# The Classification of Finite Simple Groups

Volume 1: Groups of
Noncharacteristic 2 Type

# The Classification of Finite Simple Groups

## Volume 1: Groups of Noncharacteristic 2 Type

Daniel Gorenstein

Rutgers, The State University of New Jersey
New Brunswick, New Jersey

Plenum Press · New York and London

Library of Congress Cataloging in Publication Data

Gorenstein, Daniel.
  The classification of finite simple groups.

  (The University series in mathematics)
  Bibliography: p.
  Includes index.
  Contents: v. 1. Groups of noncharacteristic 2 type.
  1. Finite simple groups. I. Title. II. Series: University series in mathematics (Plenum Press)
QA171.G638  1983                       512′.2                       83-4011
ISBN 0-306-41305-1

©1983 Plenum Press, New York
A Division of Plenum Publishing Corporation
233 Spring Street, New York, N.Y. 10013

For Helen, Diana,
Mark, Phyllis, and Julia

# Acknowledgments

I want to thank Michael Aschbacher, Enrico Bombieri, Richard Lyons, Michael O'Nan, Steven Smith, and Ronald Solomon for their valuable comments on various portions of the manuscript.

Daniel Gorenstein

# Contents

# Introduction

Never before in the history of mathematics has there been an individual theorem whose proof has required 10,000 journal pages of closely reasoned argument. Who could read such a proof, let alone communicate it to others? But the classification of all finite simple groups is such a theorem—its complete proof, developed over a 30-year period by about 100 group theorists, is the union of some 500 journal articles covering approximately 10,000 printed pages.

How then is one who has lived through it all to convey the richness and variety of this monumental achievement? Yet such an attempt must be made, for without the existence of a coherent exposition of the total proof, there is a very real danger that it will gradually become lost to the living world of mathematics, buried within the dusty pages of forgotten journals. For it is almost impossible for the uninitiated to find the way through the tangled proof without an experienced guide; even the 500 papers themselves require careful selection from among some 2,000 articles on simple group theory, which together include often attractive byways, but which serve only to delay the journey.

Nevertheless, fully cognizant of the difficulties, this is the ambitious goal we have set for ourselves: to present in some detail a global *outline* of the proof of the classification theorem, accessible to any professionally trained mathematician. We have begun the task in a previous book, entitled *Finite Simple Groups: An Introduction to Their Classification* (Plenum Press, New York, 1982), which we shall refer to as [I]. Some familiarity with [I] will be essential to read this outline, especially Chapters 1 and 2 of [I]. (However, to make this introduction self-contained, we shall repeat the definitions of the key terms from [I] that occur here.)

The primary purpose of [I] was to set the stage for the systematic outline which this book and its sequel will encompass. Indeed, in Chapters 2 and 3 of [I] we have defined each of the known finite simple groups and have given a characterization of each in terms of some set of *internal* group-theoretic conditions. Such characterizations are necessary prerequisites for completing any classification theorem concerning simple groups and allow one to ultimately identify the group under investigation. Furthermore, in Chapter 4 of [I] we have presented a detailed description of the principal techniques of simple group theory, thereby providing the fully developed language necessary for our outline.

## The Main Classification Theorems

The aim of the total endeavor is to show that the list of finite simple groups is complete and thus to establish the following result:

MAIN CLASSIFICATION THEOREM. *Every* (*nonabelian*) *finite simple group is isomorphic to one of the groups on the following list*:

(1) *A group of Lie type*;
(2) *An alternating group*; *or*
(3) *One of the following* 26 *sporadic groups*: $M_{11}$, $M_{12}$, $M_{22}$, $M_{23}$, $M_{24}$, $J_1$, $J_2$, $J_3$, $J_4$, *Mc, Ly, Hs, He, ON, Suz, Ru, .3, .2, .1, M*(22), *M*(23), *M*(24)′, $F_5$, $F_3$, $F_2$, *or* $F_1$.

(As we have said, definitions of all of these groups are given in Chapter 2 of [I].)

The statement can be reformulated in a way which makes explicit the inductive nature of the proof. One begins with a "minimal counterexample" $G$—i.e., a simple group of *least* order not isomorphic to one of the listed groups. To establish the theorem, one must derive a contradiction from the existence of such a group $G$, which will be achieved by showing that $G$ is, in fact, a *known* simple group—i.e., isomorphic to one of the listed groups.

But by the minimality of $G$, the composition factors of every proper subgroup of $G$ have lower order than $G$ and so are themselves known simple groups. For brevity, the term *K-group* is used for any group whose composition factors are among the known simple groups. In particular, every known simple group is a $K$-group.

Hence, in this terminology, the main theorem assumes the following form:

MAIN CLASSIFICATION THEOREM (*Second Form*). *If G is a finite simple group each of whose proper subgroup is a K-group, then G is a K-group.*

In Sections 1.5 and 1.6 of [I] we have given a brief overview of the proof of the main classification theorem; in particular, we have described its subdivision into four basic phases:

(a) The classification of nonconnected simple groups.
(b) The classification of connected simple groups of component type.
(c) The classification of "small" simple group of characteristic 2 type.
(d) The classification of "large" simple groups of characteristic 2 type.

(Definitions of the terms "nonconnected," "component type," and "characteristic 2 type" will be given below.)

The subdivision can be condensed into two broader categories by combining (a) and (b), and (c) and (d), respectively, which then express the fundamental breakup of simple group theory as two essentially disjoint parts:

(I) The classification of simple groups of noncharacteristic 2 type.
(II) The classification of simple groups of characteristic 2 type.

The definition of characteristic 2 type ([I, 1.43])[†] is given in terms of the *generalized Fitting* subgroup $F^*(X)$ of a group $X$, a basic group-theoretic notion [I, 1.26]. By definition

$$F^*(X) = F(X) L(X),$$

where $F(X)$ is the *Fitting* subgroup of $X$ (the unique largest normal nilpotent subgroup of $X$) and $L(X)$ is the *layer* of $X$ [the product of all subnormal quasisimple subgroups of $X$, with $L(X) = 1$ if no such subnormal subgroups exist].[‡] {A group $K$ is *quasisimple* if $K$ is perfect (i.e., $K = [K, K]$) and $K/Z(K)$ is simple, where $Z(K)$ denotes the center of $K$. A fuller discussion of these terms, including the significance of the generalized Fitting subgroup, is

---

[†] For brevity, we refer to Theorem, Proposition, Lemma, or Definition $y$ of Chapter $x$ of [I] as [I, $x \cdot y$]. To avoid confusion, we shall use a colon rather than a comma in referring to the bibliography of [I]: thus we designate Ref. 137 of [I] as [I: 137].
[‡] Many authors have adopted Bender's notation $E(X)$ for the layer of $X$, which gave harmony to the term $F^*(X) = E(X) F(X)$. We prefer the letter $L$ because it suggests an abbreviation of the term "layer."

given in Section 1.5 of [I]. In particular, the quasisimple factors of $L(X)$ are called *components* of $X$.}

Furthermore, for any group $X$, $O_2(X)$ denotes the unique largest normal 2-subgroup of $X$; and a 2-*local* subgroup of $X$ is by definition the normalizer in $X$ of a nontrivial 2-subgroup of $X$.

Now we can define characteristic 2 type:

DEFINITION. A group $X$ of even order is said to be of *characteristic* 2 type if for every 2-local subgroup $Y$ of $X$

$$F^*(Y) = O_2(Y).$$

In the contrary case, $X$ is called of *noncharacteristic* 2 type.

This division is a reflection of the fundamental dichotomy in the subgroup structure of the groups of Lie type over finite fields of, on the one hand, characteristic 2 and, on the other, odd characteristic. Indeed, in the language of Lie theory, elements of order a power of 2 are correspondingly "unipotent" or "semisimple," so it is natural to expect the 2-local structure of a group of Lie type over $GF(q)$ to differ radically according as $q$ is *even* or *odd*.

Thus, although the two parts have many common features, it is reasonable to view the main classification theorem as splitting into two separate individual theorems: (i) the classification of simple groups of noncharacteristic 2 type and (ii) the classification of those of characteristic 2 type. In fact, each of these two theorems takes up approximately half the total proof of the main classification theorem.

It should be added that over small fields and for low 2-ranks, there are coincidences between "even" and "odd" groups. For example, even though it is defined over a field of odd characteristic, the group $L_3(3)$ is of characteristic 2 type, because its subgroup $SL_2(3)$ happens to be solvable (for $q$ odd, $q > 3$, $L_3(q)$ is of noncharacteristic 2 type since $SL_2(q)$ is quasisimple). Thus, one expects some overlap in the conclusions of the classification theorem for each of the two types of groups.

In view of the length of our projected outline, we have decided to split the discussion into two volumes, one for each type of simple group. Since groups of noncharacteristic 2 type were historically investigated before those of characteristic 2 type, it is natural to treat the former type first.

Thus, we shall outline in this book a proof of the following fundamental result.

NONCHARACTERISTIC 2 THEOREM. *A minimal counterexample to the main classification theorem is of characteristic 2 type.*

Likewise, the noncharacteristic 2 theorem has a reformulation analogous to that of the main classification theorem.

NONCHARACTERISTIC 2 THEOREM (*Second Form*). *Let G be a finite simple group in which all proper subgroups are K-groups. If G is of noncharacteristic 2 type, then G is isomorphic to one of the following groups:*

(1) *A group of Lie type of odd characteristic, excluding $L_2(q)$, q a Fermat or Mersenne prime or 9, $L_3(3)$, $U_3(3)$, $U_4(3)$, $Psp_4(3)$, or $G_2(3)$.*

(2) *An alternating group of degree n, n = 7 or n $\geqslant$ 9.*

(3) *One of the following 18 sporadic groups: $M_{12}$, $J_1$, $J_2$, Mc, Ly, HS, He, ON, Suz, Ru, .3, .1, M(22), M(23), M(24)', $F_5$, $F_2$, or $F_1$.*

The excluded groups of Lie type of odd characteristic are of characteristic 2 type, as are the remaining eight sporadic groups. Of course, the groups of Lie type of characteristic 2 are all of characteristic 2 type and hence so is $A_5 \cong L_2(4)$ and $A_8 \cong L_4(2)$. Likewise $A_6 \cong L_2(9)$ is of characteristic 2 type.

## The Major Subdivision of the Noncharacteristic 2 Theorem

The proof of the second form of the noncharacteristic 2 theorem is divided into *four* major parts, the first three of which consist of independent general results about finite (simple) groups. In particular, their statements do not require the *K*-group hypothesis on proper subgroups. To state them, we recall a few further definitions from [I]. Let *X* be an arbitrary finite group.

The 2-*rank* $m_2(X)$ of *X* is the maximum rank of an abelian 2-subgroup *A* of *X* (the rank of *A* is by definition the number of factors in a direct product decomposition of *A* into cyclic subgroups).

The *sectional* 2-rank of *X* is the maximum rank of any section of *X* (by definition, a homomorphic image of a subgroup of *X* is called a *section*).

With *X* is associated a graph $\Lambda(X)$ whose vertices consist of the $Z_2 \times Z_2$ subgroups of *X* (called *four groups*), with two vertices connected by an edge if and only if the corresponding four subgroups commute elementwise. Then *X* is said to be *connected* if and only if its graph $\Gamma(X)$ is connected in the usual sense.

$O(X)$ is the unique largest normal subgroup of $X$ of odd order and is called the *core* of $X$.

Finally we say that $X$ is of *component* type if for some involution $x$ of $X$, $C_X(x)/O\big(C_X(x)\big)$ has a nontrivial layer.

Now we can state the four major results which together will yield the second form of the noncharacteristic 2 theorem:

(I) 2-RANK $\leqslant 2$ THEOREM. *If $X$ is a simple group of 2-rank $\leqslant 2$, then* $X \cong L_2(q)$, $L_3(q)$, *or* $U_3(q)$, *$q$ odd, $U_3(4)$, $A_7$, or $M_{11}$.

(II) SECTIONAL 2-RANK $\leqslant 4$ THEOREM. *If $X$ is a simple group of sectional 2-rank $\leqslant 4$, then either $X$ has 2-rank $\leqslant 2$ or $X$ is isomorphic to one of the groups on the following list*:

1. Odd characteristic. $G_2(q)$, $^3D_4(q)$, $PSp_4(q)$, $^2G_2(q)$, $q = 3^{2n+1}$, $L_4(q)$, $q \not\equiv 1 \pmod 8$, $U_4(q)$, $q \not\equiv 7 \pmod 8$, $L_5(q)$, $q \not\equiv -1 \pmod 4$, *or* $U_5(q)$, $q \equiv 1 \pmod 4$.
2. Even characteristic. $L_2(8)$, $L_2(16)$, $L_3(4)$, $U_3(4)$, $L_4(2)$, *or* $Sz(8)$.
3. Alternating. $A_8, A_9, A_{10}$, *or* $A_{11}$.
4. Sporadic. $M_{12}, M_{22}, M_{23}, J_1, J_2, J_3$, *Mc, or Ly*.

Note that the 2-rank $\leqslant 2$ and sectional 2-rank $\leqslant 4$ theorems include a number of characteristic 2 type groups: namely, $L_2(q)$, $q$ a Fermat or Mersenne prime or 9, $L_3(3)$, $U_3(3)$, $U_4(3)$, $PSp_4(3)$, $G_2(3)$, $L_2(3)$, $L_2(8)$, $L_2(16)$, $L_3(4)$, $U_3(4)$, $L_4(2)$ $(\cong A_8)$, $Sz(8)$, $M_{22}$, $M_{23}$, and $J_3$. However, the "typical" group in their conclusions is of noncharacteristic 2 type.

As described in Section 1.6 of [I], these two results yield as a corollary the classification of all simple groups with a nonconnected Sylow 2-subgroup. Indeed, by a theorem of MacWilliams [I, 1.36] such a 2-group necessarily has sectional 2-rank $\leqslant 4$, a fact which explains our interest in this class of groups. Hence, as a corollary we have the following theorem.

NONCONNECTEDNESS THEOREM. *If $X$ is a simple group with a nonconnected Sylow 2-subgroup, then $X$ is a K-group. In particular, if $X$ has 2-rank $\geqslant 3$, then $X \cong J_2$ or $J_3$.*

Hence, the second form of the noncharacteristic 2 theorem holds in the nonconnected case.

The significance of connectivity is given by the following result of John Walter and myself [I, 1.40 and 4.54] concerning balanced groups. Recall [I,

4.53] that a group $X$ is said to be *balanced* if for every pair of commuting involutions $x$, $y$ of $X$, we have

$$O(C_X(x)) \cap C_X(y) = O(C_X(y)) \cap C_X(x).$$

We note that the proof of the theorem depends on the fundamental Aschbacher–Bender classification of groups with a proper 2-generated core [I, 4.28]. Both results are fully discussed in [I].

We state the result in the following form:

BALANCE THEOREM. *If $X$ is a group with $O(X) = 1$ whose Sylow 2-subgroups are connected, then $X$ is either of characteristic 2 type or of component type.*

The point is that if $X$ is not of component type, it is immediate that $X$ is balanced [I, 4.58], whence by [I, 1.40] $X$ is of characteristic 2 type.

Thus, in view of the nonconnectedness and balance theorems, we see that it suffices to prove the second form of the noncharacteristic 2 theorem for groups of component type. The third and fourth major parts of its proof concern such groups.

The first of these is a remarkable general property of all finite groups, which has come to be called the *B-property*, and which expresses a universal fact about the centralizers of involutions in finite groups.

(III) *B*-THEOREM. *Let $X$ be a finite group, $x$ an involution of $X$, and set $C = C_X(x)$, $\bar{C} = C/O(C)$. Then we have*

$$\overline{L(C)} = L(\bar{C}).$$

[Since this equality holds trivially if $L(\bar{C}) = 1$, it is of interest only when $L(\bar{C}) \neq 1$, in which case by definition $X$ is of component type.]

We can rephrase the conclusion in the following way:

*Every component of $\bar{C}$ is the image of a component of $C$.*

Here is yet another formulation:

$$C_0 = C_C(O(C)) \text{ covers } L(\bar{C}); \ [i.e., L(\bar{C}) \leqslant \bar{C}_0].$$

We remark that the proof of the *B*-theorem uses a slightly sharpened form of the balance theorem. Furthermore, it involves several results about simple groups of low 2-rank other than those of sectional 2-rank $\leqslant 4$.

The 2-rank $\leqslant 2$, the sectional 2-rank $\leqslant 4$, and the $B$-theorems are all proved by induction; and it is for this reason that they stand as independent results about finite groups. On the other hand, the fourth major result needed for the noncharacteristic 2 theorem is a noninductive statement, involving an explicit $K$-group hypothesis. To avoid technical definitions at this point, we state it in somewhat simplified form:

(IV) STANDARD COMPONENT THEOREM. *Let $X$ be a simple group of component type with the B-property. Choose an involution $x$ of $X$ such that a component $K$ of $C_X(x)$ has maximum possible order. If $K$ is a known quasisimple group (i.e., a K-group), then $X$ is isomorphic to one of the groups listed in the conclusion of the second form of the noncharacteristic 2 theorem.*

(The term "standard component" will be defined later in the text.)

These last two results immediately yield the second form of the noncharacteristic 2 theorem. Indeed, in view of the preceding discussion, it suffices to prove that result when the group $G$ under investigation is of component type. Moreover, $G$ has the $B$ property by the $B$ theorem. Since by assumption *every* proper subgroup of $G$ is a $K$-group, the hypotheses of the standard component theorem are therefore satisfied, and so by that theorem $G$ is of one of the listed isomorphism types.

We add a remark here about the standard component theorem. In the course of the proof of the $B$-theorem, one is forced to treat a certain number of standard component problems (i.e., with certain prescribed possibilities for the component $K$). However, it turns out that a minimal counterexample to the $B$-theorem is not necessarily simple, but satisfies only the weaker condition that $F^*(X)$ is simple. Hence for those choices of $K$, the standard component theorem must be proved for this wider class of groups $X$.

## Some Comments on the Classification Proof

We conclude with a few general remarks about the proof of the main classification theorem and of our outline. First, we note that one can obtain a modest simplification in the existing proof by assuming at the very outset that every proper subgroup of the group $G$ under investigation is a $K$-group. However, in the early years when no idea of the exact statement of the classification theorem existed nor whether such a result would ever be attainable, one always strived for "compartmentalization," deriving each theorem as an independent result concerning simple groups, thus "closing off" each chapter as one progressed. (In a few situations, when that approach did not seem possible, one instead proceeded "axiomatically," imposing conditions on proper simple sections of $G$ satisfied by all the then known simple groups and hopefully remaining valid for any yet to be discovered simple groups.)

Of course, each such result "on the way" is now an immediate corollary of the full classification theorem, but the original approach, forced upon us by our ignorance of simple groups, has the decided advantage of "protecting" each chapter of simple group theory from any "disaster" which might befall a "neighbor." Furthermore, there is something esthetically appealing about an independent proof of the 2-rank $\leq 2$ theorem, say, or the wonderful $B$-property of finite groups. (The same assertion, of course, applies to the famous Schreier conjecture that the outer automorphism group of every simple group is solvable, which is now a corollary of the complete classification theorem; however, unfortunately no independent proof of this result has ever been obtained.)

To the extent possible, our outline will follow the historical, compartmentalized approach. In particular, the 2-rank $\leq 2$, sectional 2-rank $\leq 4$, and $B$-theorems will be discussed as independent results.

In saying that the classification proof occurred over a 30-year period, we mean to imply that the early results—the odd-order theorem and the classification of groups with dihedral Sylow 2-subgroups, for example—are as essential today for the final proof as they were when first established. Thus, the classification theorem is literally a structure composed of 500 "bricks," constructed slowly over these 30 years. However, in contrast to customary building practice, it was a long time before an actual "blueprint" of the simple group edifice was fully developed and, likewise, before all the "tools of the trade" were available for use.

Thus, for most of the period, work proceeded without benefit of a clearly defined plan and without the techniques that were to be developed later. Furthermore, throughout construction, there was a strong tendency to ease the task by quoting any result available at the time, in part because the individual steps were already often of excessive length and also from a desire to get on with the job. On top of this, the "work force" was spread around the world: primarily in Australia, England, Germany, Japan, and the United States. As a result, at least portions of the same job were often duplicated, and sometimes it was necessary to reshape a seemingly completed prior section to fit more neatly into the larger structure.

These factors had certain expected consequences for the total proof:

(a) bricks near the base could often be cut down and strengthened by application of later results and techniques; (b) the logical structure of the "brickwork" was undoubtedly more intricate than necessary; (c) sections of the "facade" could be improved by replacing several individual bricks by a single, better shaped one.

Of course, the somewhat haphazard evolution of the proof raised an even more basic question: how efficient is the final plan which evolved? Now that we are in a position to see the completed structure as a single entity, can we discern a much simpler way to carry out the entire construction?

Reconstruction—or as we say in the trade "revisionism"—began some 10 years ago with Bender's fundamental simplification of Feit and Thompson's basic "uniqueness" theorem of the odd-order paper and continued intermittently until the very end of the classification. However, now that the classification proof is completed, there are strong reasons for undertaking a systematic revision. I have indicated some of these in the above description of the evolutionary nature of the original proof. Moreover, many "local" errors have been uncovered in individual papers—a hardly surprising fact, considering the inordinate length of the total argument. Although these have always been quickly patched up, usually "on the spot," the only reasonable way of assuring that they have all been ferreted out is within the framework of a complete reorganization of the existing proof.

Such a reorganization is indeed already under way—in the recent work of Solomon (and also Aschbacher and Gilman) on groups with quasidihedral or wreathed Sylow 2-subgroups, [9], and of Harada, Lyons, and myself on disconnected groups [54], [44]. Furthermore, in my lectures at the 1979 Santa Cruz AMS summer conference on finite groups [39] I have put forth a possible blueprint for a "second-generation" proof.

How then is one to organize this outline? To try to keep abreast of the latest revisionist developments would be a hopeless endeavor, for this will clearly be an ongoing process, with new results regularly produced each year. Moreover, in taking that approach, one faces the possibility of losing the essential structure of the *initial* proof, which would certainly be unfortunate.

Expressed in more positive terms, it seems definitely preferable to have a historical record of the original classification of simple groups against which to measure all subsequent improvements and revisions.

This is the approach we have determined to adopt here, with the following proviso: those and only those revisions in the proof which occurred prior to the completion date of February, 1982 (primarily due to Bender) will be incorporated into the outline.

Finally, we note that the outline itself will conform rather closely to the four-part division of the second form of the noncharacteristic 2 theorem.

Thus, in Chapter 1 we outline the proof of the 2-rank $\leqslant 2$ theorem and, in particular, the celebrated Feit–Thompson proof of the solvability of groups of odd order.

Likewise, in Chapter 2 we outline the sectional 2-rank $\leqslant 4$ theorem as well as those additional low 2-rank results needed for the $B$-theorem and standard component theorem.

Chapters 3 and 4 are devoted to the $B$-theorem. In Chapter 3, we give the general picture as well as an overview of the proof. We then outline in detail two fundamental results of Aschbacher—his "component" and "classical involution" theorems, both of which are crucial for the $B$-theorem and for (the precise form of) the standard component theorem.

Finally, in Chapter 5 we outline the standard component theorem, including those results needed to complete the $B$-theorem.

## Notation

In general, our notation will conform to that of [I], and the reader is referred to Sections 1.4 and 1.5 of [I], where a good deal of notation is introduced.

We shall make a few notational shortcuts, to simplify the text. However, to avoid possible confusion, these will be limited primarily to the group $G$ under investigation.

For any element or subset $A$ of $G$, we write $C_A$ for $C_G(A)$. Furthermore, we let $\mathscr{L}(G)$ be the set of quasisimple components of $C_x/O(C_x)$ as $x$ ranges over the involutions of $G$. {In [I], we have used the notation $\mathscr{L}_2(G)$; however, there the discussion was general and included all primes, whereas in this volume our interest will be almost exclusively with the prime 2.} For the same reason, we denote by $\mathscr{I}(X)$ the set of involutions of the group $X$.

We also adopt the so-called "bar convention." Thus for any group $X$, if $\overline{X}$ is a homomorphic image of $X$, then for any subgroup or subset $Y$ of $X$, $\overline{Y}$ will denote its image in $\overline{X}$.

These notational conventions will be used throughout the book. Other notation will be introduced as we go along.

# 1

# Simple Groups of 2-Rank ⩽ 2

In this chapter we outline the classification of simple groups of 2-rank ⩽ 2, beginning with the Feit–Thompson proof of the solvability of groups of odd order [I: 93]. In particular, such a group $G$ must therefore have even order.

If $m_2(G) = 1$, then by [I, 1.35], a Sylow 2-subgroup of $G$ is either cyclic or quaternion. In the first case, Burnside's transfer theorem [I, 1.20] implies that $G$ has a normal 2-complement and so is solvable; while in the second case, the Brauer–Suzuki theorem [I, 4.88] shows that $G$ has a unique involution, which is therefore central. Hence, in either case, $G$ is not simple.

Thus a simple group with $m_2(G) \leqslant 2$ necessarily has 2-rank 2. However, in this case Alperin's theorem [I, 4.103] implies that a Sylow 2-subgroup of $G$ is either *dihedral*, *quasi-dihedral*, *wreathed* (the definitions, given in [I], will be repeated below), or isomorphic to a Sylow 2-subgroup of $U_3(4)$. Hence the 2-rank ⩽ 2 theorem is reduced to the classification of simple groups with Sylow 2-subgroups of each of these types.

As we shall see, the quasi-dihedral and wreathed cases are best treated simultaneously, so the analysis of simple groups of 2-rank 2 reduces to three separate Sylow 2-group classification theorems, which we shall discuss in succession after the odd-order theorem.

## 1.1. Groups of Odd Order; The Uniqueness Theorem

Every group of odd order is solvable. What a simple statement! Yet the reason—if there be such a single reason—is buried deep within the Feit–Thompson proof. Technical simplifications there have been since the proof first appeared in 1962, but its fundamental three-part subdivision has remained completely intact:

(a) Determine the structure and embedding of the maximal subgroups of a minimal counterexample $G$ by local analytic methods [I: 93, Chapter IV].

(b) Eliminate as many of these configurations as possible by character-theoretic and arithmetic arguments [I: 93, Chapter V].

(c) Using generators and relations, prove that the single remaining configuration leads to a contradiction [I: 93, Chapter VI]. (See [I, Section 1.1] for fuller discussion of this division.)

It has never ceased to amaze me that with appropriate interpretations of (a), (b), and (c), the classification of all simple groups has followed the identical pattern of proof; one must then regard (a) as restricted to maximal subgroups in the "neighborhood" of the centralizers of involutions; for "large" simple groups, one must replace character theory in (b) by local analysis; and in (c), the generator-relation analysis leads to a presentation of each of the known simple groups by generators and relations rather than to contradiction. Thus, despite the fact that there are no (nonabelian) simple groups of odd order, the proof of this result provides a truly quintessential insight into the nature of finite simple groups and their ultimate classification.

Moreover, the very techniques which Feit and Thompson introduced in the course of the proof underlie many of the later general developments of simple group theory: transitivity theorems, uniqueness theorems, strong $p$-embedding, the foreshadowing of signalizer functors, factorization theorems, coherent sets of characters, etc. In addition, the odd-order theorem itself has been crucial to all subsequent classification theorems—in particular, since it implies that cores of centralizers of involutions in simple groups are necessarily solvable. Thus, it is not possible to overemphasize the importance of the Feit–Thompson theorem for simple group theory.

The original proof has been closely examined by several individuals: notably Bender [I: 28] and Glauberman for the local analysis, and Dade [17] and Sibley [108] for the character theory. Indeed, motivation for Glauberman's $ZJ$-theorem [I, 4.110] lies in the odd-order proof, and it enabled him to simplify the places in the argument in which Thompson had required factorization lemmas. Shortly thereafter (and using the $ZJ$-theorem in a critical way), Bender obtained a fundamental simplification in the proof of Thompson's basic uniqueness theorem, thereby launching the "Bender method." Over the years, Glauberman and Bender further refined the local analysis portions of the proof, and Glauberman lectured on the material at the University of Chicago in 1978. The two are now (as of this writing) preparing a set of lecture notes which will provide an updated treatment of this portion of the proof.

Concerning the odd-order character theory, first Dade gave a more conceptual treatment of some of Feit's basic results on coherence; however, it was Sibley who focused on coherence in the context of the odd-order theorem and ultimately found shorter ways of eliminating some of the difficult configurations of [I: 93, Chapter V]. He, too, is in the process of preparing lecture notes for publication.

On the other hand, [I: 93, Chapter VI] has so far withstood attempts at significant simplification. This is not surprising, for it deals with a very restricted situation, which does not allow "much room to maneuver." (Glauberman, who has intensely studied this part of the proof, has managed a few minor improvements.)

Here then is the theorem.

THEOREM 1.1. *Every group of odd order is solvable.*

The proof is, of course, by contradiction. One begins with a counterexample $G$ of least order. If $G$ has a nontrivial proper normal subgroup $N$, then $N$ and $G/N$ are both solvable by the minimality of $G$, so $G$ is solvable—contradiction. Thus $G$ is simple. Furthermore, again by the minimality of $G$, every proper subgroup of $G$ is solvable. We see then that attention is focused on a simple group $G$ of odd order in which every proper subgroup is solvable; to establish the theorem, it must be shown that no such group exists. We fix such a group $G$.

The structure and embedding of the maximal subgroups of $G$ is based on a fundamental "uniqueness" theorem concerning $p$-subgroups of $G$ for any prime $p$ for which the $p$-rank $m_p(G) \geqslant 3$. The Thompson transitivity theorem represents a key step in its proof, and the uniqueness theorem itself is closely related to the notion of strong $p$-embedding. Its statement is as follows:

THEOREM 1.2. *Let $A$ be an elementary abelian p-subgroup of $G$ of order $p^3$. Then there is only one maximal subgroup of $G$ which contains $A$.*

This is a powerful result, which fairly directly yields the following important information about maximal subgroups of $G$.

PROPOSITION 1.3. *Let $P$ be a Sylow p-subgroup of $G$ with $m(P) \geqslant 3$ and let $M$ be a maximal subgroup of $G$ containing $P$. Then we have*

    (i) *M is the unique maximal subgroup of G containing P*;

    (ii) $\Gamma_{P,2}(G) \leqslant M$;

    (iii) *Either M is strongly embedded in G* [*i.e.,* $\Gamma_{P,1}(G) \leqslant M$] *or else for some element x of P of order p,* $C_P(x) = \langle x \rangle \times D$, *where D is cyclic*;

    (iv) $P \leqslant M'$; *and*

    (v) $M \cap M^g$ *has cyclic Sylow p-subgroups.*

[Recall that by definition the *k-generated p-core* $\Gamma_{P,k}(G) = \langle N_G(Q) \mid Q \leqslant P$, $m(Q) \geqslant k \rangle$].

    PROOF. Since $P$ contains a subgroup $A \cong E_{p^3}$ by assumption, (i) follows at once from Theorem 1.2.

    Suppose (ii) is false; choose a $p$-subgroup $Q$ of $M$ of maximal order with $m(Q) \geqslant 2$ such that $N = N_G(Q) \not\leqslant M$. Let $R \in \mathrm{Syl}_p(N_M(Q))$. By Sylow's theorem, we can assume without loss that $R \leqslant P$. By Theorem 1.2, $m(R) \leqslant 2$, otherwise $N \leqslant M$. In particular, $m(Q) = 2$, so $Q < P$, and hence $Q < R$ by [I, 1.10]. But then by our choice of $Q$, $N_G(R) \leqslant M$, so by definition of $R$, $R \in \mathrm{Syl}_p(N_N(R))$ and therefore $R \in \mathrm{Syl}_p(N)$, by [I, 1.11].

    Now set $Z = \Omega_1(Z(P))$ and $V = \Omega_1(Z(Q))$. Then $Z \leqslant R$. If $Z \not\leqslant Q$, then as $m(Q) = 2$, $m(ZQ) \geqslant 3$, so $m(R) \geqslant 3$—contradiction. Hence $Z \leqslant Q$ and so $Z \leqslant V$. Furthermore, as $P \leqslant N_G(Z)$, $N_G(Z) \leqslant M$ by Theorem 1.2. On the other hand, as $V$ char $Q$, $V \lhd N$, so $N \leqslant N_G(V)$. Since $N \not\leqslant M$, it follows that $Z \neq V$, whence $m(V) = m(Q) = 2$ [and $m(Z) = 1$].

    Finally, set $C = C_N(V)$. Then $C \leqslant N_G(Z) \leqslant M$. We shall argue that $CR \lhd N$, whence $N = CN_N(R)$ by the Frattini argument [I, 4.3] [as $R \in \mathrm{Syl}_p(CR)$]. Since $N_G(R) \leqslant M$, it will then follow that $N \leqslant M$—contradiction. Since $C \lhd N$, the desired assertion is clear if $R \leqslant C$, so we can assume that $R \not\leqslant C$. Set $\overline{N} = N/C$, so that $\overline{N} \leqslant \mathrm{Aut}(V) \cong GL_2(p)$, which has Sylow $p$-subgroups of order $p$. Thus, $\overline{R} \in \mathrm{Syl}_p(\overline{N})$ and $\overline{N}$ has odd order. However, it is easily checked now from the structure of $GL_2(p)$ that these conditions force $\overline{R}$ to be normal in $\overline{N}$ (cf. Lemma 1.10 below). Hence, $CR \lhd N$, as required. This proves (ii).

    Next, by (ii), either $M$ is strongly $p$-embedded in $G$ or there is $Q \leqslant P$ with $m(Q) = 1$ such that $N = N_G(Q) \not\leqslant M$. Assume the latter and choose $Q$ minimal. Since $N \leqslant N_G(\Omega_1(Q))$, it follows that $Q = \Omega_1(Q)$. But as $p$ is odd, $Q$ is cyclic by [I, 1.35], so $Q \cong Z_p$. Set $T = C_P(Q)$. Since $T \leqslant N \not\leqslant M$, again Theorem 1.2 implies that $m(T) \leqslant 2$. If $Q$ is not contained in the Frattini subgroup $\phi(T)$ of $T$, then as $T/\phi(T)$ is elementary abelian by [I, 1.12], $Q$ has a complement $D$ in $T$, whence $T = QD = Q \times D$. Since $m(T) \leqslant 2$, this forces

$m(D) \leqslant 1$, so $D$ is cyclic and (iii) holds (with $\langle x \rangle = Q$). Hence, we can suppose that $Q \leqslant \phi(T)$.

On the other hand, as $p$ is odd and $m(P) \geqslant 3$, $P$ possesses an elementary normal subgroup $E \cong E_{p^3}$ by [I, 1.36]. Set $E_0 = C_E(Q)$ and $T_0 = C_T(E)$. Then $E_0 \leqslant T$ and so $m(E_0 Q) \leqslant 2$. In particular, $E_0 < E$, so $Q \not\leqslant T_0$. But then $Q \not\leqslant E_0$, otherwise $E_0 = E$. Hence $E_0 Q = E_0 \times Q$ and so $m(E_0) \leqslant 1$. Furthermore, $\bar{T} = T/T_0 \leqslant \mathrm{Aut}(E) \cong GL_3(p)$, which has extraspecial Sylow $p$-subgroups $p^{1+2}$ of order $p^3$ and exponent $p$. Since $Q \leqslant \phi(T)$ and $Q \not\leqslant T_0$, it follows that $1 \neq \bar{Q} \leqslant \phi(\bar{T})$, where $\bar{T} \cong p^{1+2}$ and $\bar{Q} = \bar{T}'$. Thus, in the natural action of $\mathrm{Aut}(E)$ on $E$, a generator of $\bar{Q}$ can be represented by the matrix

$$\begin{pmatrix} 1 & 0 & 0 \\ 0 & 1 & 0 \\ 1 & 0 & 1 \end{pmatrix}$$

and we see that $C_E(\bar{Q}) = E_0 \cong Z_p \times Z_p$, contrary to $m(E_0) = 1$. Thus (iii) also holds.

Next, if $D \in \mathscr{D}$, where $\mathscr{D}$ denotes the Alperin–Goldschmidt conjugation family for $P$ [I, 4.86], we have $N_P(D) \in \mathrm{Syl}_p\big(N_G(D)\big)$, $D \in O_{p'p}\big(N_G(D)\big)$ and $C_P(D) \leqslant D$. Since $m(P) \geqslant 3$, these conditions immediately yield that $D$ is noncyclic, whence $m(D) \geqslant 2$ and so $N_G(D) \leqslant M$ by (ii). Thus $M$ controls the $G$-fusion of $P$; and as $G$ is simple, (iv) follows from the focal subgroup theorem [I, 1.19].

Finally, suppose that for some $g \in G - M$, $M \cap M^g$ has a noncyclic Sylow $p$-subgroup $S$; then $m(S) \geqslant 2$ and so by (ii), $N_G(S) \leqslant M \cap M^g$. Hence $M \cap M^g$ contains a Sylow 2-subgroup of $N_G(S)$ and therefore, by definition of $S$, we have $S \in \mathrm{Syl}_p\big(N_G(S)\big)$, whence $S \in \mathrm{Syl}_p(G)$. But $S^{g^{-1}} \leqslant M$ as $S \leqslant M^g$, so $S$ and $S^{g^{-1}}$ are both Sylow $p$-subgroups of $M$. Hence by Sylow's theorem, $S = S^{g^{-1}m}$ for some $m \in M$. Thus $g^{-1}m \in N_G(S) \leqslant M$ [by (ii)] and it follows that $g \in M$, which is not the case. Therefore (v) also holds.

In view of the proposition, we call $M$ a *uniqueness* subgroup for $p$.

Since the odd-order uniqueness theorem had a profound influence on the subsequent development of local analysis in simple group theory, we shall outline Bender's proof in some detail. Bender has since made further improvements using ideas from signalizer functor theory. (It is this later proof which will appear in the Bender–Glauberman lecture notes.) However, because of its historical interest and its illustraton of the way in which Bender adapted Thompson's results on maximal abelian normal $p$-subgroups, we shall follow Bender's original argument here.

If $P$ is a $p$-group, Thompson introduced the term $\mathrm{SCN}(P)$ for the set of abelian normal subgroups $B$ of $P$ maximal under inclusion, and wrote $B \in \text{SCN}_3(P)$ if $m(B) \geqslant 3$. Let $X$ be a group having $P$ as Sylow $p$-subgroup. In [I, Section 4.1], we have noted the important property of elements of $\text{SCN}(P)$:

$$\text{If } B \in \text{SCN}(P), \text{ then } C_X(B) = B \times O_{p'}\big(C_X(B)\big). \text{ In particular, } B = C_P(B) \tag{1.1}$$

The Thompson transitivity theorem [I, 4.14] asserts the following:

If all $p$-local subgroups of $X$ are solvable and we let $B \in \text{SCN}_3(P)$, then for any prime $q \neq p$, any two maximal $B$-invariant $q$-sub-  (1.2)
groups of $X$ are conjugate by an element of $C_X(B)$.

As a consequence, $P$ leaves invariant some maximal $B$-invariant $q$-subgroup of $G$. Furthermore, we also have (cf. [I, Section 4.7] and [I, 4.15]).
    If $B \in \text{SCN}(P)$ and $X$ is solvable of odd order, then

(1) $B \leqslant O_{p'p}(X)$; and
(2) $O_{p'}(X)$ contains every $B$-invariant $p'$-subgroup of $X$.  (1.3)

These are the basic results of Thompson which Bender uses. The first step in his proof of Theorem 1.2 is a key uniqueness criterion. Henceforth we fix a prime $p$ with $m_p(G) \geqslant 3$. [Also keep in mind our notational convention $C_A$ for $C_G(A)$ for any subset $A$ of $G$.]

PROPOSITION 1.4. *Let $M$ be a maximal subgroup of $G$ and $U$ a subgroup of $M$ with $U \cong Z_p \times Z_p$. If $C_u \leqslant M$ for every $u \in U^\#$, then $M$ is the unique maximal subgroup of $G$ containing $U$.*

PROOF. Expand $U$ to $P \in \text{Syl}_p(M)$. We claim first that $N_G(P) \leqslant M$, in which case $P \in \text{Syl}_p(G)$, as usual, by [I, 1.11]. If $O_{p'}(M) = 1$, then as $|M|$ is odd, $M$ is $p$-stable by [I, 4.108] and so $Z = Z(J(P)) \lhd M$ by Glauberman's $ZJ$-theorem [I, 4.110]. Since $M$ is maximal and $G$ is simple, it follows that $M = N_G(Z)$. But $N_G(P) \leqslant N_G(Z)$ as $Z$ char $P$, so $N_G(P) \leqslant M$ in this case.
    We can therefore assume that $Y = O_{p'}(M) \neq 1$. Thus $M = N_G(Y)$, again as $M$ is maximal and $G$ is simple. Let $X$ be a $P$-invariant $p'$-subgroup of $G$. Then $X$ is $U$-invariant and as $U$ is noncyclic abelian, [I, 4.13] implies that $X = \langle C_X(u) \mid u \in U^\# \rangle$, so $X \leqslant M$ by our hypothesis on $U$. But as $M$ is solvable with $P \in \text{Syl}_p(M)$, it follows at once that $X \leqslant Y$. Thus $Y$ is the unique maximal $P$-invariant $p'$-subgroup of $G$. Hence if $x \in N_G(P)$, then $Y^x$

is $P^x = P$-invariant and $Y^x$ is a $p'$-group, so $Y^x \leqslant Y$, forcing $Y^x = Y$. Since $M = N_G(Y)$, this yields $x \in M$ and we conclude that $N_G(P) \leqslant M$ is this case as well.

Now suppose the proposition false and choose a maximal subgroup $N$ of $G$ containing $U$ with $N \not\leqslant M$ such that $N \cap M$ has a Sylow $p$-subgroup $R$ of largest possible order. Without loss, we can assume $R \leqslant P$. It follows easily, as in the proof of Proposition 1.3(ii), that $R \in \mathrm{Syl}_p(N)$. By the $ZJ$-theorem, we have $N = O_{p'}(N) N_N(V)$, where $V = Z(J(R))$. But as $O_{p'}(N)$ is $U$-invariant, $O_{p'}(N) \leqslant M$, again by [I, 4.13], so $N_G(V) \not\leqslant M$. Hence, by the maximality of $N$, $R = N_P(V)$, whence, as usual, $P = R$ (and hence $V = Z$). This forces $Y = O_{p'}(M) \neq 1$, otherwise $M = N_G(Z) = N_G(V)$, contrary to $N_G(V) \not\leqslant M$.

Now we shift attention to an element $B \in \mathrm{SCN}_3(P)$. By the Thompson transitivity theorem, for each prime $q \neq p$, $C_B$ acts transitively on the maximal $B$-invariant $q$-subgroups of $G$ and $P$ leaves invariant one such subgroup $Q$. Then $U$ normalizes $Q$ and so $Q \leqslant M$. But also $U$ normalizes $O_{p'}(C_B)$, so, in addition, $C_B \leqslant M$ [using (1.1)]. We conclude that every $B$-invariant $q$-subgroup of $G$ is contained in $M$, and this in turn yields that every $B$-invariant $p'$-subgroup of $G$ is contained in $M$. But each such subgroup of $M$ is contained in $Y = O_{p'}(M)$ by (1.3)(2) and therefore $Y$ is the unique maximal $B$-invariant $p'$-subgroup of $G$.

Finally set $S = P \cap O_{p'p}(N)$, so that $S \in \mathrm{Syl}_p\big(O_{p'p}(N)\big)$ and hence $B \leqslant S$ by (1.3)(1). Since $Y$ is $S$-invariant, it follows that $Y$ is also the unique maximal $S$-invariant $p'$-subgroup of $G$. Hence as with $P$ above, $N_G(S)$ normalizes $Y$ and so $N_G(S) \leqslant M = N_G(Y)$. But $N = O_{p'}(N) N_N(S)$ by the Frattini argument, and as each factor is contained in $M$, we conclude that $N \leqslant M$, contrary to the choice of $N$.

The proposition reduces the uniqueness theorem to showing that any subgroup $A \cong E_{p^3}$ of $G$ necessarily contains a subgroup $U \cong Z_p \times Z_p$ satisfying the conditions of Proposition 1.4.

As an immediate corollary, we obtain the following:

PROPOSITION 1.5. *Let $U$, $V$ be subgroups of $G$ with $U \cong V \cong Z_p \times Z_p$ and $[U, V] = 1$. If $U$ is contained in a unique maximal subgroup $M$ of $G$, then $M$ is the unique maximal subgroup of $G$ containing $V$.*

Indeed, $U \leqslant C_v$ for each $v \in V^{\#}$, so each $C_v \leqslant M$ by our hypothesis on $U$. Thus, the assumptions of Proposition 1.4 hold for $V$ and so $M$ is the unique maximal subgroup of $G$ containing $V$.

This in turn yields a basic reduction in the proof of the uniqueness theorem.

PROPOSITION 1.6. *Let $P \in \mathrm{Syl}_p(G)$ and let $U \leqslant P$ with $U \cong Z_p \times Z_p$ be such that $m(C_P(U)) \geqslant 3$. If $U$ is contained in a unique maximal subgroup $M$ of $G$, then $M$ is the unique maximal subgroup of $G$ containing $A$ for every subgroup $A$ of $P$ with $A \cong E_{p^3}$. In particular, Theorem 1.2 holds.*

Indeed, let $Z$ be a normal subgroup of $P$ with $Z \cong Z_p \times Z_p$. Since $\mathrm{Aut}(Z) \cong GL_2(p)$ has Sylow $p$-subgroups of order $p$, $Z$ centralizes a subgroup $A_1$ of $A$ with $A_1 \cong Z_p \times Z_p$. Furthermore, by assumption $C_P(U)$ contains $B \cong E_{p^3}$ and so similarly $Z$ centralizes a subgroup $B_1$ of $B$ with $B_1 \cong Z_p \times Z_p$. But then successive terms of the chain $U, B_1, Z, A_1$ centralize each other. Since $M$ is the unique maximal subgroup of $G$ containing $U$, we conclude therefore by repeated application of Proposition 1.5 that $M$ is also the unique maximal subgroup of $G$ containing $A_1$. Since $A_1 \leqslant A$, the proposition follows.

Thus we are reduced to finding a "good" $p$-subgroup $U$. There are two special situations in which Bender shows that such a $U$ exists.

PROPOSITION 1.7. *Suppose $M$ is a maximal subgroup of $G$ such that $F(M) = O_p(M)$ [in which case $M = N_G(Z(F(P)))$ for some $P \in \mathrm{Syl}_p(G)$]. Then for any subgroup $U \cong Z_p \times Z_p$ of $P$ which is contained in an element of $SCN_3(P)$ (in particular, if $U \lhd P$), $M$ is the unique maximal subgroup of $G$ containing $U$.*

The proof is similar in spirit to that of Proposition 1.4 and again depends on the Thompson transitivity theorem. We omit the details.

PROPOSITION 1.8. *Suppose $M$ is a maximal subgroup of $G$ with $F(M) > O_p(M)$ and assume that $O_p(M)$ contains a subgroup $U \cong Z_p \times Z_p$ such that either*

(a) $m(C_{O_p(M)}(U)) \geqslant 3$; *or*
(b) *$M$ contains a Sylow $p$-subgroup $P$ of $G$ and $U \lhd P$.*

*Then $M$ is the unique maximal subgroup of $G$ containing $U$.*

It was the proof of this result which introduced the "Bender method." In [I, Sections 4.3 and 4.8] we have given two illustrations of the method; the proof of the proposition follows the same general line of argument.

For convenience, let us say that $M$ is of *restricted* type for $p$ if the assumptions of Proposition 1.7 or 1.8 hold for $p$. In particular, if $M$ is of restricted type for $p$, then $O_p(M)$ is noncyclic. Moreover, combined with Proposition 1.6, we conclude at once:

PROPOSITION 1.9. *If $G$ possesses a maximal subgroup which is of restricted type for $p$, then Theorem 1.2 holds for $p$.*

We assume henceforth that Theorem 1.2 fails for $p$. In particular, by the proposition, $G$ possesses no maximal subgroup which is of restricted type for $p$. Bender's idea in this case is to force some maximal subgroup of $G$ to be of restricted type for a suitable prime $q \neq p$ (in which case Theorem 1.2 will hold for $q$), and then to play off the prime $q$ against the prime $p$.

We need a preliminary fact about $q$-groups of rank $\leq 2$.

LEMMA 1.10. *Let $X$ be a group of odd order with $F^*(X) = O_q(X)$ for some odd prime $q$. If $O_q(X)$ has rank $\leq 2$, then we have*

(i) $\bar{X} = X/O_q(X) \leq GL_2(q)$; *and*
(ii) $\bar{X}'$ *is a $q$-group.*

Indeed, let $Q$ be a subgroup of $O_q(X)$ satisfying the conditions of [I, 4.1.6] with $O_q(X)$ in the role of $P$ (such a subgroup $Q$ is customarily called *critical*). Setting $R = \Omega_1(Q)$, it follows that $m(Q) \leq 2$ and $R$ has class at most 2 and exponent $q$, which together imply that $\bar{R} = R/\phi(R) \cong Z_q$ or $Z_q \times Z_q$. But $\bar{X}$ acts faithfully on $\bar{R}$ by [I, 4.6 and 4.7]. Since $\text{Aut}(\bar{R}) \leq GL_2(q)$, (i) follows. But now as $|\bar{X}|$ is odd, (ii) is immediate from the structure of $GL_2(q)$.

Now let $P \in \text{Syl}_p(G)$ and fix a maximal subgroup $M$ of $G$ containing $N_G(Z(P))$. In particular, $M$ is not of restricted type for $p$. Observe, first of all, the following:

LEMMA 1.11. $O_p(M)$ *is cyclic.*

Indeed, if not, then $O_p(M)$ would contain a subgroup $U \cong Z_p \times Z_p$ with $U \lhd P$, in which case $M$ would be of restricted type for $p$ by Propositions 1.7 and 1.8—contradiction.

With this information, Bender proves the following result:

PROPOSITION 1.12. (i) $M$ *is of restricted type for some prime $q$. In particular, Theorem 1.2 holds for $q$; and* (ii) $P \leq M'$.

To establish the proposition, he first proves the following lemma:

LEMMA 1.13. *If $P \cap M'$ is noncyclic, then $M$ is of restricted type for some prime $q$.*

Indeed, if not, then $O_q(M) \leqslant 2$ for all $q \neq p$, so $R = P \cap M'$ centralizes $O_q(M)$ by Lemma 1.10(ii). Furthermore, as $O_p(M)$ is cyclic, $\text{Aut}(O_p(M))$ is abelian, so $M'$ and hence $R$ centralizes $O_p(M)$. Thus $R$ centralizes $F(M)$. Since $C_M(F(M)) \leqslant F(M)$ by [I, 1.25], as $M$ is solvable, it follows that $R \leqslant F(M)$, whence $R \leqslant O_p(M)$, contrary to the fact that $R$ is noncyclic by hypotheses, while $O_p(M)$ is cyclic.

Since $m(P) \geqslant 3$, Lemma 1.13 and Proposition 1.12(ii) imply Proposition 1.12(i), so it suffices to prove (ii). The aim is to show that $M$ controls the $G$-fusion of $Z(P)$ in $P$ and then apply standard transfer arguments (cf. [I, 1.19]). (Since $G$, being simple, has no normal subgroups of index $p$, it will then follow that $M$ has no normal subgroups of index $p$, whence $P \leqslant M'$.)

The proof that $M$ controls the $G$-fusion of $Z(P)$ divides into two cases, according as $M$ is or is not restricted for some prime $q$. The first case depends on the following general fact.

LEMMA 1.14. *Let $B \in SCN_3(P)$ and let $N$ be a maximal subgroup of $G$ containing $C_B$. If $N$ is of restricted type for some prime $r$, then $N_G(T) \leqslant N$ for every subgroup $T$ of $P$ containing $B$.*

This is easily proved with the aid of the transitivity theorem for $B$ relative to maximal $B$-invariant $r$-subgroups of $G$. [Note that $B = C_P(B)$ by (1.1), so $Z(P) \leqslant B$. Hence $C_B \leqslant C_{Z(P)} \leqslant M$ (by choice of $M$). Thus Lemma 1.14 applies, in particular, to $M$ if $M$ is of restricted type for some $q$.]

In the contrary case, $P \cap M'$ and hence $P'$ is cyclic by Lemma 1.13. The proof that $M$ controls the fusion of $Z(P)$ in this case depends on the following easily established property of such $p$-groups:

LEMMA 1.15. *Let $X$ be a solvable group of odd order with Sylow $p$-subgroup $S$. If $S'$ is cyclic, then $O_{p'}(X)S \lhd X$.*

We henceforth assume that Proposition 1.12 holds. We fix a prime $q$ such that $M$ is of restricted type for $q$. We set $R = O_q(M)$ and let $Q \in \text{Syl}_q(M)$. Then $m(R) \geqslant 2$, $Q \in \text{Syl}_q(G)$, and Theorem 1.2 holds for $q$, so that $M$ is the unique maximal subgroup of $G$ containing $E$ for any

$E_{q^3} \cong E \leqslant Q$. Also by Propositions 1.7 and 1.8, $M$ is the unique maximal subgroup of $G$ containing $R$.

The following result connects $p$ and $q$.

PROPOSITION 1.16. *Let* $Z_p \times Z_p \cong U \leqslant P$. *If* $C_R(U)$ *is noncyclic, then* $M$ *is the unique maximal subgroup of* $G$ *containing* $U$.

Indeed, set $T = C_R(U)$, so that $m(T) \geqslant 2$. It will suffice to show that either $m(TC_T(R)) \geqslant 3$ or $T = R$, for then by the paragraph following Lemma 1.15, $M$ will be the unique maximal subgroup of $G$ containing $T$. Since $T \leqslant C_u$ for each $u \in U^\#$, it will then follow that each $C_u \leqslant M$, in which case the proposition will follow from Proposition 1.4.

But if $m(TC_R(T)) = 2$, then $\Omega_1(C_R(T)) \leqslant T$, and it is not difficult to show that this forces $T = R$.

We use this to prove the following result:

LEMMA 1.17. *If* $B \in SCN_3(P)$, *then* $C_R(B) = 1$.

Indeed, assume false. We need only show that $C_R(U)$ is noncyclic for some $Z_p \times Z_p \cong U \leqslant B$, for then Propositions 1.6 and 1.16 will imply that Theorem 1.2 holds for $p$, contrary to assumption.

Set $T = C_R(B)$, so that $T \neq 1$ by assumption. We can assume $T$ is cyclic, whence $T < R$. Let $S$ be a minimal $B$-invariant subgroup of $N_R(T)$ containing $T$ properly ($S$ exists by [I, 1.10]). Then $\bar{S} = S/T$ is elementary and $B$ acts irreducibly and nontrivially on $\bar{S}$. We can then find $Z_p \times Z_p \cong U \leqslant B$ such that $B$ does not centralize $\bar{V} = C_{\bar{S}}(U)$, by [I, 4.13]. But then if $V$ denotes the pre-image of $\bar{V}$ in $S$, it is immediate that $V$ is noncyclic and $V$ centralizes $U$, so $U$ has the required property.

This is the final configuration which Bender must eliminate. Set $Z = \Omega_1(Z(R))$. Then $B$ does not centralize $Z$. Again by [I, 4.13], there is $Z_p \times Z_p \cong U \leqslant B$ such that $B$ does not centralize $V = C_Z(U)$. If $C_u \leqslant M$ for every $u \in U^\#$, then Theorem 1.2 would hold for $p$ by Propositions 1.4 and 1.6, so $C_u \not\leqslant M$ for some $u \in U^\#$. Let $N$ be a maximal subgroup of $G$ containing $C_u$, so that $V \leqslant N$.

The argument splits into two cases, according as $N$ is or is not of restricted type. Bender first proves:

LEMMA 1.18. $N$ *is not of restricted type for any prime* $r$.

PROOF. Assume false. We argue that $M$ has a normal subgroup of index $p$, contrary to Proposition 1.12(ii). Set $T = P \cap O_{p'p}(M)$, so that $T \in \mathrm{Syl}_p\big(O_{p'p}(M)\big)$ and $M = O_{p'}(M)H$, where $H = N_M(T)$, by the Frattini argument. Thus it will suffice to show that $H$ has a normal subgroup of index $p$.

By (1.3)(1) we have $B \leqslant T$. However, $C_B \leqslant C_u \leqslant N$ (as $u \in B$). Since $N$ is assumed to be of restricted type, it follows therefore from Lemma 1.14 that $N_G(T)$ and hence $H$ is contained in $N$. Thus $H$ normalizes $W = Z \cap N$. Also $V \leqslant W$.

We claim that $W$ is cyclic; so assume the contrary. If $W \lhd Q$, then as $m(Q) \geqslant 3$, $m\big(C_Q(W)\big) \geqslant 3$. In the contrary case, $W < Z$ and again $m\big(C_Q(W)\big) \geqslant 3$. Thus $C_w \leqslant M$ for each $w \in W^\#$ as Theorem 1.2 holds for $q$. Hence by Proposition 1.4, applied to $q$, $M$ is the unique maximal subgroup of $G$ containing $W$. Since $W \leqslant N$, it follows that $N \leqslant M$—contradiction. This proves our claim.

Thus, $V = W$, and so $V \cong Z_q$. Hence $H$ normalizes $V$, and as $V$ is cyclic, $\bar{H} = H/C_H(V)$ is abelian. However, $B$ does not centralize $V$ by choice of $V$, so $|\bar{H}|$ is divisible by $p$, and we conclude that $\bar{H}$ and hence $H$ has a normal subgroup of index $p$, giving the desired contradiction.

Finally, Bender proves the following result:

LEMMA 1.19. *We have* $O_{q'}(N) = 1$.

Indeed, we argue first that $V \leqslant O_q(N)$. Since $N$ is not of restricted type, $O_r(N)$ has rank $\leqslant 2$ for each prime $r \neq q$. But $V = Z \cap N \cong Z_q$, as shown in the previous lemma. (That argument did not depend on the restrictedness of $N$.) Since $B$ does not centralize $V$, it follows that $V = [V, B]$. But now as $VB$ acts on $O_r(N)$, Lemma 1.10(ii) implies that $V$ centralizes $O_r(N)$, so $V$ centralizes $X = O_{q'}\big(F(N)\big)$. On the other hand, Theorem 1.2 holds for $q$, so if $m_q(N) \geqslant 3$, it would follow that $N$ is a conjugate of $M$ in $G$, which is impossible as $M$ is of restricted type for $q$, while $N$ is not. Hence, $m_q(N) \leqslant 2$. In particular, $m\big(O_q(N)\big) \leqslant 2$.

Now set $C = C_N(X)$, so that $V \leqslant C$. Since $F(N) = XO_q(N)$ and $C_N\big(F(N)\big) \leqslant F(N)$, $D = C_C\big(O_q(N)\big) \leqslant F(N)$. But now applying Lemma 1.10(ii) once again, it follows that a Sylow $q$-subgroup of $C/D$ is normal in $C/D$. Hence if we let $V \leqslant S \in \mathrm{Syl}_q(C)$, we see that $DS$ char $C$ and so $DS \lhd N$. However, as $D \leqslant F(N)$, it is immediate that $DS = (D \cap X) \times S$, and therefore $S \lhd N$, whence $S \leqslant O_q(N)$. Thus, $V \leqslant O_q(N)$, as asserted.

Next set $K = C_{O_q(N)}(V)$. We claim that $C_K \leqslant M \cap N$. We have $C_K \leqslant C_V$ as $V \leqslant K$. Also $Z \leqslant C_V$ as $V \leqslant Z$, so if $Z$ is noncyclic, then $C_V \leqslant M$, as in the previous lemma. In the contrary case, $V = \Omega_1(Z(R))$, so $R \leqslant C_V \leqslant M$ as $M$ is the unique maximal subgroup of $G$ containing $R$. In either case, $C_V$ and hence $C_K$ is contained in $M$. On the other hand, $K$ contains $Z(O_q(N))$ and so $C_K \leqslant N_G(Z(O_q(N)))$. However, the latter group is $N$ as $N$ is maximal and $1 \neq Z(O_q(N)) \lhd N$. Thus, $C_K \leqslant M \cap N$, as claimed.

Finally, set $Y = O_{q'}(N)$. Since $C_K \leqslant N$, it follows that $Y \leqslant O_{q'}(C_K)$; and as $C_K \leqslant M$, this implies that $Y \leqslant O_{q'}(C_M(K))$. But now it is an easy consequence of the $(A \times B)$-lemma [applied in $M/O_{q'}(M)$], together with the solvability of $M$, that $Y \leqslant O_{q'}(M)$. But then $Y$ centralizes $R = O_q(M)$, and so $R \leqslant C_Y$. However, if $Y \neq 1$, then as $Y \lhd N$, $N = N_G(Y)$ by the maximality of $N$, so $R \leqslant N$, forcing $M = N$—contradiction. Hence we must have $Y = 1$, and the lemma is proved.

Now Bender can derive a final contradiction at once. We have $F(N) = O_q(N)$ by Lemma 1.19, and, in addition, $m(O_q(N)) \leqslant 2$. Again by Lemma 1.10(ii), these conditions imply that $O_q(N) \in \mathrm{Syl}_q(N)$. But $N = N_G(O_q(N))$, so as usual, $O_q(N) \in \mathrm{Syl}_q(G)$, a contradiction as $m_q(G) \geqslant 3$. This completes discussion of the uniqueness theorem.

## 1.2. Solvability of Groups of Odd Order

Feit and Thompson use their uniqueness theorem as a basic tool in analyzing the maximal subgroups of $G$, meshing its conclusions for various primes simultaneously. They thus consider the set $\pi$ of primes $p$ for which $m_p(G) \geqslant 3$. On $\pi$, they define a relation $p \sim q$ for $p, q \in \pi$ by the condition that $G$ contains a $\{p, q\}$-subgroup having both $p$-rank $\geqslant 3$ and $q$-rank $\geqslant 3$; and it is an immediate consequence of the uniqueness theorem that $\sim$ is an equivalence relation on $\pi$.

In view of the uniqueness theorem, this means that with each equivalence class $\alpha$ of $\pi$ under $\sim$, there is associated a unique conjugacy class $\{M_\alpha\}$ of maximal subgroups of $G$ with the property that $M_\alpha$ is a uniqueness subgroup for $p \in \pi$ if and only if $p \in \alpha$. The uniqueness theorem and its corollary, Proposition 1.3, put severe restrictions on the structure and embedding of each $M_\alpha$, since they imply that for all $p \in \alpha$:

(a) $M_\alpha$ has no normal subgroup of index $p$;
(b) If $C_y \not\leqslant M_\alpha$ for $y \in M_\alpha^{\#}$, then $m_p(C_y) \leqslant 2$; and $\qquad$ (1.4)
(c) $M_\alpha \cap M_\alpha^g$ has cyclic Sylow $p$-subgroups.

However, one is still a long way from the exact structure and embedding of $M_\alpha$; and, in addition, one has no information yet about other maximal subgroups of $G$. Clearly sharper results will require an analysis of the embedding of $p$-subgroups of $G$ for primes $p$ for which $m_p(G) \leqslant 2$. This turns out to be extremely elaborate and the argument covers more than half of [I: 93, Chapter IV]. Recently, Bender has made significant improvements in Thompson's original analysis, but it is still a very delicate matter.

Here we mention only one important preliminary fact.

PROPOSITION 1.20. *Let $X$ be a solvable group of odd order in which* $m_p(X) \leqslant 2$ *for all primes $p$. Let $p_1 > p_2 > \cdots > p_n$ be the distinct prime divisors of $|X|$ and let $P_i \in \mathrm{Syl}_{p_i}(X)$, $1 \leqslant i \leqslant n$. Then for all $i$, $1 \leqslant i \leqslant n$,*

$$P_1 P_2 \cdots P_i \lhd X.$$

One expresses this conclusion by saying that $G$ possesses a *Sylow tower*.

We note that the proposition is an immediate consequence of Lemma 1.10(i) together with the fact that the prime divisors of $|GL_2(q)|$, $q$ odd, are all $\leqslant q$.

Under the assumption that the centralizer of every nonidentity element of $G$ is nilpotent, Feit, Hall, and Thompson [I: 91], generalizing the work of Suzuki in the centralizer abelian case [I: 275], had earlier shown that every maximal subgroup $M$ of $G$ was a Frobenius group whose kernel was a T.I. (trivial intersection) set in $G$ (i.e., $M$ is disjoint from its distinct conjugates; see [I, Section 1.1]). Now, in the general situation Feit and Thompson's objective was to approximate these conclusions as closely as possible. The exact statement is very technical, so we shall limit ourselves to an approximate version only, which requires the following terminology.

DEFINITION 1.21. A group $X$ will be called of *Frobenius type* with *kernel $K$* provided the following conditions hold:

(1) $K = F(X)$ is a Hall subgroup of $X$ (i.e., $|K|$ and $|X : K|$ are coprime);
(2) If $A$ is a complement to $K$ in $X$, then $A$ contains a subgroup $B$ of the same exponent as $A$ such that $BK$ is a Frobenius group with kernel $K$ and complement $B$; and
(3) The nonprincipal irreducible characters of $K$, when induced to $X$, have properties similar to those which hold in a Frobenius group.

(Note that $A$ exists by the Schur–Zassenhaus theorem [I, p. 51, footnote].)

DEFINITION 1.22. A group $X$ will be called a *three-step* group provided the following conditions hold:

(1) $X = X'A$, where $X' \cap A = 1$ and $A$ is a cyclic Hall subgroup of $X$;

(2) If $K = F(X)$, then $(X')' \leqslant K \leqslant X'$;

(3) If $H$ is the maximal nilpotent normal Hall subgroup of $X$ contained in $K$, then $H$ is noncyclic and $K = HC_X(H)$; and

(4) $H$ contains a cyclic subgroup $R$ such that $C_{X'}(a) = R$ for all $a \in A^{\#}$.

For uniformity of notation, we call $K$ the *kernel* of $X$ and $A$ the *complement* of $X$ in this case as well.

We give an example of a three-step group, which will help to clarify the notion and explain the terminology. Let $p$, $q$, $r$ be distinct odd primes such that $q$ divides $p - 1$ and $r$ divides $q - 1$. Then $\text{Aut}(Z_q)$ has order divisible by $r$ and so there exists a nontrivial extension $Y$ of $Z_q$ by $Z_r$. It is immediate that $Y$ is a Frobenius group with kernel $Q$ of order $q$ and complement $A$ of order $r$. Since $q \mid (p - 1)$, it is easily seen by Clifford's theorem [I, 1.5] that $Y$ can be represented faithfully on a vector space $K$ of dimension $r$ over $Z_p$ with $Q$ leaving invariant $r$ one-dimensional subspaces of $K$ that are cyclically permuted by $A$.

We let $X$ be the semidirect product of $K$ by $Y$. Then $X' = KQ$ and $X = KY = X'A$. Furthermore, $K = F(X) = (X')'$, $K$ itself is a Hall subgroup of $X$, and $K = C_X(K)$. Finally, the action of $A$ on $K$ implies that $C_K(A) = R$ has order $p$ and that $R = C_X(a)$ for all $a \in A^{\#}$. Hence, by the definition $X$ is a three-step group.

DEFINITION 1.23. A subgroup $H$ of $G$ will be said to be a *quasi-T.I.* set in $G$ provided either

(a) $H$ is a T.I. set in $G$; or

(b) If $p$ is a prime divisor of $|H|$, then either $m_p(G) \leqslant 2$ or else a Sylow $p$-subgroup $P$ of $G$ contains an element $x$ such that $C_P(x) = \langle x \rangle \times D$, where $D$ is a cyclic group.

When $H$ satisfies the conditions of (b), one cannot prove at this stage that $H$ must be a T.I. set in $G$. Furthermore, an additional (technical) requirement should be placed on the primes $p$ in (b), which for simplicity we have omitted.

DEFINITION 1.24. A maximal subgroup $M$ of $G$ will be said to be of

*F-type* or *T-type*, respectively, if $M$ is of Frobenius type or is a three-step group and if the kernel of $M$ is a quasi-T.I. set in $G$.

(In the odd-order paper itself, Feit and Thompson must distinguish four different types of three-step groups, which they designate type II, type III, type IV, and type V. Our *T*-type group combines their four categories.)

Now we can state Feit and Thompson's first main result concerning the maximal subgroups of $G$.

THEOREM 1.25. *The following conditions hold*:

   (i) *Every maximal subgroup of G is either of F-type or T-type*;
  (ii) *Either every maximal subgroup of G is of F-type or there are precisely two conjugacy classes of maximal subgroups of T-type; and*
 (iii) *If G has two conjugacy classes of maximal subgroups of T-type, we can choose representatives $M_1$, $M_2$ and complements $A_1$ in $M_1$ and $A_2$ in $M_2$ such that*
    (1) $A = A_1 A_2 = A_1 \times A_2$ *is cyclic with $A_i \neq 1$, $i = 1, 2$; and*
    (2) *If $A_0$ is a subset of $A$ with $A_0 \not\leq A_i$, $i = 1$ or $2$, then $N_G(A_0) = A$.*

The theorem shows the inherent complexity of the general odd-order problem when compared to the centralizer nilpotent special case. Even if every maximal subgroup is of *F*-type, the conclusions are weaker than before, since now the maximal subgroups only approximate Frobenius groups and, in addition, their kernels need not be disjoint from their conjugates in $G$. Nevertheless, Feit and Thompson were eventually able to treat this case along the same lines as the centralizer nilpotent case. Rather it is the existence of maximal subgroups of *T*-type which causes the major difficulty and accounts for the bulk of the subsequent analysis. Note that $A = N_G(A)$ in (iii), so that the problem is directly related to that of determining all simple groups (if any) which possess a self-normalizing cyclic subgroup.

If a maximal subgroup $M$ of $G$ is a Frobenius group whose kernel $K$ is a T.I. set in $G$, it is immediate that

$$C_x \leqslant K \qquad \text{for all } x \in K^\#, \tag{1.5}$$

a crucial requirement for smooth application of exceptional character theory. These conditions were indeed satisfied by every maximal subgroup in the centralizer nilpotent case, so in that case the method could be applied to a representative of every conjugacy class of maximal subgroups. Emulating Suzuki's calculations in the prior centralizer abelian case, Feit, Hall, and Thompson were then able to derive sufficient information about the characters of $G$ to reach an ultimate arithmetic contradiction.

Feit and Thompson were now faced with the formidable problem of extending this character-theoretic analysis to the much looser situation embodied in Theorem 1.25. If the kernel $K$ of a maximal subgroup $M$ of $G$ is a T.I. set in $G$, we at least know that $C_x \leqslant M$ for $x \in K^\#$, even though we can no longer necessarily assert that $C_x \leqslant K$ (unless $M$ is a Frobenius group). However, in the contrary case, we may well have $C_x \not\leqslant M$ for some $x \in K^\#$. Without some hold on the structure of $C_x$, it was apparent to Feit and Thompson that it would be impossible to carry through the necessary extension. Furthermore, inasmuch as $C_M(x)$ need not lie in $K$ for $x \in K^\#$, they realized they would also need information about $C_y$ for suitable elements $y \in M - K$.

The next step was therefore to determine the precise subset of elements of $M$ on which to focus and then to determine the structure of the centralizers in $G$ of these elements. The conditions are incorporated in the following two basic definitions. The key concept is that of a "tamely embedded" subset of a group, which we state here in simplified form to avoid technicalities.

DEFINITION 1.26. Let $Y$ be a subset of a group $X$ and set $N = N_X(Y)$. We say that $Y$ is *tamely embedded* in $X$ provided two elements of $Y$ are conjugate in $X$ if and only if they are conjugate in $N$ and one of the following holds;

(a) $C_X(y) \leqslant N$ for all $y \in Y^\#$; or
(b) There exists a set of nonidentity subgroups $U_i$ of $X$ with the following properties:
   (1) $U_i$ and $U_j$ have coprime orders for all $i \neq j$, $1 \leqslant i, j \leqslant n$;
   (2) If we set $N_i = N_X(U_i)$, then $U_i$ and $N_i/U_i$ have coprime orders, $1 \leqslant i \leqslant n$;
   (3) $U_i \cap Y = 1$, $1 \leqslant i \leqslant n$;
   (4) $N_i = U_i(N \cap N_i)$, $1 \leqslant i \leqslant n$; and
   (5) For every $y \in Y^\#$, there exists an $N$-conjugate $y^*$ of $y$ and an index $i$ such that

$$C_X(y^*) = C_{U_i}(y^*) \, C_N(y^*) \leqslant N_i.$$

DEFINITION 1.27. Let $M$ be a maximal subgroup of $G$ and let $K$ be the kernel of $M$. We define the *singularity* subset $M_0$ of $M$ as follows

(a) If $M$ is of $F$-type, then $M_0 = \bigcup_{x \in K^\#} C_M(x)$; and
(b) If $M$ is of $T$-type, then either $M_0 = M'$ or $M_0 = \bigcup_{x \in H^\#} C_M(x)$, where $H$ is the nilpotent normal Hall subgroup of $M$ contained in $K$, specified in the definition of a three-step group.

As we have already mentioned, Feit and Thompson distinguish four different $T$-type situations; correspondingly they specify which alternative of (b) should be taken as the definition of $M_0$.

In this terminology, their main result is as follows:

THEOREM 1.28. *Let $M$ be a maximal subgroups of $G$ and let $M_0$ be the singularity subset of $M$. Then $M = N_G(M_0)$ and $M_0$ is tamely embedded in $G$.*

This completes the local analytical portion of the proof, and now the stage is set for character theory. First, we recall some elementary general facts and terminology. For any group $X$, a *generalized* character of $X$ is by definition a $\mathbb{Z}$-linear combination of irreducible characters of $X$. The set of all such generalized characters thus forms a $\mathbb{Z}$-module $\mathscr{C}(X)$. $\mathscr{C}(X)$ is endowed with an inner product, defined for $\phi$, $\theta \in \mathscr{C}(X)$, by the rule

$$(\phi, \theta)_X = \frac{1}{|X|} \sum_{x \in X} \phi(x) \, \overline{\theta(x)}, \tag{1.6}$$

where bar denotes complex conjugation. [If we express $\phi = \sum_{i=1}^{n} a_i \chi_i$ and $\theta = \sum_{i=1}^{n} b_i \chi_i$, where the $\chi_i$ denote the distinct irreducible characters of $X$, and $a_i, b_i \in \mathbb{Z}$, $1 \leqslant i \leqslant n$, it is immediate that $(\phi, \theta)_X = \sum_{i=1}^{n} a_i b_i$.] In particular, $(\phi, \phi)_X$ is called the "norm" of $\phi$.

For any subgroup $Y$ of $X$, the induction map from characters of $Y$ into characters of $X$ extends in a natural way to a map $*$ from $\mathscr{C}(Y)$ into $\mathscr{C}(X)$ (see [I: 130, Chapter 4.4] for the definition of the induction map).

If $\mathscr{S}(Y)$ is a submodule of $\mathscr{C}(Y)$, we say that $*$ is an *isometry* on $\mathscr{S}(Y)$ if for $\phi$, $\theta \in \mathscr{S}(Y)$, we have

$$(\theta, \phi)_Y = (\theta^*, \phi^*)_X.$$

In other words, $*$ is "norm-preserving" on $\mathscr{S}(Y)$.

For any submodule $\mathscr{S}(X)$ of $\mathscr{C}(X)$, we denote by $\mathscr{S}_0(X)$ the subset of elements of $\mathscr{S}(X)$ which take the value 0 on the identity element of $X$.

Exceptional character theory deals with the following two basic questions concerning a group $X$, a subgroup $Y$, and a submodule $\mathscr{S}(Y)$ of $\mathscr{C}(Y)$:

   A. Under suitable assumptions on the structure and embedding of $Y$ in $X$ and for a suitable choice of $\mathscr{S}(Y)$, does $*$ induce an isometry of $\mathscr{S}_0(Y)$ into $\mathscr{C}_0(X)$?
   B. If $(A)$ holds for $Y$ and $\mathscr{S}(Y)$, does $*$ extend to an isometry $\tau$ of $\mathscr{S}(Y)$ into $\mathscr{C}(X)$?

If (A) and (B) both hold for $\mathscr{S}(Y)$, we say that $\mathscr{S}(Y)$ is *coherent*.

The significance of (A) and (B) is that it enables one to relate the character values $\theta(u)$ for $\theta \in \mathscr{S}_0(Y)$ and $\mathscr{S}(Y)$ and $u \in Y$ with $\theta^*(u)$ or $\theta^\tau(u)$, respectively.

A theorem of Brauer and Suzuki [I: 130, Theorem 4.5.4] asserts that if $Y$ is a Frobenius group with kernel $K$ and $K$ is a T.I. set in $X$, then indeed (A) holds for either the submodule $\mathscr{T}(Y)$ generated by the characters of $Y$ induced from nontrivial irreducible characters of $K$ or the submodule $\mathscr{T}'(Y)$ consisting of the elements of $\mathscr{C}(Y)$ which vanish on $Y - K$ [it is easily seen that, in fact, $\mathscr{T}(Y) \subseteq \mathscr{T}'(Y)$].

The whole point of tame embedding is that in its presence Feit and Thompson are able to extend the Brauer–Suzuki theory to pairs $Y, K$, where $Y$ is maximal subgroup of their minimal counterexample $G$ and $K$ is the kernel of $Y$. However, in this more general situation, they are forced to replace the induction map $*$ by a more delicately defined mapping $\sigma$ of the given submodule of $\mathscr{C}_0(Y)$ into $\mathscr{C}_0(G)$.

Concerning coherence, Feit's theorem [I: 130] gives sufficient conditions on $Y$, again in the case that $Y$ is a Frobenius group whose kernel $K$ is a T.I. set in $G$, for $\mathscr{T}(Y)$ to be coherent.

However, in the centralizer nilpotent case, Feit, Hall, and Thompson had been able to avoid the subtleties of Feit's theorem by working instead with the submodules $\mathscr{T}_d(Y)$ generated by the characters of $Y$ induced from nontrivial irreducible characters of $K$ of *degree d*. It is then automatic that each $\mathscr{T}_d(Y)$ is coherent. Since $\mathscr{T}(Y)$ can be shown to be the direct sum of its submodules $\mathscr{T}_d(Y)$, the coherence of the $\mathscr{T}_d(Y)$ is sufficient for the analysis (see [I: 130, Theorem 4.4.6]). Unfortunately this approach does not work in the general odd-order problem, and so the question of coherence must now be faced head on.

(We remark parenthetically that these coherence-type arguments have been limited to situations in which local subgroups possess Frobenius-like structures. In practice, these have arisen, only in groups of low 2-rank or with low class Sylow 2-subgroups.)

Let then $M$ be a maximal subgroup of $G$, let $K$ be the kernel of $M$, and $M_0$ the singularity subset of $M$. We define $\mathscr{I}'(M)$ to be submodule of $\mathscr{I}(M)$ of generalized characters which vanish on $M - M_0$. If $C_y \leqslant M$ for $y \in M_0^\#$ [(a) of Definition 1.26], then $*$ is again an isometry on $\mathscr{I}_0'(M)$. The difficulty arises when (b) of Definition 1.26 occurs. Let $U_i$, $N_i$, $1 \leqslant i \leqslant n$, be the corresponding subgroups specified in the definition (with $M$, $M_0$ in place of $X$, $Y$, respectively). For $\theta \in \mathscr{I}'(M)$, denote by $\theta_i$ the restriction of $\theta$ to $M \cap N$, $1 \leqslant i \leqslant n$. Since $M = N_G(M_0)$, $N_i = U_i(M \cap N_i)$ with $U_i \cap M = 1$ by Definition 1.26(b). It is clear from these conditions that $\mathscr{C}(N_i)$ contains an element $\theta_{i1}$ which is the difference of two characters of $N_i$ each having $U_i$ in its kernel with the property that $\theta_i$ is the restriction of $\theta_{i1}$ to $M \cap N_i$. On the other hand, we can induce $\theta_i$ from the subgroup $M \cap N_i$ to the group $N_i$, obtaining a character which we denote by $\theta_{i2}$. Feit and Thompson now define a mapping $\sigma$ from $\mathscr{I}_0'(M)$ into $\mathscr{C}_0(G)$ by the rule

$$\theta^\sigma = \theta^* + \sum_{i=1}^{n} (\theta_{i1} - \theta_{i2})^*, \tag{1.7}$$

and then go on to prove the following result:

PROPOSITION 1.29. $\sigma$ is an isometry of $\mathscr{I}_0'(M)$ into $\mathscr{C}_0(G)$.

Subsequently, Dade [17] gave a more conceptual treatment of this portion of their work.

Feit's theorem shows how complex the question of coherence is even for Frobenius groups, so it is hardly surprising that the problem is enormously difficult in the general odd-order context.

The easiest case for Feit and Thompson is that of maximal subgroups of $F$-type.

PROPOSITION 1.30. Let $M$ be a maximal subgroup of $G$ of $F$-type with kernel $K$ and complement $E$ and assume that $E = AB$, where $B$ is abelian, $A$ is cyclic, and $A$, $B$ have coprime order. If $\mathscr{I}(M)$ denotes the submodule of $\mathscr{C}(M)$ generated by the irreducible characters of $M$ not having $K$ in their kernel, then $\mathscr{I}(M)$ is coherent.

A large portion of Chapter V of the odd-order paper is taken up with a proof of an analogous result for maximal subgroups $M$ of $T$-type. It is for this analysis that Feit and Thompson must divide such maximal subgroups into four subtypes, which we shall here designate as $T_1$, $T_2$, $T_3$, $T_4$, respectively. To give some idea of the subdivision, we describe its principal distinguishing features.

Let $H$ be the Hall nilpotent normal subgroup of $M$ contained in the kernel $K$ of $M$. Then $H \leqslant K \leqslant M'$ and as $H$ and $M'/H$ have coprime order, $H$ possesses a complement $E$ in $M'$. Then corresponding to the four subtypes, we have

$T_1$-type:   $E$ is abelian, $E \neq 1$, and $N_G(E) \not\leqslant M$.
$T_2$-type:   $E$ is abelian, $E \neq 1$, and $N_G(E) \leqslant M$.
$T_3$-type:   $E$ is nonabelian and $N_G(E) \leqslant M$.
$T_4$-type:   $E = 1$.

Furthermore, in carrying out their analysis, Feit and Thompson must consider the *pair* of maximal subgroups $M_1$ and $M_2$ of $T$-type whose existence is specified in Theorem 1.25 and study them both *individually* and *in conjunction*. Note that, *a priori*, $M_1$ and $M_2$ may or may not be of the same subtypes. However, the precise form of Theorem 1.25 which Feit and Thompson establish yields the following additional fact.

PROPOSITION 1.31. *Either $M_1$ or $M_2$ is of $T_1$-type.*

For simplicity of notation, we put $M = M_1$ or $M_2$. Now define $\mathscr{S}(M)$, $\mathscr{T}(M)$ to be the submodules of $\mathscr{C}(M)$ generated by the characters of $M$ induced, respectively, from irreducible characters of $M'$ not having $H$ in their kernel or from all nontrivial irreducible characters of $M'$.

We now state Feit and Thompson's main moves (in simplified form) to give the reader some idea of the elaborateness of the analysis. Individual portions of the argument involve spectacular arithmetic calculations. Keep in mind that conclusions about $M$ very often involve calculations concerning both $M_1$ and $M_2$.

PROPOSITION 1.32. *If $M$ is of $T_i$-type, $1 \leqslant i \leqslant 3$, and $\mathscr{S}(M)$ is not coherent, then $i = 1$ with $H$ a 3-group and $M_j$ is of $T_4$-type for $M_j \neq M$, $j = 1$ or 2.*

PROPOSITION 1.33. *If $M$ is of $T_4$-type, then $\mathscr{T}(M)$ is coherent.*

PROPOSITION 1.34. *$M$ is not of $T_4$-type.*

Thus $M_1$ and $M_2$ are of $T_i$-type, $1 \leqslant i \leqslant 3$. In particular, it follows now from Proposition 1.32 that $\mathscr{I}(M)$ is coherent.

They next prove the following proposition.

PROPOSITION 1.35. *Every maximal subgroup of G of F-type is a Frobenius group.*

This conclusion holds whether or not there exist maximal subgroups $M_1$, $M_2$ in $G$ of $T$-type. However, in the latter case, every maximal subgroup of $G$ is a Frobenius group, in which case Feit and Thompson establish the following result by a short argument, with the aid of Burnside's normal complement theorem [I, 1.20].

PROPOSITION 1.36. *If every maximal subgroup of G is of F-type, then the centralizer of every nonidentity element of G is nilpotent.*

But now they are in a position to invoke the main result of [I: 91] to exclude this possibility. Thus, they have proved the following result:

THEOREM 1.37. *G possesses a maximal subgroup of $T_1$-type.*

In particular, the subgroups $M_1$ and $M_2$ of Theorem 1.25 exist. This is by far the most difficult case; and Feit and Thompson now wring every last drop from the character theory to establish the following culminating result.

THEOREM 1.38. *There exists a pair of primes $p$ and $q$ with $p > q$ dividing $|G|$ for which the following conditions hold*:

(i) *$(p^q - 1)/(p - 1)$ is prime to $p - 1$;*
(ii) *A Sylow p-subgroup P of G is elementary abelian of order $p^q$, a Sylow q-subgroup of G is elementary abelian of order $q^p$, and each is a T.I. set in G;*
(iii) *$M = N_G(P)$ is a maximal subgroup of G of $T_1$-type, with $M = PEQ^*$ where PE and $EQ^*$ are Frobenius groups with respective kernels P and E and cyclic complements E and $Q^*$ of respective orders $(p^q - 1)/(p - 1)$ and q;*
(iv) *If $P^* = C_P(Q^*)$, then $P^* = |p|$ and $P^*Q^*$ is a self-normalizing cyclic subgroup of G of order pq;*

(v) $C_{p^*} = PQ^*$ and $C_{Q^*} = QP^*$;

(vi) $C_E$ is a cyclic T.I. subgroup of $G$; and

(vii) $N_G(E)$ is a Frobenius group with kernel $C_E$ and cyclic complement of order $pq$.

$N_G(Q)$ satisfies similar conditions (in particular, $M_1$ and $M_2$ are both of $T_1$-type); however, explicit properties of $N_G(Q)$ are not needed for the subsequent argument. Furthermore, it should be noted that this configuration of subgroups is possible even under the assumption that every Sylow subgroup of $G$ is abelian!

Sibley's contributions to the odd-order theorem [108] consist of improved methods of establishing coherence in several of the preceding situations. This allows for some reorganization in the flow diagram of the proof and at the same time reduces the number and complexity of some of the arithmetic calculations needed to derive Theorem 1.38 from Theorems 1.25 and 1.28.

Furthermore, Thompson [123] also showed that the configuration of Theorem 2.19 has an intriguing number-theoretic consequence. Indeed, $N_G(Q)$ contains a cyclic subgroup $F$ of order $(q^p - 1)/(q - 1)$ and Thompson was able to prove that either conditions (vi) and (vii) of Theorem 1.38 hold with $F$ in place of $E$ or else $E$ is conjugate to a subgroup of $F$. Moreover, using exceptional character theory relative to both $E$ and $F$, he argued that the first possibility leads to a contradiction. However, that argument collapses in the second case as then the exceptional character theory of $E$ and $F$ coalesce. But as $E$ is then conjugate to a subgroup of $F$, Lagrange's theorem implies that $|E|$ divides $|F|$, whence

$$(p^q - 1)/(p - 1) \text{ divides } (q^p - 1)/(q - 1). \tag{1.8}$$

Remarkably, there exists no known pair of odd primes satisfying (1.8)! However, no one has yet shown that there cannot be such a pair of primes. Clearly such a result would immediately eliminate this final configuration and thus complete the proof of the odd-order theorem.

It was a full year before Thompson found a way of dealing with this remaining configuration: by getting down to the very multiplication table of the group and analyzing relations among the generators of $M = N_G(P)$. Regarding $P$ as the additive group of the field $GF(p^q)$, this generator analysis information ultimately translated into a question about the *nonexistence* of solutions of a specific system of equations over $GF(p^q)$, whose resolution immediately implied the nonexistence of a group $G$ satisfying the conditions of Theorem 1.38, thereby completing the proof of the odd-order theorem.

There is an amazingly close parallel between Thompson's approach here and the analysis of the final configurations of both the Bombieri–Thompson characterization of the Ree groups [I, pp. 164–168] and the O'Nan characterization of the unitary groups of odd characteristic [I, pp. 157–160]. Indeed, in the latter two doubly-transitive group problems, the group multiplication formulas lead to certain relations among a canonical set of generators of the group $G$ under investigation. This, in turn, translates into a question concerning the *uniqueness* of the solution of a specific system of equations over $GF(p^n)$ ($p = 3$ in the Ree case and $p$ an appropriate odd prime in the unitary case). Once uniqueness is established, it follows at once that $G$ is correspondingly isomorphic to a Ree or unitary group, giving the desired characterization.

Because of this strong parallel, finite group theorists have felt that the configuration of Theorem 1.38 was an intrinsic one, not simply a consequence of the "accidental nature" of the Feit–Thompson proof. For myself, I prefer a more colorful formulation: the given configuration of Theorem 1.38 represents a simple group in "embryonic form," and only a "miscarriage" resulting from the fact that its associated structure equations happen to possess no solutions has prevented its ultimate birth.

However one prefers to view it, there is no question that the analysis of this final case, despite its technical difficulties, represents one of the most beautiful and profound chapters of simple group theory. The total argument occupies exactly 26 pages [I: 93, Chapter VI] and is completely self-contained, so that it is quite accessible to a nonexpert in finite group theory. We shall outline it briefly here, but we hope our discussion will stimulate the reader to examine Thompson's wondrous proof in full detail.

Let $u$, $v$ be generators of $E$ and $A$, respectively, and set $e = |E|$. Then as $|u| = e = (p^q - 1)/(p - 1)$, $u$ acts irreducibly on $P$, regarded as a vector space over $Z_p$. Hence if $\omega$ is a characteristic root of $u$ on $P$, then $Z_p(\omega) = GF(p^q)$ and $\omega$ is a primitive $e$th root of unity over $Z_p$.

Thompson first uses the given conditions of Theorem 1.38 to analyze the action of $A$ on $E$ and proves the following result:

PROPOSITION 1.39. *Let* $x_1$, $x_2$, $x_3$ *be elements of* $Z_e$ *not all* 0. *Then we have*

$$vu^{x_1}v^{-2}u^{x_2}vu^{x_3} = 1$$

*if and only if*

    (i) $x_i \neq 0$, $1 \leqslant i \leqslant 3$;
    (ii) $x_1 + x_2 + x_3 = 0$; *and*
    (iii) $\omega^{x_1} + \omega^{-x_2} = 2$.

Thompson now focuses on the set $\mathcal{A}$ of all triples $(x_1, x_2, x_3)$ of elements of $Z_e$ which satisfy the three conditions of the proposition and first proves the following:

PROPOSITION 1.40. *The following conditions hold:*

    (i) $\mathcal{A}$ *is nonempty; and*
    (ii) *if* $(x_1, x_2, x_3) \in \mathcal{A}$, *then also* $(-x_2, -x_1, -x_3) \in \mathcal{A}$.

The crucial step in the argument is verification of the following single additional property of $\mathcal{A}$:

PROPOSITION 1.41. *If* $(x_1, x_2, x_3) \in \mathcal{A}$, *then there exist elements* $x_2', x_3' \in Z_e$ *such that* $(-x_1, x_2', x_3') \in \mathcal{A}$.

The proof of this result requires spectacularly elaborate manipulation of highly complex words in the elements $u, v$ and their conjugates. Underlying the argument is the following almost immediate consequence of our given conditions:

LEMMA 1.42. *For some* $y \in Q^\#$, $A = \langle v \rangle$ *normalizes* $uEu^{-1}$.

It follows that under conjugation $v$ transforms $yuy^{-1}$ into some power of itself, so for some integer $m \in Z_e$, we have

$$(yuy^{-1})^v = (yuy^{-1})^m. \tag{1.9}$$

Thompson next shows that $u, v$, and $y$ are connected by the following intricate set of identites:

LEMMA 1.43. *For all* $k \in Z_p$, *if we set* $y_k = [y, v^k]$, *then we have*

$$v^{-k} u v^k = y_k^{-1} u^{v^k} y_k.$$

From these relationships, Thompson obtains a further restriction on $\mathcal{A}$. (Here $y_1, y_2, y_3$ denote $y_k$ for $k = 1, 2, 3$, respectively.)

LEMMA 1.44. *Set* $z = yuy^{-1}$. *if* $(x_1, x_2, x_3) \in \mathcal{O}$, *then we have*

$$y_2 z^{x_1 m} y_3^{-1} y_1 z^{x_2 m^3} y_2^{-1} = y_1 z^{-x_3 m^2} y_3^{-1}.$$

It is this extremely unwieldy identity which Thompson must exploit to establish Proposition 1.41.

We hope this brief summary gives some indication of its complexity.

Now Thompson is in a position to translate the problem into a question concerning the solution of a system of equations over $GF(p^q)$. For $x \in GF(p^q)$, set

$$N(x) = x^{1 + p + \cdots + p^{q-1}}, \qquad (1.10)$$

and for $x \in GF(p^q)$ with $x \neq 2$, define

$$x^\sigma = 1/(2 - x). \qquad (1.11)$$

Note that if $x \in GF(p^q) - GF(p)$, then as $2 \in GF(p)$, it follows that $x^{\sigma^i} \in GF(p^q) - GF(p)$ for all $i$, so each $x^{\sigma^i} \neq 2$.

Thompson proves the following result:

PROPOSITION 1.45. *There exists an element* $x \in GF(p^q) - GF(p)$ *such that for all* $i \geqslant 0$,

$$N(x^{\sigma^i}) = 1.$$

Indeed, let $\mathcal{B}$ be the set of $b \in Z_e$ such that $(b, b', b'') \in \mathcal{O}$ for some $b'$, $b'' \in Z_e$ and set

$$\mathcal{C} = \{\omega^b \mid b \in \mathcal{B}\}. \qquad (1.12)$$

By the definition, each element of $\mathcal{C}$ is in the multiplicative group of $GF(p^q)$.

We shall argue that any element of $\mathcal{C}$ satisfies the conditions of the proposition. Let then $x \in \mathcal{C}$, so that $x = \omega^b$ for some $b \in \mathcal{B}$, in which case $(b, b', b'') \in \mathcal{O}$ for suitable $b', b'' \in Z_e$.

By Proposition 1.41, $-b \in \mathcal{B}$, so $\omega^{-b} \in \mathcal{C}$. Thus we have the crucial conclusion

$$x^{-1} \in \mathcal{C}. \qquad (1.13)$$

Now $b \neq 0$ by Proposition 1.39(i), so as $b \in Z_e$ and $|\omega| = e$, we have

$x \neq 1$. But $e = (p^q - 1)/(p - 1)$ is prime to $p - 1$ by Theorem 1.38(i), and consequently $x^{p-1} \neq 1$, which immediately yields that $x \notin GF(p)$. Thus

$$x \in GF(p^q) - GF(p). \qquad (1.14)$$

Furthermore, $e = (p^q - 1)/(p - 1) = 1 + p + \cdots + p^{q-1}$ and $x^e = \omega^{be} = (\omega^e)^b = 1$, so we also have

$$N(x) = 1. \qquad (1.15)$$

Finally, by Proposition 1.39(iii),

$$\omega^{-b'} = 2 - \omega^b = 2 - x, \qquad (1.16)$$

while by Proposition 1.40, $-b' \in \mathscr{B}$, so by (1.16), we get

$$2 - x \in \mathscr{C}. \qquad (1.17)$$

Hence by (1.13) and (1.17),

$$x^\sigma = 1/(2 - x) \in \mathscr{C}. \qquad (1.18)$$

Since these results apply to every $x \in \mathscr{C}$, we conclude that $x^{\sigma^i} \in \mathscr{C}$ for all $i$, whence each $x^{\sigma^i} \in GF(p^q) - GF(p)$ and $N(x^{\sigma^i}) = 1$. Therefore, $x$ satisfies the conditions of the proposition, as asserted.

But Proposition 1.45 is in direct contradiction with the following purely field-theoretic fact:

PROPOSITION 1.46. *For any $x \in GF(p^q) - GF(p)$, there exists a positive integer $i$ such that*

$$N(x^{\sigma^i}) \neq 1.$$

Thus the final configuration is eliminated and so all groups of odd order are solvable, completing our summary of Theorem 1.1.

## 1.3. Groups with Dihedral Sylow 2-Subgroups

The smallest simple group is $A_5$ $\left( \cong L_2(5) \cong L_2(4) \right)$: it has order 60 and $Z_2 \times Z_2$ as Sylow 2-subgroup. It was therefore natural for Brauer to study simple groups with such Sylow 2-subgroups, and in the early years he developed their character and 2-block theory in full detail. He soon expanded his investigations to include groups with dihedral Sylow 2-subgroups. Recall that a group $X$ is dihedral if $X$ is generated by elements $x$, $y$ subject to the relations

$$y^2 = x^n = 1 \quad \text{and} \quad x^y = y^{-1}xy = x^{-1}. \tag{1.19}$$

[Thus $Z_2 \times Z_2$ is a dihedral group of order 4.] This enabled Brauer to cover the entire family $L_2(q)$, $q$ odd [when $q \equiv 3$, 5(mod 8), $L_2(q)$ has $Z_2 \times Z_2$ as Sylow 2-subgroup]. Together with Suzuki and Wall [I: 49], he subsequently gave a beautiful characterization of these groups in terms of the structure of the centralizers of their involutions. (A proof of their result in the $Z_2 \times Z_2$ case can be found in [I: 130].) The Brauer–Suzuki–Wall theorem opened up the possibility of determining all simple groups with dihedral Sylow 2-subgroups, a problem which Walter and I attacked, while Feit and Thompson were still at work on the odd-order theorem. By good fortune, we were able to carry over Thompson's local methods to our analysis, and combining them with some of Brauer's block theory results, we were eventually able to obtain a complete classification.

THEOREM 1.47. *If $G$ is a simple group with dihedral Sylow 2-subgroups, then $G \cong L_2(q)$, $q$ odd, $q > 3$, or $A_7$.*

[$A_7$ has dihedral Sylow 2-subgroups of order 8; also $L_2(3)$ is solvable.]

A dihedral 2-group $S$ has three classes of involutions and one or two classes of four-subgroups according as $|S| = 4$ or $|S| > 4$. Using elementary properties of the focal subgroup [I, 1.19], it is very easy to determine the fusion of $S$ in any group having $S$ as Sylow 2-subgroup (see [I: 130, Theorems 7.7.1 and 7.7.3]).

PROPOSITION 1.48. *If $X$ is a group with dihedral Sylow 2-subgroup $S$, then we have*

   (i) *$X$ has a normal 2-complement if and only if $N_X(U) = C_X(U) N_S(U)$ for every four-subgroup $U$ of $S$;*
   (ii) *$X$ is fusion simple if and only if either*
      (1) *$X$ has only one conjugacy class of involutions; or*
      (2) *$N_X(U)/C_X(U) \cong A_3$ or $\Sigma_3$ (according as $|S| = 4$ or $|S| > 4$) for every four-subgroup $U$ of $S$;*

(iii) *The centralizer of every involution of $X$ has a normal 2-complement; and*

(iv) *According as $|S| = 4$ or $|S| > 4$, $X$ has one or two conjugacy classes of four-groups.*

Note that (iii) is immediate from (i); moreover, the proposition follows directly from Burnside's normal complement theorem [I, 1.20] in the case $|S| = 4$.

Now fix a Sylow 2-subgroup $S$ of $G$. Since $G$ is simple, $G$ has only one class of involutions. Let $t \in Z(S)$ and set $C = C_t$. Since $C$ has a normal 2-complement, we have

$$C = SO(C). \tag{1.20}$$

In the groups $L_2(q)$ themselves, $C$ is, in fact, a dihedral group of order $\frac{1}{2}(q - \varepsilon)$, where $\varepsilon = \pm 1$ and $q \equiv \varepsilon \pmod 4$. In particular, if $O(C) \neq 1$, then $T = C_S(O(C))$ is cyclic of index 2 in $S$ and $O(C)$ is inverted by elements of $S - T$. On the other hand, in $A_7$, $|C| = 24$, $|O(C)| = 3$ and $T = C_S(O(C)) \cong Z_2 \times Z_2$ [whence $O(C)$ is again inverted by elements of $S - T$]. Furthermore, in both cases, $N_G(Y) \leqslant C$ for any nontrivial subgroup $Y$ of $O(C)$.

The Brauer–Suzuki–Wall theorem is essentially a converse of the first set of conditions.

THEOREM 1.49. *Assume the following conditions*:

(a) *If $O(C) \neq 1$, then $T = C_S(O(C))$ is cyclic of index 2 in $S$;*

(b) *$O(C)$ is inverted by elements of $S - T$ [whence $O(C)$ is abelian]; and*

(c) *$N_G(Y) \leqslant C$ for every $1 \neq Y \leqslant O(C)$.*

*Then $G \cong L_2(q)$ for some odd $q$.*

The group $L_2(q)$ has a geometric realization as a group of fractional linear transformations of the projective line $\mathscr{L}$ over $GF(q)$ and hence as a group of permutations of the $q + 1$ points of $\mathscr{L}$. As such, it is doubly transitive and the stabilizer of the point at infinity on $\mathscr{L}$ is the subgroup of affine transformations, which is a Frobenius group of order $\frac{1}{2}q(q - 1)$ with abelian kernel of order $q$ (the translation subgroup). In particular, only the identity fixes three letters and so $L_2(q)$ is a Zassenhaus group [I, 3.19]. Moreover, Zassenhaus's theorem [I, 3.18] shows that $L_2(q)$ is the only simple doubly transitive permutation group with these properties.

The thrust of the Brauer–Suzuki–Wall argument was to force $G$ to satisfy the Zassenhaus conditions. The principal tool for their analysis was exceptional character theory, which is especially suited to the problem, since their hypothesis immediately implies that $O(C)T$ is disjoint from its conjugates in $G$. In particular, it enabled them to relate the order of $G$ to that of $C$. Indeed, if $|C| = m$, they prove

$$|G| = m(m + \varepsilon)(\tfrac{1}{2}m + \varepsilon), \tag{1.21}$$

where $\varepsilon = \pm 1$ and $m + \varepsilon$, $\tfrac{1}{2}m + \varepsilon$ are the degrees of certain irreducible characters of $G$. They then argue that $m + \varepsilon$ is necessarily an odd prime power $q = p^n$. Finally they study the permutation representation of $G$ on the cosets of $N = N_G(P)$, where $P$ is a Sylow $p$-subgroup of $G$ and prove that in this action $G$ is a Zassenhaus group with the required properties [$N$ is a Frobenius group with (elementary) abelian kernel $P$]. [For more details of their argument (in the case $|S| = 4$), see [I: 130, Theorem 15.4.1].]

Suzuki established a similar characterization of $A_7$ in the course of his classification of simple groups which contain an element of order 4 which commutes only with its own powers [119].

THEOREM 1.50.  *If $|C| = 24$ and $C_S\big(O(C)\big) \cong Z_2 \times Z_2$, then $G \cong A_7$.*

[Note that $O(C_{A_7}\big((12)(34)\big)) = \langle (567) \rangle \cong Z_3$.]

Again Suzuki's proof is character–theoretic, including block theory to completely pin down the internal structure of $G$. On the basis of this information, he first argues that $|G| = |A_7|$ and then shows by essentially the "Brauer trick" [I, p. 98] that $G$ must have a subgroup $G_0$ of index 7. Now the permutation representation of $G$ on the cosets of $G_0$ implies that $G$ is isomorphic to a subgroup of $A_7$ and as $|G| = |A_7|$, it follows that $G \cong A_7$.

Thus, the determination of all simple groups with dihedral Sylow 2-subgroups is reduced to forcing $C$ to satisfy the hypotheses of either Theorem 1.49 or Theorem 1.50. Clearly this requires an analysis of the structure of $O(C)$ and the normalizers of its subgroups. In particular, we see a basic difference between the present problem and the odd-order theorem: there it was necessary to study *every* maximal subgroup of a minimal counterexample $G$; whereas now we need only consider subgroups in the "neighborhood" of $C$. This comment paraphrases Brauer's fundamental observation that simple groups are determined from the structure and embedding of the centralizers of their involutions.

As has been already indicated, a major portion of the original proof of

Theorem 1.47 was closely modeled after the local arguments of Chapter IV of the odd-order paper. However, where Feit and Thompson constructed "large" maximal subgroups of $G$, Walter and I were able to construct a strongly embedded subgroup $M$ of $G$ with $O(M) \neq 1$, which thus led to a contradiction by Bender's theorem [I, 4.24]. The construction of $M$ involved an analysis of $p$-subgroups of $G$ invariant under a four-subgroup $U$ of $S$ for various odd primes $p$ dividing $|O(C)|$. The situation was controlled by certain general facts concerning the action of involutions and four-groups on groups of odd order (see [I: 130, Theorem 5.3.16]):

LEMMA 1.51. *If the group $X$ of odd order is acted on by the involution $u$, then*

$$X = C_X(u)Y,$$

*where $Y$ is the subset of elements of $X$ inverted by $u$.*

Using induction and Lemma 1.51, together with the solvability of groups of odd order, one easily obtains the following result:

LEMMA 1.52. *Let $X$ be a group of odd order which is acted on by a four-group $U$ and let $u_i$, $1 \leqslant i \leqslant 3$, be the involutions of $U$. Then*

$$X = C_X(u_1)\,C_X(u_2)\,C_X(u_3) = C_X(U)\,Y_1\,Y_2\,Y_3,$$

*where $Y_i$ is the subset of $C_X(u_i)$ inverted by $u_j$ for $j \neq 1$, $1 \leqslant i, j \leqslant 3$.*

The conditions under which our construction was possible were the following: for some odd prime $p$, if $P \in \mathrm{Syl}_p(O(C))$, then $P \notin \mathrm{Syl}_p(O(N_G(P)))$. Thus our analysis yielded the following basic embedding conclusions:

For any odd prime $p$, if $P \in \mathrm{Syl}_p(O(C))$, then $P \in \mathrm{Syl}_p(O(N_G(P)))$.

$$(1.22)$$

Unfortunately, our proof of (1.22) was extremely lengthy and cumbersome, in part at least because Glauberman's $ZJ$-theorem was not available at that time. However, the argument is quite easy in the special case that $S$ is a four-group and $C_S = S$ [whence the involutions of $S - \langle t \rangle$ invert $O(C)$] and is given in complete detail in [I: 130, Section 15.3]).

Condition (1.22) is very restrictive; for it immediately implies that each such $P$ has one of the following two properties:

(a) $P \in \mathrm{Syl}_p(G)$; or

(b) $P$ centralizes a four-subgroup $U$ of $C$ and $P$ is a            (1.23)
maximal $U$-invariant $p$-subgroup of $G$.

Indeed, set $N = N_G(P)$, so that $P \in \mathrm{Syl}_p(O(N))$ by (1.22). If $N$ has a normal complement, then $P \in \mathrm{Syl}_p(N)$, whence $P \in \mathrm{Syl}_p(G)$ by [I, 1.11]; so we can assume this is not the case. Since $P \in \mathrm{Syl}_p(O(C))$, a Frattini argument yields that $N$ contains a Sylow 2-subgroup of $C$, which without loss we can take to be $S$. Since $N$ does not have a normal 2-complement, Proposition 1.48(i) implies now that for some four-subgroup $U$ of $S$, $N_N(U)$ contains a 3-element cyclically permuting the involutions of $U$. But as $S$ is dihedral, clearly $t \in U$; and as $t$ centralizes $P$ with $P \lhd N$, it follows that every involution of $U$ centralizes $P$. Thus $U$ centralizes $P$.

Finally, let $Q$ be a maximal $U$-invariant $p$-subgroup of $G$ containing $P$. To establish (b), we must show that $Q = P$. If not, then again by [I, 1.11], $R = N_Q(P) > P$. Setting $\bar{N} = N/P$, it follows from Lemma 1.52 (or from [I, 4.13]) that $C_{\bar{R}}(\bar{u}) \neq 1$ for some $\bar{u} \in \bar{U}^\#$. But all involutions of $\bar{U}$ are conjugate in $\bar{N}$, so $|C_{\bar{N}}(\bar{t})|$ is divisible by $p$ and consequently $P$ is not a Sylow $p$-subgroup of $C_N(t)$, contrary to the fact that $P \in \mathrm{Syl}_p(C)$.

Playing off the conclusions of (1.23) for each $p$ dividing $|O(C)|$, a delicate local analytic argument implies that $C$ has one of *five* rather precise, but complicated structures (which we shall not explicitly list), the first two of which correspond to the hypotheses of Theorem 1.49 or 1.50. Thus at this point our analysis yielded a "loose" resemblance of $C$ to that of the centralizer of an involution in $L_2(q)$ or $A_7$; and so it remained to sharpen this resemblance by eliminating the third, fourth, and fifth possibilities. This was carried out by an intricate combination of character-theoretic and arithmetic argument. In particular, the analysis involved several formulas for the order of $G$, which Brauer had earlier derived for any fusion simple group with dihedral Sylow 2-groups, one for each 2-block of "maximal defect" (the principal 2-block being one such example.)

As an illustration, suppose $C$ happens to be isomorphic to the centralizer of an involution in an extension of $L_2(q)$ by a *nontrivial* group $E$ of odd order induced from the Galois group of the field $GF(q)$. Such a group $E$ is cyclic and centralizes a Sylow 2-subgroup of $G$. Hence replacing $E$ by a conjugate, if necessary, we can suppose $E$ centralizes $S$. Then $O(C)$ has the following structure:

$$O(C) = AE, \text{ where } A = [S, O(C)]; \tag{1.24}$$

and the conditions of Theorem 1.22 hold with $A$ in place of $O(C)$.

The strategy for eliminating the configuration (1.24) is clear: we must, in effect, show that $G$ has a normal subgroup of index $|E|$, since this is true in the "prototype" $L_2(q) \cdot E$. In other words, we must prove that the automorphisms of $C$ induced by $E$ "lift" to a suitable normal subgroup $G_0$ of $G$ [with $G_0 \cong L_2(q)$], which will, of course, contradict the simplicity of $G$.

Since $C = O(C)S = ASE$ with $AS$ normal in $C$, linear characters of $E$ determine linear characters of $C$ having $AS$ in their kernels. Moreover, as $E$ centralizes $S$, they determine 2-blocks of $G$ of maximal defect under the Brauer correspondence. Each such character $\phi$ of $C$ determines a formula for the order of $G$ which depends upon both $C$ and the degrees $f_i$ of four irreducible characters $\chi_i$ of the corresponding 2-block of $G$ together with signs $\delta_i = \pm 1$, $0 \leqslant i \leqslant 3$.

We obtain one formula when $\phi = 1_C$ ($1_C$ denoting the trivial character of $C$; this corresponds to the principal 2-block, in which case $f_0 = \delta_0 = 1$); and as $E \neq 1$, we obtain a second taking $\phi \neq 1_C$ (we denote the corresponding characters, degrees, and signs, by $\chi_i'$, $f_i'$, $\delta_i'$). Equating the two formulas for $G$ leads to the following equality:

$$\frac{\delta_0'}{f_0'} + \frac{\delta_1'}{f_2'} + \frac{\delta_2'}{f_2'} + \frac{\delta_3'}{f_3'} = 1 + \frac{\delta_1}{f_1} + \frac{\delta_2}{f_2} + \frac{\delta_3}{f_3}. \tag{1.25}$$

Now comes the punch line. The expression on the right is approximately 1. Except in a few minimal cases, the only way this can occur is for one of the fractions on the left, say $\delta_0'/f_0'$, to equal 1. Thus $f_0' = 1$ and so $\chi_0'$ is a *linear* character of $G$. However, from general principles, as this is not the principal 2-block, $\chi_0' \neq 1_G$. Since $\chi_0'$ is linear, it therefore determines a nontrivial homomorphism of $G$ onto an abelian group, which is impossible as $G$ is simple. (In fact, if we let $\phi$ vary over all such characters of $C$ and take the intersection of the kernels of the corresponding linear characters of $G$, we obtain the desired normal subgroup $G_0$ with $G = G_0 E$ and $G_0 \cap E = 1$.) Unfortunately, more delicate arithmetic estimates are required to reach contradictions in the minimal cases in which the inference $f_0' = 1$ cannot be made.

The preceding argument is typical of the method Walter and I used to eliminate the various approximate structures of $C$. In any event, this gives a brief indication of the original proof of Theorem 1.47.

About 1970, Bender decided to try his method, which had been so successful in the odd-order situation, on the dihedral Sylow 2-group problem and again made dramatic simplifications in the local analytic portions of the argument, along with some improvements in the character-theoretic parts. Over the years he refined his proof (lecturing on it in 1971 at the University of Illinois at Chicago Circle and again in 1977 at the University of Kiel). Some time after that, utilizing certain ideas of Glauberman, Bender managed to eliminate all block theory from the analysis, working solely with exceptional character theory (this part of the argument appears in a joint paper of the two of them [10]). Thus Bender has now achieved a comparatively short, elegant classification of simple groups with dihedral Sylow 2-subgroups, which we should like to describe briefly [9].

As usual, we take $G$ to be a minimal counterexample to Theorem 1.47. His analysis of the maximal subgroups of $G$ containing $C$ is based on the following key property of $C$.

PROPOSITION 1.53. *Every $C$-invariant subgroup of $G$ of odd order lies in $C$.*

As an immediate corollary, Bender obtains the following.

PROPOSITION 1.54. *If $M$ is a maximal subgroup of $G$ containing $C$, then*

  (i) *$t$ centralizes $O(M)$; and*
  (ii) *Either $t \in O_2(M)$ or $t \in L(M)$.*

Indeed, as $C$ normalizes $O(M)$, $O(M) \leqslant C$ by Proposition 1.53, so $t$ centralizes $O(M)$. If $L(M) \neq 1$, then $L(M)$ contains a four-subgroup of $S$ and hence contains $t$. In the contrary case, $F^*(M) = F(M) = F(O(M)) O_2(M)$ and as $O_2(M) \leqslant S$ with $t \in Z(S)$, it follows that $t$ centralizes $F^*(M)$. Hence by [I, 1.27], $t \in F^*(M)$, in which case $t \in O_2(M)$. Thus (i) and (ii) both hold.

We shall prove Proposition 1.53, as it gives an excellent illustration of the interplay between Bender's general techniques and specific properties of $K$-groups with dihedral Sylow 2-subgroups. In particular, we shall see how crucial the condition of *p-stability* is for the argument. {Note that by [I, 4.108] if $X$ is any group with $O_p(X) \neq 1$, $p$ an odd prime, and $X$ has dihedral Sylow 2-subgroups, then $X$ is $p$-stable; so in the present situation all $p$-locals of $G$ are $p$-stable for all odd $p$.}

It will be convenient here to extend the definition $H \rightsquigarrow K$ for two maximal subgroup $H, K$ of a group given in [I, 4.124] to the case in which $F^*(H)$ is of prime power order. Thus $H \rightsquigarrow K$ if for some $1 \neq Q \leqslant F(H)$, $N_{F^*(H)}(Q) \leqslant K$. (Of course, [I, 4.125] will still only be applicable when $|F^*(H)|$ is divisible by at least two primes.)

We separate out the argument which depends on $p$-stability.

PROPOSITION 1.55. *Let $X$ be a group with $O_p(X) = 1$ in which all $p$-locals are $p$-stable for some odd prime $p$. If $M$ and $N$ are maximal subgroups of $X$ such that $F^*(M)$ and $F^*(N)$ are both $p$-groups and if $M \rightsquigarrow N$, then $M = N$.*

PROOF. We have $F^*(M) = O_p(M) \neq 1$ and $M = N_X(O_p(M))$ [as $M$ is maximal in $X$ and $O_p(X) = 1$], so $M$ is a $p$-local subgroup of $X$. Hence by hypothesis $M$ is $p$-stable, and so if $P \in \mathrm{Syl}_p(M)$, Glauberman's $ZJ$-theorem [I, 4.110] implies that $Z(J(P)) \lhd M$, whence $M = N_X(Z(J(P)))$ by the maximality of $M$. Since $Z(J(P))$ char $P$, $N_X(P) \leqslant M$, so $P \in \mathrm{Syl}_p(X)$ by [I, 1.11]. Similarly if $R \in \mathrm{Syl}_p(N)$, we have $N = N_X(Z(J(R)))$ and $R \in \mathrm{Syl}_p(X)$. But by Sylow's theorem, $R^x = P$ for some $x \in X$. Therefore $Z(J(R))^x = Z(J(P))$ and consequently $N^x = M$.

Now set $Z = Z(O_p(M))$. Since $M \rightsquigarrow N$, $N_X(Q) \leqslant N$ for some $1 \neq Q \leqslant F(M) = O_p(M)$. Since $Z$ centralizes $Q$, it follows that $Z \leqslant N$, whence $Z^x \leqslant M$ and so $Z^{xm} \leqslant P$ for some $m \in M$. But as all $p$-locals of $X$ are $p$-stable, and $M = N_X(Z)$ by the maximality of $M$, it follows from the general form of Glauberman's theorem [I, 4.116] (see [I:109, Theorem B]) that $Z^{xm} = Z^a$ for some $a \in M$. Thus $xma^{-1} \in M$ and hence $x \in M$. Since $M^{x^{-1}} = N$, we conclude that $M = N$, as asserted.

We next list the properties of $K$-groups with dihedral Sylow 2-subgroups needed for the proof of Proposition 1.53. Part (ii) will be critical.

PROPOSITION 1.56. *If $X$ is a $K$-group with dihedral Sylow 2-subgroups, then we have the following:*

(i) *Either $X$ is solvable or else $X/O(X) \cong A_7$ or to a subgroup of $\mathrm{Aut}(L_2(r))$, $r$ odd, $r \geqslant 5$, containing $L_2(r)$. In particular, $X$ has at most one nonsolvable composition factor.*

(ii) *If $y \in \mathscr{I}(X)$ and $P$ is a $C_X(y)$-invariant $p$-subgroup of $X$, $p$ an odd prime, then $[P, y] \leqslant O_p(X)$.*

First, (i) is easily verified. Furthermore, (ii) can be directly checked in $A_7$ and in $\text{Aut}(L_2(r))$. But then, using (i) and properties of solvable groups [$O(X)$ is solvable], (ii) follows in general.

Now we prove Proposition 1.53. First of all, we have the following result:

LEMMA 1.57. *If $M$ is a maximal subgroup of $G$ containing $C$, then $N_M(U) = N_S(U)\,C_M(U)$ for some four-subgroup $U$ of $S$.*

Indeed, if not, then by Proposition 1.48, $M$ has only one class of involutions and as $C \leqslant M$, it follows that $M$ is strongly embedded in $G$. But then by Bender's theorem [I, 4.24], $G$ is not a counterexample to Theorem 1.47.

We use Proposition 1.55 and [I, 4.125] to establish the following key result.

LEMMA 1.58. *Let $M$ be a maximal subgroup of $G$ containing $C$ and suppose that $M \leadsto M^g$ for some $g \in G$. If $O(M) \neq 1$, then $M = M^g$.*

PROOF. If $F^*(M)$ is a $p$-group for some odd prime $p$, so also is $F^*(M^g)$ $\left(=(F^*(M))^g\right)$, so $M = M^g$ by Proposition 1.55 (as all $p$-locals of $G$ are $p$-stable). In the contrary case, $|F^*(M)|$ is divisible by at least two primes [as $O(M) \neq 1$], so [I, 4.125] is applicable. Hence if $\pi$ denotes the set of primes dividing $|F(M)|$, we have $O_\pi\left(F(M^g)\right) \leqslant M$. But $|F(M^g)| = |F(M)|$, so, in fact, $O_\pi\left(F(M^g)\right) = F(M^g)$, whence $F(M^g) \leqslant M$.

We claim next that $L(M) = L(M^g)$. This is clear if each is trivial, so assume not. We have $L(M) \leqslant N_G(Q) \leqslant M^g$ [as $M \leadsto M^g$ via some $1 \neq Q \leqslant F(M)$]. But $M^g/L(M^g)$ is solvable by Proposition 1.56(i), so $L(M) \leqslant L(M^g)$ and hence they must be equal, as claimed.

Thus $F^*(M^g) = L(M)\,F(M^g) \leqslant M$, which immediately implies that $M^g \leadsto M$. Hence $M = M^g$ by [I, 4.125].

Now suppose Proposition 1.53 is false, so that $C$ normalizes some subgroup $P$ of $G$ of odd oder with $P \nleqslant C$. Choose $P$ of least order. Now $t$ does not centralize $P$, and as $P$ is solvable (having odd order), it follows easily from [I, 1.25] that $t$ does not centralize $F(P)$ and hence does not centralize $O_p(P)$ for some prime $p$. Since $C$ leaves $O_p(P)$ invariant, we have $P = O_p(P)$ by our minimal choice of $P$. Thus

$$P \text{ is a } p\text{-group.} \tag{1.26}$$

Furthermore, by [I, 4.8], $t$ does not centralize $[P, t]$ (which is clearly $C$-invariant), so also

$$P = [P, t]. \tag{1.27}$$

In addition, $t$ centralizes $O_p(C)$, so by the $(A \times B)$-lemma [I, 4.9], $t$ does not centralize $C_P(O_p(C))$, so this must be $P$ by the minimality of $P$. Hence we also have

$$O_p(C) \text{ centralizes } P. \tag{1.28}$$

Now $C \leqslant N_G(P)$ and so we can choose a maximal subgroup $M$ of $G$ containing $C$ such that

$$N_G(P) \leqslant M. \tag{1.29}$$

By Lemma 1.57, $S$ contains a four-subgroup $U$ whose involutions are not conjugate in $M$. But by Proposition 1.48(ii), $N_G(U)$ contains a 3-element $g$ which cyclically permutes the involutions of $U$ (as $G$ is simple). Thus

$$g \notin M. \tag{1.30}$$

Our goal will be to contradict (1.30). First of all, as $U$ acts on $P$ with $t$ not centralizing $P$ and $t \in U$, it follows from Lemma 1.52 (or [I, 4.13]) that for some $u \in U^{\#}$,

$$R = [C_P(u), t] \neq 1.$$

Obviously $u \neq t$, so replacing $g$ by $g^{-1}$, if necessary, we can assume without loss that $u = t^g$, whence

$$R \leqslant C_u = C^g \leqslant M^g. \tag{1.31}$$

Observe next that $P = [P, t]$ is $C$-invariant and that $R = [R, t]$ (by [I, 4.8]) is invariant under $C \cap M^g = C_{M^g}(t)$, so by Proposition 1.56(ii), we have

$$P \leqslant O_p(M) \quad \text{and} \quad R \leqslant O_p(M^g). \tag{1.32}$$

Since $M^g$ contains $C_u$, (1.32) implies that $R \leqslant O_p(C_u)$. Hence conjugating (1.28) by $g$, we obtain

$$R \text{ centralizes } P^g. \tag{1.33}$$

Hence if $N$ is a maximal subgroup of $G$ containing $N_G(R)$, it follows that

$$P^g \leqslant N. \tag{1.34}$$

Now we are ready for the final argument. We compare $N$ with both $M$ and $M^g$. By (1.32), $R \leqslant F(M)$ and $R \leqslant F(M^g)$. Since $N_G(R) \leqslant N$, it follows from the definition of $\rightsquigarrow$ that

$$M \rightsquigarrow N \quad \text{and} \quad M^g \rightsquigarrow N. \tag{1.35}$$

Furthermore, we have $P^g = [P^g, t^g] = [P^g, u]$ with $P^g$ invariant under $C_u$ and, in particular, under $C_N(u)$. Since $P^g \leqslant N$, Proposition 1.56(ii), applied to $u$, yields now that

$$P^g \leqslant O_p(N). \tag{1.36}$$

But by (1.29), $N_G(P^g) \leqslant M^g$, so by (1.36) and the definition of $\rightsquigarrow$, it follows that

$$N \rightsquigarrow M^g. \tag{1.37}$$

Thus $M^g \rightsquigarrow N$ and $N \rightsquigarrow M^g$. Hence, according as $F^*(M^g)$ and $F^*(N)$ are both $p$-groups or at least one is not, Proposition 1.55 or Lemma 1.58 implies that $M^g = N$. Hence by (1.35), $M \rightsquigarrow M^g$ and we conclude now from Lemma 1.58 that $M = M^g$. Since $M$ is maximal in $G$, this yields $g \in M$, giving the desired contradiction.

Thus Propositions 1.53 and 1.54 hold. We can sharpen the latter result.

PROPOSITION 1.59. *Let $M$ be a maximal subgroup of $G$ containing $C$. If $t \in O_2(M)$, then either*

  (i)  $M = C$; *or*
  (ii) $O_2(M) \cong Z_2 \times Z_2$, $O(M) = O(C)$, $M/O(C) \cong A_4$ *or* $\Sigma_4$, *and correspondingly* $|S| = 4$ *or* $8$.

Indeed, set $R = O_2(M)$. If $R$ is cyclic or nonabelian, then $\langle t \rangle = \Omega_1(Z(R))$, whence $\langle t \rangle \lhd M$ and so $M \leqslant C$. Since $C \leqslant M$, $M = C$ and (i) holds. The only other possibility is that $R \cong Z_2 \times Z_2$, whence $M/C_M(R) \cong Z_3$ or $\Sigma_3$. In particular, $O(C) \leqslant C_M(R)$. But $C_M(R) \leqslant C$ (as $t \in R$) and so $C_M(R) = O(C)R$ [as $C_S(R) = R$]. Since $C_M(R) \lhd M$, this implies that $O(C) \lhd M$, whence $O(C) \leqslant O(M)$. On the other hand, $O(M) \leqslant O(C)$ as

$O(M)$ centralizes $R$. Thus $O(M) = O(C)$ and we conclude that $M/O(C) \cong A_4$ or $\Sigma_4$. Hence (ii) holds in this case.

We note that (i) holds in the groups $L_2(q)$, while (ii) holds in $A_7$ $\left(N_{A_7}(\langle 567 \rangle)/\langle 567 \rangle \cong \Sigma_4\right)$.

Proposition 1.53 is very strong. Indeed, a close analysis of its consequences enables Bender to pin down the possibilities for $O(C)$ and its embedding in $G$ rather tightly, again giving a loose resemblance of $C$ to the centralizer of an involution in $L_2(q)$ or $A_7$. His description is very similar to that obtained by Walter and me (but follows more quickly). We state Bender's result in a somewhat simplified (as well as modified) form, which is unfortunately still very complicated. We can view it as the analog of Proposition 1.25 of the odd-order analysis.

If $|S| = 4$, set $T = \langle t \rangle$; while if $|S| > 4$, let $T$ be the unique cyclic subgroup of $S$ of index 2. Also for any $u \in \mathcal{I}(S)$, let $I_u$ be the subset of $O(C)$ inverted by $u$. Finally set $F = F(O(C))$.

PROPOSITION 1.60. *For some maximal subgroup $M$ of $G$ containing $C$, we have*

(I) *If $t \in O_2(M)$, then one of the following holds:*
   (A) (1) *$T$ centralizes $O(C)$;*
       (2) *$I_u$ is a normal subgroup of $O(C)$ for some [and hence by (1) for every] $u \in \mathcal{I}(S)$ with $u \neq t$;*
       (3) *$N_G(Y) \leqslant M$ for every $1 \neq Y \leqslant I_u$.*
   (B) (1) *For every $u \in \mathcal{I}(S)$, $I_u$ is a normal subgroup of $O(C)$ of index relatively prime to its order;*
       (2) *$I_u \neq I_v$ for some $u, v \in \mathcal{I}(S)$ with $u \neq t \neq v$; and*
       (3) *$N_G(Y) \leqslant M$ for every $1 \neq Y \leqslant F$.*
   (C) (1) *For some $u \in \mathcal{I}(S)$ with $u \neq t$, $F_u = C_F(u)$ is a nontrivial cyclic group;*
       (2) *$N_G(Y) \leqslant M$ for every $1 \neq Y \leqslant F_u$; and*
       (3) *One of the following holds:*
           (a) *$I_u$ is normal in $O(C)$ and $T \neq O_2(C)$;*
           (b) *For some cyclic subgroup $K_u$ of $I_u$ of order a multiple of $|F_u|$, we have $|O(C): C_{O(C)}(u)K_u| \leqslant |K_u|$; or*
           (c) *$F_u$ has prime order $p > 3$, $O(C)/F$ is abelian and $|O(C): C_{O(C)}(u)F| \leqslant \frac{1}{2}(p + 1)$.*
(II) *If $t \in L(M)$, then $L(M) \cong L_2(r)$ and $L(M)S \cong PGL_2(r)$ for some odd $r$. In particular, $S \not\leqslant L(M)$.*

[Note that by Proposition 1.54(ii), either $t \in O_2(M)$ or $t \in L(M)$.]

Setting $A = I_u$ in (I)(A), we see that we have exactly the conditions of (1.24). Hence the goal of the analysis must be to force $A = I_u = O(C)$, in which case the hypotheses of the Brauer–Suzuki–Wall theorem (Theorem 1.49) will be satisfied, and it will follow that $G \cong L_2(q)$ for some odd $q$. Furthermore, in (I)(B), Bender must force $|O(C)| = 3$ and $|S| = 8$ with $I_u = O(C)$ and $I_v = 1$ for suitable $u, v \in \mathscr{I}(S)$ with $u \neq t \neq v$, for then the hypotheses of Suzuki's theorem (Theorem 1.50) will be satisfied and it will follow that $G \cong A_7$. Likewise, he must show that cases (I)(C) and (II) lead to contradictions.

It is now that character theory enters the analysis. In [10], Bender and Glauberman establish two results, which play a crucial role in Bender's arithmetic analysis of these various configurations. The first is specifically designed to deal with (I)(A). [Actually they establish a slight extension which Bender uses to eliminate a minimal configuration of Proposition 1.60(I)(B).] Furthermore, the second result is an extension of the Brauer–Suzuki–Wall group order formula (1.21).

Underlying their proof is a basic correspondence relating irreducible characters of $O(C)T$ and generalized characters of $G$. This "coherence" result gives powerful arithmetic information which is then effectively exploited. Here are their results:

PROPOSITION 1.61. *The following conditions hold*:

   (i) *If Proposition* 1.60(I)(A) *holds, then* $I_u = O(C)$; *and*
  (ii) *Assume the following conditions hold*:
      (a) *$T$ centralizes $O(C)$*;
      (b) *$m = |SO(C) : C_{O(C)}(S)|$ is prime to $|G : C|$*;
      (c) *There exists $C_{O(C)}(S) \leqslant Y \leqslant O(C)$ with $Y$ normal in $C$ of index at least 12 such that the set $TO(C) - Y$ is disjoint from its conjugates in $G$.*
*Then for a suitable sign $\varepsilon = \pm 1$, we have*

$$|G : C| = (m + \varepsilon)(\tfrac{1}{2}m + \varepsilon).$$

Thus in Case (I)(A) of Proposition 1.60, Proposition 1.61(i) reduces us immediately to the Brauer–Suzuki–Wall hypothesis and so $G \cong L_2(q)$ for some odd $q$ by Theorem 1.49.

If case (I)(B) holds, Bender splits the analysis into two parts, according as $O_2(M)$ is noncyclic or cyclic. In the first case, he forces $O_2(M) \cong Z_2 \times Z_2$ (whence $|S| = 8$) and then carries out a delicate, but elementary, count of the involutions of $G$, which ultimately yields the exact order of $G$: $|G| = 2^3 \cdot 3^2 \cdot 5 \cdot 6 \cdot 7 = |A_7|$. In particular, $|O(C)| = 3$ and so Suzuki's Theorem 3.4 implies that $G \cong A_7$.

On the other hand, if $O_2(M)$ is cyclic, a purely local-theoretic argument (together with the use of transfer at one point) yields the following properties of $C$:

> (a) $O(C) = AB \times P$, where $|B| = |P| = 3$, $|A| \geqslant 13$, and
>     $AB$ is a Frobenius group with kernel $A$;
> (b) $B = C_{O(C)}(S)$ and $AB = C_{O(C)}(T)$; and             (1.38)
> (c) $TAP$ is disjoint from its conjugates in $G$.

These are the precise hypotheses of the Bender–Glauberman extension of Proposition 1.61(i), and their result yields that $G$ has a normal subgroup of index 3, contrary to the simplicity of $G$.

Thus it remains to eliminate cases (I)(C) and (II) of Proposition 3.14. In the course of establishing that proposition [when (I)(A) and (I)(B) fail], Bender constructs a second maximal subgroup $N$ of $G$ with $t \in L(N)$ and has very precise information concerning $N \cap C$. In particular, if $C \not\leqslant N$, one of the alternatives of (I)(C) holds; while if $C \leqslant N$, (II) holds with $N = M$. This maximal subgroup $N$ plays a key role in eliminating each of the remaining possibilities; the analysis is very delicate.

Bender first argues that $L(N) \cong L_2(r)$ for some $r$; and then counting involutions of $G - N$, he forces $S \leqslant L(N)$. This eliminates case (II) of Proposition 1.60 since then $S \not\leqslant L(N)$ (as $N = M$). Thus (I)(C) holds. It also follows that $C \not\leqslant N$ (otherwise $N$ would be strongly embedded in $G$, giving the usual contradiction). The argument also yields that $T = O_2(C)$, so (I)(C)(3)(a) does not hold. Again a count of involutions of $G - N$ leads to an arithmetic contradiction if (I)(C)(3)(b) holds.

Finally to eliminate case (I)(C)(3)(c), Bender invokes Proposition 1.61(ii), arguing first that its hypotheses are satisfied with $Y = C_{O(C)}(u)F = C_{O(C)}(S)F$ [as $T$ centralizes $O(C)$]. Thus $|G : C| = (m + \varepsilon)(\frac{1}{2}m + \varepsilon)$, where $m = |SO(C) : C_{O(C)}(S)|$ and $\varepsilon = \pm 1$. Now a further arithmetic analysis forces $F_u$ to have order $p = 3$, contrary to (I)(C)(3)(c).

This completes the discussion of Theorem 1.47.

REMARK. The theorem easily yields the structure of an arbitrary core-free group $X$ with dihedral Sylow 2-subgroup $S$, a result often needed for the applications. Indeed, suppose first that $O_2(X) \neq 1$. If $O_2(X)$ is cyclic, nonabelian, or if $X \cong Z_2 \times Z_2$, then $X$ has a central involution $z$, in which case $X = C_X(z) = SO(X) = S$ by Proposition 1.48(iii). The only other possibility is that $O_2(X) \cong Z_2 \times Z_2$, in which case $X \cong A_4$ or $\Sigma_4$. On the other hand, if $O_2(X) = 1$, then $L = L(X) = F^*(X)$ is a nontrivial direct product of simple groups with $T = S \cap L \in \mathrm{Syl}_2(L)$. Since $S$ is dihedral, so also is $T$ and hence $L$ must be simple. The structure of $L$ is therefore given by Theorem 1.47 and as $X \leqslant \mathrm{Aut}(L)$, the possible structures of $X$ are determined in this case as well.

In particular, it follows that every group $X$ with dihedral Sylow 2-subgroups (core-free or not) is a $K$-group, so Proposition 1.56 applies to $X$. Likewise as a consequence of the preceding paragraph, if $X$ is perfect, then necessarily $X/O(X) \cong L_2(q)$, $q$ odd, $q > 3$, or $A_7$.

Combining this last observation with the Brauer–Suzuki theorem [I, 4.88], we obtain the following basic result on groups of 2-rank 1:

THEOREM 1.62. *If $X$ is a perfect group of 2-rank 1, then $X/O(X) \cong SL_2(q)$, $q$ odd, $q > 3$, or $\hat{A}_7$.*

Indeed, without loss, we can assume that $O(X) = 1$. Since $X$ is perfect, Burnside's theorem [I, 1.20] implies that a Sylow 2-subgroup $S$ of $X$ is not cyclic, so as $m(S) = 1$, $S$ is quaternion by [I, 1.35]. Furthermore, by the Brauer–Suzuki theorem, $Z(S) \leqslant Z(X)$. But then $\bar{X} = X/Z(S)$ is a perfect group with dihedral Sylow 2-subgroup $\bar{S} = S/Z(S)$. Also, as $O(X) = 1$, it is immediate that $O(\bar{X}) = 1$, so $\bar{X} \cong L_2(q)$, $q$ odd, $q > 3$, or $A_7$. Since $Z(S)$ is cyclic, $X$ does not split over $Z(S)$ and we conclude at once that $X \cong SL_2(q)$ or $\hat{A}_7$, as asserted.

## 1.4. Groups with Quasi-Dihedral or Wreathed Sylow 2-Subgroups

Recall that a 2-group $S$ is *quasi-dihedral* if $S$ is generated by elements $x$, $y$ subject to the relations

$$y^2 = x^{2^n} = 1 \quad \text{and} \quad x^y = x^{-1+2^{n-1}}, \qquad n \geqslant 3;$$

and $S$ is *wreathed* if it is generated by elements $x$, $y$, $z$ subject to the relations

$$x^{2^n} = y^{2^n} = z^2 = 1, \quad [x, y] = 1, \quad \text{and} \quad x^z = y, \qquad n \geqslant 2.$$

The main classification theorem, due to Alperin, Brauer, and me [I: 3, 4], asserts the following:

THEOREM 1.63. *If G is a simple group with quasi-dihedral or wreathed Sylow 2-subgroups, then $G \cong L_3(q)$ or $U_3(q)$ for some odd q or $M_{11}$.*

Despite their quite distinct defining relations, it is natural to consider both types of Sylow 2-groups simultaneously since the Sylow 2-subgroups of $L_3(q)$, $q \equiv \varepsilon$ (mod 4) and $U_3(q)$, $q \equiv -\varepsilon$ (mod 4) are quasi-dihedral when $\varepsilon = -1$ and wreathed when $\varepsilon = 1$ ($M_{11}$ has quasi-dihedral Sylow 2-subgroups of order 16).

The theorem has a long history, for Brauer's first centralizer of an involution characterization theorem [I: 40] dealt with the quasi-dihedral situation, characterizing the groups $L_3(q)$, $q \equiv -1$ (mod 4), and $M_{11}$ in terms of the *exact* structure and embedding of the centralizers of their involutions. Moreover, this result represented the first major application of block theory (for the prime 2) to classification problems; there is no doubt that it strongly shaped Brauer's ideas concerning the intimate connections between a group and the centralizers of its involutions. Over the years, Brauer systematically worked out the general theory of modular characters of arbitrary groups with quasi-dihedral and wreathed Sylow 2-subgroups (a portion of the latter jointly with W. Wong [11]), not only for the principal 2-block, but for other 2-blocks of maximal defect. It is an extremely elaborate theory. Moreover, the existing proof of Theorem 1.63 uses every last bit of information that Brauer was able to extract from the character theory. Indeed, the use of block theory here is far greater than in any other single classification result concerning simple groups, even reaching into the local analytic portions of the argument.

It is surprising to me that in the roughly ten-year interval from the publication of the quasi-dihedral, wreathed paper in 1970 to the completion of the classification of simple groups, no attempt was made to simplify the original local analysis by the Bender method. The presence of $SL_2(p^m)$ sections and the resulting possible failure of *p*-stability was undoubtedly the strongest deterrent. However, since then, as part of the "revisionist" approach to the classification, Solomon (together with Aschbacher and Gilman) have indeed carried this out [109]. However, it is difficult to measure the extent of their simplification, for they have not yet succeeded in reducing the character theory needed for other portions of the proof; and until that is accomplished, not much is to be gained by elimination of Brauer's character-theoretic results from the local analysis.

In any event, we shall follow the original proof here, for several reasons: first, historical; second, because it splendidly reveals the power of Brauer's methods; and third, because the local analysis introduces the notion of a "covering $p$-local," which was to be used in several subsequent Sylow 2-group characterization theorems.

We begin by summarizing a few of the principal features of the 2-block theory of groups with quasi-dihedral or wreathed Sylow 2-subgroups. However, first we describe the involution fusion pattern, by analogy with Proposition 1.48 in the dihedral case.

PROPOSITION 1.64. *If $X$ is a group with quasi-dihedral or wreathed Sylow 2-subgroup $S$, then we have*

(i) *The following statements are equivalent*:
    (1) *$X$ has only one conjugacy class of involutions*; *and*
    (2) *$S$ contains no involution which is isolated in $X$.*

(ii) *If $X$ is fusion simple, then*
    (1) *If $T$ is a four-subgroup of $S$, with $T \lhd S$ if $S$ is wreathed, then $N_X(T)/C_X(T) \cong \Sigma_3$*; *and*
    (2) *If $Q$ is a quaternion subgroup of $S$ of order 8, then $N_X(Q)/QC_X(Q) \cong \Sigma_3$.*

(iii) *If some involution of $S$ is isolated in $X$, then $O(X) Z(S) \lhd X$.*

The proposition is easily established with the aid of Glauberman's $Z^*$-theorem [I, 4.95], Thompson's transfer lemma [I, 1.21], and the Alperin–Goldschmidt conjugation family for $S$ [I, 4.86].

We limit our discussion of the character theory to the case in which our group $G$ has only one class of involutions. Let $t \in \mathcal{I}(G)$ and set $C = C_t$. Using the dihedral Sylow 2-group classification theorem, we shall see below (Proposition 1.73) that one obtains a rather sharp picture of the group $\bar{C} = C/O(C)$. In particular, it will follow that $\bar{C}$ has a uniquely determined normal subgroup $\bar{L} \cong SL_2(q)$ for some odd prime power $q = p^n$. Since $G$ has only one class of involution, the integer $q$ is independent of the choice of $t$ and will be called the *characteristic power* of $G$.

However, Brauer's character theory development of $G$ proceeds independently of the structure of $C/O(C)$. The integer $q$ appears instead as the degree of an irreducible character in the principal 2-block of $C$, and he can take this as the definition of the characteristic power of $G$. [Using the structure of $C/O(C)$, Brauer indeed shows that these two definitions of the characteristic power of $G$ coincide.]

By analyzing the principal 2-block of $G$, Brauer establishes a basic formula for the order of $G$.

PROPOSITION 1.65. *If $T$ is a four-subgroup of $G$ containing $t$ (with $T$ normal in a Sylow 2-subgroup of $G$ in the wreathed case), then*

$$|G| = \frac{|C|^3}{|C_T|^2} \frac{f_1(f_1 + \delta_1)}{(f_1 - \varepsilon\delta_1 q)^2} \frac{(q + \varepsilon)}{q} = \frac{|C|^3}{|C_T|^2} \frac{f_2(f_2 + \delta_2)}{(f_2 + \varepsilon\delta_2 q)^2} \frac{(q - \varepsilon)}{q},$$

*where $f_1, f_2$ are degrees of certain irreducible characters in the principal 2-block of $G$, $q \equiv \varepsilon$ (mod 4), and $\delta_1$, $\delta_2 = \pm 1$.*

Note that if $f_1$ or $f_2$ is "large" relative to $q$, then $|G|$ is approximately $|C|^3/|C_T|^2$. It turns out that much sharper results on $f_1, f_2$, and $|G|$ can be obtained under an additional assumption, relating the integer $|C|^3/|C_T|^2$ with the prime $p$ ($q = p^n$), which is the basis of a notion that is fundamental for the character theory of groups with quasi-dihedral or wreathed Sylow 2-subgroups.

DEFINITION 1.66. *$G$ is said to be regular* if the power of $p$ dividing $|G|$ is at least as great as the power with which it divides $|C|^3/|C_T|^2$; otherwise $G$ is said to be *irregular*.

Indeed, the exceptional $M_{11}$ conclusion of Theorem 1.47 can be explained in terms of this notion, for $M_{11}$ is an irregular group, while the groups $L_3(q)$, $U_3(q)$ are regular for all odd $q$.

Brauer proves the following result:

PROPOSITION 1.67. *If $G$ is regular, then*

(i) *In a suitable order the pair $(f_1, f_2) = (q^3, q^2 - \varepsilon q + 1)$, where $\varepsilon = \pm 1$; and*

(ii) $|G| = \dfrac{|C|^3}{|C_T|^2} \dfrac{(q^2 - \varepsilon q + 1)}{(q - \varepsilon)^2} = rq^3(q + 1)(q - 1)(q^2 - \varepsilon q + 1)$

*for some integer $r$ depending on the structure of $C$. [In particular, the first equality of (ii) implies that $p$ divides $|G|$ with exactly the same exponent as it divides $|C|^3/|C_T|^2$.]*

The exact value of $r$ in (ii) is very important for the analysis. Brauer shows that

$$r = 2ab^3e, \qquad (1.39)$$

where we have $a = |C_T \cap O(C)|$, $b = |O(C): C_T \cap O(C)|$, and $e = |O(C_S): O(C_S) \cap O(C)|$, $S \in \mathrm{Syl}_2(C)$.

The character theory analysis in the wreathed case is more difficult than in the quasi-dihedral, primarily because a wreathed 2-group has a more involved structure. On the other hand, when it is completed, the conclusions are stronger. Indeed, Brauer proves the following:

PROPOSITION 1.68. *If $G$ has wreathed Sylow 2-subgroups, then $G$ is regular.*

The analysis of other 2-blocks of maximal defect is extremely elaborate and involves some of Brauer's deepest character-theoretic ideas. Eventually they lead to group order formulas for $G$ similar to those of Proposition 1.65. We shall not attempt to describe these, but we do mention one important general consequence.

PROPOSITION 1.69. *Let $S$ be a Sylow 2-subgroup of $C$ containing $T$, with $T \lhd S$ if $S$ is wreathed, and set $E = O(C_S)$. Then*

$$|G : G'| = |E : E \cap [O(C), y]C'|,$$

*where $y$ is a 3-element of $N_G(T)$ cyclically permuting the involutions of $T$.*

The index on the right may look artificial, but it is not. Indeed, if we let $G$ be an extension of $G_0 \cong L_3(q)$ or $U_3(q)$, $q$ odd, by a (cyclic) group $E_0$ of field automorphisms of $G_0$ of odd order, then $G$ is a group with quasi-dihedral or wreathed Sylow 2-subgroups and one class of involutions. Moreover, $G' = G_0$, so that $|G : G'| = |E_0|$. Furthermore, if we choose the involution $t$ to centralize $E_0$, then $E_0$ is, in fact, a complement to $E \cap [O(C), y]C'$ in $E$. Thus Proposition 1.69 amounts, in effect, to a general formulation of the situation in this "prototype" example.

In addition, Brauer established an important sufficient condition for the group $G$ to be regular, in terms of degrees of the characters in the principal 2-block of $G$ (the degrees had to be coprime to $p$) and the nontriviality of $O(C_S)$.

This summary will give some idea of Brauer's general character theory developments. We should also mention that for many of their applications concerning the action of four-groups on groups of odd order, Lemma 1.52 is an important auxiliary tool (as it is throughout the proof of Theorem 1.63).

As remarked above, Brauer's early interest focused on the application of the character theory to obtain a characterization of the groups $L_3(q)$ in terms of the precise form of the centralizers of their involutions; and it is clear from his work that much of the motivation for his development of the general 2-block theory of groups with quasi-dihedral or wreathed Sylow 2-subgroups was aimed at achieving such a result. We state his theorem only in the simple case [I: 42].

THEOREM 1.70. *Let $G$ be a simple group having an involution $t$ such that the group $C = C_t$ is isomorphic to a factor group of $GL_2(q)$ for some odd $q$ by a central subgroup of odd order. Then $G \cong L_3(q)$ or $M_{11}$.*

In the regular case, Brauer knows that $q^3$ divides the order of $G$. He is then able to show that $C_t$ normalizes two distinct Sylow $p$-subgroups $P$ and $P^*$ of $G$ with $P \cap P^* = 1$ (here $q = p^n$, $p$ an odd prime). He also knows the exact index of $PC_t$ in $G$—namely, $q^2 - \varepsilon q + 1$, where $q \equiv \varepsilon$ (mod 4). This information is so strong that he can construct *a Desarguesian* projective plane $\mathscr{P}$ from the internal structure of $G$, on which $G$ acts as a group of projective transformations. It then follows that $G \cong L_3(q)$.

The points of $\mathscr{P}$ are the $G$-conjugates of $tP$ and its lines are the $G$-conjugates of $tP^*$. It is clear from the way the geometry of $G$ is built up that Brauer could equally well have used his information to construct a $(B, N)$-pair $G_0$ in $G$ with $G_0 \cong L_3(q)$ and then shown that $G_0 = G$. The latter approach was more common in subsequent characterizations of individual families of groups of Lie type of odd characteristic [however, Aschbacher's classical involution proof [I: 13] (see Chapter 3 below) follows a more geometric path].

In the irregular case [whence necessarily $q \equiv -1$ (mod 4) and $G$ has quasi-dihedral Sylow 2-subgroups], Brauer first shows that $O(C) = 1$ by further character analysis along the lines of his general criterion for regularity. [Note that under the hypothesis of Theorem 1.70, $O(C) \leqslant O(C_S)$, where $S$ is a Sylow 2-subgroup of $G$ contained in $C$.] The balance of the argument is quite delicate: character-theoretic and arithmetic calculations show first that $\frac{1}{2}(q - 1)$ is divisible only by the primes 3 and/or 5, and then that it is not divisible by either 9 or 25, so that the only possibilities are

$$q = 3, 7, 11, \text{ or } 31. \tag{1.40}$$

Moreover, in each case Brauer has precise information about the characters in the principal 2-block of $G$ (there are two subcases when $q = 11$). If $q = 7$, he uses the 11-blocks of $G$ to derive a contradiction; if $q = 11$, he uses 5-blocks or 41-blocks (corresponding to the respective subcases), and if $q = 31$, he uses 991-blocks.

The idea is to derive additional arithmetic relations from the appropriate odd prime blocks, which are eventually shown to be incompatible with the already developed 2-block information. The calculations are very subtle.

This leaves only the possibility that $q = 3$. In this case the information yields the precise order of $G$: $7920 = |M_{11}|$; and so $G \cong M_{11}$ by [I, 3.44].

Because the groups $L_3(q)$ have Lie rank 2, their internal structure is "richer" than that of the groups $U_3(q)$ which have only Lie rank 1 (it is undoubtedly for this reason that Brauer considered the former family first). As we know [I, Section 2.1], the latter groups are thus doubly transitive on the cosets of a Borel subgroup. Moreover, in [I, 3.32], we have described O'Nan's characterization of $U_3(q)$, $q$ odd, as such a doubly transitive permutation group. (We note that chronologically his theorem constituted the final step in the proof of Theorem 1.63.)

However, it was fully evident at the outset that Theorem 1.63 would require a centralizer of involution characterization for the groups $U_3(q)$ similar to that of Theorem 1.70. Moreover, we knew that this characterization would entail a reduction to double transitivity. Brauer carried out the reduction—but within the context of Theorem 1.63 itself rather than as an independent effort. Combined with O'Nan's theorem, his result can be stated as follows:

THEOREM 1.71. *Let $G$ be a simple group having an involution $t$ such that the group $C = C_t$ is isomorphic to a factor group of $GU_2(q)$ for some odd $q$ by a central subgroup of odd order. Then $G \cong U_3(q)$.*

The cases $q = 3$ or 5 require somewhat exceptional treatment. Moreover, these cases cause special problems at various other points in the proof of Theorem 1.63. Because of this, Brauer decided to deal with them once and for all in complete generality. Hence, these two cases will be covered by Theorem 1.72 below; the discussion of Theorem 1.71 will thus be limited to the case $q \geqslant 7$.

Brauer's goal is to show that $G$ is regular. In view of Proposition 1.68, it suffices to treat the case that $G$ has quasi-dihedral Sylow 2-subgroups. The most difficult part of the argument occurs when $O(C) = 1$, and the proof is closely patterned after that of the corresponding portion of Theorem 1.70. However, this time Brauer works with 5-blocks, ultimately obtaining arithmetic relations incompatible with previously derived 2-block information.

Once $O(C) \neq 1$, Brauer is able to apply the sufficiency criterion for regularity mentioned above. However, the argument breaks down for $q = 7$; the proof of regularity in this case is similar in spirit to the analyses when $q = 3$ or 5.

At this point Brauer knows that a Sylow $p$-subgroup $P$ of $G$ has order $q^3 = p^{3n}$ and the next step is to shown that $G$ is, is fact, doubly transitive on the cosets of $N = N_G(P)$. This follows easily once it is proved that $|G:N| = q^3 + 1$. Brauer's proof of this equality is again character-theoretic.

To apply O'Nan's theorem, it remains only to pin down the precise structure of $N/C$. Indeed, one must prove that

$$N/C \text{ is cyclic of order } (q^2 - 1)/d, \text{ where } d = \text{g.c.d.}(q + 1, 3). \quad (1.41)$$

If $2^k$ is the exact power of 2 dividing $q + 1$, it is easily seen that $(1.41)$ is equivalent to the assertion:

$$|O(C)| = (q + 1)/2^k d. \qquad (1.42)$$

To establish $(1.42)$, Brauer studies $p_0$-blocks of $G$ for each odd prime $p_0$ dividing $q + 1$. The case $p_0 = 3$ is very difficult. In particular, Brauer must pin down the exact structure of a Sylow 3-subgroup of $G$. Moreover, the argument in this case involves both local analysis and character theory.

Brauer's $q = 3$ or 5 result is as follows.

THEOREM 1.72. *If $G$ is a simple group with quasi-dihedral or wreathed Sylow 2-subgroups of characteristic power $q = 3$ or 5, then $G \cong L_3(q)$, $U_3(q)$, or $M_{11}$.*

If $q = 3$ and a Sylow 2-subgroup $S$ of $G$ is quasi-dihedral, Brauer shows that there are precisely two possibilities for the characters in the principal 2-block of $G$ with $G$ having the corresponding order:

$$|G| = ab^3 |M_{11}| \quad \text{or} \quad |G| = ab^3 |L_3(3)|, \qquad (1.43)$$

where $a$ and $b$ have the same meanings as in $(1.39)$.

If $b = 1$, Brauer argues on the basis of his general character-theoretic results that $a = 1$. It follows immediately now, using Proposition 1.73 below, that $C = C_t \cong GL_2(3)$ for any $t \in \mathcal{T}(G)$. But then $G \cong L_3(3)$ or $M_{11}$ by Theorem 1.70. On the other hand, if $b \neq 1$, Brauer obtains a contradiction by analyzing the 11-blocks or 13-blocks of $G$, respectively [ $M_{11}$ is divisible by 11 to the first power and $|L_3(3)|$ by 13 to the first power]. For example, in the first case, he obtains

$$ab^3 \mid 3^4 \cdot 5 \quad \text{and} \quad ab^3 \equiv 1 \ (\text{mod } 11), \tag{1.44}$$

which cannot be satisfied simultaneously (with $b \neq 1$). [The first condition of (1.44) is a consequence of a general result of Schur [104] concerning *rational-valued* characters. Schur's theorem is used several times in the proof of Theorem 1.72 (as well as in the proof of Theorem 1.70 when $q$ is small).]

On the other hand, if $q = 3$ and $S$ is wreathed (whence $G$ is regular), Brauer shows that one of the characters in the principal 2-block of $G$ has degree 7, so that $G$ has a 7-dimensional irreducible complex representation. All such finite groups have been completely analyzed by Lindsey [76] and Wales [124]; and their results yield that $G \cong U_3(3)$.

The case $q = 5$ is similar, but more difficult. If $S$ is quasi-dihedral, there are now three possibilities for degrees of the characters in the principal 2-block with corresponding order of $G$:

$$|G| = ab^3 |U_3(5)|, \ |G| = ab^3 \cdot 2^4 \cdot 3^2 \cdot 5^3 \cdot 7, \ \text{or}$$
$$|G| = ab^3 \cdot 2^4 \cdot 3^2 \cdot 5 \cdot 7 \cdot 17. \tag{1.45}$$

The first case is the most involved. Here Brauer works with 7-blocks [$|U_3(5)|$ is divisible by 7 to the first power]. Eventually he is able to show that there are exactly two possibilities for $a$ and $b$:

$$a = 1, b = 1, \ \text{or} \ a = 3, \qquad b = 1. \tag{1.46}$$

In the first case of (1.46) $|G| = |U_3(5)|$ and so $G \cong U_3(5)$ by a theorem of Harada [I: 161]. In the second case of (1.46), Brauer argues that $G$ has a normal subgroup of index 3—contradiction.

In the second case of (1.45) Brauer works with 19-blocks when $a$ or $b \neq 1$ and with 3-blocks when $a = b = 1$. In each case, he ultimately reaches an arithmetic contradiction.

On the other hand, in the third case of (1.45), he obtains a similar contradiction, using 17-blocks.

Finally if $q = 5$ and $S$ is wreathed (whence again $G$ is regular), Brauer proves

$$|G| = ab^3 |L_3(5)|, \tag{1.47}$$

and determines the degrees of the characters in the principal 2-block. If $a = b = 1$, then (again using Proposition 1.73 below) one checks that $C \cong GL_2(5)$. But then $G \cong L_3(5)$ by Theorem 1.70. The argument which forces $a = b = 1$ uses the same methods as in the quasi-dihedral case, and again is very delicate.

This is by no means the end of Brauer's contributions to the classification of simple groups with quasi-dihedral or wreathed Sylow 2-subgroups; but it will be best to discuss his remaining results in their proper context.

It was likewise abundantly clear at the very outset that the thrust of the proof of Theorem 1.63 would involve a reduction of the structure of the centralizer of an involution to conditions of the type *assumed* in Theorems 1.70 and 1.71. It will therefore be helpful to reformulate their assumptions in a more uniform way. This is possible because the groups $SL_2(q)$ and $SU_2(q)$ are isomorphic. Hence in both theorems, $C$ satisfies the following conditions:

(a) $L = [C, C] \cong SL_2(q)$;
(b) $Y = C_C(L)$ is cyclic of order $(q - \varepsilon)/d$, where $q \equiv \varepsilon \pmod 4$, $\varepsilon = \pm 1$, and $d$ is odd;
(c) $C = LYS$, where $S \in \mathrm{Syl}_2(C)$ [and hence $S \in \mathrm{Syl}_2(G)$] and $|S : S \cap LY| = 2$; $\tag{1.48}$
(d) $C/Y \cong PGL_2(q)$; and
(e) $O(Y) = O(C)$ and $C = (LS) \times O(C)$.

For brevity, we shall refer to these as the *Brauer structure conditions*.

Let us turn at last to arbitrary groups with quasi-dihedral or wreathed Sylow 2-subgroups. First of all, as we have already remarked, the dihedral classification theorem can be used to obtain the essential structure of the group $C/O(C)$. It is this structure which represents the starting point for the proof of Theorem 1.63.

PROPOSITION 1.73. *Let $G$ be a group with quasi-dihedral or wreathed Sylow 2-subgroups having only one class of involutions. For $t \in \mathscr{I}(G)$, if we set $C = C_t$ and $\bar{C} = C/O(C)$, then the following conditions hold:*

(i) $\bar{L} = [\bar{C}, \bar{C}] \cong SL_2(q)$ *for some odd prime power $q = p^n$; and*
(ii) *If $S \in \mathrm{Syl}_2(C)$, then $\bar{C} = \bar{L}\bar{S}\bar{E}$, where*

(1) $\overline{LS} \lhd \overline{C}$;

(2) $\overline{E}$ is cyclic of odd order dividing n and induces a group of field automorphisms on $\overline{L}$;

(3) $O_2(\overline{C}) = Z(\overline{S}) \leqslant Z(\overline{C})$ is cyclic $\big(Z(\overline{S}) = Z(\overline{L})$ when $S$ is quasi-dihedral$\big)$;

(4) $|\overline{S}: (\overline{S} \cap \overline{L}) Z(\overline{S})| = 2$; and

(5) $\overline{LS}/Z(\overline{S}) \cong PGL_2(q)$.

Indeed, as $G$ has only one class of involutions, $t$ is 2-central and so $C$ contains a Sylow 2-subgroup $S$ of $G$. Applying Proposition 1.64 to both $G$ and $C$ ($t$ is, of course, isolated in $C$), we easily obtain

(a) $Z(\overline{S}) \lhd \overline{C}$;

(b) $S/Z(S) \cong \overline{S}/Z(\overline{S})$ is a dihedral group and hence $\tilde{C} = \overline{C}/Z(\overline{S})$ has dihedral Sylow 2-subgroups; and

(c) $\tilde{C}$ contains a fusion-simple normal subgroup $\tilde{C}_0$ of index 2.

The dihedral classification theorem now yields that either $\tilde{C}_0 \cong A_7$ or else $\tilde{C}_0 = \tilde{L}\tilde{E}$, where $\tilde{L} \cong L_2(q)$ for some $q = p^n$, $p$ an odd prime, and $\tilde{E}$ is cyclic of odd order dividing $n$ and inducing a group of field automorphisms on $\tilde{L}$. However, the first case is excluded here, for then $\tilde{C} \cong \Sigma_7$ or $A_7 \times Z_2$, both of which have Sylow 2-subgroups of the form $D_8 \times Z_2$, contrary to the fact that $\tilde{S}$ is dihedral. In the second case, the pre-image of $\tilde{L}$ in $\overline{C}$ cannot split over $Z(\overline{S})$ as $Z(\overline{S})$ is cyclic, which implies that $\overline{C}$ contains a normal subgroup $\overline{L} \cong SL_2(q)$ mapping on $\tilde{L}$. The remaining parts of the proposition follow at once.

If we compare Proposition 1.73 with the Brauer structure conditions (1.48), we see that the resemblance between $C/O(C)$ in the two cases is quite sharp (indeed, it becomes an actual isomorphism if $\overline{E} = 1$). Moreover, we can state very precisely what is required to show that the Brauer conditions in fact hold in $C$:

A. Prove that $O(C)$ is cyclic.
B. Prove that $\overline{C} = \overline{LS}$ (equivalently, that $\overline{E} = 1$).                         (1.49)
C. Prove that $O(C) \leqslant Z(C)$.

Since $G$ has only one conjugacy class of involutions, the integer $q = p^n$ $\big(\overline{L} \cong SL_2(q)\big)$ is determined independently of the choice of $t$, and is called the *characteristic power* of $G$. As noted earlier, Brauer shows that this integer is identical to that determined from the degree of a certain character in the principal 2-block of $G$.

In operational terms, we have thus reduced the proof of Theorem 1.63

to verification of assertions (1.49) a, b, and c. Despite their simple form, one should not be lulled into believing they are necessarily easy to establish!

In fact, Alperin, Brauer, and I were able to do so only for a minimal counterexample to the theorem; so henceforth we shall assume that $G$ is such a minimal counterexample. (In particular, $q \geqslant 7$ by Theorem 1.72.) The significance of this assumption is that if $X$ is a proper simple section of $G$ with quasi-dihedral or wreathed Sylow 2-subgroups, we shall know that $X \cong L_3(r)$, $U_3(r)$, $r$ odd, or $M_{11}$. This fact, again combined with the dihedral classification theorem, enables us to describe all proper subgroups of $G$:

PROPOSITION 1.74. *If $H$ is a proper subgroup of $G$ and we set $\bar{H} = H/O(H)$, then either $\bar{H}$ is solvable (and the possibilities are easily specified) or*

$$\bar{H}' \cong L_2(r) \text{ or } SL_2(r) \text{ } (r \text{ odd, } r \geqslant 5), \text{ } L_3(r) \text{ or } U_3(r) \text{ } (r \text{ odd}), A_7, \text{ or } M_{11}.$$

*In particular, every proper subgroup of $G$ is a K-group.*

Obviously if (1.49)A holds, then $C_C(O(L))$ covers $\bar{L}$ (as the automorphism group of a cyclic group is abelian). Thus when Alperin and I began our work, we knew that as a major first step in establishing (1.49)A, we had to derive this covering property of $O(C)$. Moreover, we knew from the dihedral and abelian Sylow 2-group classification proofs that, arguing by contradiction, the goal would be to construct a strongly embedded subgroup in $G$ by local methods and then apply Bender's classification theorem to obtain a contradiction. However, try as we would, we were never able to achieve this objective. The best we could do, and this after a great effort, was to construct *weakly* embedded subgroups, "one prime at a time."

To describe these, we make the following definition:

DEFINITION 1.75. Call a prime $r$ dividing $|O(C)|$ *ordinary* provided an $S$-invariant Sylow $r$-subgroup $R$ of $O(C)$ ($R$ exists by [I, 4.2]), satisfies the following two conditions:

(a) $R \in \text{Syl}_r(O(N_G(R)))$; and
(b) $|S : C_S(R)| \leqslant 2$; and if equality holds, then elements of $S - C_S(R)$ invert $R$.

Call the remaining primes dividing $|O(C)|$ *exceptional*.

To establish the desired covering property, it suffices to show that every prime dividing $|(O)|$ is ordinary. Indeed, then if $r$ is any prime dividing $|O(C)|$ and $R$ is an $S$-invariant Sylow $r$-subgroup of $C$, $S'$ will centralize $R$, so, in fact, $S'$ then centralizes $O(C)$. Hence if we set $D = C_C(O(C))$, then $\bar{S}' \cap \bar{L} \leqslant \bar{D} \cap \bar{L}$. One checks directly that $\bar{S}' \cap \bar{L} > Z(\bar{L})$, so $\bar{D} \cap \bar{L} > Z(\bar{L})$. But $D \lhd C$, so $\bar{D} \cap \bar{L} \lhd \bar{L}$. Since $\bar{L}/Z(\bar{L})$ is simple (as $q \geqslant 7$), this forces $\bar{D} \cap \bar{L} = \bar{L}$. Thus $D = C_C(O(C))$ covers $\bar{L}$, as claimed.

Concerning exceptional primes, Alperin and I were able to prove the following result:

PROPOSITION 1.76. *For any exceptional prime $r$ dividing $|O(C)|$, there exists an $r$-local subgroup $M$ of $G$ with the following properties*:

(i) $S \leqslant M$;
(ii) *$M$ has only one class of involutions; and*
(iii) $C = O(C)(M \cap C)$.

Thus $M$ is "nearly" strongly embedded. Indeed, if $O(C) \leqslant M$, then $C \leqslant M$ and it follows that $M$ is strongly embedded. Because $M$ covers $\bar{C} = C/O(C)$ (in particular, covers $\bar{L}$), we refer to $M$ as a *covering $r$-local subgroup* of $G$. Since this covering property holds for every involution of $M$ (as all involutions of $M$ are $M$-conjugate), $M$ is an example of a *weakly embedded* subgroup.

Our proof of Proposition 1.76 involved a very complicated process of pushing-up in "two directions." Furthermore, because of the presence of $SL_2(p^m)$ sections in $r$-local subgroups of $G$, the case $r = p$ was especially difficult as then $p$-stability might well fail, in which case Glauberman's $ZJ$-theorem would not be applicable. To get around this, Alperin and I developed a "relativized" $ZJ$-theorem for $p$-constrained $p$-local subgroups $K$ of $G$ containing $D = N_S(T)$, where $T$ is a four-subgroup of $S$, with $T \lhd S$ if $S$ is wreathed (thus $D \cong D_8$ or $D = S$ according as $S$ is quasi-dihedral or wreathed). We then limited ourselves to maximal $D$-invariant $p$-subgroups of $G$ rather than to Sylow $p$-subgroups. On the other hand, when $r \neq p$, we were able to use the $ZJ$-theorem itself.

Our procedure was as follows: we first pushed-up $C$ "as far as possible," obtaining an $r$-constrained $r$-local subgroup $H$ which covered $C/O(C)$ and according as $r \neq p$ or $r = p$ contained a Sylow $r$-subgroup $Q$ of $G$ or a maximal $D$-invariant $p$-subgroup $Q$ of $G$. Then we turned to $N_G(T)$, which by our prior fusion analysis necessarily contained a 3-element cyclically permuting the involutions of $T$. Since by assumption $r$ is exceptional, it was easy to show that $|O(C_T)|$ was necessarily divisible by $r$, so that by a Frattini argument, $N_G(P)$ covered $N_G(T)/C_G(T)$, where $P$ is a $D$-invariant Sylow $r$-subgroup of $O(C_T)$. We then pushed-up $P$ as far as possible, preserving this covering property, and eventually obtained an $r$-constrained $r$-local subgroup $H^*$, which again contained either a Sylow $r$-subgroup $Q^*$ of $G$ or a maximal $D$-invariant $p$-subgroup $Q^*$ of $G$, with the additional property that all involutions of $T$ were conjugate in $H^*$.

Finally, using the $r$-locals $H$ and $H^*$, we were able to show that $M = N_G(Z(J(Q)))$ was the desired covering $r$-local subgroup. Indeed, using the appropriate form of the $ZJ$-theorem, it followed that $M$ covered $C/O(C)$. Similarly all involutions of $T$ are conjugate in $N_{H^*}(Z(J(Q^*)))$. But if $r \neq p$, $Q^*$ and $Q$, being Sylow $r$-subgroups of $G$, are conjugate, so all involutions of some four-subgroup of $M$ must then be conjugate in $M$, which immediately implies [given the structure of $C/O(C)$] that $M$ has only one class of involutions. To achieve the same conclusion in the case $r = p$, we first had to derive a suitable conjugacy theorem for maximal $D$-invariant $p$-subgroups of $G$ (limited to those containing a fixed maximal $D$-invariant $p$-subgroups of $C$), as a replacement for Sylow's theorem.

Having constructed these covering $r$-local subgroups, Alperin and I were stymied as to how to use them to reach a contradiction, so we asked Brauer whether he had any suggestions. Indeed, he did! For, using the powerful general character theory results he had already established in the quasi-dihedral, wreathed cases, he was able to prove the following remarkable result:

THEOREM 1.77. *Let $X$ be a group with quasi-dihedral or wreathed Sylow 2-subgroups, and suppose $X$ contains a weakly embedded subgroup $H$ [i.e., $H$ has one class of involutions and for any $y \in \mathscr{I}(H)$, $H$ covers $C_X(y)/O(C_X(y))$]. If $H$ is a $K$-group with $H/O(H) \not\cong M_{11}$, then $O(H) \leqslant O(X)$.*

Since our group $G$ is simple and $q \geqslant 7$, the theorem gives an immediate contradiction to the existence of covering $r$-locals. Hence, combined with Proposition 4.14, the theorem yielded the desired covering property of $O(C)$.

PROPOSITION 1.78. *The following conditions hold*:

(i) *Every prime dividing $|O(C)|$ is ordinary; and*
(ii) $C_C(O(C))$ *covers $\bar{L}$.*

Note that because of the covering assumption of Theorem 1.77, $X$ and $H$ have the same characteristic powers. Furthermore, as $M_{11}$ is excluded and

$H$ is a $K$-group, $H$ is, in fact, regular. Brauer easily concludes from this that also $X$ is regular. Since $X$ and $H$ are both regular of the *same* characteristic power, there is a close connection between the degrees of the irreducible characters in their principal 2-blocks, which Brauer was able to exploit. In fact, by a very delicate arithmetic analysis, he was able to show that one of these characters $\chi$ of $X$ *remained irreducible* upon restriction to $H$ (and hence is in the principal 2-block of $H$). But then by [I, 4.89], $O(H)$ is in the kernel of the restriction of $\chi$ to $H$, which in turn yields the desired conclusion $O(H) \leqslant O(X)$.

Here then was a beautiful marriage between local analysis and character theory. It was so effective that Alperin, Brauer, and I decided at that very point to join forces and attack the full classification together. (Indeed, it was subsequent to this decision that Brauer established Theorem 1.72 and completed his work on Theorem 1.71 in the wreathed case.)

Returning now to Proposition 1.78(ii), let $L_0$ be the pre-image of $\bar{L}$ in $C$ and put $L_1 = C_{L_0}(O(C))$. Then $\bar{L}_1 = \bar{L}$ by the proposition. Also clearly $L_1 \lhd C$. Now set $L = L_1^{(\infty)}$, the ultimate term of the derived series of $L_1$. In particular, $L = [L, L]$, so $L$ is perfect. Since $\bar{L}_1 = \bar{L}$ and $\bar{L} = [\bar{L}, \bar{L}]$, it follows easily by taking successive commutators that $L$ itself maps on $\bar{L}$. Clearly this forces $O(L) \leqslant O(C)$. But $L_1$ and hence $L$ centralizes $O(C)$, so $O(L) \leqslant Z(L)$. Thus by definition, $L$ is quasisimple and so is a covering group of $SL_2(q)$ and hence a covering group of $L_2(q)$. Also $L$ char $L_1$ and so is normal in $C$. Now, from the known structure of the Schur multiplier of $L_2(q)$ [I, Table 4.1], we obtain the following basic conclusion:

PROPOSITION 1.79. *C contains a quasi-simple normal subgroup $L$ which maps on $\bar{L}$ and either $L \cong \bar{L} \cong SL_2(q)$ or else $q = 9$ and $L \cong SL_2(9)/Z_3$, the 3-fold cover of $SL_2(9)$.*

Thus we have made a major push towards the Brauer conditions. But we have not yet used the full force of the ordinary property of the primes in $|O(C)|$. Indeed, this condition yields very strong information about the structure and embedding of $O(C)$.

PROPOSITION 1.80. *If we set $A = C_{O(C)}(S)$ and $B = [O(C), S]$ (whence $O(C) = AB$), then we have*

    (i) *$A$ is cyclic of order dividing $q \pm 1$;*

    (ii) *$B$ is abelian and inverted by an involution of $S$;*

(iii) *If A and B are both nontrivial, then AB is a Frobenius group with complement A and kernel B; in particular, A and B have coprime orders;*

(iv) *If $B = 1$, then $L \cong SL_2(q)$ [i.e., the $SL_2(9)/Z_3$ case is excluded];* *and*

(v) *$C_u \leqslant C$ for every $u \in O(C)^{\#}$.*

The argument is delicate, but not difficult, and is entirely local. One first reduces (i), (ii), and (iii) to verification of (v). Lemma 1.52 is especially critical to this reduction, as is the easily verified fact that for any $x \in \mathscr{I}(S)$ with $x \neq t$, $C_L(x)$ has a *cyclic* normal 2-complement of order dividing $q \pm 1$.

But now, if (v) fails, it fails for some element $u$ of $O(C)$ of prime order $r$. For such a choice of $r$, one first constructs an $r$-local subgroup $H$ of $G$ containing $LS$ such that $C_H(O_r(H)) \leqslant O(H)$ (in particular, $H$ is $r$-constrained). But then if $x \in \mathscr{I}(S)$ with $x \neq t$, $x$ neither centralizes nor inverts $O_r(H)$. Since $x$ and $t$ are conjugate in $N_G(\langle t, x \rangle)$, one is able to produce a nontrivial $r$-subgroup $Q$ of $O(C)$ such that $t$ does not centralize a Sylow $r$-subgroup of $O(N_G(Q))$. Then pushing-up $Q$, one eventually shows that $t$ does not centralize a Sylow $r$-subgroup of $O(N_G(R))$, where $R$ is an $S$-invariant Sylow $r$-subgroup of $O(C)$. Since $t$ centralizes $R$, clearly $R \notin \mathrm{Syl}_r(O(N_G(R)))$, contrary to the fact that $r$ is ordinary by Proposition 1.78. As would be anticipated, this pushing-up argument is more involved if $r$ happens to be the characteristic prime $p$ of $G$. On the other hand, since we are pushing-up in only a *single* direction, the complete argument is consideraly easier than that of Proposition 1.76.

Finally, if $L \cong SL_2(9)/Z_3$, it is easy to show that the involution $x$ must invert $O(L) \cong Z_3$, whence $O(L) \leqslant B$ and so $B \neq 1$, proving (iv).

As an immediate corollary of Propositions 1.79 and 1.80, we get the following.

PROPOSITION 1.81. *The Brauer structure conditions hold provided:*

(a) *$\bar{E} = 1$; and*
(b) *$B = 1$.*

Indeed, as $A$ centralizes $S$, it will then follow that $C = (LS) \times O(C)$ with $O(C)$ cyclic of order dividing $q + 1$ or $q - 1$, as the case may be, which are the Brauer conditions.

The proof that $\bar{E} = 1$ follows immediately from Proposition 1.69. [Indeed, it is clear that $\bar{E}$ in some way "wants to lift" to a group $E_0 (\cong \bar{E})$ of field automorphisms of a subgroup $G_0 \cong L_3(q)$ or $U_3(q)$ of $G$.] Thus we have the following result:

PROPOSITION 1.82. *We have $\bar{E} = 1$.*

Hence the proof of Theorem 1.63 is reduced finally to establishing the following single assertion:

PROPOSITION 1.83. *We have $B = 1$.*

The proof proceeds in three stages. First, if $G$ is regular, it is immediate from Proposition 1.65 and (1.39) that for any prime $p_0$ dividing $|B|$, a Sylow $p_0$-subgroup of $C$ is not a Sylow $p_0$-subgroup of $G$. [One checks easily that $|C| = 2abe(q-1)(q+1)$.] On the other hand, it is a direct consequence of Proposition 1.80(v) that for such a prime $p_0$, $C$ contains a Sylow $p_0$-subgroup of $G$—contradiction.

Next, if $A \neq 1$, Brauer again uses his sufficiency criterion (as in the proof of Theorem 1.71) to show that $G$ is regular. Hence in view of Proposition 1.68, Proposition 1.83 is reduced to the quasi-dihedral case with $A = 1$ and $G$ irregular.

This final case is very delicate; the proof involves both character-theoretic and local group-theoretic arguments.

There you have an outline of the quasi-dihedral, wreathed classification theorem. Surely Brauer's spectacular use of modular character theory in this theorem represents one of the high points of the classification of finite simple groups!

## 1.5. Simple Groups of 2-Rank $\leqslant 2$

In view of the introductory discussion of the chapter, to complete the classification of simple groups of 2-rank $\leqslant 2$, it remains only to consider groups with a Sylow 2-subgroup of *type* $U_3(4)$ [i.e., isomorphic to a Sylow 2-subgroup of $U_3(4)$].

The problem was studied by Lyons in his doctoral thesis under Thompson [78]. He proved the following theorem.

THEOREM 1.84. *If $G$ is a simple group with Sylow 2-subgroup $S$ of type $U_3(4)$, then $G \cong U_3(4)$.*

One easily checks that such a 2-group $S$ has the following properties:

(1) $|S| = 64$;
(2) $Z(S) = \Omega_1(S) = [S, S] \cong Z_2 \times Z_2$; and                          (1.50)
(3) $S/Z(S) \cong E_{16}$, an elementary group of order 16.

Using Glauberman's $Z^*$-theorem and the Alperin–Goldschmidt conjugation family, one easily obtains the following fusion information:

PROPOSITION 1.85.  (i) *All involutions of G are conjugate*;
(ii) *If $N = N_G\big(Z(S)\big)$, then $N/C_{Z(S)} \cong Z_3$; and*
(iii) *N controls the G-fusion in S.*

In particular, if $t \in \mathscr{I}(S)$ [whence $t \in Z(S)$], it follows that $Z(S)/\langle t \rangle$ is isolated in $C_t/\langle t \rangle$, so by the $Z^*$-theorem and the Frattini argument, we see that

$$C_t = O(C_t)(N \cap C_t).$$                          (1.51)

Thus, $N$ is another example of a weakly embedded subgroup. To prove that $N$ is actually strongly embedded in $G$, it therefore suffices to show that $O(C_t) \leqslant N$—equivalently, that

$$Z(S) \text{ centralizes } O(C_t).$$                          (1.52)

Once (1.52) is proved, Theorem 1.84 follows immediately from Bender's classification theorem [since $U_3(4)$ is the only group among $L_2(2^n)$, $Sz(2^n)$, $U_3(2^n)$ with Sylow 2-subgroups of type $U_3(4)$].

Hence Theorem 1.84 is reduced to verification of the single assertion (1.52). Lyons's proof of (1.52) is based entirely on block-theoretic methods (for the prime 2), à la Brauer. The resulting computations are, however, extremely elaborate. On the other hand, because of the involved nature of $S$, there turn out to be several independent relations among the irreducible characters of $G$, and so the final conclusions are very sharp. Indeed, the upshot of the argument is to force $G$ to have a *rational* representation of degree 12.

Now Schur's theorem [104] applies to yield the following restriction on the order of $G$:

$$|G| \text{ divides } 2^6 \cdot 3^8 \cdot 5^3 \cdot 7^2 \cdot 11 \cdot 13.$$                          (1.53)

Suppose now that $Z(S)$ does not centralize $O(C_t)$. Then $Z(S)$ normalizes, but does not centralize, a Sylow $p$-subgroup of $O(C_t)$ for some prime $p$. However, the character theory analysis implies that for any such prime $p$, $p^{12}$ must divide $|G|$, contrary to (1.53).

We note that the conclusion that $p^{12}$ divides $|G|$ should not be surprising, for it can be shown that the minimal dimension of a faithful representation of $U_3(4)$ on a vector space over the prime field $GF(p)$, $p$ an odd prime, is 12; and it is this fact which, in essence, underlies the given assertion.

As in the quasi-dihedral, wreathed case, it would be of interest to know whether the $U_3(4)$-type case is amenable to resolution by the Bender method. It would appear in this latter case that critical $p$-locals of $G$ will turn out to be $p$-stable for all odd $p$ [for $p \geqslant 5$, it is immediate from (1.51) that all $p$-locals are $p$-stable, so only the prime 3 could possibly cause difficulty]; one should therefore be able to emulate the argument of the dihedral case.[†]

Finally, combining Theorem 1.84 with previous results, we obtain the 2-rank $\leqslant 2$ theorem itself:

THEOREM 1.86. *If $G$ is a simple group of 2-rank at most 2, then $G \cong L_2(q)$ ($q$ odd, $q \geqslant 5$), $L_3(q)$ or $U_3(q)$ ($q$ odd), $U_3(4)$, $A_7$, or $M_{11}$.*

---

[†] (Added in proof) Solomon has indeed produced a proof of Theorem 1.84 along the lines of the Bender–Glauberman dihedral analysis.

# 2

# Simple Groups of Low 2-Rank

With the completion of the classification of simple groups with quasi-dihedral or wreathed Sylow 2-subgroups (and with it of the 2-rank ≤2 theorem), a veritable "cottage industry" developed, having as its purpose the classification of simple groups having a Sylow 2-subgroup $S$ isomorphic to that of some known simple group $X$ (or family of simple groups). (For brevity, we say that $S$ is of *type X*.) At first these were limited to cases in which $S$ had 2-rank 3 or 4, but eventually led to groups or families in which $S$ had higher rank—in some cases, even arbitrary rank. Indeed, there was a period in which it looked as though we were heading towards a characterization of every known simple group in terms of the structure of its Sylow 2-subgroups.

However, as the analysis of simple groups progressed, it became gradually evident that some of these Sylow 2-type theorems were to play a critical role in the general classification theorem, while others were to be of value only for themselves. The one topic which provided a focus for many of these theorems was the classification of simple groups with a *nonconnected* Sylow 2-subgroup. (The importance of connectivity for the effectiveness of the signalizer functor method has been fully stressed in [I] (cf., for example, [I, 4.35 and 4.36]). Moreover, as noted in the general introduction, nonconnected 2-groups always have sectional rank ≤4.

Thus these low 2-rank Sylow 2-type theorems in turn provided a basis for attacking the classification of arbitrary simple groups of sectional 2-rank ≤4 and thereby the nonconnected Sylow 2-group problem. Because the sectional 2-rank condition carries over to all subgroups and homomorphic images, it lends itself to an inductive argument; at the time it seemed to be the only way of dealing with nonconnectivity (which is not an inductive notion).

However, subsequently it turned out that other low 2-rank problems (including some of 2-rank exceeding 4) were required for the proof of the noncharacteristic 2 theorem, their solutions following the same lines of argument as used in the sectional 2-rank $\leqslant 4$ classification.

In this chapter we describe the principal results concerning simple groups of low 2-rank, but limiting ourselves to those which have been used at later places in the classification proof.

Since we shall be focusing henceforth on the prime 2, we shall write $m(X)$ for the 2-rank $m_2(X)$ for any group $X$.

## 2.1. Simple Groups with Specified Sylow 2-Subgroups; General Considerations

In the next three sections we shall discuss the classification of simple groups of low 2-rank (at least 3) in terms of the structure of their Sylow 2-subgroups, limiting ourselves, as we have said, to those results which have been needed for the classification proof.

The following table lists the set of such simple groups:

A. *Odd Characteristic.* $G_2(q)$, $^2G_2(q)$   $q = 3^n$, $n$ odd, $^3D_4(q)$, $PSp_4(q)$, $L_4(q)$, $U_4(q)$, $L_5(q)$, $U_5(q)$, or $P\Omega_7(q)$ with $q \equiv 3,5 \pmod 8$ in this last case.

B. *Characteristic 2.* $L_2(2^n)$,   $U_3(2^n)$,   $Sz(2^n)$,   $L_3(2^n)$,   $PSp_4(2^n)$ (although these groups have 2-ranks $n$, $n$, $n$, $2n$, and $3n$, respectively, they are easily handled uniformly for all values of $n$; see Theorem 2.2 below), $L_4(2)$ ($\cong A_8$), $L_5(2)$, $U_5(2)$, or $PSp_6(2)$.

C. *Alternating.* $A_n$, $8 \leqslant n \leqslant 13$.

D. *Sporadic.* $M_{12}, M_{22}, M_{23}, M_{24}, J_1, J_2, J_3, Mc, Ly, HS$, or $He$.

For convenience, we denote by $\mathscr{F}$ the individual groups and families of groups listed above. Moreover, if a group $X$ has a Sylow 2-subgroup isomorphic to that of some member of $\mathscr{F}$, we shall say that $X$ is of *type $\mathscr{F}$*.

The principal result is as follows:

THEOREM 2.1. *If $G$ is a simple group of type $\mathscr{F}$, then $G$ is isomorphic to a member of $\mathscr{F}$.*

In several cases, a Sylow 2-subgroup of $G$ has class at most 2, so the result follows from that classification theorem and hence from the Goldschmidt–Rowley strongly closed class at most 2 theorem [I, 4.129].

However, we note that some of these cases have also received independent treatment in the literature (Bender [I: 27], Collins [14, 15], Griess [45], and Walter [I: 314, 315]).

THEOREM 2.2. *If $G$ is a simple group with Sylow 2-subgroup of type $L_2(2^n)$, $Sz(2^n)$, $U_3(2^n)$, $L_3(2^n)$, or $PSp_4(2^n)$, then $G \cong {}^2G_2(3^n)$, $L_2(2^n)$, $Sz(2^n)$, $U_3(2^n)$, $L_3(2^n)$, $PSp_4(2^n)$, or $J_1$.*

Note that the groups ${}^2G_2(3^n)$ and $J_1$ have elementary Sylow 2-subgroups of order 8, and so they are of type $L_2(8)$.

As we have described in [I, 4.129], the Goldschmidt–Rowley theorem follows the Bender method. However, the remaining cases of Theorem 2.1 (with one character-theoretic exception) have been treated entirely by means of signalizer functors, which enable one to avoid the troublesome problem of non-$p$-stable $p$-locals encountered in applying the Bender method (the difficulty does not occur under the assumptions of Theorem 2.2).

In this section, we shall note some of the principal similarities and differences among the remaining elements of $\mathscr{F}$ to give some feeling of the problems to be faced in establishing Theorem 2.1. We shall then give an overview of the main steps of the analysis; however, the more detailed discussion will be deferred to Sections 2.2 and 2.3.

## Similarities

(1) The groups $G_2(q)$ and ${}^3D_4(q)$ have isomorphic Sylow 2-subgroups for a given value of $q$ ($q$ odd); moreover, if $q \equiv 3, 5 \pmod 8$, these Sylow 2-subgroups are isomorphic to those of $M_{12}$. Thus $G_2(q)$, ${}^3D_4(q)$, $q \equiv 3, 5 \pmod 8$, and $M_{12}$ are all of the same type. Many other groups and families on our list have the same type. We note only the following cases:

(2) $A_8, A_9$, and $PSp_4(q)$, $q \equiv 3 \pmod 8$.

(3) $A_{10}, A_{11}$.

(4) $A_{12}, A_{13}, PSp_6(2)$, and $P\Omega_7(q)$, $q \equiv 3 \pmod 8$.

(5) $M_{22}, M_{23}, Mc, L_4(q)$, $q \equiv 3 \pmod 8$, and $U_4(q)$, $q \equiv 5 \pmod 8$.

(6) $M_{24}, He$, and $L_5(2)$.

(7) $J_2$ and $J_3$.

In the final two cases, there exists an even greater similarity:

(8) Centralizers of 2-*central* involutions (i.e., involutions in the center of a Sylow 2-subgroup) are isomorphic in $M_{24}, He$, and $L_5(2)$ and are isomorphic in $J_2$ and $J_3$.

Here are some of the key differences in the internal structures of members of $\mathscr{F}$.

## Differences

(1) $G_2(q)$, $^3D_4(q)$, $q$ odd, and $M_{12}$ have 2-rank 3; the remaining elements of $\mathscr{F}$ have 2-rank $\geqslant 4$.

(2) $J_2$ and $J_3$ have nonconnected Sylow 2-subgroups; all other elements of $\mathscr{F}$ have connected Sylow 2-subgroups.

The number of conjugacy classes of involutions in the members of $\mathscr{F}$ varies from one to four. For example:

(3) $M_{22}$, $M_{23}$, $Mc$, $Ly$, and $J_3$ have only one conjugacy class of involutions.

(4) $L_5(2)$, $U_5(2)$, $A_n$, $8 \leqslant n \leqslant 11$, $M_{12}$, $M_{24}$, $J_2$, $HS$, and $He$ have two conjugacy classes of involutions.

(5) For elements of $\mathscr{F}$ of Lie type of odd characteristic $q$, the number of classes of involutions usually depends on some congruence on $q$. However, the number is one for the groups $G_2(q)$ and $^3D_4(q)$ for all values of $q$.

Likewise there are differences in the degree of balance which holds in members of $\mathscr{F}$ (see [I, 4.60] for the definition, also cf. [I, 4.56 and 4.61]). For example:

(6) $G_2(q)$, $^3D_4(q)$, $q$ odd, $A_n$, $n$ even, and every sporadic group in $\mathscr{F}$ are locally balanced groups.

(7) With the exception of $A_{11}$, every member of $\mathscr{F}$ is locally 2-balanced; however, $A_{11}$ is locally 4-balanced, but not locally 3-balanced.

Despite these various differences, there is considerable uniformity in the pattern of the analysis in each case except for Sylow 2-groups of type $M_{12}$, a particular subcase of which was treated by Brauer and Fong [I: 45] entirely by character-theoretic methods (see Theorem 1.3 below). Furthermore, even though signalizer functors are used crucially in proving Theorem 2.1, this pattern is remarkably similar to that of the quasi-dihedral, wreathed problem. Let me describe its main features.

## Global Outline of Proof

(1) Determine the possible $G$-fusion patterns for the involutions of a Sylow 2-subgroup $S$ of $G$ and related 2-local information. (In any individual case, the number of such fusion patterns varies from one to five.)

(2) For each fusion pattern under (1), and with the aid of prior classification theorems, determine the structure of $C/O(C)$, where $C = C_t$ and $t \in Z(S)$. This analysis will yield that

$$C/O(C) \text{ resembles } C^*/O(C^*),$$

where $C^* = C_{G^*}(t^*)$ with $G^*$ an appropriate simple group in $\mathscr{F}$ and $t^*$ a 2-central involution of $G^*$.

(3) At this point, one assumes that $G$ is a minimal counterexample to Theorem 2.1, which allows identification of proper simple sections of $G$ whose Sylow 2-subgroups are of type $\mathscr{F}$. This information is necessary for verification of appropriate $k$-balance conditions in $G$. Such verification may involve either or both of the following additional steps:

(4)  If $G$ has more than one conjugacy class of involutions, determine the structure of $C_y/O(C_y)$ for involutions $y$ of $G$ which are not $G$-conjugate to $t$.

(5) Construct "covering $p$-local" subgroups of $G$ for suitable odd primes $p$, and use their existence to verify appropriate $k$-balance properties of $G$.

(6) On the basis of the established balance properties of $G$, and with the aid of the signalizer functor theorem, limit the structure of $O(C)$; more specifically, prove that one of the following holds:

(a)  $O(C) = 1$; or
(b)  $L(C/O(C)) \neq 1$ and $L(C)$ maps onto $L(C/O(C))$. (In particular, show that $O(C) = 1$ if $C$ is 2-constrained.)

Hence the effect of (6) [in conjunction with (2)] is to assert:

(7) $C$ resembles $C^*$.

(8) In those cases in which $G$ has more than one class of involutions, the analysis will similarly yield that

$$C_y \text{ resembles } C_{G^*}(y^*)$$

for suitable involutions $y$ of $G$ not $G$-conjugate to $t$, where $y^*$ is an appropriate involution of $G^*$.

In particular, at this stage we have obtained a rather precise description of the centralizer $C$ of a 2-central involution of $G$, and so we are left with the following task:

(9) Solve the resulting centralizer of an involution problem; in other words, show that $G \cong G^*$ for an appropriate $G^* \in \mathscr{F}$

There are, however, some important distinctions in the treatment of specific cases, so we add a few further comments, each relevant to only certain elements of $\mathcal{F}$ and hence to the appropriate known groups $G^*$.

## Groups of Lie Type

(a) First, if $G^*$ is of Lie type of odd characteristic, then just as in the quasi-dihedral, wreathed problem, $C$ may actually resemble the centralizer of an involution in some nontrivial *extension* of $G^*$ by a group of outer automorphisms of odd order [the case $\bar{E} \neq 1$ of (1.73)]. However, in contrast to that situation, where this possibility was eliminated by Brauer's character theory results, such automorphisms arising from $C$ are simply "carried along" and are not eliminated until the final step of the proof.

(b) In some cases, the structure of $C$ obtained in (6) gives $L(C) \cong L(C^*)$; however, in others, the resemblance of $C$ to $C^*$ may not be quite so exact. Thus further work may be required to prove that

$$L(C) \cong L(C^*).$$

(c) Next, using the structure of $C$ and the involution fusion pattern in $G$ (and possibly also the structure of $C_y$ for other involutions $y$ not $G$-conjugate to $t$), determine the structure of a Sylow $p$-subgroup of $G$, where $p$ is the characteristic of $G^*$.

(d) With this information, construct a subgroup $G_0$ of $G$ containing $C$ which is a split $(B, N)$-pair for appropriate subgroups $B$ and $N$ of $G$.

(e) Next show that the multiplication table of $G_0$ is uniquely determined: specifically, show that $G_0$ possesses a normal subgroup $G_1$ of odd index with $G_1 \cong G^*$.

(f) Finally prove that $G = G_1$. (Otherwise $G_0 < G$, as $G$ is simple, and it follows that $G_0$ is strongly embedded in $G$, in which case Bender's theorem yields the usual contradiction.)

## $G^*$ an Alternating Group

(a) Prove that $C \cong C^*$.

(b) For an appropriate involution $y$ of $C$, show that $C_y \cong C_{G^*}(y^*)$, where $y^*$ is a product of two transpositions in the natural representation of $G^*$ (thus $y^*$ is a "short" involution).

(c) With this information, construct a subgroup $G_0$ of $G$ having a presentation by generators and relations identical to that of $G^*$ [this is a direct analog of constructing a $(B, N)$-pair in $G$ in the Lie type case].

(d) Invoke [I, 3.12 or 3.14] to conclude that $G_0 \cong G^*$.

(e) Finally prove that $G = G_1$. (This can either be achieved by Bender's strong embedding theorem or by the Thompson order formula [I, 2.43], which will yield that $|G| = |G_0|$.)

## $G^*$ a Sporadic Group

(a) Prove that $C \cong C^*$.

(b) Invoke [I, 3.50], which characterizes the group $G^*$ in terms of the structure of the centralizer of a 2-central involution, to conclude that $G \cong G^*$.

It will be valuable for the reader to reexamine the discussion of [I, Sections 2.1 and 3.1]. Since the sporadic groups have no "natural" presentation by generators and relations, the construction of an "intermediary" subgroup $G_0$ is always bypassed. In its place, one pins down the exact order of $G$ (which, in effect, means that one knows in advance that this hypothetical subgroup $G_0$ is $G$ itself). However, the remainder of the analysis can be viewed as forcing a presentation for $G$ by generators and relations, whence $G$ is uniquely determined from the structure of $C$ and hence $G \cong G^*$. On the other hand, because this presentation is not "visible," it is much more difficult to achieve than in the Lie type and alternating group cases.

As mentioned above, the one example which does not conform to this outline is Brauer and Fong's characterization of $M_{12}$ by the structure of its Sylow 2-subgroup [I: 45]. Although it is clear that one could have followed the above procedure in this case, too, Brauer and Fong instead used character-theoretic methods to force the order of their group $G$ to equal that of $M_{12}$ [in particular, it then follows that $O(C) = 1$]. Their argument is entirely analogous to that of Lyons in the $U_3(4)$ case. Once they obtained the order of $G$, they were then able to quote Stanton's theorem [I, 3.44] to conclude that $G \cong M_{12}$.

Since their result is used in the proof of Theorem 2.1, we state it explicitly.

THEOREM 2.3. *If $G$ is a simple group with Sylow 2-subgroups of type $M_{12}$ having more than one conjugacy class of involutions, then $G \cong M_{12}$.*

[As we shall see in the next section, the case that $G$ has Sylow 2-subgroups of type $M_{12}$ and only one class of involutions will lead to the groups $G_2(q)$ and $^3D_4(q)$ for suitable odd $q$.]

## 2.2. Simple Groups with Sylow 2-Subgroups of Type $G_2(q)$ or $PSp_4(q)$, $q$ Odd

Despite the uniformity of the outline given in the previous section, individual cases involve sufficiently distinct technical difficulties that, in fact, Theorem 2.1 was established on a case-by-case basis.

We shall now describe two representative cases in some detail; then in the next section we shall limit ourselves to brief comments concerning each of the remaining cases.

It is natural to consider the $G_2(q)$ and $PSp_4(q)$, $q$ odd, cases together, for 2-groups $S$ of these types have very similar structures. Indeed, in both cases

$$S = (Q_1 * Q_2)\langle u \rangle, \tag{2.1}$$

where $Q_1, Q_2$ are isomorphic quaternion groups (whose orders depend on the value of $q$), $u$ is an involution, and

(a) If $S$ is of type $G_2(q)$, $u$ normalizes $Q_1$ and $Q_2$, and $Q_i\langle u \rangle$ is quasi-dihedral, $i = 1, 2$.

(b) If $S$ is of type $PSp_4(q)$, then $u$ interchanges $Q_1, Q_2$ under conjugation.

Despite these similarities in structure, we shall see that there are sufficient differences in the analysis to make it preferable to treat even these two cases separately.

We first discuss the $G_2(q)$ case, which was treated by Harada and me [I: 135], our argument consisting of a reduction to a centralizer of an involution problem, the latter having been previously treated by Fong and W. Wong [I: 103] and Harris [55].

THEOREM 2.4. *If $G$ is a simple group with Sylow 2-subgroup of type $G_2(q)$, $q$ odd, then $G \cong G_2(r)$, $^3D_4(r)$, for some odd $r$, or $M_{12}$.*

We fix a Sylow 2-subgroup $S$ of $G$ and let $Q_1, Q_2, u$ be as in (2.1) above. Again the initial involution fusion analysis uses Glauberman's $Z^*$-

theorem [I, 4.95], Thompson's transfer lemma [I: 21], and the Alperin–Goldschmidt conjugacy family for $S$ [I, 4.85]. In the present case, the analysis yields:

PROPOSITION 2.5. *One of the following holds*:
  (i) *$G$ has only one conjugacy class of involutions*; or
  (ii) *$|S| = 64$ and $G$ has exactly two conjugacy classes of involutions*.

Using the Brauer–Fong theorem (Theorem 2.3), $G \cong M_{12}$ in the latter case. Thus we can assume henceforth that

$$G \text{ has only one class of involutions} \qquad (2.2)$$

Next let $t$ be the involution of $Z(S)$ and set $C = C_t$. Our goal is to show that $C/O(C)$ resembles the centralizer $C^*$ of a 2-central involution $t^*$ of $G_2(r)$ or ${}^3D_4(r)$, $r$ odd. Specifically we prove the following:

PROPOSITION 2.6. *If we set $\bar{C} = C/O(C)$, then $\bar{C}$ possesses normal subgroups $\bar{L}_1, \bar{L}_2$ with the following properties*:

  (i) *$\bar{L}_i \cong SL_2(q_i)$, $q_i$ odd, $i = 1, 2$;*
  (ii) *For a suitable choice of the quaternion subgroups $Q_i$ of $S$, $\bar{Q}_i \in \mathrm{Syl}_2(\bar{L}_i)$, $i = 1, 2$; and*
  (iii) *$[\bar{C}, \bar{C}] = \bar{L}_1 \bar{L}_2 = \bar{L}_1 * \bar{L}_2$.*

We indicate the proof. First, $S \leqslant C$ as $t \in Z(S)$. Set $\tilde{C} = \bar{C}/\langle \bar{t} \rangle$, so that $\tilde{S} = (\tilde{Q}_1 \times \tilde{Q}_2)\langle \tilde{u} \rangle$, where $\tilde{Q}_1 \cong \tilde{Q}_2$ is dihedral and also $\tilde{Q}_1\langle \tilde{u} \rangle \cong \tilde{Q}_2\langle \tilde{u} \rangle$ is dihedral. The initial fusion analysis not only yields the stated conclusions of Proposition 2.5, but also gives strong information concerning the portion of that $G$-fusion which is "realized" in $C$ (and hence concerning the fusion of elements of $\tilde{S}$ in $\tilde{C}$), to be precise:

  (a) $\tilde{C}$ has a normal subgroup $\tilde{C}_0$ with Sylow 2-subgroup $\tilde{S}_0 = \tilde{Q}_1 \times \tilde{Q}_2$;
  (b) All involutions of $\tilde{Q}_i$ are conjugate in $\tilde{C}_0$, $i = 1, 2$; and
  (c) $\tilde{Q}_i$ is strongly closed in $\tilde{S}_0$ with respect to $\tilde{C}_0$, $i = 1, 2$ $\qquad (2.3)$
      (see [I, 4.127] for definition).

Thus Goldschmidt's product fusion theorem [I, 4.148] is applicable and together with the dihedral Sylow 2-group theorem (Theorem 1.47) easily yields that $\tilde{C}_0$ contains normal subgroups $\tilde{L}_1, \tilde{L}_2$ with the following properties:

$$\begin{aligned}
&\text{(a)} \ \ \tilde{L}_i \cong L_2(q_i), q_i \text{ odd, or } A_7, i = 1, 2; \\
&\text{(b)} \ \ \tilde{Q}_i \in \mathrm{Syl}_2(\tilde{L}_i), i = 1, 2; \\
&\text{(c)} \ \ \tilde{L}_1 \tilde{L}_2 = \tilde{L}_1 \times \tilde{L}_2; \text{ and} \\
&\text{(d)} \ \ \tilde{C}_0/\tilde{L}_1 \tilde{L}_2 \text{ is abelian of odd order.}
\end{aligned} \tag{2.4}$$

Again the $A_7$ case is excluded. Indeed, as $\tilde{u}$ normalizes $\tilde{Q}_1$ and $\tilde{Q}_2$, it leaves both $\tilde{L}_1$ and $\tilde{L}_2$ invariant. But then if $\tilde{L}_i \cong A_7$, $\tilde{L}_i \langle \tilde{u} \rangle \cong \Sigma_7$ or $A_7 \times Z_2$ and so has Sylow 2-subgroups of type $D_8 \times Z_2$, contrary to the fact that $\tilde{Q}_i \langle \tilde{u} \rangle$ is a dihedral group.

Since $\tilde{u}$ leaves both $\tilde{L}_1$ and $\tilde{L}_2$ invariant, it also follows from the structure of $\mathrm{Aut}\big(L_2(q_i)\big)$ that $\tilde{C}/\tilde{L}_1 \tilde{L}_2$ is abelian. Now taking pre-images in $\bar{C}$, we immediately obtain the conclusions of Proposition 2.6.

What then is required to turn the resemblance of Proposition 2.6 into an isomorphism of $C$ with $C^*$? In other words, what is the exact structure of $C^*$? The answer is as follows (assume here, for definiteness, that $q_1 \geqslant q_2$):

$$\begin{aligned}
&\text{A.} \ \ \text{Prove that } O(C) = 1. \\
&\text{B.} \ \ \text{Prove that } \bar{C} = \bar{L}_1 \bar{L}_2 \bar{S}. \\
&\text{C.} \ \ \text{Prove that either } q_2 = q_1 \ [\text{the } G_2(q_2)\text{-case}] \text{ or } q_1 = q_2^3 \\
&\ \ \ \ \ \ [\text{the } {}^3D_4(q_2)\text{-case}].
\end{aligned} \tag{2.5}$$

These conditions should be compared with those of equation (1.49) for the quasi-dihedral, wreathed problem. There we had to show that $O(C)$ was cyclic and in the center of $C$; here we need the stronger conclusion that $O(C)$ is trivial. There we had to prove $\bar{C} = \overline{LS}$; here we need the analogous conclusion $\bar{C} = \bar{L}_1 \bar{L}_2 \bar{S}$. But now there is an additional problem, having no analog in the previous situation—namely, determining the precise relationship between the characteristic powers $q_1, q_2$ of the "components" $\bar{L}_1, \bar{L}_2$ of $\bar{C}$.

To establish (2.5)A, we must therefore "kill" the cores of the centralizers of all involutions of $G$ [keep in mind that as $G$ has only one class of involutions by (2.2), killing the core of any one such centralizer accomplishes this for every centralizer of an involution].

As stated in the outline of the previous section, this is accomplished by the signalizer functor method. In the present case, our aim will be to show that [I, 4.54] concerning balanced groups with a connected Sylow 2-subgroup is applicable, for then it will follow that $O(C_x) = 1$ for every $x \in \mathscr{I}(G)$—in particular, that $O(C) = 1$. Furthermore, it is immediate from the structure of $S$ given in (2.1) that $S$ is connected. Thus we have the following result:

PROPOSITION 2.7. *If G is a balanced group, then $O(C) = 1$.*

How easy is it to establish the balance property of $G$? Recall that by [I, 4.58], $G$ is balanced if $K/Z(K)$ is locally balanced for every $K \in \mathscr{L}(G)$. As $G$ has only one class of involutions, Proposition 2.6 implies that $\mathscr{L}(G) = \{SL_2(q_1), SL_2(q_2)\}$ [with the proviso that the corresponding $SL_2(q_i)$ term is omitted if $q_i = 3$]. Thus if $K \in \mathscr{L}(G)$, then $K/Z(K) \cong L_2(q)$ for some odd $q > 3$. But it is easily verified that the centralizer of an involution in $L_2(q)$ is a dihedral group of order $q + \varepsilon$, where $\varepsilon = \pm 1$ and $4 | (q + \varepsilon)$. Hence, in general, the core of this centralizer is nontrivial, so, in general, $L_2(q)$, q odd, is *not* locally balanced.

Thus the direct approach for establishing balance will not work here, and so we must follow a more roundabout procedure. The method we employ is that of *covering p-locals*—the same technique which worked so well in the quasi-dihedral, wreathed problem. As there, we must restrict ourselves henceforth to a minimal counterexample (to Theorem 2.4), for under this assumption, we shall again be able to describe the structure of a proper fusion simple subgroup $H$ which contains a Sylow 2-subgroup of $G$. Namely, if we set $Y = F^*(H/O(H))$, it follows easily then that either $Y$ is a 2-group (in which case $H$ is 2-constrained), or else $Y$ is simple and contains the image of a Sylow 2-subgroup of $H$, so that $Y$ satisfies the hypothesis and hence, by the minimality of $G$, the conclusion of Theorem 2.3 [whence $Y \cong G_2(r)$, $^3D_4(r)$, $r$ odd, or $M_{12}$]. Furthermore, in the constrained case, it is not difficult to show that $H/O(H)$ is necessarily an extension of $E_8$ by $GL_3(2)$. This description will enable us to establish properties of covering $p$-locals which we shall need for the verification of the balance condition in $G$.

DEFINITION 2.8. *If $p$ is a prime divisor of $|O(C)|$, we say that a $p$-local subgroup $H$ of $G$ is a covering p-local provided:*

(a) $H$ covers $C/O(C)$;
(b) $H$ contains both $S$ and an $S$-invariant Sylow $p$-subgroup of $O(C)$; and
(c) $H$ is fusion simple.

We prove the following result:

PROPOSITION 2.9. *G possesses a covering p-local subgroup for every prime p dividing $|O(C)|$.*

The proof is similar to (but somewhat easier than) the corresponding assertion in the quasi-dihedral, wreathed problem (Proposition 1.76). Again it involves a pushing up argument in "two directions" and depends on a suitable extended form of Glauberman's $ZJ$-theorem.

The above discussion also easily yields the following:

PROPOSITION 2.10. *If $H$ is a covering $p$-local subgroup of $G$ for some prime $p$, then*

  (i) $F^*(H/O(H)) \cong G_2(r)$ *or* $^3D_4(r)$ *for some odd $r$;*
  (ii) *All involutions of $S$ are conjugate in $H$;*
  (iii) *If $y \in \mathcal{I}(S)$, then $H$ covers $C_y/O(C_y)$; and*
  (iv) $O(C) \cap H \leqslant O(H)$.

The point of this is that as $H$ is a covering $p$-local, it involves $SL(2, q_1) * SL(2, q_2)$ as $C/O(C)$ does. But neither $M_{12}$ nor an extension of $E_8$ by $GL(3, 2)$ involves a subgroup of this shape, so $F^*(H/O(H))$ must be isomorphic to one of the groups listed in (i) (and must also contain the image of $S$).

The remaining parts of the proposition follow from (i). Indeed, as the groups $G_2(r)$, $^3D_4(r)$, $r$ odd, have only one class of involutions, so therefore does $H$, proving (ii). Furthermore, if $y \in \mathcal{I}(S)$, $y$ is $H$-conjugate to $t$ by (ii), so as $H$ covers $C/O(C) = C_t/O(C_t)$, it covers $C_y/O(C_y)$ as well, proving (iii).

Finally, to establish the crucial conclusion (iv), set $D = O(C) \cap H$. Since $D \vartriangleleft C \cap H = C_H(t)$, it follows that $D \leqslant O(C_H(t))$. Then by general principles, if we set $\bar{H} = H/O(H)$, the image of $O(C_H(T))$ is contained in $O(C_{\bar{H}}(\bar{t}))$. Hence $\bar{D} \leqslant O(C_{\bar{H}}(\bar{t}))$. But $\bar{H}$ is isomorphic to a subgroup of $\mathrm{Aut}(Y)$, where $Y \cong G_2(r)$ or $^3D_4(r)$. However, it is easily checked that these groups are *locally balanced* (this is the reason why covering locals work!), so $O(C_{\bar{H}}(\bar{t})) = 1$ by definition of this term. Thus $\bar{D} = 1$ and so $D \leqslant O(H)$, proving (iv).

Now we can immediately obtain our objective.

PROPOSITION 2.11. *$G$ is a balanced group.*

Indeed, we must show that $O(C_x) \cap C_y = O(C_y) \cap C_x$ for every pair of commuting involutions $x, y \in G$. As $G$ has only one class of involutions, we can suppose without loss that $x = t$ and $y \in S$. By symmetry, it suffices to show that $C_{O(C)}(y) \leqslant O(C_y)$. Moreover, this inclusion will clearly follow if we show that for each prime $p$, a Sylow $p$-subgroup $Q$ of $C_{O(C)}(y)$ is contained in $O(C_y)$. We can also assume that $Q \neq 1$, so that $p$ divides $|O(C)|$.

Expand $Q$ to a $y$-invariant Sylow $p$-subgroup $P$ of $O(C)$. By [I, 4.2], any two such $y$-invariant Sylow $p$-subgroups are conjugate in $C_y$, so as $y \in S$, we can assume without loss that $P$ is $S$-invariant. We now let $H$ be a covering $p$-local subgroup of $G$ containing $SP$. Then $P \leqslant O(C) \cap H$, so by Proposition 2.10(iv), $P \leqslant O(H)$. In particular, $Q \leqslant O(H)$, so certain $Q \leqslant O(C_H(y))$. However, by Proposition 2.10(iii), $H$ and hence $C_H(y)$ covers $C_y/O(C_y)$, which immediately forces $O(C_H(y))$ to lie in $O(C_y)$. Thus $Q \leqslant O(C_y)$, as required.

Hence combined with Proposition 2.7, we obtain (2.5)A as a corollary.

PROPOSITION 2.12. *We have $O(C) = 1$.*

We can therefore write $L_1, L_2$ for $\bar{L}_1, \bar{L}_2$. It follows now from Propositions 2.6(iii) and 2.12 that

$$C = L_1 L_2 SE, \tag{2.6}$$

where $E$ is abelian of odd order and acts as a group of field automorphisms on $L_1$ and $L_2$ [note that for a given $i$, some element of $E^\#$ may act trivially on $L_i$, $i = 1$ or 2; however, as $O(C) = 1$, $E$ acts faithfully on $L_1 L_2$]. Thus we have shown at this point that $C$ "resembles" the centralizer of an involution in an odd order extension of $G_2(q)$ or $^3D_4(q)$, $q$ odd, and so have reduced the proof of Theorem 2.1 to the solution of a centralizer of an involution problem. This reduction completes Harada's and my contribution; and now we turn to the work of Fong and Wong [I: 103] and Harris [55].

Fong and Wong treated the case in which $E = 1$ (i.e., $C = L_1 L_2 S$); Harris has shown that their argument extends to the case $E \neq 1$. It is certainly conceivable, perhaps even likely, that a character-theoretic approach similar to Brauer's in the quasi-dihedral, wreathed case (Proposition 1.82) could be used here to force $E = 1$. However, this has not been attempted, since Harris has shown that the Fong–Wong argument can be extended in its entirety to the case $E \neq 1$. For simplicity, we shall limit the discussion here to their special case, adding a few remarks at the end concerning Harris's extension.

Thus we assume through Proposition 2.19 that

$$C = L_1 L_2 S. \tag{2.7}$$

As we shall see, large portions of the argument have the same flavor as the corresponding parts of the quasi-dihedral, wreathed analysis and involve considerable character theory, including an analysis of the principal 2-block of $G$. Likewise, ultimately $G$ is identified by means of its $(B, N)$-pair structure.

On the other hand, the resemblance (2.7) is not as precise as at the corresponding stage of the quasi-dihedral, wreathed problem, for as yet there is no relationship between the integers $q_1$ and $q_2$. Moreover, a large part of the argument is devoted to showing that, in fact, they must either be equal or one is the cube of the other. As a first step, Fong and Wong prove the following result:

PROPOSITION 2.13. $q_1$ and $q_2$ are powers of the same (odd) prime p.

Suppose false and for definiteness, assume $q_1 > q_2$. Then $q_2 = p^n$ for some prime $p$ and a Sylow $p$-subgroup $X_2$ of $L_2$ is elementary of rank $n$. On the other hand, as $q_1$ is not a power of $p$, a Sylow $p$-subgroup $X_1$ of $L_1$ is cyclic (possibly trivial). In particular, $X_1 X_2 = X_1 \times X_2$ is abelian.

Using results of Brauer from the quasi-dihedral and wreathed problem, they first argue that $X_1 X_2$ is not a Sylow $p$-subgroup of $G$. This in turn implies that $X_1 X_2$ is not a Sylow $p$-subgroup of $C_{X_1 X_2}$ and then that $X_2$ is not a Sylow $p$-subgroup of $O(C_{X_2})$. But $L_1 \leqslant C_{X_2}$ and so by a Frattini argument $L_1$ acts on some Sylow $p$-subgroup of $O(C_{X_2})$. Eventually this leads to the following inequalities:

$$\tfrac{1}{2}(q_2 + 1) \leqslant \tfrac{1}{2}(q_1 - 1) \quad \text{and} \quad p^{(q_2 - 3)/4} \leqslant q_2. \tag{2.8}$$

Since also $q_1 \geqslant q_2 + 2$ (as $q_1, q_2$ are both odd), (2.8) yields a contradiction except in the single case:

$$q_1 = 5, \qquad q_2 = 3, \tag{2.9}$$

which requires a more detailed analysis to eliminate.

Henceforth we let $p$ be the prime divisor of $q_1$ and $q_2$.

The preceding argument actually provides some initial information in this case. Indeed, it yields the following result:

PROPOSITION 2.14. Let $X_2 \in \mathrm{Syl}_p(L_2)$ and let $Y_2 \in \mathrm{Syl}_p(O(C_{X_2}))$. If $q_1 \geqslant q_2$, then one of the following holds:

(i) $q_1 = q_2$ and $|G|$ is divisible by $(q_1 q_2)^3$;
(ii) $q_1 = q_2^3$ and $|G|$ is divisible by $(q_1 q_2)^3$; or
(iii) Either $Y_2 = X_2$ or $|Y_2/X_2| = q_1^2$.

Thus the desired relations between $q_1$ and $q_2$ will follow if we can show that, in fact, $q_1 \geqslant q_2$, and (iii) does not hold.

The next step is crucial both for determining the exact relationship between $q_1$ and $q_2$ and for showing that $G$ is a $(B, N)$-pair.

PROPOSITION 2.15. *The order of $G$ is divisible by $(q_1 q_2)^3$.*

The proof is similar to that of the corresponding portion of the quasi-dihedral, wreathed problem. Again there are "regular" and "irregular" cases. In the first, an analysis of the principal 2-block of $G$ yields the following formula for the order of $G$:

$$|G| = (q_1 q_2)^3 (q_1 + \varepsilon)(q_2 + \varepsilon) f, \tag{2.10}$$

where $\varepsilon = \pm 1$ and $f$ is the degree of an irreducible character of the principal 2-block of $G$. In particular, $(q_1 q_2)^3$ divides $|G|$ in this case.

Fong and Wong then show that, in fact, either (2.10) holds or else $q_1, q_2$ satisfy one of the following:

$$\begin{aligned} &\text{(a) } q_1 = q_2 = 11; \\ &\text{(b) } q_1 = q_2 = 9; \text{ or} \\ &\text{(c) } \min\{q_1, q_2\} = 3, 5, \text{ or } 7. \end{aligned} \tag{2.11}$$

Next they analyze (2.11)(c) in the case $q_1 \neq q_2$ and argue with the aid of Proposition 2.14 that again $(q_1 q_2)^3$ divides $|G|$. Hence they are left with the following five cases:

$$q_1 = q_2 = 3, 5, 7, 9, \text{ or } 11. \tag{2.12}$$

Thus $q_1 = q_2 = p$ and $(q_1 q_2)^3 = p^6$, $p = 3, 5, 7, 9$, or 11. A purely local argument along the lines of Proposition 2.14 yields:

$$\text{Either } p^6 \text{ divides } |G| \text{ or else } X_1 X_2 \in \mathrm{Syl}_p(G). \tag{2.13}$$

On the other hand, an argument similar to that which derived (2.12) from (2.11) shows that $X_1 X_2 \notin \mathrm{Syl}_p(G)$ when $p = 5$ or 7. This leaves the cases $p = 3, 9$, and 11, which require character theory to eliminate.

The case $p = 3$ was, in fact, treated by Janko [70], who characterized $G_2(3)$ in terms of the structure of the centralizer of one of its involutions. His proof is interesting since he shows that the group $G$ under investigation satisfies the 2-local and 3-local conditions under which Thompson characterizes $G_2(3)$ as a $(B, N)$-pair in Section 8 of the $N$-group paper [I: 289]. However, we do not need Janko's complete argument here since his initial character-theoretic analysis produces formulas for the order of $G$, in each of which 27 divides $|G|$, so that then $3^6$ divides $|G|$ by (2.13).

Similarly when $p = 9$ or 11, formulas for $|G|$ are obtained by character theory which yield that $X_1 X_2 \notin \mathrm{Syl}_p(G)$, whence by (2.13) $p^6$ divides $|G|$ in these cases as well.

Using this result, Fong and Wong push Proposition 2.14 one step further to obtain the following sharper result:

PROPOSITION 2.16. *Let* $X_i \in \mathrm{Syl}_p(L_i)$, *set* $C_i = C_{X_i}$, *and let* $Y_i \in \mathrm{Syl}_p(O(C_i))$, $i = 1, 2$. *Then for either* $i = 1$ *or* 2, *we have*

   (i) $Y_i/X_i$ *is abelian of order* $(q_1 q_2)^2$;
   (ii) *If* $Y_i$ *is chosen to be* $X_j$-*invariant*, $j \neq i$, *then* $P = X_j Y_i$ *has order* $(q_1 q_2)^3$ *and is a Sylow p-subgroup of* $G$; *and*
   (iii) $X_i = Z(P)$.

Note that as $L_j \leqslant C_i$ and $L_j \cap O(C_i) = 1$, $X_j \cap Y_i = 1$, so by (i) $P$ has the order specified in (ii). Fong and Wong establish sufficient commutator relations in $P$ to conclude that $X_i = Z(P)$. But then $N_G(P) \leqslant N_i = N_G(X_i)$. Since $C_i \lhd N_i$, it follows by a Frattini argument that, in fact, $N_i = C_i C_{N_i}(t)$. Since $C = C_t$ and $X_1 X_2 \in \mathrm{Syl}_2(C)$ with $X_1 X_2 \leqslant C_i$, this in turn implies that $P \in \mathrm{Syl}_p(N_i)$. Thus $P \in \mathrm{Syl}_p(N_G(P))$ and so $P \in \mathrm{Syl}_p(G)$ by [I, 1.11].

Now Fong and Wong are in a position to determine the relation between the values of $q_1$ and $q_2$ and to prove that $G$ is a split $(B, N)$-pair of rank 2. Observe that for both $k = 1$ and 2, $N_{L_k}(X_k)$ contains a cyclic subgroup $H_k$ of order $q_k - 1$, so $C_i$ is invariant under the abelian group $H = H_1 H_2$ of order $\frac{1}{2}(q_1 - 1)(q_2 - 1)$. Furthermore, it is not difficult to show that $P$ can be chosen to be invariant under $H$. Note that $t \in H$, so that $C_H \leqslant C$, whence by the structure of $C$, $C_H = H$. Set $N = N_G(H)$ and $W = N/H$, so that $W = N_G(H)/C_H$. From the initial 2-fusion analysis, they easily conclude that

$$W \text{ is a dihedral group of order 12.} \qquad (2.14)$$

Thus $W$ is a Coxeter group of rank 2 (see [I, 2.14]). At this point, Fong and Wong can obtain a precise picture of $P \cap P^w$ for $w \in N$ and can determine the commutator relations among the six "root" subgroups of $P$. In fact, this information is strong enough to allow them to construct *two* $(B, N)$-pair subgroups of $G$, one of which is ultimately shown to be $G$ itself, while the other is used to force the exact relation between $q_1$ and $q_2$.

We first consider the second of these $(B, N)$-pairs. Fong and Wong work with an appropriate $H$-invariant subgroup $P^*$ of $P$, with $P^*$ a product of precisely three "root" subgroups, each of order $q_i$, where $i$ is determined as in Proposition 2.16, and they set $B^* = P^*H$ and $W^* = N^*/H$, where $N^*$ is a subgroup of $N$, containing $H$ with $N^*/H \cong \Sigma_3$. They prove the following proposition:

PROPOSITION 2.17. *The following conditions hold*:

   (i) $G^* = \langle B^*, N^* \rangle = B^*N^*B^*$ *is a split* $(B, N)$-*pair*;
  (ii) $|G^*| = q_i^3(q_i^3 - 1)(q_i + 1)(q_j - 1)$, *where* $j \neq i$; *and*
 (iii) $C_{G^*}(t) = L_jH$. *In particular*, $G^*$ *has quasi-dihedral or wreathed Sylow* 2-*subgroups*.

Now they can quickly prove:

PROPOSITION 2.18. *Either* $q_1$ *and* $q_2$ *are equal or one is the cube of the other*.

For definiteness, suppose $q_1 \geqslant q_2$. We need only show that the distinguished value of $i$ in Proposition 2.16 can be taken to be 2, for then as $q_1 \geqslant q_2$, Proposition 2.14 will be applicable. Moreover, in view of Proposition 2.16(ii), Proposition 2.14(iii) will be excluded, so that Proposition 2.18 will follow from Proposition 2.16(i), (ii).

We use Proposition 2.17 and the quasi-dihedral, wreathed theorem (Theorem 1.63). Since $W^* \cong \Sigma_3$, it follows from the latter result that $G^*/O(G^*)$ contains a normal subgroup of index 1 or 3 isomorphic to $L_3(q_i)$. Hence $L_3(q_i)$ divides $|G^*|$, and consequently

$$\frac{1}{d} q_i^3(q_i^3 - 1)(q_i^2 - 1) \text{ divides } |G^*|, \qquad (2.15)$$

where $d = 1$ or 3 and if $d = 3$, then $q_i \equiv 1 \pmod 3$. But $|G^*|$ is given by Proposition 2.17(ii), and so by (2.15), it follows that

$$d(q_j - 1)/(q_i - 1) \text{ is an integer.} \qquad (2.16)$$

However, this is possible only if $q_j \geqslant q_i$. Since the index 2 is specified solely by the requirement $q_1 \geqslant q_2$, we conclude that we *can* take $i = 2$, as required.

They next prove the following result:

PROPOSITION 2.19. *G is a split* $(B, N)$-*pair of rank* 2 *with Weyl group of order* 12, *and G has order*

$$(q_1 q_2)^3 (q_1^2 - 1)(q_2^2 - 1)(1 + q_1 q_2 + q_1^2 q_2^2).$$

To establish the proposition, they set $B = PH$ and first show that the subgroup $G_0 = \langle B, N \rangle = BNB$ is a split $(B, N)$-pair of rank 2 of the specified order. Thus they need only prove that $G = G_0$. But it is immediate that $C \leqslant G_0$ and that $G_0$ has only one class of involutions. Hence if $G_0 < G$, then $G_0$ would be strongly embedded in $G$ and Bender's classification theorem would yield a contradiction.

It thus remains only to identify the split $(B, N)$-pair $G$. The Fong–Seitz classification theorem [I, 3.14] implies that $G \cong G_2(q)$ or $^3D_4(q)$. However, unfortunately, as the present Fong–Wong theorem was available at the time Fong and Seitz established their result, they invoked it, for convenience, to handle the corresponding situation, so their theorem cannot be used here. However, the pattern of proof is essentially the same as in their general case, the goal being to show that the multiplication table of $G$ is uniquely determined by $q_1$ and $q_2$. This is achieved by proving the following:

(a) The multiplication table of $P$ is uniquely determined;
(b) The action of $H$ on $P$ is uniquely determined (and hence the structure of $B$ is uniquely determined); and     (2.17)
(c) The structure of $N$ is uniquely determined.

Although the various moves are forced, the details are fairly elaborate.

Thus, there is at most one simple group satisfying the specified conditions.

Finally assuming $q_1 \geqslant q_2$, then according as $q_1 = q_2$ or $q_1 = q_2^3$, the group $G_2(q_2)$ or $^3D_4(q_2)$ satisfy the specified conditions. Thus we reach the desired conclusion.

PROPOSITION 2.20. *Assume* $C = L_1 L_2 S$ *and suppose, for definiteness, that* $q_1 \geqslant q_2$. *Then either* $q_1 = q_2$ *and* $G \cong G_2(q_2)$ *or* $q_1 = q_2^3$ *and* $G \cong {}^3D_4(q_2)$.

Finally a few words about Harris's extension to the case $E \neq 1$. Surprisingly, this assumption actually leads to a simplification of certain parts of the argument, because one is now able to apply induction at certain key places. Namely, set $K_i = C_{L_i}(E)$, $i = 1, 2$, $M = C_E$, and $\bar{M} = M/E$. Since $E$ induces field automorphisms on $L_i$, we have $K_i \cong SL_2(r_i)$ with $q_i = r_i^{m_i}$ for some $m_i$, $i = 1, 2$. Furthermore, from the initial fusion analysis, it is not difficult to show that $\bar{M}$ is fusion simple (and, in fact, simple) with $C_{\bar{M}}(\bar{t}) = \bar{K}_1 \bar{K}_2 \bar{S}$. Since $|\bar{M}| < |G|$, induction then yields that $\bar{M} \cong G_2(r_2)$ or $^3D_4(r_2)$ (assuming, for definiteness, that $r_1 \geqslant r_2$). In particular, $r_1$ and $r_2$ are powers of the *same* prime $p$, so that $q_1$ and $q_2$ are also powers of $p$, thus proving Proposition 2.13. Given the structure of a Sylow $p$-subgroup of $G_2(r)$ or $^3D_4(r)$, it likewise follows that $C$ does not contain a Sylow $p$-subgroup of $G$, which in turn implies that the first alternative of Proposition 2.14(iii) does not hold.

Similarly, it is now no longer necessary to determine in advance the exact relationship between $q_1$ and $q_2$, for once one knows that $Y_i/X_i$ is abelian of order $(q_1 q_2)^2$ [Proposition 2.16(i)], one can construct the $(B, N)$-pair $G_0$ exactly as before, but with $B = PHE$. It is then easily shown that $E$ centralizes $W = N/H$ and that $G_0$ possesses a normal subgroup $G_1$, which is itself a $(B, N)$-pair with $C_{G_1}(t) = C \cap G = L_1 L_2 S$. Thus $G_1 < G$, so that again by induction $G_1 \cong G_2(q_2)$ or $^3D_4(q_2)$ (assuming $q_1 \geqslant q_2$), which in turn yields the conclusion $q_1 = q_2$ or $q_1 = q_2^3$. Of course, the structure of $G_0$ actually leads to an immediate contradiction, since by Bender's strong embedding theorem we must have $G = G_0$, whence $G_1 \lhd G$, contrary to the simplicity of $G$.

Harris's proof is based on the observation that the various subgroups which Fong and Wong must consider in their analysis are all $E$-invariant, so that one can keep track of the effect of $E$ on their argument. The most difficult step for Harris is verification of Proposition 2.16(i). The point is that if we set $N_i = N_G(X_i)$, then $E \leqslant N_i$ and we may have $E \cap O(N_i) \neq 1$, so that $Y_i$ need not be a Sylow $p$-subgroup of $O(N_i)$, which is a complicating factor for the analysis.

This completes the discussion of Theorem 2.4 and we turn now to groups with Sylow 2-subgroups of type $PSp_4(q)$, $q$ odd, which we treat more sketchily.

THEOREM 2.21. *If $G$ is a simple group with Sylow 2-subgroup of type $PSp_4(q)$, $q$ odd, then $G \cong PSp_4(r)$ for some odd $r$, $A_8$, or $A_9$.*

We note that as in the $G_2(q)$ case, analysis of the subgroup structure of a minimal counterexample to Theorem 2.21 requires a general description of fusion simple groups $X$ with Sylow 2-subgroups of type $PSp_4(q)$. If

$O(X) = 1$, then as in the $G_2(q)$ case, either $F^*(X)$ is simple or $X$ is 2-constrained. In the latter case, it is not difficult to show here that $X$ is either a split extension of $E_8$ by $GL_3(2)$ or a split extension of $E_{16}$ by $A_5$ (acting nontransitively on the involutions of the $E_{16}$). [We note that there also exists a transitive such split extension, but it has Sylow 2-subgroups of type $L_3(4)$].

Let $S \in \mathrm{Syl}_2(G)$, with $S = (Q_1 * Q_2)\langle u \rangle$, as in (2.1) above, and let $\langle t \rangle = Z(S)$. Observe that as $u$ interchanges $Q_1$ and $Q_2$, $C_S(u) = \langle u \rangle \times \langle t \rangle \times D$, where $D$ is a dihedral group. Thus $m(C_S(u)) = 4$ and it follows that $S$ has rank 4.

For convenience we shall say that $G$ has the *involution fusion pattern* of $G^* = A_8$ or $\mathrm{PSp}_4(q)$, $q$ odd, provided $S$ is isomorphic to a Sylow 2-subgroup $S^*$ of $G^*$ under an isomorphism $\alpha$ and for $a, b \in \mathscr{I}(S)$, $a$ is conjugate to $b$ in $G$ if and only if $\alpha(a)$ is conjugate to $\alpha(b)$ in $G^*$. This terminology makes it easier to describe the involution fusion patterns in $G$.

As before, we set $C = C_t$ and $\overline{C} = C/O(C)$. Again it was Harada and I who carried out the reduction to a centralizer of an involution problem [I: 134]. To describe the results of the initial involution fusion analysis, we let $A$ be an elementary subgroup of $S$ of order 16 (note that any two such subgroups are conjugate in $S$) and set $X = N_G(A)/C_A$.

PROPOSITION 2.22. *One of the following holds:*

(A) *$G$ has two classes of involutions and the involution fusion pattern of $PSp_4(q)$, $q$ odd, and $X \cong A_5$;*

*or $S$ has order 64 and either:*

(B) *$G$ has two classes of involutions and the involution fusion pattern of $A_8$, and $X \cong \Sigma_3 \times \Sigma_3$;*

(C) *$G$ has three classes of involutions, and $X \cong Z_2 \times \Sigma_3$; or*

(D) *$A$ is strongly closed in $S$ with respect to $G$ (and $X \cong A_5$).*

Applying Goldschmidt's strongly closed abelian theorem [I, 4.128], we, of course, immediately eliminate (D).

PROPOSITION 2.23. *Case (D) of Proposition 2.22 does not hold.*

[We note that if we had allowed $G$ to be core-free fusion simple, (D) would lead to the extension of $E_{16}$ by $A_5$.]

We shall refer to the three remaining possibilities of Proposition 2.22 as cases $A$, $B$, $C$, respectively.

As we shall see, also case $C$ leads to a strongly closed elementary subgroup of $S$ of order 8 and is thus likewise eliminated by Goldschmidt's theorem [and likewise leads to the split extension of $E_8$ by $GL_3(2)$ in the core-free fusion simple case].

In each case, we obtain the structure of $\bar{C}$ by the usual procedure.

PROPOSITION 2.24. (i) *If case A holds, then $\bar{C}$ contains isomorphic subgroups $\bar{L}_i \cong SL_2(q)$, for some odd $r$, $i = 1, 2$, with the following properties*:

(1) $\bar{L}_1 \bar{L}_2 = \bar{L}_1 * \bar{L}_2$ *and* $\bar{L}_1 \bar{L}_2 \lhd C$;
(2) *For a suitable choice of $Q_1, Q_2$, $\bar{Q}_i \in Syl_2(\bar{L}_i)$, $i = 1, 2$;*
(3) $\bar{u}$ *interchanges $\bar{L}_1$ and $\bar{L}_2$ under conjugation; and*
(4) $\bar{C} = \bar{L}_1 \bar{L}_2 \langle \bar{u} \rangle \bar{E}$, *where $\bar{E}$ is an abelian group of odd order inducing field automorphisms on $\bar{L}_1$ and $\bar{L}_2$.*

(ii) *If case B or case C holds, then $\bar{C} \cong C_{A_8}((12)(34)(56)(78))$.*

Since $u$ interchanges $Q_1$ and $Q_2$, (i)(3) is a consequence of (i)(2). Thus, in contrast to the previous $G_2(q)$ case, we know at the outset of the $PSp_4(q)$ case that the "components" $\bar{L}_1, \bar{L}_2$ are isomorphic and hence defined over the same field. In particular, we have

$$\bar{L}_1 \bar{L}_2 \langle \bar{u} \rangle \cong C_{PSp_4(r)}(t^*) \tag{2.18}$$

for any 2-central involution $t^*$ of $PSp_4(r)$.

We shall first treat case $A$, which, of course, will lead to the groups $PSp_4(r)$. Thus we assume in Propositions 2.25–2.35 that

$$\text{Case } A \text{ holds.} \tag{2.19}$$

Let $U$ be a normal four-subgroup of $S$, so that $U = \langle t, y \rangle$ for some involution $y$. The initial fusion analysis shows that $y$ is a representative of the second conjugacy class of involutions of $G$. Then we have

PROPOSITION 2.25. *If we set $\bar{C}_y = C_y/O(C_y)$, then $\bar{C}_y$ has a normal subgroup $\bar{L} \cong L_2(h)$ for some odd $h$, with $C_{\bar{C}_y}(\bar{L})$ a cyclic 2-group.*

With this information, we again use signalizer functors to prove that $O(C) = 1$. We note that covering $p$-locals might possibly be avoided in the present situation by working with Goldschmidt-type signalizer functors rather than with balanced and 2-balanced functors as Harada and I have done. Covering $p$-locals are defined here as in the $G_2(q)$ case. Assuming that $G$ is a minimal counterexample, we prove the following result:

PROPOSITION 2.26. *G possesses a covering p-local subgroup for every prime p dividing O(C).*

The existence of covering $p$-locals enables one to relate the characteristic power $r$ of $\bar{C}$ with the characteristic power $h$ of $\bar{C}_y$.

PROPOSITION 2.27. *If H is a covering p-local subgroup of G for some prime p dividing $|O(C)|$, then*

(i) $F^*(H/O(H)) \cong PSp_4(r)$; *and*
(ii) *H covers $C_y/O(C_y)$. In particular, $h = r$.*

One cannot expect to use Proposition 2.27 to prove that $G$ is balanced, for, in general, the group $PSp_4(r)$ is not locally balanced. Indeed, the centralizer of a non-2-central involution of $PSp_4(r)$ contains a cyclic normal subgroup of order $r - \varepsilon$, where $\varepsilon = \pm 1$ and $r \equiv \varepsilon \pmod 4$ and so, in general, has a nontrivial core—indeed, this core is nontrivial unless $r$ is a Fermat or Mersenne prime or 9. Thus Harada and I were able to prove the following:

PROPOSITION 2.28. (i) *If r is a Fermat or Mersenne prime or 9, then G is balanced; and*
(ii) *In the contrary case, G is 2-balanced.*

Since $S$ is easily seen to be connected, [I, 4.54] yields:

PROPOSITION 2.29. *If r is a Fermat or Mersenne prime or 9, then $O(C) = 1$.*

The nonbalanced case is a little more difficult. Moreover, the fact that $r > 9$ under this assumption is crucial to the argument. Indeed, let $A$ be an elementary subgroup of $S$ of order 16 containing $U$. Using properties of the groups $SL_2(r)$ and $L_2(h)$, we easily prove the following:

PROPOSITION 2.30. (i) $C = \Gamma_{A,3}(C) = \langle N_C(B)|B \leqslant A, m(B) \geqslant 3\rangle$; *and*
(ii) *If $h > 9$, then $C_y = \Gamma_{A,3}(C_y)$.*

Thus $C$ is "3-generated with respect to $A$" for all $r$, but $C_y$ has this property only if $h > 9$. We use these facts as follows. Let $\theta$ be the 2-balanced $A$-signalizer functor on $G$ [I, 4.65]. By the signalizer functor theorem [I, 4.38], its completion $\theta(G; A)$ has odd order. Assuming $O(C) \neq 1$, it follows that $N_G(\theta(G; A))$ is a proper subgroup of $G$.

On the other hand, we claim that the 2-balanced functor has the following general property:

$$N_G(B) \leqslant N_G\big(\theta(G;A)\big) \tag{2.20}$$

for every subgroup $B$ of $A$ of rank at least 3.

Since $N_G(B)$ clearly normalizes $\theta(G;B)$, it will suffices to show that $\theta(G;B) = \theta(G;A)$. Now $\theta(G;A)$ is generated by its subgroups $\varDelta_G(D)$ for $D \leqslant A$, and $m(D) = 2$ by definition of the 2-balanced functor, so we need only prove that each such $\varDelta_G(D) \leqslant \theta(G;B)$ [as the reverse inclusion $\theta(G;B) \leqslant \theta(G;A)$ obviously holds]. But $\varDelta_G(D)$ is $A$- and hence $B$-invariant and as $|\varDelta_G(D)|$ is odd, [I, 4.13] implies that $\varDelta_G(D)$ is generated by its subgroups $\varDelta_G(D) \cap C_F$ as $F$ ranges over the set of $Z_2 \times Z_2$ subgroups of $B$. However, as $G$ is 2-balanced (with respect to $A$),

$$\varDelta_G(D) \cap C_f \leqslant O(C_f) \tag{2.21}$$

for each $f \in F^\#$, and consequently

$$\varDelta_G(D) \cap C_F \leqslant \bigcap_{f \in F^\#} O(C_f) = \varDelta_G(F) \leqslant \theta(G;B). \tag{2.22}$$

Thus each $\varDelta_G(D) \leqslant \theta(G;B)$ and our claim is proved.

We use (2.20) here in conjunction with Proposition 2.30. Since $h = r$ by Propositions 2.28 and 2.29, as $O(C) \neq 1$, we conclude:

PROPOSITION 2.31. *If* $O(C) \neq 1$, *and* $r > 9$, *then*

$$\langle C, C_y \rangle \leqslant N_G\big(\theta(G;A)\big).$$

Since $G$ has precisely two conjugacy classes of involutions, it is immediate now, under the assumptions of Proposition 2.31, that $N_G\big(\theta(G;A)\big)$ is strongly embedded in $G$. Since the existence of a strongly embedded subgroup implies that $G$ has only one class of involutions [I: 130, Theorem 9.2.1], we obtain a contradiction without recourse to Bender's theorem. Thus also $O(C) = 1$ when $r > 9$, and so we have proved the following result:

PROPOSITION 2.32. *We have* $O(C) = 1$.

Again this completes Harada's and my reduction to the centralizer of an involution problem. The balance of the analysis of case $A$ was carried out by W. Wong [131] (the case $\bar{E} = 1$) and Harris [55] (the case $\bar{E} \neq 1$).

As $\bar{C} = C$, we may again drop the bars on $\bar{L}_1, \bar{L}_2$, and $\bar{E}$.

The initial fusion analysis shows that the involution $u$ of $S$ is not $G$-conjugate to $t$. As $u$ interchanges $L_1$ and $L_2$, it follows easily that $C_{L_1 L_2}(u) = \langle t \rangle \times K$, where $K \cong L_2(r) [= PSL_2(r)]$.

The next step in the analysis is to determine the exact structure of $C_u$. First of all, it is not difficult to show that $N_C(E)$ contains a Sylow 2-subgroup of $C$, which without loss we can assume to be $S$. Thus $u$ normalizes $E$. However, as $u$ interchanges $L_1$ and $L_2$, it is no longer automatic [as it was in the $G_2(q)$ case] that $u$ must centralize $E$; this conclusion requires further argument, involving the centralizers of various subgroups of $S$. In the end, one obtains the following:

PROPOSITION 2.33. *Let* $r \equiv \varepsilon \pmod 4$, *where* $\varepsilon = \pm 1$ *and let* $2^m$ *be the exact power of* 2 *dividing* $r - \varepsilon$. *Then* $C_u$ *contains a normal subgroup* $D_u$ *of index* $2n$ *with* $D_u = O(C_u) KE$, $O(C_u) K \cap E = 1$, *and* $O(C_u)$ *cyclic of order* $(r - \varepsilon)/2^m$.

At this point, one has pinned down the precise structure of $C = C_t$ and $C_u$. Since $G$ has exactly two classes of involutions, Thompson's order formula [I, 2.43] can now be used to obtain the following result:

PROPOSITION 2.34. *We have* $|G| = |PSp_4(r)| |E| = \frac{1}{2} r^4 (r^4 - 1) |E|$.

Now one is in a position to construct a $(B, N)$-pair in $G$, as in the $G_2(r)$ case. If $r = p^n$, $n$ a prime, one first determines the structure of a Sylow $p$-subgroup of $G$, then defines the Borel subgroup $B = HPE$, sets $N = N_G(H)$ and $W = N/H$, and argues here that $W$ is dihedral of order 8. Using this information, one then shows that $G_0 = \langle B, N \rangle = BNB$ is a $(B, N)$-pair of rank 2. As before, the simplicity of $G$ and Bender's strong embedding theorem force $G = G_0$ and $E = 1$. Finally one must prove that the multiplication table of $G$ is uniquely determined, from which it follows that $G \cong PSp_4(r)$.

Thus we conclude:

PROPOSITION 2.35. *If case $A$ holds, then* $G \cong PSp_4(r)$.

We can therefore assume that case $B$ or $C$ holds. In particular, $|S| = 64$.

Furthermore, it is immediate from Proposition 2.24(ii) that $\overline{C}$ and hence $C$ is solvable. Again the goal is to prove that $O(C) = 1$. Using the exact involution fusion pattern, we first prove:

PROPOSITION 2.36.   *In case C, the centralizer of every involution of G is solvable.*

If the centralizer of every involution of $G$ is solvable, then $G$ is balanced by [I, 4.52] and as $S$ is connected, [I, 4.54] thus yields the following result:

PROPOSITION 2.37. *If the centralizer of every involution of G is solvable, then $O(C) = 1$. In particular, $O(C) = 1$ if case C holds.*

Thus we can suppose we are in case $B$ with the centralizer of some involution nonsolvable. Let $U \leqslant A$ with $U \cong Z_2 \times Z_2$ and $U \lhd S$ ($U$ exists as $A \lhd S$). Then $U = \langle t, y \rangle$ and again the initial fusion analysis shows that $y$ is not conjugate to $t$ in $G$. Since $G$ has precisely two classes of involutions in case $B$, it follows that $C_y$ is nonsolvable. To pin down the structure of $C_y/O(C_y)$, we again resort to covering $p$-local subgroups of $G$. Likewise, we again assume that $G$ is a minimal counterexample. With the aid of such covering $p$-locals, we prove the following result:

PROPOSITION 2.38. *If we set $\overline{C}_y = C_y/O(C_y)$, then*

$$\overline{C}_y \cong C_{A_9}\big((12)(34)(56)(78)\big).$$

Thus $\overline{C}_y$ contains a normal subgroup of index 2 isomorphic to $Z_2 \times Z_2 \times A_5$.

Observe next that the group $A_5$ is not locally balanced; however, nonbalance arises only from *outer* automorphisms [more precisely, if $x \in \mathscr{S}(\Sigma_5)$, then $C_{\Sigma_5}(x)$ is a 2-group if $x \in A_5$ and $C_{\Sigma_5}(x) \cong Z_2 \times \Sigma_3$ if $x \notin A_5$ (whence $x$ is a transposition); correspondingly $O\big(C_{\Sigma_5}(x)\big) \cong 1$ or $Z_3$].

The point is that since some involution of $C_S(y)$ induces an outer automorphism on $\overline{K} = L(\overline{C}_y)(\cong A_5)$, we cannot prove that our group $G$ is necessarily balanced. On the other hand, $A \leqslant C_y$ and from the structure of $C_y$, $\overline{A}$ induces *inner* automorphisms on $\overline{K}$, so $\overline{K}$ is locally balanced "with respect to $\overline{A}$". Using this observation together with the fact that $C$ is solvable, it is not difficult to prove the following result:

PROPOSITION 2.39. *G is balanced with respect to A.*

But now if $O(C) \neq 1$ and we let $\theta$ be the 1-balanced $A$-signalizer functor on $G$ [i.e., $\theta(C_a) = O(C_a)$ for $a \in A^{\#}$], we conclude, as usual, that the completion $\theta(G; A)$ of $\theta$ is nontrivial of odd order [whence $M = N_G(\theta(G; A))$ is proper in $G$] and also that

$$N_G(B) \leqslant M \tag{2.23}$$

for all noncyclic subgroups of $A$.

This is sufficient information to prove that $M$ is strongly embedded in $G$. Indeed, $\langle O(C), O(C_y) \rangle \leqslant \Gamma_{A,2}(G) \leqslant M$ by [I, 4.13]; and $S \leqslant M$ as $A \lhd S$. Also as $\bar{K} = L(\bar{C}_y)$ centralizes a four-subgroup of $A$, it follows that $M$ covers $\bar{K}$, so, in fact, $C_y \leqslant M$. But now we conclude easily from the minimality of $M$ that $M/O(M) \cong A_9$, which in turn implies that $M$ covers $C/O(C)$. Thus also $C \leqslant M$ and so $M$ is indeed strongly embedded in $G$, giving the usual contradiction. Thus $O(C) = 1$ in this case as well.

PROPOSITION 2.40. *We have $O(C) = 1$.*

Thus $C$ is isomorphic to the centralizer of a "long" involution in $A_8$. The resulting centralizer of an involution problem was treated by Held [63].

Pursuing the fusion of involutions of case $C$, he argues that $S$ contains an elementary subgroup of order 8 which is strongly closed in $S$ with respect to $B$. Again [I, 4.128] yields a contradiction. Thus we have the following:

PROPOSITION 2.41. *Case B holds.*

The goal of Held's analysis is, as expected, to prove the following result:

PROPOSITION 2.42. *We have $C_y \cong C_{A_n}((12)(34))$, $n = 8$ or 9.*

Once this is established, we know the exact structure of the centralizers of all involutions of $G$, and hence, using the Thompson order formula, we conclude that $|G| = |A_n|$. Therefore to complete the analysis, it remains only to construct a subgroup $G_0$ of $G$ with $G_0 \cong A_n$. When $n = 9$, Held constructs $G_0$ in terms of a presentation by generators and relations identical to that which holds in $A_9$. Thus $G_0 \cong A_9$ by [I, 3.42]. Undoubtedly a similar argument would work when $n = 8$, or alternatively, as $A_8 \cong L_4(2)$, one could construct $G_0$ as a $(B, N)$-pair. However, as W. Wong [122] characterized $A_8$ from the structure of these two centralizers, Held quotes that result instead. We remark that Wong's proof (1963), carried out prior to many of the later developments (in particular, before the Thompson order formula) consists of an elaborate character-theoretic argument, using block theory. Eventually he determines the degrees of the irreducible characters in the principal 2-block of $G$, the order of $G(=|A_8|)$, and representatives of every conjugacy class of elements of $G$. Finally he uses the Brauer trick to show that $G$ contains a subgroup $G_1 \cong A_7$. Thus $G$ has a permutation representation of degree 8 on the cosets of $G_1$. Since $G$ is simple with $|G| = |A_8|$, this forces the desired conclusion $G \cong A_8$.

This completes the discussion of Theorem 2.21.

## 2.3. Other Types of Sylow 2-Subgroups

We shall now briefly describe the principal features of the classification theorems for the remaining 2-groups of type $\mathscr{F}$. The proofs follow the same outline sketched at the end of Section 2.1. In particular, they consist of reductions to appropriate centralizer of involution problems. In the case of sporadic groups, these centralizer of involution problems have been discussed in [I], and the results summarized in [I, 3.50]. We note also that some of these problems possess fusion simple, 2-constrained solutions, which give rise to additional cases that have to be considered in the analysis.

We let $S, t, C$ have the same meanings as above and again we set $\bar{C} = C/O(C)$.

First, Harada and I treated the $J_2$ case [40], with Janko handling the resulting centralizer of involution problem [I: 188].

THEOREM 2.43. *If $G$ is a simple group with Sylow 2-subgroups of type $J_2$, then $G \cong J_2$ or $J_3$.*

We note that $S = T\langle u \rangle$, where $T$ is of type $L_3(4)$, and $u$ is an involution such that $C_T(u) \cong Q_8$.

The initial fusion analysis yields that $G$ has either one or two conjugacy classes of involutions, that $\bar{C}$ is a split extension of $Q_8 * D_8$ by $A_5$ in both cases, and in the latter case if $y$ is a non-2-central involution, then $C_y$ has elementary Sylow 2-subgroups of order 16 and either $C_y$ is solvable or $L(C_y/O(C_y)) \cong L_2(q)$ for some $q \equiv 3,5 \pmod 8$. (Of course, in $J_2$ itself we have $q = 5$.) At this point we need only prove that $O(C) = 1$, for then the desired conclusion follows from [I: 188]. This is achieved by showing that $G$ is balanced and connected, whence $O(C) = 1$ by [I, 4.54]. Assuming $O(C) \neq 1$, we used the existence of covering locals $H$ to establish both balance and connectedness (when $G$ has only one class of involutions, centralizers of involutions are 2-constrained, so balance follows from [I, 4.52] without the use of covering locals). Note that $S$ itself is not connected; however, as $H/O(H) \cong J_2$ or $J_3$, which are connected groups, $H$ is connected and this enabled us to prove that $G$ is connected.

Harada and I also treated the $Ly$ case [40], with Lyons responsible for the corresponding centralizer of involution problem [78].

THEOREM 2.44. *If $G$ is a simple group with Sylow 2-subgroups of type $Ly$, then $G \cong Ly$.*

In this case $S = T\langle u, v \rangle$, where $T$ is of type $L_3(4)$, $\langle u, v \rangle \cong Z_2 \times Z_2$, $T\langle u \rangle$ is of type $J_2$, and $C_T(v) \cong D_8$.

This time we show that $G$ has only one class of involutions and that $\bar{C} \cong \hat{A}_{10}$ or $\hat{A}_{11}$. We then use covering 3-locals $H$ to prove that $G$ is balanced. [The point is that centralizers of long involutions in $A_{11}$ have cores of order 3, so that if $C/O(C) \cong \hat{A}_{11}$, *a priori* $G$ need not be balanced "relative to 3-subgroups." On the other hand, $Ly$ is a balanced group and as $H/O(H) \cong Ly$, we can conclude that $G$ is balanced.] Since $S$ is connected, [I, 4.54] again yields that $O(C) = 1$. Now [I, 2.32 and 2.50] give the desired conclusion $G \cong Ly$.

Harada and I [I: 134] and D. Mason [80, 81] treated the $L_4(q)$, $q$ odd, case, which gives rise to several distinct centralizer of involution problems: Phan [96, 99] and Mason handled those leading to $L_4(q)$, $U_4(q)$, Held [64] and Kondo [74], respectively, those leading to $A_{10}$, $A_{11}$, Janko [I: 190] those leading to $M_{22}, M_{23}$, and Janko and S. K. Wong [72] and Janko [I: 189] those leading to $Mc$ (Janko's latter result is needed to rule out the possibility $C \cong \hat{A}_9$).

THEOREM 2.45. *If $G$ is a simple group with Sylow 2-subgroups of type $L_4(q)$, $q$ odd, then $G \cong L_4(r)$, $U_4(r)$, $A_{10}, A_{11}, M_{22}, M_{23}$, or $Mc$.*

The structure of $S$ depends upon the congruence of $q(\mathrm{mod}\ 8)$. Moreover, in each case there are several (often not completely obvious) equivalent descriptions of $S$. We give the ones used in the actual analysis:

(a) If $q \equiv 3, 7(\mathrm{mod}\ 8)$, then $S = D_{2^n} \wr Z_2, n \geq 3$;
(b) If $q \equiv 5(\mathrm{mod}\ 8)$, then $S = T\langle u \rangle$, where $T$ is of type $L_3(4)$ and $C_T(u) \cong D_8$; or
(c) If $q \equiv 1(\mathrm{mod}\ 8)$, then $S = TD$, where $T \cong Z_{2^n} \times Z_{2^n} \times Z_{2^{n-2}}, n \geq 3$, and $D \cong D_8$ acts in a suitably prescribed fashion on $T$.                    (2.24)

The initial fusion analysis was carried out separately for each congruence on $q \pmod 8$ [the groups $A_{10}, A_{11}$ occur when $q \equiv 3 \pmod 8$; $M_{22}, M_{23}, Mc$ when $q \equiv 5 \pmod 8$]. Harada and I treated these two congruences, Mason the remaining two. The final result is as follows:

PROPOSITION 2.46. *One of the following holds*:

(1) *G has the involution fusion pattern of $L_4(q)$ and $\bar{C}$ contains a normal subgroup $\bar{L} \cong SL_2(r) * SL_2(r)$ for some $r \equiv q \pmod 8$;*
(2) *$q \equiv 3 \pmod 8$, G has two classes of involutions, and $\bar{C} \cong C_{A_{10}}\big((12)(34)(56)(78)\big)$ (whence $\bar{C}$ is a split extension of $E_{16}$ by $\Sigma_4$);*
(3) *$q \equiv 5 \pmod 8$, G has one class of involutions, and $\bar{C} \cong \hat{A}_8$ or $\hat{A}_9$; or*
(4) *$q \equiv 5 \pmod 8$, G has one or two classes of involutions and $\bar{C}$ is isomorphic to the centralizer of an involution in $M_{22}$ or $M_{23}$ [whence correspondingly C is a split extension of $E_{16}$ by $\Sigma_4$ or $L_3(2)$].*

A few remarks about the proposition. When $q \equiv 3 \pmod 8$, in which case $S$ is of type $A_{10}$, much of the fusion analysis was carried out by Kondo as part of his prior centralizer of involution characterization of $A_{11}$ [74].

Furthermore, in case 1 of the proposition, $G$ has one conjugacy class of involutions when $q \equiv 5 \pmod 8$ and two classes for each of the other three congruences of $q \pmod 8$.

Note also that the groups $M_{22}$ and $M_{23}$ have only one conjugacy class of involutions, so that the possibility of $G$ having two classes of involutions in case 4 must be shown to lead to a contradiction.

Finally the split extensions of $E_{16}$ by $\Sigma_4$ in cases 2 and 4 are different. The centralizer of $(12)(34)(56)(78)$ in $A_{10}$ is easily computed, while the centralizer $C^*$ of an involution in $M_{22}$ can be described as follows: there exists a unique nontrivial split extension $H$ of $E_{16}$ by $A_6$. Then $A_6$ acts transitively on $E_{16}^{\#}$ and $C^* \cong C_H(t^*)$ for any involution $t^*$ of $E_{16}$.

Harada and I treated cases 2, 3, and 4, while Mason treated case 1. We describe the former analysis first.

*Case 2.* The first step is to show that if $y$ is a non-2-central involution, then

$$\bar{C}_y = C_y/O(C_y) \cong C_{A_n}\big((12)(34)\big) \text{ for } n = 10 \text{ or } 11, \qquad (2.25)$$

which is equivalent here to the assertion that $\bar{C}_y$ contains a normal subgroup $\bar{K}$ isomorphic to $A_6$ of $A_7$. Interestingly enough, this conclusion does not require covering locals. Indeed, one argues at the outset that $\bar{K}$ is simple with dihedral Sylow 2-subgroups of order 8, whence $\bar{K} \cong A_7$ or $L_2(r)$ for suitable $r \equiv 1,7 \pmod 8$. However, in the latter case, one can show that some 2-central involution $z$ with $z \in C_y$ induces a field automorphism on $K$, so that either $r = 9$ (whence $\bar{K} \cong A_6$) or $C_{\bar{K}}(\bar{z}) \cong PGL_2(r^{1/2})$ [whence $C_{\bar{K}}(\bar{z})$ and hence $C_z$ is nonsolvable]. But the latter is impossible as $z \sim t$ in $G$ and $C = C_t$ is solvable.

When $K \cong A_6$, it is immediate that $G$ is balanced. Since $S$ is also connected, [I, 4.54] implies, as usual, that $O(C) = O(C_y) = 1$. Thus $C$ and $C_y$ are isomorphic to the centralizers of long and short involutions in $A_{10}$, respectively, and it follows from Held's result [64] that $G \cong A_{10}$. His characterization of $A_{10}$ is proved in much the same way as his characterization of $A_9$, which we have described earlier.

On the other hand, if $\bar{K} \cong A_7$, the proof is considerably more difficult since $A_7$ is not locally 2-balanced, so that a priori $G$ need not be 2-balanced. Moreover, this time the possibility cannot be eliminated by using covering locals, since the group $A_{11}$ is also not locally 2-balanced (in fact, is not locally 3-balanced). To get around the problem, we make use of a special signalizer functor $\theta$, designed solely for the present situation.

First of all, the goal is to show that $O(C_y) = 1$ and $O(C) \cong Z_3$ with $O(C)$ inverted by an involution of $C$, for then $C$ and $C_y$ will be isomorphic, respectively, to the centralizers of long and short involutions in $A_{11}$, in which case Kondo's theorem [74] (which is established by the same generator-relation method as Held's) will imply that $G \cong A_{11}$.

We have the following preliminary fact:

LEMMA 2.47. *If* $D \in Syl_3(O(C))$, *then* $|D| \geqslant 3$ *and equality holds if and only if* $O(C_y)$ *is a* $3'$-*group.*

We now choose an appropriate $E_{16}$-subgroup $A$ of $S$. In defining $\theta$, we must distinguish the cases that $O(C_y)$ is or is not a $3'$-group. If $O(C_y)$ is a $3'$-group, for $a \in A^\#$, we set

$$\theta(C_a) = O^3\big(O(C_a)\big). \tag{2.26}$$

[Recall that for any group $X$, $O^3(X)$ is the subgroup of $X$ generated by the elements of $X$ of order prime to 3.] We use covering locals to prove that $\theta$ is an $A$-signalizer functor, and then apply the signalizer functor theorem in the standard way to force $\theta$ to be trivial. Now with the aid of Lemma 2.47, we easily conclude that $C$ and $C_y$ have the required structures.

On the other hand, if $O(C_y)$ is not a $3'$-group, we first prove the existence of a covering 3-local subgroup $H$, which is then used in the very definition of $\theta$. Indeed, for $a \in A^{\#}$, we set

$$\theta(C_a) = O_{3'}\big(O(C_a)\big)\big(O(H) \cap C_a\big) \text{ if } a \sim t \text{ in } G; \text{ and}$$
$$\theta(C_a) = O(C_a) \text{ if } a \sim y \text{ in } G. \tag{2.27}$$

Again $\theta$ is shown to be an $A$-signalizer functor on $G$ and the proof goes in the same way as before.

*Case* 3. Once again the goal is to show that $O(C) = 1$, for then [I, 3.50] will imply that $C \cong \hat{A}_8$ and $G \cong Mc$. However, as $G$ has only one class of involutions and $\bar{C} \cong \hat{A}_8$ or $\hat{A}_9$, it is immediate that $G$ is balanced and connected, so that again the desired conclusion follows from [I, 4.54].

*Case* 4. Once more we argue that $O(C) = 1$, in which case $C$ is isomorphic to the centralizer of an involution of $M_{22}$ or $M_{23}$. But then $G \cong M_{22}$ or $M_{23}$ by [I, 3.50]. We need only show that $G$ is balanced, for then as $S$ is connected, it will follow, as usual, that $O(C) = 1$. Since $C$ is 2-constrained, we can assume that $G$ has more than one class of involutions (whence $G$ has precisely two classes of involutions), otherwise $G$ is balanced by [I, 4.52].

Let $y \in \mathcal{I}(G)$ with $y \nsim t$ in $G$. If $\bar{C} \cong \Sigma_4/E_{16}$, our fusion information easily yields that $C_y$ has a normal 2-complement, so again $G$ is balanced. Thus we can assume that $\bar{C} \cong L_3(2)/E_{16}$. Setting $\bar{C}_y = C_y/O(C_y)$, the fusion information implies this time that $\bar{C}_y$ contains a normal subgroup $\bar{K} \cong L_2(q)$ for some $q \equiv 3,5 \pmod 8$. If $q > 5$, it follows directly that an $E_{16}$-subgroup of $C_y$ normalizes a subgroup of $\bar{K}$ of order $p$ for some odd prime $p$. But now using covering $p$-locals, we derive a contradiction. Hence, we must have $q \leqslant 5$, in which case balance is easily verified.

Finally we turn to Case 1.

*Case* 1. In those cases in which $G$ has two classes of involutions [$q \not\equiv 5 \pmod 8$], Mason first determines the structure of the centralizer of a non-2-central involution $y$.

PROPOSITION 2.48. *Setting* $\bar{C}_y = C_y/O(C_y)$ *and* $\bar{K} = L(\bar{C}_y)$, *we have either*

(i) $\bar{K} \cong L_2(r^2)$ *with* $r \equiv 3,7 \pmod 8$; *or*
(ii) $\bar{K} \cong L_3(r), r \equiv 1 \pmod 8$.

The initial fusion information yields only that $\bar{K} \cong L_2(d^2)$ or $A_7$ in (i) and $\bar{K} \cong L_3(d)$ for some odd $d$ in (ii), so that to establish the proposition, one must eliminate the $A_7$ possibility and show that $d = r$. The basis of the argument is the following observation: $S$ contains an involution $u$ which interchanges $\bar{L}_1$ and $\bar{L}_2$ under conjugation, whence $C_{\bar{L}}(\bar{u}) \cong Z_2 \times L_2(r)$. Hence if we set $H = C_{\langle t, u \rangle}$ and $\bar{H} = H/O(H)$, then $\bar{H}$ contains a normal subgroup isomorphic to $L_2(r)$. Working with an appropriate conjugate of $y$, this last conclusion is played off against the structure of $C_y$.

Next, Mason uses covering locals to prove that $G$ is 2-balanced and then with the aid of the 2-balanced $A$-signalizer functor with respect to an appropriate $E_{16}$-subgroup $A$ of $S$, he forces $O(C)$ and $O(C_y)$ to be cyclic. Thus at this point, he has reduced to two centralizer of involution problems [according as Proposition 2.48(i) or (ii) holds].

These problems are entirely analogous to that of the $PSp_4(r)$ case of the preceding section. Phan treated the special case in which $C$ is isomorphic to the exact centralizer of a 2-central involution of $L_4(r)$ or $U_4(r)$ [in which case $O(C)$ has a prescribed order with $O(C) \leqslant Z(C)$ and $\bar{C} = \overline{LS}$]. He constructs a $(B, N)$-pair $G_0$ in $G$ with $G_0 \cong L_4(r)$ or $U_4(r)$ and then argues that $G = G_0$. Mason shows that Phan's argument carries over to the general case. (In particular, he handles the "Harris" problem of odd-order field automorphisms on $\bar{L}$ occurring within $\bar{C}$.)

Mason also treated the $L_5(q)$, $q$ odd, case [82]. This time the resulting centralizer of involution problem was handled by Collins and Solomon [16] and Phan [99, 100].

THEOREM 2.49.  *If $G$ is a simple group with Sylow 2-subgroups of type $L_5(q)$, $q$ odd, then $G \cong L_5(r)$ or $U_5(r)$, $r$ odd.*

This time there are two possibilities for the structure of $S$:

$$
\begin{aligned}
&\text{(a) If } q \equiv -1 \,(\mathrm{mod}\ 4),\ S \cong D \wr Z_2,\ \text{where } D \text{ is quasi-dihedral;} \\
&\text{(b) If } q \equiv 1 \,(\mathrm{mod}\ 4),\ S \cong (D_{2^n} \wr Z_2) \wr Z_2,\ n \geqslant 2.
\end{aligned}
\tag{2.28}
$$

However, now in both cases the initial fusion analysis leads to the same conclusions. In particular, $G$ has two classes of involutions, represented by $t$ and $y$, and again we set $\bar{C}_y = C_y/O(C_y)$. Mason uses his fusion information to prove the following:

PROPOSITION 2.50.  (i) $\bar{L} = L(\bar{C}) \cong SL_4(r)$    *or*    $SU_4(r)$    *for    some*
$r \equiv q (\mathrm{mod}\ 4);$ *and*

(ii) $\overline{C}_y$ contains a normal subgroup of the form $\overline{K}_1 \times \overline{K}_2$, where $\overline{K}_1 \cong SL_2(r)$ and $\overline{K}_2 \cong L_3(r)$ or $U_3(r)$, respectively.

The proof that $\overline{K}_1$ and $\overline{K}_2$ are defined over the same field as $\overline{L}$ is similar to that of Proposition 2.48.

This time 2-balance is established without recourse to covering locals, so that with the aid of an appropriate 2-balanced signalizer functor, Mason is again able to prove that $O(C)$ and $O(C_y)$ are cyclic (of the same order), so again he has reduced the theorem to a centralizer of an involution problem.

Again it was Phan who treated the case in which $C$ is isomorphic to the exact centralizer of a 2-central involution of $L_5(r)$ or $U_5(r)$; however, this time Collins and Solomon were the ones who showed that Phan's arguments carry over to the general case, forcing $G \cong L_5(r)$ or $U_5(r)$.

At this point, we have covered those elements of $\mathscr{F}$ needed for the classification of groups of sectional 2-rank $\leqslant 4$. The remaining possibilities [types $L_5(2)$, $U_5(2)$, $HS$, and $A_{12}$] are needed for the classification of groups in which the centralizer of some involution (mod core) has a component isomorphic to $L_2(q)$, $q$ odd (see Section 2.8 below).

Harris and I treated the Higman–Sims case [I: 137]. Janko and S. K. Wong studied the resulting centralizer of involution problem [I: 193] and determined the order of $G$ by the Thompson order formula as equal to that of $HS$, thus reducing the problem to a prior characterization by Parrott and S. K. Wong of $HS$ by its order [I: 234].

THEOREM 2.51. *If $G$ is a simple group with Sylow 2-subgroups of type HS, then $G \cong HS$.*

In this case, $S = TD$, where $T \cong Z_4 \times Z_4 \times Z_4$, $D \cong D_8$, and $T \cap D = 1$. Moreover, the action of $D$ on $T$ is determined from that of the unique nontrivial split extension of $T$ by $L_3(2)$, taking $D$ to be a Sylow 2-subgroup of $L_3(2)$. (Indeed, as already noted in [I, 2.34 and 2.35], HS contains a 2-local of this shape.) We fix this notation for $S$.

The initial fusion analysis yields:

PROPOSITION 2.52. *One of the following holds:*

(i) *$G$ has two classes of involutions and $\overline{C}$ is isomorphic to the centralizer of a 2-central involution of HS (whence $\overline{C} \cong \Sigma_5 / Q_8 * Q_8 * Z_4$, 2-constrained); or*

(ii) *G has three classes of involutions and $\Omega_1(T)$ is strongly closed in S with respect to G.*

As $G$ is simple, Goldschmidt's theorem [I, 4.128] eliminates the second possibility. (This case leads to the solution $G \cong L_3(2)/(Z_4 \times Z_4 \times Z_4)$, which is fusion simple, but not simple.)

Thus $G$ has two classes of involutions, represented by $t$ and $y$; and we next prove the following result:

PROPOSITION 2.53. *Setting $\bar{C}_y = C_y/O(C_y)$, we then have $\bar{C}_y \cong Z_2 \times \mathrm{Aut}(L_2(9))$.*

The initial fusion analysis yields only that $\bar{C}_y$ has a normal subgroup $\bar{K} \cong L_2(q)$, where $q$ is a square and some involution $u$ of $S$ induces a nontrivial field automorphism on $\bar{K}$. But $\langle y, u \rangle$ contains a 2-central involution, and it follows from the structure of $C$ that $C_{\langle y,u \rangle}$ is solvable, whence $C_{\bar{K}}(\bar{u})$ is solvable, thus forcing $q = 9$.

Since $HS$ is a balanced group, we could have used covering locals to prove that $G$ is balanced. However, this is unnecessary, for one can show directly that $G$ is balanced with respect to any $E_{16}$-subgroup, which suffices here for application of the standard signalizer functor arguments. [The point is that as $\mathrm{Aut}(L_2(9))$ contains an involution whose centralizer on $L_2(9)$ has a core, $L_2(9)$ is not locally balanced, so it is not automatic that $G$ is a balanced group.] In any event, we conclude, as usual, that $O(C) = O(C_y) = 1$.

Now the Thompson order formula yields that $|G| = |HS|$. Parrott and Wong take up the problem at this point, employing the same methods as in the corresponding order characterizations of the Mathieu groups which we have described in [I, 3.44]. On the basis of local information which they establish, they eventually determine the full character table of $G$ and, in particular, show that $G$ possesses an irreducible character $\chi$ of degree 22. Then, using the "Brauer trick" [I, p. 98] with $\chi$, they produce a simple subgroup $G_0$ of $G$ in which the centralizer of an involution is isomorphic to that of an involution of $M_{22}$. Hence by Janko's theorem [I: 190], $G_0 \cong M_{22}$.

Since $|G| = |HS|$, it follows now that $|G : G_0| = 100$. Finally Parrott and Wong show that the permutation representation of $G$ on the right cosets of $G_0$ is of rank 3 and in this representation $G_0$ has orbits of lengths 1, 22, and 77. Now they are in a position to invoke a theorem of Wales [117] which asserts that the associated graph of this representation is isomorphic to the one Higman and Sims used in their construction of $HS$ [I, p. 107, 108], and it follows from their results that $G \cong HS$.

Schoenwaelder treated the $L_5(2)$ case [103], with Held handling the resulting centralizer of involution problems [I: 165].

THEOREM 2.54. *If $G$ is a simple group with Sylow 2-subgroups of type $L_5(2)$, then $G \cong L_5(2)$, $M_{24}$, or He.*

In this case $S = TD$, where $T \cong D_8 * D_8 * D_8$, $D \cong D_8$, and $T \cap D = 1$. Moreover, the action of $D$ on $T$ can be determined from the structure of the centralizer of a 2-central involution of $L_5(2)$, which is a split extension of $D_8 * D_8 * D_8$ by $L_3(2)$. Of course, $S$ can also be described as the group of nonsingular $5 \times 5$ triangular matrices over $GF(2)$. We fix this notation.

The initial portion of Schoenwaelder's proof is organized in a somewhat different fashion from that described in the outline of Section 2.1. Indeed, he first focuses on the structure of $\bar{C}$, independent of its embedding in $G$, and studies the fusion in $\bar{C}$ of elements of $\bar{S}$, eventually pinning down the possibilities for $\bar{C}$. The structure of the subgroup $N_{\bar{C}}(\bar{T})/\bar{T}C_{\bar{C}}(\bar{T})$ plays a key role in this analysis. The final result is as follows:

PROPOSITION 2.55. *We have $\bar{T} \lhd \bar{C}$ and $\bar{C}/\bar{T} \cong L_3(2)$, $\Sigma_4$, or $D_8$.*

Schoenwaelder now considers $C$ as a subgroup of $G$ and uses Thompson's transfer lemma to show that $G$ has a normal subgroup of index 2 when $\bar{C}/\bar{T} \cong \Sigma_4$ or $D_8$. Thus $\bar{C}/\bar{T} \cong L_3(2)$, and he concludes:

PROPOSITION 2.56. *$\bar{C}$ is isomorphic to the centralizer of a 2-central involution of $L_5(2)$.*

Now Schoenwaelder is in a position to pin down the $G$-fusion of $S$. First of all, it is immediate from the description of $S$ in terms of $5 \times 5$ matrices over $GF(2)$ that $S$ contains exactly two self-centralizing normal $E_{16}$-subgroups $E_1$ and $E_2$.

Using the fusion simplicity of $G$ and the previously determined structure of $C$, he next proves the following result:

PROPOSITION 2.57. *One of the following holds*:

(i) *$G$ has exactly two classes of involutions; or*
(ii) *For a suitable choice of $i = 1$ or 2, $E_i$ is strongly closed in $S$ with respect to $G$.*

Since $G$ is simple, (ii) is, as usual, excluded by [I, 4.128]. [Under the assumption of fusion simplicity, (ii) leads to the solution $G \cong L_4(2)/E_{16}$, which is the structure of a maximal parabolic in $L_5(2)$.]

As usual, let $y$ be an involution of $G$ not conjugate to $t$. Schoenwaelder's analysis together with the class 2 Sylow 2-subgroup classification theorem [I, 4.129] now yields the following additional result:

PROPOSITION 2.58. *Either* $C_y$ *is 2-constrained or else* $C_y/O(C_y) \cong L_3(4)/(Z_2 \times Z_2)$.

In either case, it is immediate that $G$ is balanced, and as $S$ is connected, [I, 4.54] once again implies that $O(C) = O(C_y) = 1$. Thus, Schoenwaelder has reduced the theorem to Held's centralizer of an involution problem.

It is the structure of the groups $N_i = N_G(E_i)$, $i = 1, 2$, which distinguish $L_5(2)$, $M_{24}$, and *He*. Indeed, using additional fusion analysis, Held proves:

PROPOSITION 2.59. *For* $i = 1$ *and* 2, *either* $N_i \leqslant C$ *or* $N_i$ *is a split extension of* $E_{16}$ *by* $L_4(2)$.

There are thus three possibilities, according as $N_i \leqslant C$ for zero, one, or two values of $i$.

Held now develops sufficient local information to apply the Thompson order formula, and obtains the following result:

PROPOSITION 2.60.  (i) *If* $N_i \cong L_4(2)/E_{16}$ *for both values of* $i = 1, 2$,
                            *then* $|G| = |L_5(2)|$;
                     (ii) *If* $N_i \cong L_4(2)/E_{16}$ *for exactly one value of*
                            $i = 1, 2$, *then* $|G| = |M_{24}|$; *and*
                    (iii) *If* $N_i \cong L_4(2)/E_{16}$ *for no values of* $i = 1, 2$
                            (*equivalently* $N_i \leqslant C$ *for both* $i = 1, 2$), *then*
                            $|G| = |He|$.

In case (i), we have two of the maximal parabolics of $L_5(2)$, and it is clear that one can use these to construct a $(B, N)$-pair $G_0$ in $G$ with $G_0 \cong L_5(2)$, forcing $G = G_0$ as each has the same order. However, instead, Held works directly with the permutation representation on the set $\Omega$ of cosets of $N_1$ in $G$, which is doubly transitive. He argues that $N_2$ is not $G$-conjugate to $N_1$, so that $N_2$ is not a one-point stabilizer. He then shows that $N_2$ is not transitive on $\Omega$ and that $N_2$ is not faithful on one of its orbits, at which point he can invoke a theorem of Ito on doubly transitive groups which contain a subgroup with these two properties [68] and conclude that $G$ is necessarily of Lie type, which immediately yields that $G \cong L_5(2)$.

On the other hand, in case (ii), [I, 3.44] applies directly to yield that $G \cong M_{24}$.

Finally in case (iii), Held determines the complete local structure of $G$, on the basis of which Higman and McKay subsequently proved the existence and uniqueness of $He$ [I: 171].

We come next to the $U_5(2)$ case. When we began this book, the only existing classification of groups with such Sylow 2-subgroups occurred as part of the much more general problem of determining all simple groups with Sylow 2-subgroups of order at most $2^{10}$ [2-groups of type $U_5(2)$ have order $2^{10}$], an elaborate undertaking initiated by Held and carried out by the combined efforts of several of his students at the University of Mainz—Beisiegel, Fritz, Nah, and Stingl—using the same techniques as in the classification of simple groups of sectional 2-rank $\leqslant 4$, which we shall outline in the next three sections.

We mentioned to Aschbacher in passing the unfortunate fact that there was no independent characterization of $U_5(2)$ by its Sylow 2-subgroup; and he quickly produced a very short proof of the desired result (unpublished), which we shall now outline.

THEOREM 2.61.   *If $G$ is a simple group with Sylow 2-subgroups of type $U_5(2)$, then $G \cong U_5(2)$.*

The structure of $S$ in this case can be read off from the structure of the centralizer $C^*$ of a 2-central involution of $U_5(2)$:

$$C^* \cong GU_3(2)/Q_8 * Q_8 * Q_8, \text{ 2-constrained.} \tag{2.29}$$

Furthermore, $GU_3(2)$ is a split, extension of $3^{1+2}$ (an extraspecial group of order $3^3$ and exponent 3) by $SL_2(3)$.

Thus $S = TD$, where $T \cong Q_8 * Q_8 * Q_8, D \cong Q_8$, and the action of $D$ on $T$ is determined from (2.29).

As expected, Aschbacher studies the involution fusion pattern of $G$ and ultimately shows that $C \cong C^*$. He then goes on to argue that $t$ is a class of 3-transpositions and then to invoke Fischer's theorem [I, 2.58] [this could also have done for $L_5(2)$ above].

Using the action of $D$ on $T$ determined by $C^*$, Aschbacher easily establishes the following properties of $S$:

LEMMA 2.62.   *Set* $\langle y \rangle = Z(D)$ *and* $\tilde{C} = C/\langle t \rangle$. *Then we have*

   (i) *There are exactly two conjugacy classes of involutions in* $S - T$, *represented by* $y$ *and* $yt$, *each of cardinality* 4;
  (ii) $C_T(y) = E * Q$, *where* $E \cong E_8$ *and* $Q \cong Q_8$;
 (iii) $\tilde{E} = Z(\tilde{S}) \cong Z_2 \times Z_2$;
 (iv) $\langle y^S \rangle = \langle (yt)^S \rangle = A \cong E_{16}$;
  (v) $\tilde{T} = J(\tilde{S})$ *is the unique elementary subgroup of* $\tilde{S}$ *of order* 64; *and*
 (vi) $C_{\tilde{T}}(\tilde{y}) \cong E_{16}$ *and* $[\tilde{T}, \tilde{y}] \cong Z_2 \times Z_2$.

Lemma 2.62(v) immediately yields the following result:

LEMMA 2.63.   $\tilde{T}$ *is weakly closed in* $\tilde{S}$ *with respect to* $\tilde{C}$ *and* $T$ *is weakly closed in* $S$ *with respect to* $C$.

As a corollary of his work on strongly closed abelian 2-subgroups, Goldschmidt established a criterion for a weakly closed abelian subgroup to be strongly closed [I: 124]. His precise conditions are satisfied by $\tilde{T}$ since $\tilde{S}/\tilde{T}$ has rank 1, while $[T, y]$ has rank exceeding 1. Hence we have the following further result:

LEMMA 2.64.   $\tilde{T}$ *is strongly closed in* $\tilde{S}$ *with respect to* $\tilde{C}$.

Aschbacher next proves the following:

LEMMA 2.65.   $t$ *is not conjugate in* $G$ *to an involution of* $T - \langle t \rangle$.

Indeed, if $t^g = u$ for some $u \in T - \langle t \rangle$, we can assume that $C_T(u) \leqslant S^g$, whence $C_T(u)$ normalizes $T^g$. But $C_T(u) \cong Z_2 \times D_8 * Q_8$ with $\langle t \rangle = C_T(u)'$. Since $S^g/T^g \cong Q_8$, $t \in R = C_T(u) \cap T^g$, so $C_T(u)/R$ is abelian. On the other hand, it is easy to show that $C_T(u)$ cover $S^g/T^g$, whence $C_T(u)/R \cong Q_8$—contradiction.

Now let $\varDelta$ denote the set of $G$-conjugates of $t$ contained in $S$. By the $Z^*$-theorem, $t$ is $G$-conjugate to some involution of $S - \langle t \rangle$. We conclude at once therefore from Lemmas 2.62(iii) and 2.65 (the second assertion follows immediately from the first as $A$ is abelian):

LEMMA 2.66.   (i) $\langle \varDelta \rangle = A$; *and*
            (ii) $N = N_G(A)$ *acts transitively on* $\varDelta$.

In $U_5(2)$, there is a maximal parabolic $P$ with $P/O_2(P) \cong A_5$ and $O_2(P)$ a special 2-group with center $Z \cong E_{16}$ and $O_2(P)/Z \cong E_{16}$. Using his fusion information, Aschbacher next proves that $N/O(N)$ has essentially this structure:

PROPOSITION 2.67.  (i) $N/C_A \cong A_5$ with $A$ the natural $A_5$-module; and
(ii) $R = C_S(A)$ is special of order $2^8$ with $A = \Omega_1(R) = Z(R)$.

In particular, the proposition has the following consequence.

LEMMA 2.68. $N$ contains a 3-element $x$ with the following properties:

(i) $x$ centralizes $t$ and normalizes $E$ $(= Z(C_T(y)))$ with $E = [E, x] \times \langle t \rangle$; and
(ii) $S = T[S, x]$.

Combining Lemmas 2.64 and 2.68, Aschbacher can now prove:

PROPOSITION 2.69. We have $\bar{T} \lhd \bar{C}$.

Indeed, as $\bar{C} = C/O(C)$, it will suffice to show that $\tilde{T}O(\tilde{C}) \lhd \tilde{C}$ and hence that $T^* \lhd C^*$, where $C^* = \tilde{C}/O(\tilde{C})$. But by Lemma 2.64, $T^*$ is strongly closed in $S^*$ with respect to $C^*$. Hence if we set $Y^* = \langle T^{*C^*} \rangle$, the general form of Goldschmidt's strongly closed abelian theorem [I: 123] (without the assumption of simplicity) yields

$$T^* \in \mathrm{Syl}_2(Y^*) \quad \text{and} \quad Y^* = O_2(Y^*) \times L(Y^*). \tag{2.30}$$

On the other hand, if $x$ is as in Lemma 2.68, then $x^* \in C^*$ and $E^* = [E^*, x^*] \cong Z_2 \times Z_2$ with $E^* = Z(S^*)$ [Lemma 2.62(iii)]. Since $x^*$ acts transitively on $(E^*)^\#$, it follows that either $E^* \leqslant O_2(Y^*)$ or $E^* \leqslant L(Y^*)$. Since $E^* = Z(S^*)$, this forces correspondingly $L(Y^*) = 1$ [whence $T^* = O_2(Y^*) \lhd C^*$, as required] or $O_2(Y^*) = 1$. So assume the latter, in which case $T^* \in \mathrm{Syl}_2(L^*)$, where $L^* = L(Y^*)$, by (2.30).

Now as $T^* \cong E_{64}$, $L^*$ is the direct product of $r \leqslant 3$ components $L_i^*$, $1 \leqslant i \leqslant r$, with abelian Sylow 2-subgroups (whence $L_i^* \cong L_2(q)$, $^2G_2(3^n)$, or $J_1$ by [I, 4.126]). In particular, $\mathrm{Out}(L_i^*)$ is abelian, $1 \leqslant i \leqslant 3$. But as $S^* = T^*[S^*, x^*]$, we see that $S^*$ leaves each $L_i^*$ invariant, and consequently $(S^*)'$ induces inner automorphisms on each $L_i^*$ and hence on $L^*$. In particular, $y^*$ induces an inner automorphism on $L^*$ and so $y^*$ centralizes $T^*$, which is not the case. We conclude that $T^* \lhd C^*$, as required.

Note that $\mathrm{Out}(T) \cong O_6^-(2) \left(\cong Sp_4(3)\right)$ and so $\overline{C}/\overline{T}$ is isomorphic to a subgroup of $Sp_4(3)$ with $Q_8$ Sylow 2-subgroups. Using the existence of the element $x$, Aschbacher easily concludes from these conditions:

PROPOSITION 2.70. *One of the following holds*:

(i) $\overline{C}$ *is the semidirect product of* $\overline{T}$ *and* $SL_2(3)/3^{1+2}$; *or*
(ii) $\overline{A} \lhd \overline{C}$.

By Proposition 2.67, $t$ and $v = yt$ are representatives of the classes of involutions of $A$ under the action of $N$. Moreover, it follows easily from the existing fusion information that $C_v = O(C_v)(N \cap C_v)$, so $C_v$ is solvable. Since $C = C_t$ is also solvable, $G$ is therefore balanced with respect to $A$. Now Aschbacher concludes, as usual:

PROPOSITION 2.71. $O(C_a) = 1$ *for every* $a \in A^\#$.

Indeed, in the contrary case, if we set $M = N_G\big(\theta(G;A)\big)$ with $\theta(G:A)$ the completion of the $A$-signalizer functor $\theta = O$, it is not difficult to show that $M/O(M)$ is fusion simple, whence by induction $M/O(M) \cong U_5(2)$. But then it follows that $M$ is strongly embedded in $G$, contrary to Bender's theorem.

As remarked above, Aschbacher is now in a position to complete the proof of Theorem 2.61 by showing that $G$ is a 3-transposition group.

We come finally to Sylow 2-subgroups of type $A_{12}$, which is primarily the work of Solomon [111, 112] with Yamaki [140, 141] and Olsson [90] treating the resulting centralizer of involution problems. The complete result is as follows:

THEOREM 2.72. *If* $G$ *is a simple group with Sylow 2-subgroups of type* $A_{12}$, *then* $G \cong A_{12}, A_{13}, PSp_6(2)$, *or* $P\Omega_7(q), q \equiv 3,5 \pmod 8$.

In this case $S$ possesses a unique (and hence normal) elementary abelian subgroup $T$ of order 64 and $S = TD$, where $D$ is dihedral of order 8 with $T \cap D = 1$ and a uniquely determined action of $D$ on $T$. Solomon's initial fusion analysis determines not only the possible involution fusion patterns, but also the corresponding structure of $N = N_G(T)$. Specifically he proves the following:

PROPOSITION 2.73. *One of the following holds*:

    (i) $N/T \cong \Sigma_4$, *G has three conjugacy classes of involutions, and the involution fusion pattern of $A_{12}$*;

    (ii) $N/T \cong L_3(2)$, *G has four conjugacy classes of involutions, and the involution fusion pattern of $PSp_6(2)$; or*

    (iii) $N/T \cong A_7$, *G has three classes of involutions, and the involution fusion pattern of $P\Omega_7(q)$, $q \equiv 3,5 (\bmod 8)$.*

*Moreover, in each case T contains a representative of each conjugacy class of involutions.*

The proof of the proposition divides into four parts. Using the fusion-simplicity of $G$, Solomon obtains initial information about the $G$-fusion of involutions of $S$ by studying the Alperin–Goldschmidt conjugation family [I, 4.85, 4.86]. In the present case, this is rather elaborate since the group $S$ itself possesses 28 conjugacy classes of involutions.

The uniqueness of $T$ implies that $N$ controls the $G$-fusion of $T$. But then on the basis of his initial fusion analysis, Solomon concludes that $\bar{N} = N/T$ must act indecomposably on $T$, regarded as a six-dimensional vector space over $GF(2)$. (This differs from the situation of Proposition 2.55, where $\bar{C}/\bar{T}$ acts decomposably on $\bar{T} \cong E_{64}$.) Since $\bar{N}$ is a subgroup of $GL_6(2)$ with $D_8$ Sylow 2-subgroups, he knows that either $\bar{N}$ has one of the required three forms or else $\bar{N} \cong D_8, \Sigma_5$, or $A_6$, and he argues that the latter three possibilities are incompatible with the indecomposable action of $\bar{N}$ on $T$. Thus

$$\bar{N} \cong \Sigma_4, L_3(2), \text{ or } A_7. \tag{2.31}$$

Since $S$ splits over $T$, so does $N$ by Gaschütz's theorem [I: 154, Theorem 15.8.6]. Hence $N = TK$, where $K \cong \Sigma_4, L_3(2)$, or $A_7$. Solomon argues next that the action of $K$ on $T$ is uniquely determined. We note that in dealing with the $\Sigma_4$ and $L_3(2)$ cases, he makes use of Green's general results on modules (cf. [I: 88, Chapter III, Section 5]). On the other hand, in the $A_7$ case, he easily forces $T$ to be the standard module for $K \cong A_7$.

Finally, Goldschmidt's strongly closed abelian theorem [I, 4.128] implies that $T$ is not strongly closed in $S$ with respect to $G$ (otherwise $T$ would be normal in $G$, contrary to the simplicity of $G$). Now with this information, Solomon can show that the involution pattern of $G$ is uniquely determined in each case and is correspondingly that of $A_{12}$, $PSp_6(2)$, or $P\Omega_7(q)$, $q \equiv 3,5 (\bmod 8)$. In particular, $T$ contains a representative of every conjugacy class of involutions of $G$ in each case.

We refer to the three possibilities of the proposition as *cases* 1, 2, and 3, respectively.

Solomon consider cases 1 and 2 together. Set $Z = Z(S)$, so that $Z \cong Z_2 \times Z_2$. Note that in case 1, all involutions of $Z$ are conjugate, while in case 2, $Z$ contains two conjugacy classes of involutions. However, in both cases, he proves the following result:

PROPOSITION 2.74. *In cases 1 and 2, we have $C_Z/ZO(C_Z) \cong X \times Z_2$, where $X$ is a nontrivial split extension of $E_8$ by $\Sigma_4$ containing a normal four-subgroup.*

Hence for any involution $t$ of $Z$, the proposition gives some information about the fusion of involutions in $C_t$.

Utilizing his full knowledge of the fusion of involutions, Solomon next proves:

PROPOSITION 2.75. *For some choice of the involution $t \in Z$,*

$$C_t/O(C_t) \cong C_{A_{12}}\big((12)(34)(56)(78)(910)(11\ 12)\big).$$

Thus, if we set $\bar{C}_t = C_t/O(C_t)$, then $\bar{C}_t/\langle \bar{t} \rangle$ has a normal subgroup of index 2 which is a nontrivial split extension of $E_{16}$ by $A_6$. In particular, Solomon uses the sectional 2-rank $\leqslant 4$ theorem to identify this normal subgroup of index 2.

To reduce to Yamaki's characterization of the groups $A_{12}, A_{13}$, and $PSp_6(2)$, it thus remains for Solomon to show that $O(C_t) = 1$. As usual, to accomplish this, he must first pin down the centralizers of the remaining involutions of $G$.

PROPOSITION 2.76. *If $u$ is an involution of $G$ not conjugate to $t$, then either*

(i) $C_u$ *is solvable; or*
(ii) *Case 1 holds and $L\big(C_u/O(C_u)\big) \cong A_8$ or $A_9$.*

Now Solomon can quickly obtain the desired conclusion $O(C_t) = 1$. Note that $C_t$ is 2-constrained by Proposition 2.75. Hence if case 2 holds, it follows at once from Proposition 2.76(i) and [I, 4.52] that $G$ is balanced. Since $S$ is connected, we conclude therefore from [I: 4.54] that $O(C_t) = 1$.

On the other hand, in case 1 it is not difficult to show with the aid of Proposition 2.76 that $G$ is balanced *with respect to $T$*. But then if $\theta$ denotes the balanced $T$-signalizer functor on $G$ and we set $M = N_G\big(\theta(G; T)\big)$, where, as usual, $\theta(G; T)$ is the completion of $\theta$, it follows easily by standard signalizer functor arguments, using the structure of the centralizers of involutions of $G$, that either $\theta$ is trivial or $M$ is strongly embedded in $G$. However, as $G$ has more than one class of involutions, the latter possibility is excluded, and we conclude that $O(C_t) = 1$ in this case as well.

This is the starting point for Yamaki's characterization of the groups $A_{12}, A_{13}$, and $PSp_6(2)$. Of course, Solomon's results allow us to dispense with Yamaki's initial involution fusion analysis and move directly to his characterization of these three groups.

In case 1, he first pins down the exact structure of the centralizers of representatives $x, y$ in $T$ of the conjugacy classes of non-2-central involutions of $G$, arguing that $C_x$ and $C_y$ are isomorphic to the centralizers in $A_k$, $k = 12$ or 13, of the product of two or six transpositions, respectively. In addition, he shows that the three intersections $C_t \cap C_x, C_t \cap C_y$, and $C_x \cap C_y$ are isomorphic to the intersections of the corresponding centralizers in $A_k$. At this point he invokes a theorem of Kondo [73] to conclude that $G \cong A_k$.

Kondo's proof consists in showing that the subgroup $G_0 = \langle C_t, C_x, C_y \rangle$ is generated by elements satisfying the canonical defining relations for $A_k$[I, 3.42]. It is also immediate that $C_u \leqslant G_0$ for every $u \in \mathscr{I}(G_0)$. This forces $G = G_0$; otherwise $G_0$ would be strongly embedded in $G$, whence $G_0$ would possess only one conjugacy class of involutions, which is not the case. Thus $G \cong A_k$, $k = 12$ or 13.

On the other hand, in case 2 Yamaki constructs a subgroup $G_0 \cong PSp_6(2)$ of $G$ before determining the structure of the centralizers of all involutions of $G$. Indeed, he argues that there exist natural isomorphisms of $C_t$ and $N_G(T)$ onto the corresponding 2-local subgroups of $PSp_6(2)$ such that $C_t \cap N_G(T)$ maps onto the corresponding intersection in $PSp_6(2)$. Once this is established, it follows easily that $G_0 = \langle C_t, N_G(T) \rangle$ is a split $(B, N)$-pair, identifiable as $PSp_6(2)$. Now he determines the structure of the centralizer of every involution of $G$; in particular, it again follows that $C_u \leqslant G_0$ for every $u \in \mathscr{I}(G_0)$. As in case 1, this forces $G = G_0 \cong PSp_6(2)$.

For the proof of the noncharacteristic 2 theorem, one requires only the above characterizations of the groups $A_{12}, A_{13}$, and $PSp_6(2)$. However, for completeness, we add a few comments concerning Solomon's analysis in case 3, which follows the same pattern of argument as Mason in his characterization of the groups $L_5(q)$, $U_5(q)$ (Theorem 2.49 above). Indeed, using the 2-balanced $T$-signalizer functor, Solomon is able to prove the following:

PROPOSITION 2.77. *In case* 3, *the centralizer of every involution of $G$ is isomorphic to the centralizer of an involution in a suitable extension of*

$P\Omega_7(q)$ *for some* $q \equiv 3,5 \pmod 8$ *by a cyclic group of field automorphisms of odd order. In particular,* $L(C) \cong Sp_4(q)$ *and* $O(C) = 1$.

At this point, it remains only to construct a $(B, N)$-pair subgroup $G_0$ of $G$, identifiable as $P\Omega_7(q)$, for then the usual argument will force $G_0 = F^*(G) \cong P\Omega_7(q)$. This last step was carried out by Olsson [90] as part of a centralizer of an involution characterization of the orthogonal groups of odd degree $n$. [Although Olsson's published argument requires $n \geqslant 11$, he points out that with the aid of a theorem of W. Wong [131], which provides a very nice presentation of the orthogonal groups, the proof can be extended to the case $n = 7$. Of course, by Proposition 2.77, the centralizer of some involution of $G$ possesses a subnormal $SL_2(q)$-subgroup, so $F^*(G)$ can in any case be identified from Aschbacher's classical involution theorem (Theorem 3.64 below).]

## 2.4. Groups of Sectional 2-Rank at Most 4; General Considerations

When Harada and I began our characterizations of the known simple groups of 2-rank 3 or 4 in terms of the structure of their Sylow 2-subgroups (the $J_2$ case being the first), we were motivated primarily by the prior classifications of groups with dihedral, quasi-dihedral, and wreathed Sylow 2-subgroups and had no broader strategy in mind. Only later did we realize that our accumulated results could provide a jumping off point for the sectional 2-rank $\leqslant 4$ theorem, thereby settling the nonconnected Sylow 2-subgroup problem as a corollary. However, when I suggested to Harada that we attack the general problem, it was with considerable naiveté—I had no idea how elaborate the ultimate proof would be, how intricate a case subdivision it would require, nor how complex the accompanying 2-fusion and 2-local analysis. In the end, the bulk of the proof was the creation of Harada, and it vividly reveals his consummate mastery of the 2-local structure of simple groups of low 2-rank.

Because of the length of the statement of the sectional 2-rank $\leqslant 4$ theorem, we shall not repeat it here, but refer the reader to the Introduction, where it appears in full.

Furthermore, as noted there, it follows as a corollary that $J_2$ and $J_3$ are the only simple groups of 2-rank $\geqslant 3$ with a nonconnected Sylow 2-subgroup.

The proof of the sectional 2-rank $\leqslant 4$ theorem is long and tortuous. As a result, we break up its outline into several parts, the present section devoted to general considerations, and, in particular, to a division of the proof into six major cases.

One begins with a minimal counterexample $G$, which we fix for the entire discussion. Since the hypothesis carries over to every section of $G$, it follows that the nonsolvable composition factors of every proper subgroup of $G$ are isomorphic to one of the groups in the conclusion of the theorem; thus

$$\text{Every proper subgroup of } G \text{ is a } K\text{-group.} \qquad (2.32)$$

The basic thrust of the analysis is 2-local in character, its aim being to force a Sylow 2-subgroup $S$ of $G$ to be isomorphic to that of one of the listed groups. It will then follow that $S$ is of type $\mathscr{F}$, in which case $G$ will be isomorphic to one of the groups in the conclusion of the sectional 2-rank $\leqslant 4$ theorem by Theorem 2.1, contrary to the fact that $G$ is a counterexample by assumption. In other words, the operative assumption throughout the proof is the following:

$$\text{A Sylow 2-subgroup } S \text{ of } G \text{ is not of type } \mathscr{F}. \qquad (2.33)$$

On the basis of prior classification theorems (specifically the 2-rank $\leqslant 2$ theorem and [I, 4.128]), we can eliminate certain possibilities at once. Thus we have the following result:

PROPOSITION 2.78.  (i) *S has 2-rank 3 or 4; and*
  (ii) *No nontrivial abelian subgroup of S is strongly closed in S with respect to G.*

The automorphism group of a 2-group of sectional 2-rank at most 4 is very restricted, a fact which is extremely important throughout the proof. We state the result in the following form:

PROPOSITION 2.79. *If $X$ is a 2-constrained core-free group such that $O_2(X)$ has sectional 2-rank $\leqslant 4$, then $\bar{X} = X/O_2(X)$ is isomorphic to a subgroup of $A_8$. In particular,*

  (i) *A Sylow 3-subgroup of $\bar{X}$ is elementary of order at most 9;*
  (ii) *A Sylow p-subgroup of $\bar{X}$ has order 1 or p for $p \geqslant 5$ and is trivial for $p > 7$;*

(iii) *If $X$ is nonsolvable, then $\bar{X} \cong L_3(2)[=GL_3(2)]$, $A_n$, $5 \leqslant n \leqslant 8$, $\Sigma_n$, $5 \leqslant n \leqslant 7$, or else $\bar{X} \cong (Z_3 \times A_5) \cdot Z_2$ with $\bar{X}/O_3(\bar{X}) \cong \Sigma_5$; and*

(iv) *If $X$ is nonsolvable and $X$ has sectional 2-rank $\leqslant 4$, then $\bar{X} \not\cong \Sigma_6$, $\Sigma_7$, or $A_8$.*

Indeed, set $Y = O_2(X)$ and $X^* = X/\phi(Y)$, so that $Y^*$ is elementary of rank at most 4 (as $Y$ has sectional 2-rank $\leqslant 4$). Furthermore, as $X$ is 2-constrained with $O(X) = 1$, $C_X(Y) \leqslant Y$. But by [I, 1.12] any element of $X$ of odd order which acts trivially on $Y^*$ acts trivially on $Y$, so by the previous inclusion $C_{X^*}(Y^*)$ is a 2-group. Since $Y = O_2(X)$, clearly $Y^* = O_2(X^*)$. Thus, $X^*/Y^*$ and hence $X/Y = \bar{X}$ is isomorphic to a subgroup of $\text{Aut}(Y^*)$ $\big(\cong GL_4(2)\big)$ with no nontrivial normal 2-subgroups. Since $GL_4(2) = L_4(2) \cong A_8$, all parts of the proposition except possibly (iv) follow. [We shall prove (iv) below, following the statement of Lemma 2.90.]

In her work on 2-groups having no normal ementary abelian subgroups of order 8, MacWilliams [I: 207] treated certain special cases of the sectional 2-rank $\leqslant 4$ theorem. Moreover, one of the families of 2-groups which she obtained as possible candidates for Sylow 2-subgroups of a (fusion) simple group was subsequently eliminated by Harada [52]. Combining his result with MacWilliams's work, we obtain the following theorem:

THEOREM 2.80. *Let $X$ be a fusion simple group with Sylow 2-subgroup $S$ of rank at least 3 which contains no elementary normal subgroup of order 8. If for some normal four-subgroup $U$ of $S$, the involutions of $U$ are conjugate in $X$, then $S$ is either of type $G_2(q)$, $q \equiv 1$, $7(\text{mod } 8)$, $J_2$, or $Ly$.*

(Note that by [I, 1.35], a 2-group of rank $\geqslant 3$ necessarily possesses a normal four-subgroup).

We shall give the initial arguments of Theorem 2.80 as they are indicative of a certain type of reduction which occurs frequently in the general sectional 2-rank $\leqslant 4$ proof. Set $R = C_S(U)$, $C = C_X(U)$, $N = N_X(U)$, and $\bar{N} = N/C$. Then $R \lhd S$, $S \leqslant N$, and as $C$ is $S$-invariant, $R = S \cap C \in \text{Syl}_2(C)$. Furthermore, $\bar{N} \leqslant \text{Aut}(U) \cong GL_2(2)$. Since $GL_2(2) \cong \Sigma_3$, $|\bar{S}| \leqslant 2$. Thus we have

$$|S/R| \leqslant 2. \tag{2.34}$$

LEMMA 2.81. *We have $U = \Omega_1\big(Z(R)\big)$.*

Indeed, set $Z = \Omega_1(Z(R))$, so that $U \leqslant Z$ char $R$. Hence $Z \lhd S$ and as $S$ contains no normal $E_8$-subgroups, we must have $m(Z) \leqslant 2$, whence $U = Z$, as asserted.

We next prove the following result:

LEMMA 2.82. $N_N(R)$ contains an element $y$ which cyclically permutes the involutions of $U$.

Indeed, let $z \in Z(S) \cap U$, so that $U = \langle z, u \rangle$ for some $u \in U^\#$. By hypothesis, $u = z^x$ for some $x \in X$. Setting $C_z = C_X(z)$ and $C_u = C_X(u)$, we have $C_u = (C_z)^x$, $S \leqslant C_z$, and $\langle R, S^x \rangle \leqslant C_u$.

Consider first the case $R = S$ [whence $U \leqslant Z(S)$]. Then $S$ and $S^x$ are both Sylow 2-subgroups of $C_u$, so by Sylow's theorem $S^{xa} = S$ for some $a \in C_u$. Putting $y = xa$, it follows that $y$ normalizes $R$ and $z^y = (z^x)^a = u^a = u$. But $y$ normalizes $U$ as $U$ char $R$ by Lemma 2.81, so $y \in N$. Since $z^y \neq z$, $y \notin C$. Furthermore, as $S = R$, $\bar{N} \cong Z_3$ and so $\langle \bar{y} \rangle = \bar{N}$. Thus $y$ induces an automorphism of $U$ of order 3 and hence cyclically permutes the involutions of $U$, as required.

Suppose then that $|S/R| = 2$, whence $U \not\leqslant Z(S)$ and $Z(S)$ is cyclic. Moreover, if we let $s \in S - R$, we have

$$z^s = z, \qquad u^s = zu, \qquad \text{and} \qquad (zu)^s = u. \tag{2.35}$$

On the other hand, as $S^x \leqslant C_u$, $R \notin \mathrm{Syl}_2(C_u)$, so $R$ has index 2 in a Sylow 2-subgroup $T$ of $C_u$ [by (2.34)]. We have $u \in Z(T)$ and $Z(T)$ is cyclic as $T \in \mathrm{Syl}_2(X)$ and $Z(S)$ is cyclic. Thus $R = C_T(U)$ and $U \lhd T$ (by Lemma 2.81). Hence as with $S$, if we let $t \in T - R$, we have

$$u^t = u, \qquad z^t = zu, \qquad \text{and} \qquad (zu)^t = z. \tag{2.36}$$

But now if we set $y = st$, (2.35) and (2.36) imply that $y$ cyclically permutes the involutions of $U$. In particular, $y \in N$. But also as $R \lhd \langle S, T \rangle$, $y$ normalizes $R$, so the desired conclusion holds in this case as well.

LEMMA 2.83. Set $H = N_N(R)$ and $\bar{H} = H/O(H)$. Then we can choose $y$ so that

   (i) $|\bar{y}| = 3$;
  (ii) $\langle \bar{y} \rangle$ is permutable with $\bar{S}$; and
 (iii) $\bar{S}\langle \bar{y} \rangle / \bar{R} \cong Z_3$ or $\Sigma_3$.

Indeed, as $y$ induces an automorphism of $U$ of order 3, $|y| = 3^a \cdot b$, where $a > 1$ and $b$ is prime to 3. Then $\langle y \rangle = \langle y_1 \rangle \times \langle y_2 \rangle$, where $|y_1| = 3^a$ and $|y_2| = b$. (cf. [I: 130, Theorem 1.3.1]). Since $y_2$ is prime to 3, $y_2$ cannot cyclically permute the involutions of $U$, so $U_2 = C_U(y_2) \neq 1$. But $U_2$ is $y_1$-invariant as $y_1$ centralizes $y_2$, so $U_2$ is $y$-invariant. However, $y$ fixes no involution of $U$, so we must have $U_2 = U$. Thus $y_2$ centralizes $U$ and so $y_1$ has the same properties as $y$. Hence without loss, $y$ is a 3-element.

Observe next that as $S$ and hence $R$ has sectional 2-rank $\leqslant 4$, $y^3$ centralizes $R$ by Proposition 2.79. Hence $y^3 \in K = C_H(R)$. But as $K \cap R = Z(R) \in \mathrm{Syl}_2(K)$, $K = O(K) Z(R)$ by Burnside's theorem [I, 1.20]. Thus $y^3 \in O(K)$. Since $K \lhd H$, it is immediate that $O(K) = O(H)$, so $\bar{y}$ has order 3. In particular, all parts of the lemma hold with this choice of $y$ if $R = S$, so we can suppose that $|S/R| = 2$.

To treat this case, set $H^* = \bar{H}/\bar{R}$. Since $O(\bar{H}) = 1$ and $C_{\bar{H}}(\bar{R}) \leqslant \bar{R}$, $H^* \leqslant A_8$, again by Proposition 2.79. But as $S^* \cong Z_2$, $H^*$ has a normal 2-complement, again by Burnside's theorem. Since $H^* \leqslant A_8$, it follows now that a Sylow 3-subgroup $D^*$ of $H^*$ is normal in $H^*$ and $D^*$ is elementary of order at most 9. In particular, $y^* \in D^*$. Now $S^*$ does not centralize $y^*$, otherwise $y^*$ would leave $\langle z \rangle = C_U(S^*)$ invariant, which is not the case, so if $S^*$ leaves $\langle y^* \rangle$ invariant, then necessarily $S^* \langle y^* \rangle \cong \Sigma_3$ and again the lemma holds.

Thus we can assume that $S^*$ does not leave $\langle y^* \rangle$ invariant, in which case $D^* \cong Z_3 \times Z_3$ and $S^*$ does not invert $D^*$. Hence by Lemma 1.51, $D^* = D_1^* \times D_2^*$, where $S^*$ inverts $D_1^*$ and centralizes $D_2^*$, and $D_i^* \cong Z_3, i = 1, 2$. Then $D_2^*$ leaves $\langle z \rangle = C_U(S^*)$ invariant, forcing $D_2^*$ to centralize $U$. Since $y^* \in D^*$, it follows that a generator $y_1^*$ of $D_1^*$ cyclically permutes the involutions of $U$. But then we can choose $y$ so that $y^* = y_1^*$, in which case $\langle y^* \rangle S^* \cong \Sigma_3$ and the lemma holds in this case as well.

In effect, Lemma 2.83 reduces Theorem 2.80 to a purely local problem:

What are the possible structures of the group
$\bar{S} \langle \bar{y} \rangle$ and, in particular, those of $\bar{S}$?

Once a list of possibilities for $\bar{S}$ and hence of $S$ is obtained, fusion arguments will enable us to eliminate those which do not appear in the conclusion of Theorem 2.80.

We make one further reduction. Let $Y$ be the preimage of $\bar{S} \langle \bar{y} \rangle$ in $H$, so that $O(H) = O(Y)$ centralizes $R$ and $Y/O(Y) = \bar{S} \langle \bar{y} \rangle$. Let $A$ be an abelian subgroup of $R$ of largest order such that $A \lhd Y$. Since $U \leqslant Z(R)$, clearly

$U \leqslant A$. Since $A \lhd S$ and $S$ has no elementary $E_8$-subgroups, $m(A) \leqslant 2$. Thus $U = \Omega_1(A)$ and as $A$ admits $y$, we conclude that

$$A \cong Z_{2^n} \times Z_{2^n}. \tag{2.37}$$

We claim $A > U$. Indeed, suppose $A = U$. If $R = U$, it is immediate that $S \cong Z_2 \times Z_2$ or $D_8$, contrary to $m(S) \geqslant 3$, so $R > A$. Set $\tilde{Y} = Y/U$ and let $1 \neq \tilde{Q} \leqslant \Omega_1(Z(\tilde{R}))$ with $\tilde{Q} \lhd \tilde{Y}$. Since $O(\tilde{Y})\tilde{R}$ centralizes $\tilde{Q}$ and $\tilde{Y}/O(\tilde{Y})\tilde{R} \cong Z_3$ or $\Sigma_3$, we can assume $\tilde{Q} \cong Z_2$ or $Z_2 \times Z_2$. Let $Q$ be the pre-image of $\tilde{Q}$ in $Y$, so that $Q \lhd Y$ and $|Q| = 8$ or 16. Since $Q/U$ is elementary of order 2 or 4, and $U \leqslant Z(Q)$, $Q'$ is cyclic and $Q' \leqslant U$. But $Q$ is $y$-invariant and hence so is $Q'$. Since $y$ acts irreducibly on $U$, this forces $Q' = 1$, whence $Q$ is abelian, and our maximal choice of $A$ is contradicted. Thus we have

$$A > U. \tag{2.38}$$

We now let $B$ be the subgroup $\Omega_2(A)$ generated by the elements of $A$ of order 4. Thus

$$B \cong Z_4 \times Z_4, \tag{2.39}$$

and as $B$ char $A \lhd Y$, also $B \lhd Y$.

We next prove:

LEMMA 2.84. *We have* $C_{\bar{Y}}(\bar{B}) = \bar{A}$. *In particular,* $\bar{Y}/\bar{A}$ *is isomorphic to a subgroup of* $\mathrm{Aut}(Z_4 \times Z_4)$.

For simplicity of notation, we drop the bars [equivalently we assume $O(Y) = 1$]. Set $P = C_Y(B)$, so that $A \leqslant P$ and $P \lhd Y$. Also $y$ does not centralize $B$ as $U = \Omega_1(B)$, so $P \leqslant R$. We are done if $A = P$, so we can assume that $A < P$.

Again set $\tilde{Y} = Y/A$. Then as $\tilde{P} \lhd \tilde{Y}$, we can choose $\tilde{Q}$ as before with $\tilde{Q} \leqslant \tilde{P} \cap \Omega_1(Z(\tilde{R}))$. Letting $Q$ be the pre-image of $\tilde{Q}$ in $P$, we have that $A < Q \leqslant P$, $Q \lhd Y$ and $Q/A \cong Z_2$ or $Z_2 \times Z_2$ (according as $y$ centralizes or does not centralize $Q/A$). Suppose $U = \Omega_1(Q)$, in which case $y$ induces an automorphism of $Q$ cyclically permuting its involutions. Hence Higman's theorem [I: 170] (also cf. [I, p. 150, 151]) is applicable and yields that either $Q \cong Z_{2^{n+1}} \times Z_{2^{n+1}}$ or $|Q| = 2^6$ with $U = Z(Q)$ (as $|U| = 4$). The first case is excluded by the maximality of $A$ and the second as $B \leqslant Z(Q)$ [since $Q \leqslant P = C_Y(B)$].

Thus $\Omega_1(Q) > U$ and so $Q - A$ contains an involution $w$. We claim that $E = U\langle w \rangle = \Omega_1(A\langle w \rangle)$. Indeed, if not, then $A\langle w \rangle$ contains an involution of the form $aw$, where $a \in A$ and $|a| = 2^m$, $m \geqslant 2$. But $(aw)^2 = awaw = 1$, so $a^w = a^{-1}$. Now $b = a^{2^{m-2}}$ has order 4 and so $b \in B = \Omega_2(A)$. Furthermore, $w$ inverts $b$ as it inverts $a$, so certainly $w$ does not centralize $b$ (as $|b| = 4$). This is a contradiction as $w \in Q$ centralizes $B$.

Since $w \in Q \leqslant R$, $E = U\langle w \rangle \cong E_8$. Also $E$ char $A\langle w \rangle$; so if $A\langle w \rangle = Q$, then $E \lhd Y$ and $S$ has a normal $E_8$-subgroup—contradiction. Hence $Q/A \cong Z_2 \times Z_2$, and it follows that $C_Q(y) = 1$.

Since $Q/A \leqslant Z(R/A)$, $A\langle w \rangle \lhd R$ and hence $E \lhd R$. Set $F = \langle E, E^y, E^{y^2} \rangle$. Then $F$ is $y$-invariant and it is immediate that $F \lhd Y$. Furthermore, $E^y$ and $E^{y^2}$ are each normal in $R$ and hence in $F$. Since $U \leqslant E \cap E^y \cap E^{y^2}$, it follows that $F/U$ is elementary of order 2, 4, or 8. But $F/U$ admits $y$ and $C_Q(y) = 1$, so we must have $F/U \cong Z_2 \times Z_2$. But then, as above, $F' = 1$ and $F$ is abelian. Thus $F \cong Z_4 \times Z_4$ or $E_{16}$. But as $w \in F$ and $w$ is an involution of $F - U$, the first possibility is excluded. Hence $F \cong E_{16}$ and as $F \lhd S$, this contradicts our assumption that $S$ contains no normal $E_8$-subgroups and completes the proof of the lemma.

The structure of $\text{Aut}(Z_4 \times Z_4)$ is easily determined and yields the following properties of $\overline{Y}/\overline{A}$.

LEMMA 2.85. Set $Y^* = \overline{Y}/\overline{A}$ and $T^* = O_2(Y^*)$. Then we have

(i) $T^*$ is elementary of order $\leqslant 16$;
(ii) $T^*$ centralizes both $U$ and $B/U$;
(iii) $|C_{T^*}(y^*)| \leqslant 4$; and
(iv) $Y^*/T^* \cong Z_3$ or $\Sigma_3$.

Thus at this point we have obtained a rather precise description of $\overline{Y}$. Indeed, it is sufficiently tight that MacWilliams can now pin down those possible structures in which $\overline{S}$ contains no normal $E_8$-subgroups [and $m(\overline{S}) \geqslant 3$]. Her result is as follows:

PROPOSITION 2.86. If we set $T = [R, \bar{y}]$, then $T \lhd Y$ and one of the following holds:

(i) $T$ is of type $L_3(4)$ and $S$ is of type $J_2$ or $Ly$; or
(ii) (1) $T \cong Z_{2^m} \times Z_{2^m}$, $m \geqslant 3$;
  (2) $S = TF$, where $|F| = 2$ or 4, $T \cap F = 1$, and $C_F(T) = 1$;
  (3) $F_1 = C_F(\bar{y}) \cong Z_2$ and $R = TF_1$; and
  (4) If $T_1$ is a cyclic subgroup of $T$ of order $2^m$, then $T_1 F_1$ is a dihedral or quasi-dihedral group; moreover, if $|F| = 4$, then $T = T_1 T_1^f$ for any $f \in F - F_1$.

Note that if (ii) holds with $F \cong Z_2 \times Z_2$ and $T_1 F_1$ dihedral, then $S$ is, in fact, of type $G_2(q)$, $q \equiv \pm 1 \pmod{2^m}$. Thus the proposition shows that Theorem 2.80 will be proved once we rule out the other alternatives in (ii): namely,

$$
\begin{array}{ll}
\text{(a)} \ F \cong Z_2 \ \text{or} \ Z_4; \ \text{or} \\
\text{(b)} \ F \cong Z_2 \times Z_2 \ \text{and} \ T_1 F_1 \ \text{is quasi-dihedral.}
\end{array}
\qquad (2.40)
$$

The two cases of (2.40)(a) were treated by MacWilliams (and also within the general sectional 2-rank $\leqslant 4$ analysis) by straightforward fusion analysis; while Harada eliminated the more difficult quasi-dihedral case of (2.40)(b) in [52]. To state their combined result, it will be convenient to say that a 2-group satisfying the conditions of Proposition 2.85(ii) is of type $(Z_{2^m} \times Z_{2^m})^F$.

PROPOSITION 2.87. *If $X$ is a fusion simple group with Sylow 2-subgroup $S$ of type $(Z_{2^m} \times Z_{2^m})^F$, $m \geqslant 3$, then $S$ is of type $G_2(q)$, $q \equiv 1,7 \pmod{8}$.*

Together, Propositions 2.86 and 2.87 complete the proof of Theorem 2.80.

Finally, we briefly describe Harada's proof, which is very pretty and depends on the classification of groups with dihedral Sylow 2-subgroups. First write $S = TF$ as in Proposition 2.86 and set $T = \langle a \rangle \times \langle b \rangle$, where $|a| = |b| = 2^m$, $z_1 = a^{2^{m-1}}$, $z_2 = b^{2^{m-1}}$, and $z = z_1 z_2$ with $z \in Z(S)$. We can assume that $F = \langle f, f_1 \rangle$, where $f$ is an involution interchanging $a$ and $b$, $\langle f_1 \rangle = F_1$, and $\langle a, f_1 \rangle$, $\langle b, f_1 \rangle$ are each quasi-dihedral of order $2^{m+1}$. Also set $w = a^{2^{m-2}} b^{2^{m-2}}$, so that $\langle w \rangle \lhd S$.

Using elementary fusion analysis, Harada establishes in succession the following results (the first also depends on Burnside's transfer theorem):

$$
\begin{array}{ll}
\text{(a)} \ C_X(\langle z, z_1 \rangle) \ \text{has a normal 2-complement.} \\
\text{(b)} \ X \ \text{has only one conjugacy class of involutions.} \\
\text{(c)} \ z_1 \sim f \sim ff_1 \sim f_1 \ \text{in} \ C_X(z). \\
\text{(d)} \ f \sim z_1 \ \text{in} \ C_X(w).
\end{array}
\qquad (2.41)
$$

Set $C = C_X(z)$. By (2.41)(a), $C_C(z_1)$ has a normal 2-complement; while by (2.41)(b), $z_1 \sim f_1$ in $C$, so we obtain

$$
C_C(f_1) \ \text{has a normal 2-complement.} \qquad (2.42)
$$

Harada now uses (2.41)(d) to contradict (2.42). Indeed, set $H = C_X(w)$.

Since $f_1$ inverts $w$, $f_1$ acts on $H$. Also $H \leqslant C$ as $w^2 = z$. Hence if we set $\bar{H}\langle \bar{f}_1 \rangle = H\langle f_1 \rangle / O(H)$, it will suffice to show that $C_{\bar{H}}(\bar{f}_1)$ does not possess a normal 2-complement.

To prove this, observe that as $\langle w \rangle \lhd S$, $T\langle f \rangle \in \text{Syl}_2(H)$, so $H$ has wreathed Sylow 2-subgroups. Furthermore, $\bar{H}$ does not have a normal 2-complement by (2.41)(d). Since $\bar{z} \in Z(\bar{H})$ and $O(\bar{H}) = 1$, we conclude easily from Proposition 1.64 and Theorem 1.47 that $\bar{H}$ contains a normal subgroup $\bar{L} \cong SL_2(q)$ for some odd $q$ (it is this assertion which depends on the dihedral classification theorem). It is easy to check now that $\bar{f}_1$ centralizes $\bar{T}\langle \bar{f} \rangle \cap \bar{L}/\langle \bar{z} \rangle$, which implies that $\bar{f}_1$ necessarily induces a field automorphism on $\bar{L}$ (possibly trivial). Thus $C_{\bar{L}}(\bar{f}_1) \cong SL_2(q_1)$, where $q_1 = q$ or $q^{1/2}$, so $C_L(f_1)$ and hence also $C_H(f_1)$ does not possess a normal 2-complement, giving the desired contradiction.

MacWilliams's theorem illustrates another feature of the sectional 2-rank $\leqslant 4$ proof: namely, in a given situation, the possibilities for a Sylow 2-subgroup $S$ of $G$ may well include cases which do not occur in any simple group. We state an omnibus result which covers all such cases (apart from those already included in Proposition 2.85).

We denote by $L_3(4)^f$, $U_3(4)^f$, and $U_3(4)^{f'}$ the split extension of $L_3(4)$, $U_3(4)$, and $U_3(4)$, by a field automorphism of order 2,2, and 4, respectively.

PROPOSITION 2.88. *If $X$ is a simple group of 2-rank $\geqslant 3$, then a Sylow 2-subgroup $S$ of $X$ does not have any of the following forms:*

(i) *$S$ of type $L_3(4)^f$, $U_3(4)^f$, or $U_3(4)^{f'}$;*

(ii) *$S = (Q_1 * Q_2)\langle t \rangle$, where $Q_1, Q_2$ are quasi-dihedral of order 16 and $t$ is an involution interchanging $Q_1$ and $Q_2$;*

(iii) *$\Omega_1(S) \cong (Z_2 \times Z_2) \wr Z_2$ and $|S : \Omega_1(S)| \leqslant 2$;*

(iv) *$S$ contains a cyclic normal subgroup of index 4; or*

(v) *$S = Q_1 \times Q_2$, where $Q_1$ is dihedral of order at least 8, quasi-dihedral, or quaternion, and $Q_2$ is either cyclic, dihedral, quasi-dihedral, or quaternion.*

Apart from (v) with $Q_1$ and $Q_2$ dihedral or quasi-dihedral, the proposition, in fact, holds under the weaker assumption that $X$ is fusion simple, On the other hand, in the case of (v), one argues that $X$ must have product fusion and then quotes Goldschmidt's theorem [I, 4.148] to contradict the simplicity of $G$. In the remaining cases, the principal tools are, as expected, Glauberman's $Z^*$-theorem [I, 4.95] and Thompson's transfer lemma [I, 1.21], which are used to derive a fusion contradiction. Apart from the case in which $S$ contains a cyclic normal subgroup of index 4, which was treated by W. Wong in [I: 323], the remaining cases are covered within the general sectional 2-rank $\leqslant 4$ analysis.

Proposition 2.86 depends heavily upon the structure of the automorphism group of 2-groups of type $Z_4 \times Z_4$ and of type $L_3(4)$. Likewise the general sectional 2-rank $\leqslant 4$ argument depends upon detailed information about the automorphism groups of these and other 2-groups [e.g., 2-groups of type $E_{16}$ or $U_3(4)$].

Consider, for example, a 2-group $T$ of type $L_3(4)$. One can give precise generators and relations for $T$. In particular, $T$ has order 64 and contains two elementary normal subgroups $E$, $F$ of order 16, $T = EF$, $E \cap F = Z(T) = T' \cong Z_2 \times Z_2$, and $T/Z(T) \cong E_{16}$. Here are some of the properties of Aut($T$).

LEMMA 2.89. *Let $T$ be a 2-group of type $L_3(4)$, set $A = Aut(T)$, let $B$ be the subgroup of $A$ acting trivially on $Z(T)$ and on $T/Z(T)$, and let $D$ be a Sylow 3-subgroup of $A$. Then we have*

   (i) $|A| = 2^{10} \cdot 3^2$;
   (ii) $B \cong E_{28}$ and $D \cong Z_3 \times Z_3$;
   (iii) $A/B \cong \Sigma_3 \times \Sigma_3$;
   (iv) *If $V$ is a Sylow 2-subgroup of $N_A(D)$, then $V \cong Z_2 \times Z_2$ and $V \cap B = 1$; moreover, the semidirect product of $T$ and $V$ is of type $Ly$;*
   (v) $D = D_1 \times D_2$, *where $D_1 \cong D_2 \cong Z_3$ and $C_T(D_1) = 1$, $C_T(D_2) = Z(T)$; and*
   (vi) *If $v$ is an involution of $N_A(D_i)$, $i = 1$ or 2, with $v \notin B$, then the semidirect product $T\langle v \rangle$ is either of type $J_2$, Mc, or $L_3(4)'$.*

Parts (iv) and (vi) can be used in the proof of Proposition 2.86(i).

Here is another 2-group result, which is very important for the analysis.

LEMMA 2.90. *Let $X$ be a core-free 2-constrained group with $O_2(X) \cong E_{16}$ and $\bar{X} = X/O_2(X) \cong A_5$ or $\Sigma_5$ and let $T \in \mathrm{Syl}_2(X)$. Then we have*

   (i) *$X$ splits over $O_2(X)$;*
   (ii) *If $\bar{X} \cong A_5$, then $X$ has two possible structures, in one of which $X$ acts transitively on $O_2(X)^\#$ and in the other intransitively; correspondingly $T$ is of type $L_3(4)$ or $A_8$;*

(iii) *If $\bar{X} \cong \Sigma_5$ and $\bar{X}$ acts intransitively on $O_2(X)^{\#}$, then the structure of $X$ is uniquely determined and $T$ is of type $A_{10}$; and*

(iv) *If $\bar{X} \cong \Sigma_5$ and $\bar{X}$ acts transitively on $O_2(X)^{\#}$, then $X$ has three possible structures and correspondingly $T$ is of type $J_2, Mc$, or $L_3(4)^f$.*

We note that the proof of (iv) depends on the previous lemma.

Lemmas 2.89 and 2.90 are very effective in many situations for pushing-up 2-groups, primarily because of the following fact:

LEMMA 2.91. *A 2-group $R$ of type $J_2, Mc, L_3(4)^f, A_{10}$, or $Ly$ possesses a unique normal 2-subgroup $T$ of type $L_3(4)$. In particular, $T$ char $R$.*

Let me illustrate the power of these results by proving Proposition 2.79(iv). Assume false. Then with the notation there, $Y^* \cong E_{16}$ and $X^*/Y^* \cong \Sigma_6, \Sigma_7$, or $A_8$. We check in each case that $X^*/Y^*$ contains a subgroup isomorphic to $A_5$ which acts *transitively* on $(Y^*)^{\#}$. By Lemma 2.89, its pre-image in $X^*$ splits [and so has the form $Y^*L^*$, where $L^* \cong A_5$ and $L^*$ acts transitively on $(Y^*)^{\#}$], and, in addition, a Sylow 2-subgroup $T^*$ of $Y^*L^*$ is of type $L_3(4)$. We can assume without loss that $T^* \cap L^* \in \mathrm{Syl}_2(L^*)$, so that $T^* \cap L^* \cong Z_2 \times Z_2$. Then $N_{L^*}(T^* \cap L^*) \cong A_4$ and it follows from the transitive action of $L^*$ that $N_{L^*}(T^*)$ contains an element $d^*$ of order 3 with $C_{T^*}(d^*) = 1$.

Finally expand $T^*$ to $S^* \in \mathrm{Syl}_2(X^*)$ and set $R^* = N_{S^*}(T^*)$. Since $C_{X^*}(T^*) \leqslant C_{X^*}(Y^*) \leqslant Y^*$, we have $C_{R^*}(T^*) \leqslant Y^* \leqslant T^*$. By our hypothesis, $S^*$ has sectional rank $\leqslant 4$. Using this fact, we can now argue that Lemma 2.89(vi) is applicable and conclude that $T^*$ is type $J_2, Mc$, or $L_3(4)^f$. But then $T^*$ char $R^*$ by Lemma 2.91 and so $R^*$ is a Sylow 2-subgroup of $N_{S^*}(R^*)$. Hence $R^* = S^*$ by [I, 1.11] and consequently $|S^*| = |R^*| = 2^7$, whence $|S^*/Y^*| = 8$. However, this is a contradiction as $S^*/Y^* \in \mathrm{Syl}_2(X^*/Y^*)$ and $X^*/Y^* \cong \Sigma_6, \Sigma_7$, or $A_8$ has Sylow 2-subgroups of order at least 16.

Proposition 2.79 gives a partial description of any 2-constrained proper section of our minimal counterexample $G$. Since the Schur multipliers of all simple $K$-groups are known [I, Tables 4.1 and 4.2], we can obtain a comparable picture of a non-2-constrained proper section of $G$. We are interested primarily in the case of centralizers of involutions. The main result is as follows:

PROPOSITION 2.92. *Let $t \in \mathscr{I}(G)$ and set $C = C_t, \bar{C} = C/O(C)$. If $C$ is* not 2-constrained and $\bar{K}$ is a component of $\bar{C}$, then one of the following holds:

(i) $\bar{K}$ is simple with $\bar{K} \cong L_2(q)$, $L_3(q)$, $U_3(q)$, $q$ odd, $^2G_2(3^{2n+1})$, $L_2(8)$, $Sz(8)$, $A_7$, $M_{11}$, or $J_1$; or

(ii) $\bar{K}$ is nonsimple with $\bar{K} \cong SL_2(q)$, $q$ odd, $SL_4(q)$, $q \equiv -1 \pmod 4$, $SU_4(q)$, $q \equiv 1 \pmod 4$, $Sp_4(q)$, $q$ odd, $Sz(8)/Z_2$, $\hat{A}_n$, $7 \leqslant n \leqslant 11$, or $\hat{M}_{12}/Z_2$.

Note that the Schur multiplier of $Sz(8)$ is $Z_2 \times Z_2$ [however, the full cover of $Sz(8)$ has 2-rank 5 and so is excluded here]. Since this full cover admits an automorphism of order 3 cyclically permuting its three central involutions, the extension $Sz(8)/Z_2$ of $Sz(8)$ by $Z_2$ is unique up to isomorphism. We also observe that $J_2$ has a covering group by $Z_2$; however, it can be shown to have sectional 2-rank 5, so it, too, is excluded here. Likewise in (i), $\bar{K}\langle t \rangle = \bar{K} \times \langle t \rangle$, so in this case $\bar{K}$ has sectional 2-rank $\leqslant 3$. Indeed, our sectional 2-rank condition yields the following corollary of the proposition.

PROPOSITION 2.93. *Let* $t \in \mathscr{I}(G)$ *with* $C = C_t$ *not 2-constrained and set* $\bar{C} = C/O(C)$ *and* $\bar{L} = L(\bar{C})$. *Then we have the following:*

(i) *Either* $\bar{L}$ *is quasisimple (and determined from Proposition 2.92) or else* $\bar{L}$ *is a central product of two components* $\bar{L}_1, \bar{L}_2$ *of sectional 2-rank 2 with* $\bar{L}_1$ *nonsimple; thus*

  (1) $\bar{L}_1 \cong SL_2(q_1)$, $q_1$ *odd, or* $\hat{A}_7$; *and*

  (2) $\bar{L}_2 \cong SL_2(q_2)$, $q_2$ *odd*, $\hat{A}_7$, $L_2(q_2)$, $q$ *odd*, $L_3(q_2)$, $q_2 \equiv -1 \pmod 4$, $U_3(q_2)$, $q_2 \equiv 1 \pmod 4$, $A_7$, *or* $M_{11}$;

(ii) *If* $\bar{L}$ *has sectional 2-rank 4, then* $C_{\bar{C}}(\bar{L}) = Z(\bar{L})$;

(iii) *If* $\bar{L}$ *has sectional 2-rank 3, then* $C_{\bar{C}}(\bar{L})$ *is a cyclic 2-group; and*

(iv) *If* $\bar{L}$ *has sectional 2-rank 2, then a Sylow 2-subgroup of* $C_{\bar{C}}(\bar{L})$ *has sectional 2-rank* $\leqslant 2$.

Observe that $\tilde{C} = \bar{C}/C_{\bar{C}}(\bar{L}) \leqslant \text{Aut}(\tilde{L})$ with $\tilde{L} \cong \bar{L}/Z(\bar{L})$ simple and so if $R \in \text{Syl}_2(C)$, then the possibilities for $\tilde{R}$ are completely determined. If $\bar{L}$ has sectional 2-rank 3 or 4, then in view of (ii) and (iii), the structure of $\bar{R}$ and hence of $R$ is essentially also determined. On the other hand, if $\bar{L}$ has sectional 2-rank 2, the structure of $\bar{R}$ is determined only up to that of a Sylow 2-subgroup of $C_{\bar{C}}(\bar{L})$, an as yet unspecified group of 2-rank $\leqslant 2$.

In the course of the analysis, we must frequently consider normalizers of subgroups $H$ of $G$ with $\bar{H} = H/O(H)$ nonabelian. It is immediate that for such a subgroup $H$, $\bar{H}/Z(\bar{H})$ has sectional 2-rank at least 2 and hence so does $H/K$, where $K$ is the pre-image of $Z(\bar{H})$ in $H$. But if $N$ denotes the subgroup of $N_G(H)$ which acts trivially on $\bar{H}$ (in particular, $N = C_H$ if $H$ is a 2-group), then $H \cap N = K$ and as $HN$ has sectional 2-rank $\leqslant 4$, it follows that $N/K$ has sectional 2-rank $\leqslant 2$. Thus the general structure of groups of sectional 2-rank at most 2 will be important to us.

The following two results are easily established; the second, of course, depends on the classification of simple groups of 2-rank 2.

LEMMA 2.94. *If $X$ is 2-group of sectional 2-rank $\leqslant 2$, the either $\Omega_1(X) \cong Z_2 \times Z_2$ or $X$ is cyclic, dihedral, quaternion, or quasi-dihedral.*

PROPOSITION 2.95. *Let $X$ be a core-free group of sectional 2-rank $\leqslant 2$, let $S \in Syl_2(X)$, and set $Y = O^{2'}(X)$. Then one of the following holds:*

  (i) $S = X$;
  (ii) $S = O_2(X)$ is homocyclic abelian of rank 2 and $|X/S| = 3$;
  (iii) $S$ is quaternion and $Y \cong SL_2(q)$, $q$ odd, $\hat{A}_7$, or a uniquely determined nonsplit extension of $SL_2(q)$, $q$ odd, by $Z_2$;
  (iv) $S$ is dihedral and $Y \cong L_2(q)$ or $PGL_2(q)$, $q$ odd, or $A_7$; or
  (v) $S$ is quasi-dihedral and $Y \cong L_3(q)$, $q \equiv -1 \pmod 4$, $U_3(q)$, $q \equiv 1 \pmod 4$, $M_{11}$, or a uniquely determined extension of $L_2(q)$ or $SL_2(q)$, $q$ odd, by $Z_2$, respectively, nonsplit or split.

The various results we have listed above can be viewed as the underlying technical machinery needed for the proof of the sectional 2-rank $\leqslant 4$ theorem.

We conclude the section with a description of the major subdivision of the analysis into six principal cases. Since many of the groups or families of groups listed in the theorem's conclusion have widely disparate internal 2-structures, it is only natural that this fact be reflected in the analysis. Not only does it account for the six-part division, but also helps to explain the extreme length of the total proof.

It is to be understood that the assumptions of each case listed below include the stipulation that the conditions of none of the prior cases hold.

### Major Case Division of the Proof of the Sectional 2-Rank $\leqslant 4$ Theorem

  I. All 2-local subgroups of $G$ are solvable.
  II. One of the following holds:
      (a) The centralizer of every 2-central involution of $G$ is solvable and some 2-local subgroup of $G$ is nonsolvable and 2-constrained; or

(b) The centralizer of some involution of $G$ is nonsolvable and 2-constrained.

[In particular, (I) or (II) holds if all 2-locals are 2-constrained.]

III. For some involution $t$ of $G$, $L(C_t/O(C_t))$ has sectional 2-rank 4 and each of its components is nonsimple.

IV. For some 2-central involution $t$ of $G$, $L(C_t/O(C_t))$ has sectional 2-rank 3.

V. For some 2-central involution $t$ of $G$, $L(C_t/O(C_t))$ is nontrivial.

VI. The centralizer of every 2-central involution of $G$ is solvable and the centralizer of some involution of $G$ is not 2-constrained.

[Since $C_t$ is 2-constrained if and only if $L(C_t/O(C_t)) = 1$, we see that if the assumptions of (I)–(V) fail, then those of (VI) necessarily hold.]

We shall refer to these as *cases* I, II, III, IV, V, and VI, respectively.

## 2.5. An Extension of the Balanced, Connected Theorem

Harada and I completed the sectional 2-rank $\leqslant 4$ proof after Thompson's classification of $N$-groups [I: 289], but prior to its extension to the classification of groups with solvable 2-locals [43, 69, 109] (the *thin* group case—i.e., when $G$ has 2-local $p$-rank $\leqslant 1$ for all odd primes $p$—had by then been treated by Janko [69]). As a result, we were forced to provide a separate analysis in case I when $G$ had 2-local 3-rank 2, eventually forcing $G$ to be an $N$-group and invoking Thompson's theorem.

It seems preferable to avoid a discussion of those special arguments and instead to subsume case I within the general classification of simple groups of characteristic 2 type with solvable 2-local subgroups, which we shall fully describe in the sequel to this volume. Thus we have only to show that $G$ is, in fact, of characteristic 2 type, under the assumptions of case I.

Our original proof of this assertion went as follows. Since $G$ has solvable 2-locals, it is balanced by [I, 4.52]. Hence if a Sylow 2-subgroup $S$ of $G$ is connected, the desired conclusion follows from [I, 4.54 and 1.40]. In the contrary case, we invoked a result of Janko and Thompson [71], which by a rather elaborate 2-fusion analysis showed in effect that the hypothesis of MacWilliams's theorem (Theorem 2.80) was necessarily satisfied, forcing $S$ to be of type $\mathscr{F}$, contrary to (2.33).

However, there is an argument in the $N$-group paper (Theorem 10.7 of [I: 289]) which achieves the same objective and at the same time foreshadows the signalizer functor method. We shall present it here in its entirety, adapting it to the language of signalizer functors, for not only does it provide further insight into the way signalizer functors are used in practice, but it is also capable of further generalization, leading eventually to an extension of Aschbacher's proper 2-generated core theorem [I, 4.28].

Thus we shall prove the following theorem:

THEOREM 2.96. *If $X$ is a core-free fusion simple group of 2-rank $\geqslant 3$ in which the centralizer of every involution is 2-constrained (in particular, solvable), then $X$ is of characteristic 2 type.*

We begin with the following general consequence of Thompson's transfer lemma.

LEMMA 2.97. *If $Y$ is a group of 2-rank $\geqslant 3$, then either $Y$ has a normal subgroup of index 2 or $m\big(C_Y(y)\big) \geqslant 3$ for every involution $y$ of $Y$.*

PROOF. We may assume that $Y$ has no normal subgroups of index 2. Let $T \in \mathrm{Syl}_2(Y)$. As $m(T) \geqslant 3$, $T$ contains a normal four-subgroup $V$ by [I, 1.35]. Set $R = C_T(V)$, so that $|T:R| \leqslant 2$. One easily checks that $m(R) = m(T)$, so that $m(R) \geqslant 3$. Without loss $y \in T$. Since $Y$ has no normal subgroups of index 2, Thompson's transfer lemma implies that $y$ is conjugate in $Y$ to an involution $x$ of $R$, so we need only show that $m\big(C_R(x)\big) \geqslant 3$ to conclude that $m\big(C_y(y)\big) \geqslant 3$. But $V \leqslant Z(R)$ and $m(R) \geqslant 3$. Hence $V \leqslant C_R(x)$ and so either $x \in V$ or $\langle V, x \rangle \cong E_8$. However, in the first case $R = C_R(x)$, so $m\big(C_R(x)\big) \geqslant 3$ in either case, as required.

We proceed by contradiction, fixing a Sylow 2-subgroup $S$ of $G$. We immediately obtain the following result:

LEMMA 2.98.  (i) $O\big(C_X(y)\big) \neq 1$ for some $y \in \mathscr{I}(S)$;
  (ii) $S$ is not connected;
  (iii) $Z(S)$ is cyclic; and
  (iv) $S$ contains no normal elementary subgroups of order 8.

PROOF. If (i) is false, then $O\big(C_X(x)\big) = 1$ for every $x \in \mathscr{I}(X)$ by Sylow's theorem. Since all 2-locals of $X$ are 2-constrained by assumption, it follows from [I: 131] (see also [I, 1.40]) that $X$ is of characteristic 2 type. Since we are proceeding by contradiction, (i) holds. But $X$ is balanced by [I, 4.52], so [I, 4.54] in turn forces $S$ to be nonconnected, proving (ii). But (ii) implies (iii) and (iv), so all parts of the lemma thus hold.

We fix a normal four-subgroup $U$ of $S$. For each noncyclic elementary subgroup $A$ of $S$, set

$$W_A = \langle O(C_X(a)) \mid a \in A^\# \rangle. \tag{2.43}$$

It is immediate from the definitions that

$$W_{A^x} = (W_A)^x \tag{2.44}$$

for all $x \in X$, and that

$$N_X(A) \leqslant N_X(W_A). \tag{2.45}$$

We next prove:

LEMMA 2.99. $W_U$ is nontrivial of odd order.

PROOF. By Lemma 2.98(i), there is $y \in \mathscr{S}(X)$ such that $O(C_X(y)) \neq 1$. Since $X$ is fusion simple, $m(C_X(y)) \geqslant 3$ by Lemma 2.97, so $y \in A \leqslant X$ with $A \cong E_8$. Replacing $y$ by a suitable conjugate, we can assume without loss that $A \leqslant S$. Since $X$ is balanced, $O$ is an $A$-signalizer functor on $G$, so by the signalizer functor theorem [I, 4.38], $W_A$ is of odd order. Since $y \in A$ with $O(C_X(y)) \neq 1$, also $W_A \neq 1$.

But using the balance property of $X$, it follows as in [I, Section 4.3] that

$$W_B = W_A \tag{2.46}$$

for any four-subgroup $B$ of $X$ which is connected to $A$. However, as $A \leqslant S$ with $Z_2 \times Z_2 \cong U \lhd S$, $U$ centralizes a four-subgroup of $A$, so $U$ is connected to $A$. Thus

$$W_U = W_A, \tag{2.47}$$

and we conclude that $W_U$ is nontrivial of odd order.

We set $M = N_X(W_U)$ and immediately obtain the following result:

LEMMA 2.100. (i) $M < X$; and

(ii) If $E$ is a noncyclic elementary subgroup $S$ which is connected to $U$, then $N_X(E) \leqslant M$; in particular, $S \leqslant N_X(U) \leqslant M$.

PROOF. Since $X$ is core-free by hypothesis and $W_U$ is nontrivial of odd order, (i) holds. If $E$ is as in (ii), then as with (2.46) and (2.47) above, we have $W_E = W_U$, so $N_X(E) \leqslant M$ by (2.45). Since $U$ is obviously connected to itself and $U \lhd S$, both assertions of (ii) thus hold.

Our aim is to prove that $M$ contains a normal subgroup of index 2. We need a preliminary result.

LEMMA 2.101. *Let $B$ be a four-subgroup of $M$ and $D$ a Sylow 2-subgroup of $C_M(B)$. If $m(D) = 2$, then $N_M(D)$ does not contain a 3-element cyclically permuting the involutions of $B$.*

PROOF. Since $S \leqslant M$ and $S \in \mathrm{Syl}_2(G)$, $S \in \mathrm{Syl}_2(M)$. Hence without loss $D \leqslant S$. If $D = B$, then $B \in \mathrm{Syl}_2(C_M(B))$ and it follows from [I: 130, Theorem 5.4.5] that $S$ is dihedral or quasi-dihedral, contrary to $m(S) \geqslant 3$. Thus $D > B$. Since $m(D) = 2$, this forces $D$ to be nonelementary, so $\phi(D) \neq 1$ by [I, 1.12].

Suppose now that the lemma is false, so that $N_M(D)$ contains a 3-element $y$ cyclically permuting the involutions of $B$. We have $m(D) = 2$ with $B$ central in $D$, so as $\phi(D) \neq 1$, this forces $B \cap \phi(D) \neq 1$. But $\phi(D)$ char $D$ and so $y$ leaves $\phi(D)$ invariant. Given the action of $y$ on $B$, this is possible only if $B \leqslant \phi(D)$. We therefore conclude that $B \leqslant \phi(S)$.

Finally set $R = C_S(U)$, so that again $|S : R| \leqslant 2$ and $m(R) = m(S)$. Since $|S : R| \leqslant 2$, clearly $\phi(S) \leqslant R$, so $B \leqslant R$. Thus $B$ centralizes $U$, so $U \leqslant C_S(B) = D$. Since $\langle U, B \rangle$ is elementary and $m(D) = 2$, this forces $U = B$. But then $D = R$, contrary to the fact that $m(R) = m(S) \geqslant 3$.

Now we can prove:

LEMMA 2.102. *$M$ has a normal subgroup of index 2.*

PROOF. Assume false. We shall argue that $\Gamma_{S,2}(X) \leqslant M$, in which case Aschbacher's proper 2-generated core theorem [I, 4.28] will yield the usual contradiction. It will suffice to prove the following assertion:

> If $B$ is a four-subgroup of $M$, then $B^v$ is connected to
> $U$ for some $v \in M$.                                                               (2.48)

Indeed, assume (2.48) and let $Q \leqslant S$ with $m(Q) \geqslant 2$. We argue that

$N_X(Q) \leqslant M$. Let then $x \in N_X(Q)$ and let $B$ be a four-subgroup of $Q$. Then also $B^x \leqslant Q$, so by (2.48), there are $v$, $v' \in M$ such that

$$B^v \text{ and } B^{xv'} \text{ are connected to } U \text{ in } M. \tag{2.49}$$

Again as with (2.46), this implies that

$$W_{B^v} = (W_B)^v \quad \text{and} \quad W_{B^{xv'}} = (W_B)^{xv'}, \tag{2.50}$$

whence by (2.49),

$$(W_U)^{v^{-1}} = (W_U)^{(xv')^{-1}}. \tag{2.51}$$

Thus $v^{-1}xv' \in N_X(U) \leqslant M$ and as $v$, $v' \in M$, we conclude that $x \in M$. Hence $N_X(Q) \leqslant M$, as required.

Now let $B$ be a four-subgroup of $M$. If $m(C_M(B)) \geqslant 3$, there is $B \leqslant E \leqslant M$ with $E \cong E_8$. Then $E^v \leqslant S$ for some $v \in M$ by Sylow's theorem [as $S \in \mathrm{Syl}_2(M)$]. Since $E^v \cong E_8$, $E^v$ is connected to $U$ and consequently $B^v \leqslant E^v$ is as well, so (2.48) holds in this case.

Hence we can assume that $m(C_M(B)) = 2$. We shall derive a contradiction with the aid of the previous lemma. Replacing $B$ by an $M$-conjugate, we can assume without loss that $R = N_S(B) \in \mathrm{Syl}_2(N_X(B))$ Then $D = C_S(B) \in \mathrm{Syl}_2(C_X(B))$. In particular, $m(D) = 2$ and so $B = \Omega_1(D)$. But $Z = \Omega_1(Z(S)) \leqslant D$, so $Z \leqslant B$. Since $D < S$ [as $m(S) \geqslant 3$], $Z < B$, so $Z = \langle z \rangle$ and $B = \langle z, b \rangle$ for suitable involutions $z, b$. By [I, 1.11], $R > D$ as $S > D$, so if we take $r \in R - D$, we have

$$z^r = z, \qquad b^r = bz, \quad \text{and} \quad (bz)^r = b. \tag{2.52}$$

On the other hand, let $T \in C_M(b)$ with $D \leqslant T$. Since $M$ has no normal subgroups of index 2, Lemma 2.97 implies that $m(T) \geqslant 3$, so $T > D$ and consequently also $N_T(D) > D$. But $B = \Omega_1(D)$ char $D$ and $C_T(D) \leqslant D$ [as $D \in \mathrm{Syl}_2(C_M(B))$]. Hence if we choose $t \in N_T(D) - D$, we have

$$b^t = b, \qquad z^t = bz, \quad \text{and} \quad (bz)^t = z. \tag{2.53}$$

We see then from (2.52) and (2.53) that the element $tr \in N_T(D)$ cyclically permutes the involutions of $B$. Hence $N_T(D)$ contains a 3-element with this property, contrary to the previous lemma.

The final step of the proof is to show that $M$ "controls" the $X$-fusion of $S$, so that by the previous lemma and the focal subgroup theorem [I, 1.19], $X$ will have a normal subgroup of index 2, contrary to assumption.

We let $\mathcal{D}$ be the Alperin–Goldschmidt conjugation family for $S$ [I, 4.85]. We must show that $N_X(D) \leqslant M$ for every $D \in \mathcal{D}$. Let $z$ be the unique involution of $Z(S)$ [by Lemma 2.98(iii), $Z(S)$ is cyclic]. By definition of $\mathcal{D}$, $z \in D$ for every $D \in \mathcal{D}$. We first prove:

LEMMA 2.103. *Let* $D \in \mathcal{D}$ *and set* $N = N_X(D)$. *If* $z$ *is* $N$-*conjugate to an involution of* $D - \langle z \rangle$, *then* $N \leqslant M$.

PROOF. We have $z \in E = \Omega_1(Z(D))$. Since $E$ char $D$ and $D \lhd N$, $E \lhd N$. Hence by our hypothesis on $z$, $\langle z \rangle < E$. Thus $m(E) \geqslant 2$. Since $E \lhd N$, $N \leqslant N_X(E)$, so we have only to show that $N_X(E) \leqslant M$. This follows from Lemma 2.99(ii) if $E$ is connected to $U$ in $X$; and, in particular, if $m(E) \geqslant 3$. Thus we can assume that $m(E) = 2$ (whence $E$ is a four-group) and $E$ is not connected to $U$ in $X$. In particular, this forces $m(D) = 2$.

Since $E$ is a four-group, our hypothesis on $z$ implies now that $N$ contains a 3-element $y$ cyclically permuting the involutions of $E$. But now we can repeat the argument of Lemma 2.101. Indeed, if $D = E$, again $S$ is dihedral or quasi-dihedral, contrary to $m(X) \geqslant 3$. Thus $\phi(D) \neq 1$ and the action of $y$ on $E$ together with $m(D) = 2$ forces $E \leqslant \phi(D) \leqslant \phi(S) \leqslant R = C_S(U)$, so $E$ centralizes $U$ and hence $E$ is connected to $U$, contrary to assumption.

Thus, it remains to show that $N = N_X(D) \leqslant M$ for those $D \in \mathcal{D}$ for which $z$ is isolated in $N$. But by definition of $\mathcal{D}$, $D$ is a Sylow 2-subgroup of $O_{2'2}(N)$, so $D$ and hence $z$ centralize $O(N)$. Therefore $z \in Z(N)$ for any such $N$ and so $N \leqslant C_X(z)$. Hence to complete the proof that $M$ controls the $X$-fusion of $S$ and thereby the proof of Theorem 2.96, it will suffice to show that $C_X(z) \leqslant M$.

So far we have not used the basic hypothesis that the centralizer of every involution of $X$ is 2-constrained. We use it now to prove the following result:

LEMMA 2.104. *For* $u \in U^{\#}$, $U \leqslant O_{2'2}(C_X(u))$.

PROOF. Let $u \in U^{\#}$ and set $C = C_X(u)$. Again put $R = C_S(U)$, so that $|S : R| \leqslant 2$ and $R \leqslant C$. Since $S \in \mathrm{Syl}_2(X)$, it follows that $R$ has index at most 2 in a Sylow 2-subgroup $P$ of $C$, and either $P = R$ or $P \in \mathrm{Syl}_2(X)$.

We claim that $U \triangleleft P$. If $P = R$, then $U \leqslant Z(P)$ and the assertion is clear. Hence, we can assume that $P \in \mathrm{Syl}_2(X)$. Set $E = \Omega_1(Z(R))$, so that $U \leqslant E$ and $E$, being characteristic in $R$, is normal in $P$. But by Lemma 2.98(iv), $S$ and hence $P$ contains no normal elementary subgroups of order 8, so $m(E) \leqslant 2$, forcing $U = E$. Thus $U \triangleleft P$ in this case as well.

Now set $\bar{C} = C/O(C)$, $\bar{T} = O_2(\bar{C})$, and $\tilde{C} = \bar{C}/\langle \bar{u} \rangle$. Clearly $\tilde{T} = O_2(\tilde{C})$. Also, $\tilde{U} \triangleleft \tilde{P}$ and $|\tilde{U}| = 2$, so $\tilde{U} \leqslant Z(\tilde{P})$. Hence $\tilde{U} \leqslant C_{\tilde{C}}(\tilde{T})$. It will therefore suffice to show that $C_{\tilde{C}}(\tilde{T}) \leqslant \tilde{T}$, for then $\tilde{U} \leqslant \tilde{T}$, whence $\tilde{U} \leqslant \tilde{T}$ and consequently $U \leqslant O_{2'2}(C)$, as required.

Let $\tilde{x} \in C_{\tilde{C}}(\tilde{T})$ with $\tilde{x}$ of odd order and let $\bar{x}$ be an element of odd order in $\bar{C}$ which maps on $\tilde{x}$ (it is easy to see that such an $\bar{x}$ exists). Then $\bar{x}$ acts on each term of the chain $\bar{T} > \langle \bar{u} \rangle > 1$ and centralizes the factors $\tilde{T} = \bar{T}/\langle \bar{u} \rangle$ and $\langle \bar{u} \rangle = \langle \bar{u} \rangle / 1$, so $\bar{x}$ "stabilizes" the given chain, whence $\bar{x}$ centralizes $\bar{T}$ by [I, 4.10]. But as $\bar{C}$ is 2-constrained by assumption and $O(\bar{C}) = 1$, $C_{\bar{C}}(\bar{T}) \leqslant \bar{T}$. Thus $\bar{x} \in \bar{T}$ and as $|\bar{x}|$ is odd, it follows that $\bar{x} = 1$, whence also $\tilde{x} = 1$. Hence $C_{\tilde{C}}(\tilde{T})$ is a 2-group. But $C_{\tilde{C}}(\tilde{T}) \triangleleft \tilde{C}$ as $\tilde{T} \triangleleft \tilde{C}$, so $C_{\tilde{C}}(\tilde{T}) \leqslant O_2(\tilde{C}) = \tilde{T}$, and the lemma is proved.

Finally we prove:

LEMMA 2.105. *We have* $C_X(z) \leqslant M$.

PROOF. Set $C = C_X(z)$ and $T = S \cap O_{2'2}(C)$, so that $U \leqslant T$ by the previous lemma. Also as $Z(S)$ is cyclic and $U \triangleleft S$, we have $z \in U$, so $O(C) = O(C_X(z)) \leqslant W_U \leqslant M$. But by a Frattini argument, $C = O(C) N_C(T)$. Thus to establish the lemma, it will suffice to prove that $N_X(T) \leqslant M$.

Note that as $C$ is 2-constrained with $T \in \mathrm{Syl}_2(O_{2'2}(C))$, it is immediate that any $T$-invariant subgroup of $C$ of odd order is necessarily contained in $O(C)$, an observation which we shall need for the proof.

Let $x \in N_X(T)$. We argue that $x$ leaves $W_U$ invariant, in which case $x \in M = N_X(W_U)$, as required. Set $Y = (W_U)^x$. We must show that $Y = W_U$. Since $S \leqslant M$ and $W_U \triangleleft M$, $W_U$ is $T$-invariant. Since $x$ normalizes $T$, also $Y$ is $T$-invariant. Furthermore, $|Y|$ is odd as $|W_U|$ is odd. Since $U \leqslant T$, $U$ thus acts on $Y$ and so by Lemma 1.52, we have

$$Y = \langle C_Y(U), [C_Y(u), U] \, | \, u \in U^\# \rangle. \qquad (2.54)$$

Now $C_Y(U) \leqslant C$ and $C_Y(U)$ is $T$-invariant as $U \triangleleft T$. Since $C_Y(U)$ has odd order, it follows that $C_Y(U) \leqslant O(C)$ by the observation of the second paragraph. Hence $C_Y(U) \leqslant W_U$. Furthermore, setting $Y_u = [C_Y(u), U]$, for $u \in U^\#$, we have $Y_u = [Y_u, U]$ by [I, 4.8]. But $U \leqslant O_{2'2}(C_X(u))$ by the previous lemma. Since $|Y_u|$ is odd, the preceding equality immediately implies that $Y_u \leqslant O(C_X(u))$. Thus $Y_u \leqslant W_U$ for each $u \in U^{\#}$. Since also $C_Y(U) \leqslant W_U$, we conclude from (2.54) that $Y \leqslant W_U$.

But $Y$, being conjugate to $W_U$, has the same order as $W_U$, so we must have $Y = W_U$, as required.

This completes the proof of Theorem 2.96.

Thus, leaving case I aside until the classification of groups of characteristic 2 type, we shall now outline the sectional 2-rank 4 proof in cases II–VI. (The definitions of these cases are given at the end of Section 2.4.)

## 2.6. Specified 2-Subgroups of Small Index

In cases II, III, and IV, the analysis leads fairly quickly to a 2-subgroup of $S$ of specified type, having index at most 8 in $S$. In contrast, the initial information in cases V and VI leads only to a specified subgroup of $S$ which, *a priori*, may have arbitrarily large index in $S$. As a result, the methods of pinning down the exact structure of $S$ differ considerably in the two situations. We shall discuss cases II, III, and IV in this section and cases V and VI in the next.

In each case, we shall list every possibility for $S$ forced by the argument (i.e., the possible elements of $\mathscr{F}$) even though some of these can be excluded on the basis of prior results. For example, in case II (Theorem 2.107 below), a possible answer is $S$ of type $Ly$. But in this case, $G$ is necessarily isomorphic to $Ly$ by Theorem 2.1, a group in which the centralizer of every involution is non-2-constrained. Hence if $S$ is of type $Ly$, $G$, in fact, does not satisfy the assumptions of case II.

The analysis in case II is based on a more precise form of Proposition 2.79, due to Harada.

THEOREM 2.106. *Let $X$ be a core-free nonsolvable group of sectional 2-rank $\leqslant 4$. Then we have*
  (i) *If $X/O_2(X) \cong L_3(2)$, then $O_2(X) \cong E_8, E_{16}$, or $Z_4 \times E_8$; and*
  (ii) *If $X/O_2(X)$ contains a subgroup isomorphic to $A_5$, then $O_2(X) \cong E_{16}$ or $Q_8 * D_8$.*

The proof is by induction on $|X|$. We set $Y = O_2(X)$. We can assume that $X/Y$ is perfect (and hence simple by Proposition 2.79); otherwise we can apply induction to the pre-image of $(X/Y)'$ in $X$ and conclude that $O_2(X)$ has one of the required structures. Thus $X/Y \cong L_3(2)$, $A_5, A_6$, or $A_7$. Since $X/Y$ contains an $A_5$-subgroup in the latter three cases, we can similarly reduce to the case in which $X/Y \cong L_3(2)$ or $A_5$. We can also assume that $Y$ is not elementary or again $Y$ has the required form.

The last assumption implies that $\phi(Y) \neq 1$. We choose a nontrivial subgroup $Z$ of $Z(Y) \cap \phi(Y)$, minimal subject to being normal in $X$, and set $\bar{X} = X/Z$. Then $\bar{X}$ has sectional 2-rank $\leqslant 4$ and again using [I, 1.12], we see that $\bar{X}$ is 2-constrained with trivial core. Hence by induction, $\bar{Y}$ has one of the listed structures.

There are two distinct cases according as $C_X(Z) = Y$ or $C_X(Z) > Y$ [each having two subcases corresponding to $X/Y \cong L_3(2)$ or $A_5$]. In the first case, $X/Y$ acts faithfully on $Z$ and by the minimality of $Z$, $Z \cong E_8$ or $E_{16}$, according to the corresponding subcase. On the other hand, in the second case, the minimality of $Z$ (together with the simplicity of $X/Y$) implies that $Z \cong Z_2$ and $Z \leqslant Z(X)$. Thus $X$ satisfies the following conditions:

(1)  $X$ has sectional 2-rank $\leqslant 4$;
(2)  $X/Y \cong L_3(2)$ or $A_5$;
(3)  If $X/Y \cong L_3(2)$, then $Y/Z \cong E_8, E_{16}$, or $Z_4 \times E_8$ and $Z \cong Z_2$   (2.55)
     or $E_8$; and
(4)  If $X/Y \cong A_5$, then $Y/Z \cong E_{16}$ or $Q_8 * D_8$ and $Z \cong Z_2$ or $E_{16}$.

Harada must deduce from these conditions that $Y$ has one of the required structures. Because $L_3(2)$ $[\cong GL_3(2)]$ acts faithfully on a three-dimensional vector space over $GF(2)$, while $A_5$ does not, the $L_3(2)$ case is considerably more difficult.

The basic idea of the proof is to show that if $Y$ does not have the desired form, then (2), (3), and (4) force $X$ to have sectional 2-rank $\geqslant 5$, contradiction. For example, suppose $Y/Z \cong Q_8 * D_8$ (an easy case), whence $Z \cong Z_2$ or $E_{16}$ and $X/Y \cong A_5$. In the latter case, one argues easily that the pre-image of $(Y/Z)'$ must split over $Z$, in which case, $Y$ contains an $E_{32}$ subgroup—contradiction. Similarly in the first case $Y$ does not split over $Z$ (or else $Y$ has sectional 2-rank $\geqslant 5$), and it follows readily that $Y$ must be of type $U_3(4)$. However, it is not difficult to show that the automorphism group of such a 2-group is solvable, contrary to the fact that $\text{Aut}(Y)$ involves $A_5$.

Theorem 2.106 is used to prove the following result, which together with (2.33) eliminates case II.

THEOREM 2.107.  *If case II holds, then a Sylow 2-subgroup of G is of type $L_4(q)$, $U_4(q)$, $PSp_4(q)$, $G_2(q)$ [$q \equiv 3,5 \pmod 8$ in each case], $L_3(4)$, $A_8, A_{10}, J_2, Mc$, or $Ly$.*

By the hypothesis of case II, $G$ contains a 2-constrained 2-local subgroup $H$ and either $H$ can be taken to be the centralizer of an involution [case II(b)] or else the centralizer of every 2-central involution is solvable [case II(a)]. Proposition 2.79 and Theorem 2.106 give the possible structures of $\bar{H} = H/O(H)$ and $\bar{Y} = O_2(\bar{H})$ and, in particular, a rather precise description of a Sylow 2-subgroup $R$ of $H$. For example, if $R \in \mathrm{Syl}_2(G)$, then with some additional 2-fusion analysis, one can prove the following result:

PROPOSITION 2.108. *If* $R \in \mathrm{Syl}_2(G)$, *then* $R$ *is either of type* $PSp_4(q)$, $G_2(q)$, $q \equiv 3, 5 \pmod 8$, $L_3(4)$, $A_8$, $A_{10}$, $J_2$, *Mc, or Ly.*

Indeed, set $\tilde{H} = \bar{H}/\bar{Y}$, and suppose first that $\bar{Y}$ is abelian and $\tilde{H}$ contains a subgroup isomorphic to $A_5$. Then by Lemma 2.90 and Proposition 2.79, $\tilde{H}$ contains a subgroup $\tilde{L} \cong A_5$ or $\Sigma_5$ with $\bar{R} \leqslant \bar{Y}\bar{L}$. Moreover, Lemmas 2.89 and 2.90 then give the possibilities for $R(\cong \bar{R})$ and we see that either the proposition holds or else $R \cong L_3(4)^f$. However, as $G$ is simple, the latter case is excluded by Proposition 2.88(i).

Assume next that $\tilde{H} \cong L_3(2)$, in which case $\bar{Y} \cong E_8, E_{16}$, or $Z_4 \times E_8$. (In particular, $\bar{Y}$ is abelian.) If $\bar{Y} \cong E_8$, then according as $\bar{H}$ splits or does not split over $\bar{Y}$, we check that $R$ is necessarily of type $A_8$ or $G_2(q)$, $q \equiv 3,5 \pmod 8$ [it is known that $G_2(q)$, $q$ odd, contains a nonsplit extension of $E_8$ by $L_3(2)$].

If $\bar{Y} \cong E_{16}$ and $\bar{H}$ splits over $\bar{Y}$, then $R$ is necessarily of type *Mc*. In the contrary case, it can be shown that $\bar{Y}$ is the unique $E_{16}$-subgroup of $\bar{R}$; in particular, the pre-image $Y$ of $\bar{Y}$ in $R$ is weakly closed in $R$ with respect to $G$. Combining this conclusion with Thompson's transfer lemma, a rather involved argument yields that $G$ has a normal subgroup of index 2—contradiction.

Finally if $Y \cong Z_4 \times E_8$, with $Y$ again the pre-image of $\bar{Y}$ in $R$, the sectional 2-rank hypothesis is used to force $Y_1 = \Omega_1(Y) = \Omega_1(R)$. so that now $Y_1$ is weakly closed in $R$ with respect to $G$. But if $z$ denotes the unique involution of $Y_1$ which is a square in $Y$, clearly $\bar{z} \in Z(\bar{H})$, so that $z$ is isolated in $R$ with respect to $H$. Now, using the weak closure of $Y_1$, one can argue that, in fact, $z$ is isolated in $R$ with respect to $G$, so that by the $Z^*$-theorem $G$ is not simple—contradiction.

Thus the proposition holds if $\bar{Y}$ is abelian. In the remaining case, $\bar{Y} \cong Q_8 * D_8$ and $\tilde{H} \cong A_5$ or $\Sigma_5$ ($Q_8 * D_8$ has $\Sigma_5$ as its outer automorphism group). Setting $\tilde{H} = \bar{H}/Z(\bar{Y})$, Lemma 2.90 implies that $\tilde{H}$ is a split extension of $\tilde{Y}(\cong E_{16})$ by $\tilde{L} \cong A_5$ or $\Sigma_5$. Furthermore, as some elements of $\tilde{Y}^\#$ are images of involutions of $\bar{Y}$, while others are images of elements of $\bar{Y}$ of order 4, we see that $\tilde{L}$ must act intransitively on $\tilde{Y}^{\#}$. We let $\bar{L}$ be the pre-image $\tilde{L}$ in $\bar{H}$.

Suppose first that $\tilde{L} \cong A_5$, in which case $\tilde{R}$ is of type $A_8$ by Lemma 2.90(ii) and the structure of $\tilde{H}$ is uniquely determined. Moreover, either $\bar{L}$ splits over $Z(\bar{Y})$ (whence $\bar{L} \cong Z_2 \times A_5$) or it does not [whence $\bar{L} \cong \hat{A}_5 \cong SL_2(5)$]. Thus there are exactly two possible structures for $\bar{H}$ and correspondingly we find that $R$ is of type $J_2$ or $Mc$.

On the other hand, if $\bar{L} \cong \Sigma_5$, a fusion analysis shows that necessarily $\bar{L}' \cong Z_2 \times A_5$, whence $\bar{R}_1 = \bar{R} \cap \bar{H}'$ is of type $J_2$. In particular, $\bar{R}_1$ has a unique normal subgroup $\bar{T}$ of index 2 of type $L_3(4)$. Then $\bar{T} \lhd \bar{R}$ of index 4. If Lemma 2.89(iv) is applicable to $N_{\bar{H}}(\bar{T})$, $R$ is of type $Ly$ and we are done. In the contrary case, one argues that the structure of $R$ is again uniquely determined; and, in particular, $\bar{R}$ does not split over $\bar{R}_1$ [$\bar{R} = \bar{R}_1 \langle \bar{v} \rangle$, where $\langle \bar{v}^2 \rangle = Z(\bar{Y})$]. This case leads to a contradiction by further 2-fusion analysis.

As an immediate corollary we have the following result:

PROPOSITION 2.109. *Theorem* 6.2 *holds in case* II($b$).

Indeed, if we can take $H = C_z$ for some involution $z$ of $G$, then by Proposition 2.79 and Theorem 2.106, either $Y \cong E_{16}$ or $Z_4 \times E_8$ [with $\bar{H}/\bar{Y} \cong L_3(2)$] or $\bar{Y} \cong Q_8 * D_8$. Let $R \in \mathrm{Syl}_2(H)$. Since $C_{\bar{H}}(\bar{Y}) \leqslant \bar{Y}$, it follows in each case that $\langle \bar{z} \rangle = \Omega_1(Z(\bar{R}))$. Thus $\langle z \rangle$ char $R$ and, as usual, [I, 1.11] now yields that $R \in \mathrm{Syl}_2(G)$. Hence the desired conclusion follows from the preceding proposition.

Thus it remains to treat case II($a$) [with $R \notin \mathrm{Syl}_2(G)$]. Since the involution $z$ of the preceding proposition was 2-central in each case, we see that the only possibilities for $Y$ in case II($a$) are

$$\bar{Y} \cong E_8 \quad \text{or} \quad E_{16}. \tag{2.56}$$

The analysis depends in part on the following result, which will also be needed in dealing with case VI.

PROPOSITION 2.110. *Let* $X$ *be a fusion simple group of sectional 2-rank* $\leqslant 4$ *in which the centralizer of every 2-central involution is solvable. Let* $S \in \mathrm{Syl}_2(X)$ *and assume the following conditions hold:*

(a) $S$ *contains a subgroup* $Q \cong Q_8 * Q_8$; *and*

(b) $N_X(Q)$ *contains a 3-element which acts fixed-point-free on* $Q/Z(Q)$. *Then* $Q \lhd S$ *and* $|S| \leqslant 2^8$.

*If, in addition, we assume*

(c) $Q \lhd T \leqslant S$, *where* $T$ *is of type* $A_{10}$ *and* $T$ *splits over* $Q$.
*Then either* $S = T$ *is of type* $A_{10}$ *or* $S$ *is of type* $PSp_4(q)$, $q \equiv 1,7 (\mathrm{mod}\ 8)$.

The sectional 2-rank assumption implies that $Z(Q) \in \mathrm{Syl}_2\big(C_X(Q)\big)$. Hence $Z(S) = Z(Q) = \langle z \rangle$ and $z$ is a 2-central involution, so $C = C_X(z)$ is solvable by hypothesis. Using (b), we easily force $Q \in \mathrm{Syl}_2\big(O_{2'2}(C)\big)$, whence $Q \lhd S$ and this in turn yields $|S| \leqslant 2^8$ (using the solvability of $C$).

Now assume (c). Then we can pin down the structure of $S$. Indeed, assuming $T < S$, one argues that $S = (Q_1 * Q_2)\langle y \rangle$, where $Q_1, Q_2$ are either quaternion or quasi-dihedral of order 16 and $y$ is an involution interchanging $Q_1$ and $Q_2$ under conjugation. However, Proposition 2.88(ii) shows that $X$ is not fusion simple in the latter case, so $Q_1, Q_2$ are quaternion of order 16, in which case $S$ is of type $PSp_4(q)$, $q \equiv 1,7 (\mathrm{mod}\ 8)$ [recall that the centralizer of a 2-central involution of $PSp_4(q)$, $q$ odd, has the form $(L_1 * L_2)\langle y \rangle$, where $L_1 \cong L_2 \cong SL_2(q)$ and $y$ is an involution interchanging $L_1, L_2$].

Now we can prove:

PROPOSITION 2.111. *If* $R \notin \mathrm{Syl}_2(G)$, *then* $R$ *is of type* $PSp_4(q)$, $q \equiv 1,7 (\mathrm{mod}\ 8)$, $A_{10}$, *or* $Mc$.

As Case II(a) holds, the centralizer of every 2-central involution of $G$ is solvable and $\bar{Y} \cong E_8$ or $E_{16}$ [(2.56)]. Let $Y$ be the pre-image of $\bar{Y}$ in $R$, so that $Y \cong \bar{Y}$ and $Y \geqslant O_2(H)$ with $O_2(H) \neq 1$ as $H$ is a 2-local. If $\tilde{H} = H/\bar{Y}$ acts irreducibly on $\bar{Y}$, then clearly $Y = O_2(H)$. The only other possibility is that $\bar{Y} \cong E_{16}$ and $\tilde{H} \cong L_3(2)$. Since the centralizer of every 2-central involution is solvable, we have $O_2(H) \cong E_8$ in this case. Setting $H_1 = N_G(Y)$, a Frattini argument yields that $H_1$ covers $\tilde{H}$ and so is nonsolvable. Using the sectional 2-rank condition, it also follows that $H_1$ is 2-constrained, so we could have worked throughout with $H_1$ in place of $H$. Thus without loss we can assume that $H_1 = H$, in which case $Y = O_2(H)$ in this case as well.

Thus $Y \lhd H$ and so $Y$ is not characteristic in $R$, otherwise $R \in \mathrm{Syl}_2(G)$, contrary to assumption. In particular, $R$ is not of type $A_8$ and $Y \neq \Omega_1(R)$.

Suppose first that $\tilde{H}$ contains an $A_5$-subgroup and $\tilde{H}$ acts intransitively on $\bar{Y}^\#$, in which case necessarily $\tilde{H} \cong A_5$ or $\Sigma_5$. The first alternative is excluded here as then $R$ would be of type $A_8$ by Lemma 2.90(ii). Hence by Lemma 2.90(iii), $\bar{H} = \bar{Y}\bar{L}$, where $\bar{L} \cong \Sigma_5$ and $R$ is of type $A_{10}$. Furthermore, $R \cap H'$ is of type $A_8$ (as $\bar{H}' = \bar{Y}\bar{L}'$). We check now from the structure of $\bar{H}$ that $R$ possesses a normal subgroup $Q \cong Q_8 * Q_8$ such that $R$ splits over $Q$ with $N_{H'}(Q)$ containing a 3-element which acts fixed-point-free on $Q/Z(Q)$.

Since $R \notin \mathrm{Syl}_2(G)$, the previous proposition now yields that $G$ has Sylow 2-subgroups of type $PSp_4(q)$, $q \equiv 1,7(\mathrm{mod}\ 8)$.

On the other hand, if $\tilde{H}$ is transitive on $Y^{\#}$, then again by Lemma 2.90(ii), $R$ contains a subgroup $T$ of type $L_3(4)$, and with the aid of Lemma 2.89(vi), we argue that $R$ must be of type $J_2$, $M_c$, or $L_3(4)^f$. Since then $T$ char $R$ by Lemma 2.91, it follows that $R \in \mathrm{Syl}_2(G)$—contradiction.

Thus we can assume that $\tilde{H} \cong L_3(2)$, so that $\overline{Y} \cong E_8$ or $E_{16}$. In the first case, $R$ is of type $A_8$ or $G_2(q)$, $q \equiv 3,5(\mathrm{mod}\ 8)$ according as $H$ splits or does not split over $Y$. In both cases, we find that $R$ possesses a subgroup $Q \cong Q_8 * Q_8$ such that $N_H(Q)$ contains a 3-element acting fixed-point-free on $Q/Z(Q)$. Hence if we expand $R$ to $S \in \mathrm{Syl}_2(G)$, the preceding proposition yields that $Q \lhd S$ and that $|S| \leqslant 2^8$, whence $|S| = 2^7$ or $2^8$. This is very tight information, and with some further fusion analysis, one can eliminate the nonsplit case and in the split case force $S$ to be of type $A_{10}$ or $Mc$.

Finally if $Y \cong E_{16}$, one argues that $Y = \Omega_1(R)$ (using the sectional 2-rank condition), a possibility we have excluded above.

This summary should give the reader the spirit and delicate flavor of the analysis in case II. The argument in case III has a similar character.

THEOREM 2.112. *In case III, a Sylow 2-subgroup of $G$ is of type $L_4(q)$, $q \equiv 7(\mathrm{mod}\ 8)$, $PSp_4(q)$, $q \equiv 1,7(\mathrm{mod}\ 8)$, $G_2(q)$, $q$ odd, $L_5(q)$, $q \equiv -1(\mathrm{mod}\ 4)$, $A_8$, $A_{10}$, $J_2$, $Mc$, or $Ly$.*

First of all, Propositions 2.92 and 2.93 give the following picture of the centralizer of a suitable involution of $G$ in case III.

PROPOSITION 2.113. *For some $t \in \mathscr{I}(G)$, if we set $C = C_t$, $\overline{C} = C/O(C)$, and $\overline{L} = L(C)$, then one of the following holds*:

(1) $\overline{L} \cong Sz(8)/Z_2$;
(2) $\overline{L} \cong \hat{A}_n$, $8 \leqslant n \leqslant 11$;
(3) $\overline{L} \cong M_{12}/Z_2$;
(4) $\overline{L} \cong SL_4(q)$, $q \equiv -1(\mathrm{mod}\ 4)$, $SU_4(q)$, $q \equiv 1(\mathrm{mod}\ 4)$, or $Sp_4(q)$, $q$ odd;
(5) $\overline{L} \cong \overline{L}_1 \times \overline{L}_2$, where $\overline{L}_i \cong SL_2(q_i)$, $q_i$ odd, or $\hat{A}_7$, $i = 1, 2$; or
(6) $\overline{L} = \overline{L}_1 * \overline{L}_2$, where $\overline{L}_i \cong SL_2(q_i)$, $q_i$ odd, or $\hat{A}_7$, $i = 1, 2$.

*Furthermore, in all cases, $C_{\overline{C}}(\overline{L}) = Z(\overline{L})$.*

We refer to these as subcases (1)–(6) (of case III). Let $R \in \mathrm{Syl}_2(C)$ and expand $R$ to $S \in \mathrm{Syl}_2(G)$. As pointed out in Section 2.4, the possibilities for

$R$ are completely determined and easily described. Using these, we immediately obtain the following result:

LEMMA 2.114.  (i) *In subcases* 1, 2, 4, *and* 5, $R = S$ *and* $Z(R) = \langle t \rangle$; *and*

(ii) *In subcases* 3 *and* 6, $|S:R| \leqslant 2$ *and* $Z(R) \cong Z_2 \times Z_2$.

We consider the individual subcases separately.

PROPOSITION 2.115. *If subcases* 1, 2, *or* 3 *hold, then* $S$ *is of type Mc or Ly.*

Indeed, in subcase 1, $A = \Omega_1(S) \cong E_{16}$, and from the structure of $N_{\bar{L}}(\bar{A})$, $N = N_G(A)$ contains a 7-element which centralizes $t$ and acts nontrivially on $A/\langle t \rangle$. On the other hand, from the $Z^*$-theorem, it follows that $C$ does not cover $N/C_A$, and we conclude easily that $S \notin \mathrm{Syl}_2(N)$. But $S \leqslant N$ as $A \triangleleft S$—contradiction.

In subcase 2, the goal is to show that $\bar{C} = \bar{L}$, for then $S$ will be of type $\hat{A}_n$, in which case $S$ is of type $Mc$ if $n = 8$ or 9 and of type $Ly$ if $n = 10$ or 11 (recall that $\hat{A}_8$ and $\hat{A}_{11}$ are the centralizers of involutions in $Mc$ and $Ly$, respectively). In the contrary case, if we set $\tilde{C} = \bar{C}/\langle \bar{x} \rangle$, then $\tilde{C} \cong \Sigma_n$. However, if $n = 10$ or 11, we find that $S$ is of sectional 2-rank 5—contradiction—so $n = 8$ or 9. Furthermore, it is known that there are two possible structures for $\bar{S}$ (and hence for $S$) according as the pre-image in $C$ of the involution of $\tilde{C}$ corresponding to a transposition or to a product of three transpositions in $\Sigma_n$, respectively, can be represented by an involution in $\bar{C}$. In each case a fusion analysis leads to a contradiction of the simplicity of $G$.

The $M_{12}/Z_2$ case is the most complex of these three subcases. First, as $|\mathrm{Aut}(M_{12}): \mathrm{Inn}(M_{12})| = 2$, either $\bar{C} = \bar{L}$ or $|\bar{C}: \bar{L}| = 2$. Moreover, in the latter case there are two possibilities for the structure of $R$ according as $\bar{C} - \bar{L}$ contains or does not contain an involution. Thus there are three possible structures for $R$; and, in addition, either $S = R$ or $S > R$.

The case $S = R$ is eliminated by 2-fusion analysis. It is shown first that $t$ is not $G$-conjugate to any involution of $(R \cap C') - \langle t \rangle$, so that by the $Z^*$-theorem, and the simplicity of $G$, $R > R \cap C'$. Thus $|\bar{C}: \bar{L}| = 2$ and again by the $Z^*$-theorem, $S = R \langle y \rangle$ for some involution $y$ which is $G$-conjugate to $t$. Using the structure of $R$, it is shown that no such $y$ exists.

On the other hand, elimination of the case $S > R$ depends on an analysis of the group $N = N_G(Q)$, where $Q$ is a uniquely determined special

2-group of order 64 with $Q$ char $R$. (Indeed, if $\tilde{C} = \bar{C}/\langle \bar{t} \rangle$, then $\tilde{R} \cap \tilde{L}$ contains a unique subgroup $\tilde{Q}$ of index 2 with $\tilde{Q} \cong Q_8 * Q_8$, and $Q$ is defined to be the pre-image of $\tilde{Q}$ in $R$.) Thus $S \leqslant N$ and $S > R$. Ultimately, the structure of $N$ forces $S$ to have sectional 2-rank 5—contradiction.

PROPOSITION 2.116. *If the subcases* 4 *or* 5 *hold, then* $S$ *is of type* $L_5(q), q \equiv -1 (\mathrm{mod}\ 4)$.

We note that the groups $L_5(q)$, $q \equiv -1 (\mathrm{mod}\ 4)$ and $U_5(q)$, $q \equiv 1 (\mathrm{mod}\ 4)$ have Sylow 2-subgroups of type $D \int Z_2$, where $D$ is quasi-dihedral of order $2^m$ for some $m$, so the goal of the analysis is to show that $S$ must have this form.

The following lemma, which is a direct consequence of the structure of $SL_4(q)$, $SU_4(q)$, $Sp_4(q)$, shows that subcases 4 and 5 have considerable similarity.

LEMMA 2.117. *In subcase* 4, *there exists a subgroup* $\bar{C}_0$ *of* $\bar{C}$ *containing* $\bar{R}$ *with the following properties*:

    (i) $\bar{C}_0$ *possesses a normal subgroup* $\bar{K}$ *with* $\bar{K} = \bar{K}_1 \times \bar{K}_2$, *where* $\bar{K}_1 \cong \bar{K}_2 \cong SL_2(q)$;
    (ii) $\bar{C}_0 = C_{\bar{C}}\big(Z(\bar{K})\big)$ *and* $C_{\bar{C}_0}(\bar{K}) = Z(\bar{K})$; *and*
    (iii) *An involution of* $\bar{R}$ *interchanges* $\bar{K}_1$ *and* $\bar{K}_2$ *under conjugation*.

Thus $\bar{C}_0$ has essentially the same structure as $\bar{C}$ in subcase 5. (For uniformity, we set $\bar{C} = \bar{C}_0, \bar{L} = \bar{K}$, and $\bar{L}_i = \bar{K}_i$, $i = 1, 2$ in subcase 5.) Note also that no element of $\bar{R}$ induces a nontrivial field automorphism on both $\bar{K}_1$ and $\bar{K}_2$, as then $R$ would have sectional 2-rank $\geqslant 5$. Using this fact, we obtain the following description of $R$.

LEMMA 2.118. *Let* $T, Q$ *be the pre-image in* $R$ *of* $\bar{R} \cap \bar{L}$ *and* $\bar{R} \cap \bar{K}$, *respectively. Then we have*

    (i) $Q = Q_1 \times Q_2$, *where* $Q_1, Q_2$ *are quaternion*;
    (ii) $Q \lhd T$ *and* $T/Q \cong Z_2 \times Z_2, Z_2 \times Z_2, Z_2$ *or* 1 *according as* $\bar{L} \cong SL_4(q), SU_4(q), Sp_4(q)$, *or* $\bar{L} = \bar{L}_1 \times \bar{L}_2$; *and*
    (iii) $T \lhd R$ *and* $R/T$ *is isomorphic to a subgroup of* $D_8$.

Set $\langle z_i \rangle = Z(Q_i)$, $i = 1, 2$. Note that $t \in \langle Z_1, Z_2 \rangle$.
In subcase 4 a careful fusion analysis enables us to show that $t$ is not

isolated in $C_{z_i}$, $i = 1, 2$. [In the course of this analysis, the conclusion $T/Q \cong Z_2 \times Z_2$ is forced, thus excluding the possibility $\bar{L} \cong \mathrm{Sp}_4(q)$.]

Now we are able to pin down the layer of $\tilde{C}_{z_i} = C_{z_i}/O(C_{z_i})$. Indeed, a preliminary picture of $L(\tilde{C}_{z_i})$ is already determined from $C$. Let $K_1, K_2$ be subgroups of $C_C(z_i)(= C_{\langle z_1, z_2 \rangle})$, chosen minimal subject to mapping on $\bar{K}_1, \bar{K}_2$ and with $K_1 K_2$ invariant under $R$. If $q > 3$, $\bar{K}_1, \bar{K}_2$ are quasisimple and so $K_1, K_2$ are 2-components of $C_C(z_i)$; in fact, $K_1 K_2 = L_{2'}(C_C(z_i))$. Hence by $L$-balance [I, 4.73], we have

$$\bar{K}_1 \bar{K}_2 \geqslant L(\tilde{C}_{z_i}), \qquad i = 1, 2. \tag{2.57}$$

The information from $C$ is not quite so strong when $q = 3$, so this case takes a little more care. The final result is as follows:

LEMMA 2.119. *In subcase* 4, *if we set* $\tilde{C}_{z_i} = C_z/O(C_{z_i})$, *then* $C_{z_i}$ *possesses a component* $\tilde{J}_i \cong L_3(q)$, $q \equiv -1 \pmod 4$, $U_3(q)$, $q \equiv 1 \pmod 4$, *or* $M_{11}$(*with* $q = 3$) *such that* $\tilde{Q}_i \leqslant \tilde{J}_i$ *and* $\tilde{J}_i$ *centralizes* $\tilde{Q}_j$, $i \neq j$, $1 \leqslant i, j \leqslant 2$.

In particular, $J_i$ has quasi-dihedral Sylow 2-subgroups. It follows easily now that $S$ ($=R$ in subcase 4) contains involutions $y_i$ such that $Q_i^*/Q_i \langle y_i \rangle$ is quasi-dihedral [with $\tilde{Q}_i^* \in \mathrm{Syl}_2(\tilde{J}_i)$], $i = 1, 2$ and $Q_1^* Q_2^* = Q_1^* \times Q_2^*$; and, in addition, that $S = Q_1^* Q_2^* \langle y \rangle$, where $y$ is an involution interchanging $Q_1^*$ and $Q_2^*$. Thus we have the following result:

LEMMA 2.120. *In subcase* 4, $S \cong D \int Z_2$, *where* $D$ *is quasi-dihedral.*

In subcase 5, $Z(R) = \langle z_1, z_2 \rangle$ and we see that either $S = R$ or $\langle z \rangle = Z(S)$, where $z = z_1 z_2$. Considering $C_z$ and applying $L$-balance together with Propositions 2.92 and 2.93, it follows that subcase 4 or 5 must hold for $C_z/O(C_z)$. Since $S$ has the required structure in the first case by the previous lemma, we can assume that subcase 5 holds for $C_z$. Hence without loss, we can take $t = z$, so that $t$ is 2-central and $R = S$.

In this case, a fusion analysis shows that neither $z_1$ nor $z_2$ is 2-central, so again $\bar{S}$ contains an element interhanging $\bar{L}_1 = \bar{K}_1$ and $\bar{L}_2 = \bar{K}_2$. But now if $t$ is not isolated in $C_{z_i}$, $i = 1$ or 2, the argument of subcase 4 applies with no essential change. Thus we have:

LEMMA 2.121. *If* $t$ *is not isolated in* $C_{z_i}$, $i = 1$ *or* 2, *then* $S \cong D \int Z_2$, *where* $D$ *is quasi-dihedral.*

In the contrary case, we have $\tilde{K}_1\tilde{K}_2 = L(\tilde{C}_{z_i})$, $i = 1, 2$, so the structure of $\tilde{C}_{z_i}$ is precisely specified. Using the $Z^*$-theorem, we now force $t$ to be $G$-conjugate to an involution $y$ such that $Q_i\langle y\rangle$ is quasi-dihedral for both $i = 1, 2$ (in particular, this eliminates the possibility that $\bar{L}_i \cong \hat{A}_7$ for $i = 1$ or 2). $C_S(y)$ is then shown to contain a unique subgroup $A \cong E_{16}$ and an analysis of $N = N_G(A)$ yields that $C_N(t)$ contains a 3-element acting nontrivially on $A$, which is impossible by the structure of $C$. Hence this case leads to a contradiction.

Finally we describe subcase 6.

PROPOSITION 2.122. *If subcase 6 holds, then $S$ is of type $L_4(q)$, $q \equiv 7(\mathrm{mod}\ 8)$, $PSp_4(q)$, $q \equiv 1,7(\mathrm{mod}\ 8)$, $G_2(q)$, $q \equiv 1,7(\mathrm{mod}\ 8)$, $A_8, A_{10}$, $Mc$, $J_2$, or $Ly$.*

The most difficult fusion analysis occurs in this subcase. We have $R = S$ and we let $Q$ be the pre-image in $S$ of $\bar{S} \cap \bar{L}$, so that $Q = Q_1 * Q_2$, where $Q_i \cong Q_{2^{m_i}}$ and $\bar{Q}_i \in \mathrm{Syl}_2(\bar{L}_i)$, $i = 1, 2$. Again one has $S/Q$ isomorphic to a subgroup of $D_8$.

If $S$ contains a normal $E_8$-subgroup, so must $Q$, as is easily seen, in which case $Q \cong Q_8 * Q_8$. (In particular, $\bar{L}_i \not\cong \hat{A}_7$, $i = 1, 2$). A short fusion analysis eliminates the $S/Q \cong 1$ or $D_8$ cases (whence $|S| = 2^6$ or $2^7$). The exact structure of $S$ is then forced by further fusion analysis (the case $|S| = 2^7$ is handled by earlier work of Harada [51]). The final result is as follows:

LEMMA 2.123. *If $S$ contains an elementary normal subgroup of order 8, then $S$ is of type $G_2(q)$, $q \equiv 3,5(\mathrm{mod}\ 8)$, $A_8, A_{10}$, or $Mc$.*

Thus we can assume that $S$ contains no such elementary subgroup. Let $U$ be a normal four-subgroup of $S$, so that $U = \langle t, u\rangle$ for some involution $u$. If the involutions of $U$ are conjugate in $G$, then $S$ is determined by MacWilliams's theorem (Theorem 2.80) and Proposition 2.122 holds, so we can assume that $u$ is not $G$-conjugate to $t$. But then by the structure of $\bar{L}$, we have

$$\text{No element of } Q - \langle t\rangle \text{ is } G\text{-conjugate to } t. \qquad (2.58)$$

In particular, by the $Z^*$-theorem and the simplicity of $G$, $t$ is $G$-conjugate to an involution $y$ of $S - Q$. (In particular, $S \neq Q$.) We fix such an involution $y$. The first portion of the fusion analysis yields the following result:

LEMMA 2.124. *Some element of $S$ interchanges $L_1$ and $L_2$ under conjugation.*

But now if $|S/Q| = 2$, then $S = Q\langle y \rangle$ and $\bar{y}$ must interchange $\bar{L}_1$ and $\bar{L}_2$. Thus $S = (Q_1 * Q_2)\langle y \rangle$ with $y$ interchanging $Q_1$ and $Q_2$, so that $S$ is, in fact, of type $PSp_4(q)$, $q \equiv 1,7 \pmod 8$ and hence Proposition 2.122 holds. Therefore we can assume

$$|S/Q| = 4 \text{ or } 8. \tag{2.59}$$

This case is very delicate. Ultimately the fusion analysis yields the following conclusion:

LEMMA 2.125. *We have $S/Q \cong Z_2 \times Z_2$ and $S$ is of type $L_4(q)$, $q \equiv 7 \pmod 8$.*

Thus, in effect, the argument proves that $S$ is of the form $(Q_1 * Q_2) V$, where $V = \langle v, v' \rangle$ is a four-group with $v$ interchanging $Q_1$ and $Q_2$ and $Q_i \langle v' \rangle$ quasi-dihedral, $i = 1$ and $2$.

This should give some feeling for the nature of the analysis in case III. In contrast, case IV is easily eliminated.

THEOREM 2.126. *Case IV does not hold.*

Suppose false. Using Propositions 2.120 and 2.121 this time we obtain the following result:

PROPOSITION 2.127. *For some $t \in \mathscr{I}(G)$, if we set $C = C_t$, $\bar{C} = C/O(C)$, and $\bar{L} = L(C)$, then $\bar{L} \cong {}^2G_2(3^{2n+1})$, $L_2(8)$, $Sz(8)$, or $J_1$.*

Again we let $R \in \mathrm{Syl}_2(C)$, expand $R$ to $S \in \mathrm{Syl}_2(G)$, and let $Q$ be the pre-image of $\bar{R} \cap \bar{L}$ in $R$. Also let $T$ be the pre-image of $C_{\bar{R}}(\bar{L})$ in $R$. The sectional 2-rank condition implies that $T$ is cyclic. Also $|\mathrm{Aut}(L)/\mathrm{Inn}(L)|$ is odd in each case, so $R = Q \times T$. Furthermore, either $Q \cong E_8$ or $\bar{L} \cong Sz(8)$ and $|Q| = 2^6$ with $Z(Q) \cong E_8$. Hence $A = \Omega_1(Z(R)) = \Omega_1(R) \cong E_{16}$ and $A = B \times \langle t \rangle$, where $B = \Omega_1(Q)$. Also by the structure of $L$, $N_C(A)$ contains a 7-element $x$ which acts nontrivially on $B$. We set $N = N_G(A)$ and $\tilde{N} = N/C_A$ and analyze the structure of $\tilde{N}$.

If $S = R$, then $|\tilde{N}|$ is odd. However, the $Z^*$-theorem implies that $N \not\leqslant C$, which together with the existence of $x$, forces $|\tilde{N}|$ to be even—contradiction.

Hence $S > R$ and this immediately gives $T = \langle t \rangle$. Since $A$ char $R$, $S \cap N > R$ and we find that the only possibility is $O_2(\tilde{N}) \cong E_8$ with $\tilde{N}/O_2(\tilde{N}) \cong Z_7$ or Frobenius of order 21. Using the sectional 2-rank assumption, we argue first that $A = \Omega_1(S \cap N)$ (whence $S \leqslant N$) and then that $R > A$. Thus $\bar{L} \cong Sz(8)$ and $S$ is a precisely determined 2-group of order $2^{10}$. The final step is to show that there exists no 2-group with the specified properties.

This summary of cases II, III, IV should amply illustrate the painful detail and intricate case division required in the analysis.

## 2.7. Specified 2-Subgroups of Indeterminate Index

We are left with cases V and VI, which require an analysis of a somewhat different texture. Indeed, the initial conditions in these two cases provide relatively weak information about the structure of a Sylow 2-subgroup $S$ of $G$, and the work required to determine $S$ precisely accounts for more than half the total section 2-rank $\leqslant 4$ classification proof.

In case V, we do start with the centralizer $C$ of an involution $t$ of $Z(S)$, so that $S \leqslant C$: however, the exact shape of only a (possibly) very small subgroup of $S$ is known at the outset. In case VI, the situation is even worse, for now the initial involution $t$ is not 2-central, so that $S$ is not even "in the picture," but must be "built up" from a Sylow 2-subgroup $R$ of $C$ and the fusion information we can deduce from the structure of $C$ (without loss, $R \leqslant S$). We note that except for a few subcases, it does ultimately turn out that $R$ has small index in $S$; however, the pushing-up process which yields this conclusion is considerably more difficult than any previously encountered. Moreover, there exist subcases in which the index of $R$ in $S$ is arbitrarily large, and even one subcase in which it is arbitrarily large relative to the order of $R$ itself, since there is no restriction on the length of the pushing-up process of this last subcase.

We first treat case V.

THEOREM 2.128. *If case V holds, then a Sylow 2-subgroup $S$ of $G$ is of type $G_2(q)$, $q$ odd, $J_2$, or $Ly$.*

Since cases III and IV do not hold, Propositions 2.92 and 2.93 yield in this case the following conclusion:

PROPOSITION 2.129. *For some involution $t \in Z(S)$, if we set $C = C_t$, $\bar{C} = C/O(C)$, and $\bar{L} = L(\bar{C})$, then one of the following holds:*

(i) $\bar{L} \cong SL_2(q)$, $L_2(q)$, $U_3(q)$, $q$ odd, $A_7$, $\hat{A}_7$, or $M_{11}$; or

(ii) $\bar{L} = \bar{L}_1 \times \bar{L}_2$, where $\bar{L}_1 \cong L_2(q_1)$, $L_3(q_1)$, $U_3(q_1)$, $q_1$ odd, $A_7$, or $M_{11}$ and $\bar{L}_2 \cong SL_2(q_2)$, $q_2$ odd, or $\hat{A}_7$.

The point is that as cases III and IV fail, each component of $\bar{L}$ has sectional 2-rank 2 and if $\bar{L}$ has two components, they cannot both be nonsimple, so we are left with the listed possibilities.

We can make a slight reduction in the case $\bar{L} \cong SL_2(q)$, $q$ odd, or $\hat{A}_7$; namely, we can then assume that $\langle \bar{t} \rangle = Z(\bar{L})$. Indeed, suppose $\langle \bar{t} \rangle \neq Z(\bar{L})$ and let $z$ be the involution of $S$ such that $\langle \bar{z} \rangle = Z(\bar{L})$. Then as $L \lhd C$, also $z \in Z(S)$. Setting $L = L_2\big(C_C(z)\big)$, it follows that $L$ maps on $\bar{L}$ and $z \in L$. Now set $\tilde{C}_z = C_z/O(C_z)$ and $\tilde{L}_z = L(\tilde{C}_z)$, so that by $L$-balance $\tilde{L} \leqslant \tilde{L}_z$. Thus $C_z$ is not 2-constrained and so Proposition 2.129 applies equally well to $C_z$. But now if $\tilde{L}_z$ consists of a single component, then as $\tilde{L} \leqslant \tilde{L}_z$, we must have $\tilde{L}_z \cong SL_2(q_z)$, $q_z$ odd, or $\hat{A}_7$. Moreover, and this is the point of the argument, as $\tilde{z} \in \tilde{L}$, $\langle \tilde{z} \rangle = Z(\tilde{L}_z)$. Hence either $\tilde{L}_z$ is a single component with $\tilde{z} \in \tilde{L}_z$ or $\tilde{L}_z$ consists of two components as in Proposition 2.129(ii). However, as $z$ is 2-central, we could have chosen $z$ in place of $t$. Thus we have the following result:

LEMMA 2.130. *We can choose the involution $t$ so that if $\bar{L} \cong SL_2(q)$, $q$ odd, or $\hat{A}_7$, then $\langle \bar{t} \rangle = Z(\bar{L})$.*

We assume henceforth that $t$ is chosen to satisfy the condition of Lemma 2.130.

There turns out to be no advantage in treating Proposition 2.129(ii) separately; rather it is preferable to split the argument into the following two subcases:

*Subcase 1.* $\bar{L} \cong SL_2(q)$, $q$ odd, or $\hat{A}_7$ [and $\langle \bar{t} \rangle = Z(\bar{L})$].

*Subcase 2.* $\bar{L}$ contains a component $\bar{L}_1 \cong L_2(q_1)$, $L_3(q_1)$, $U_3(q_1)$, $q_1$ odd, $A_7$, or $M_{11}$.

(Thus in subcase 2, either $\bar{L} = \bar{L}_1$ or $\bar{L} = \bar{L}_1 \times \bar{L}_2$ with $\bar{L}_2 \cong SL_2(q_2)$, $q_2$ odd, or $\hat{A}_7$.)

For uniformity of notation, set $\bar{L} = \bar{L}_1$ (and $q = q_1$) in subcase 1 and let $Q_1, Q_2$ be the pre-images of $\bar{S} \cap \bar{L}_1$ and $C_{\bar{S}}(\bar{L}_1)$ in $S$, respectively. Then $Q_1$ and $Q_2$ are each normal in $S$ with $Q_1$ centralizing $Q_2$. Also $Q_1$ is either quaternion (subcase 1) or dihedral, quasi-dihedral, or wreathed (subcase 2), so $Q_1$ is essentially determined. On the other hand, according as we are in subcase 1 or 2, we see that the group $\tilde{Q}_2 = Q_2/\langle t \rangle$ or $\tilde{Q}_2 = Q_2$ has sectional 2-rank $\leqslant 2$.

Hence by Lemma 2.94,

$$\text{Either } \Omega_1(\tilde{Q}_2) \cong Z_2 \times Z_2 \text{ or } \tilde{Q}_2 \text{ is cyclic,}$$
$$\text{quaternion, dihedral, or quasi-dihedral.} \tag{2.60}$$

Thus we have partial information about the structure of $Q_2$, but no information about its order (in comparison with that of $Q_1$). Hence even though $S$ is "visible" at the outset (i.e., $S \leqslant C$), we will not have a good hold on its structure until $Q_2$ has been pinned down more closely. Once this is achieved, fusion arguments similar to those in case III can be used to force the exact structure of $S$. It sounds simple enough, but the process turns out to be surprisingly long. Furthermore, the arguments in the two subcases are quite distinct, so we must treat them separately.

THEOREM 2.131. *In subcase* 1 *of case* V, S *is of type* $G_2(q)$, q *odd*, $J_2$, *or* Ly.

We continue the above notation [thus $\tilde{Q}_2 = Q_2/\langle t \rangle$ in (2.60)]. The first step in the proof is a sharpening of (2.60), which is obtained by the proverbial fusion analysis.

PROPOSITION 2.132. *Either* $Q_2$ *contains a normal four-subgroup of* S *or* $Q_2$ *is cyclic, quaternion, dihedral, or quasi-dihedral.*

In the present situation, we cannot immediately rule out the possibility that some element of $S$ induces a nontrivial field automorphism on $\bar{L}$ [or in the $A_7$ case that $\bar{C}/C_{\bar{C}}(\bar{L}) \cong \Sigma_7$]. We are thus forced to work with the (unique) largest subgroup $R_2$ of $S$ which centralizes a quaternion subgroup of $Q_1$. We have $Q_2 \leqslant R_2$ (as $Q_2$ centralizes $Q_1$), but the inequality may be strict. Also it is easy to see that $Q_1 R_2$ is of index at most 2 in $S$.

A further fusion analysis allows us to relate the $G$-fusion of $t$ to the subgroup $R_2$; the result is basic for the subsequent argument.

PROPOSITION 2.133. *If for* $g \in G$, $t^g \in S - \langle t \rangle$, *then* $R_2 \cap R_2^g = 1$.

Combined with the $Z^*$-theorem, the proposition can be used in certain cases to force $S$ to contain a subgroup of the form $R_2 \times R_2^g$ for some $g \in G$, a conclusion which has strong implications for the structure of $S$.

We break up the analysis according to the possibilities for $Q_2$ in Proposition 2.132, first proving the following result:

PROPOSITION 2.134. $Q_2$ *does not contain a normal four-subgroup of* $S$.

Assuming false, we focus on an element $D$ of the Alperin–Goldschmidt conjugation family for $S$ [I, 4.85] with the property that $t$ is not isolated in $N = N_G(D)$ (such a $D$ exists by the $Z^*$-theorem). A lengthy argument eventually pins down the structure of $D$ and forces $S$ to lie in $N$. Moreover, in the course of the argument, it is shown that $N$ and hence $S$ contains a normal subgroup $A \cong E_{16}$ with $t \in A$. By the structure of $C$, every elementary normal subgroup of $S$ lies in $Q_1 R_2$, so $A \leqslant Q_1 R_2$. On the other hand, as $t$ is not isolated in $N$, there is $y \in N$ such that $t^y \in A - \langle t \rangle$. But now one argues first that $|A \cap R_2| \geqslant 8$ and then that $R_2 \cap R_2^y \neq 1$, contrary to the previous proposition.

PROPOSITION 2.135. *If* $Q_2$ *is cyclic, then* $S$ *is of type* $G_2(q)$, $q \equiv$ $1, 7 \pmod 8$, $J_2$, *or* $Ly$.

Indeed, if $U$ is a normal four-subgroup of $S$, a fusion analysis forces the involution of $U$ to be $G$-conjugate. Hence if $S$ has no normal $E_8$-subgroups, the proposition follows from MacWilliams's theorem (Theorem 2.80). In the contrary case, it is shown first that $S$ must be a split extension of a wreathed 2-group of order $2^{2n+1}$ by a cyclic group, where $|Q_2| = 2^n$; and then that $G$ is not fusion simple.

Finally one has the following result:

PROPOSITION 2.136. *If* $Q_2$ *is quaternion, dihedral, or quasi-dihedral, then* $S$ *is of type* $G_2(q)$, $q \equiv 3, 5 \pmod 8$ (*whence* $Q_1 \cong Q_2 \cong Q_8$ *and* $|S| = 2^6$).

If false, it follows easily that either $Q_1$ or $Q_2$ has order at least 16. We derive a contradiction by showing, on the one hand, that $t$ must be conjugate in $G$ to an involution of $S - Q_1 Q_2$ and, on the other hand, that no such conjugacy is possible.

This describes the proof of Theorem 2.131 in outline.

We turn now to subcase 2. However, we note first that if $Z(S)$ is noncyclic, our choice of $t$ is not unique and so there may or may not exist a central involution of $S$ for which subcase 1 holds. Indeed, we prove the following:

THEOREM 2.137. *If subcase 2 of case* V *holds, then subcase 1 holds for some involution of* $Z(S)$.

The theorem will therefore reduce us to subcase 1, and so combined with Theorem 2.131 will complete the proof of Theorem 2.128.

In proving Theorem 2.137, we can assume

$$\text{Subcase 1 does not hold for any involution of } Z(S). \qquad (2.61)$$

This assumption reduces us to the following possibilities:

PROPOSITION 2.138. *We have* $\bar{L}_1 \cong L_2(q_1)$, $q_1$ *odd*, $L_3(3)$, $U_3(3)$, $A_7$, *or* $M_{11}$.

Indeed, if false, then by definition of subcase 2, $\bar{L}_1 \cong L_3(q_1)$ or $U_3(q_1)$, $q_1$ odd, $q_1 \geqslant 5$. Since $\bar{L}_1 \lhd \bar{C}$, $Z(\bar{S}) \cap \bar{L}_1 \neq 1$, so $\bar{z} \in \bar{L}_1$ for some involution $z \in Z(S)$. By the structure of $L_3(q_1)$ and $U_3(q_1)$, $\bar{K} = C_{\bar{L}_1}(\bar{z})' \cong SL_2(q_1)$ with $\langle \bar{z} \rangle = Z(\bar{K})$. Moreover, as $q_1 \geqslant 5$, $\bar{K}$ is quasisimple, so $L_{2'}(C_C(z))$ possesses a 2-component $K$ with $z \in K$ and $K$ mapping on $\bar{K}$. But now if we repeat the proof of Lemma 2.130 with $K$ in place of $L$, we see that subcase 1 holds for $z$, contrary to (2.61).

By the proposition, $Q_1$ is either dihedral $(\bar{L}_1 \cong L_2(q_1)$ or $A_7)$, quasi-dihedral of order 16 $(\bar{L}_1 \cong L_3(3)$ or $M_{11})$, or wreathed of order 32 $(\bar{L}_1 \cong U_3(3))$. Furthermore, (2.60) holds now for $Q_2$ and $Q_1 Q_2 = Q_1 \times Q_2$.

We first dispose of the wreathed case.

PROPOSITION 2.139. $Q_1$ *is not wreathed.*

A wreathed 2-group contains the central product of a quaternion group and a cyclic group of order 4 and so has sectional 2-rank 3. Since $S$ has sectional 2-rank at most 4 and $Q_1 Q_2 = Q_1 \times Q_2$, $Q_2$ is necessarily cyclic if $Q_1$ is wreathed. Furthermore, a Sylow 2-subgroup of $\mathrm{Aut}(U_3(3))/\mathrm{Inn}(U_3(3))$ has order 2, so the possible structures of $S$ are precisely determined at the outset in this case. In particular,

$$\Omega_1(S) \leqslant Q_1 Q_2, \qquad (2.62)$$

for if $S - Q_1 Q_2$ contains an involution $x$, one can show that $\bar{L}_1 \langle \bar{x} \rangle$ has sectional 2-rank 4, whence $S$ has sectional 2-rank 5—contradiction.

As $Q_1$ is wreathed of order 32, it contains a $Z_4 \times Z_4$ subgroup $A_1$ of index 2, in which case $A = A_1 \times Q_2$ is abelian of rank 3 and $A \lhd S$. Using the $Z^*$-theorem, one derives an easy fusion contradiction by analyzing the structure of $N_G(A)$.

Thus $Q_1$ is either dihedral or quasi-dihedral (of order 16).

As in subcase 1, we now define $R_2$ to be the (unique) largest subgroup of $S$ disjoint from $Q_1$ such that $C_{Q_1}(R_2)$ contains a four-group. Then $Q_2 \leqslant R_2$ and $Q_1 R_2$ is of index at most 2 in $S$. Furthermore, as $Q_1 \cap R_2 = 1$, $R_2$ has sectional 2-rank at most 2. We also set $Z = \Omega_1(Z(S))$. Since $\bar{Z} \cap \bar{L}_1 \neq 1$ and $t \in Z$, we see that $|Z| \geqslant 4$.

A lengthy fusion analysis pins the situation down more tightly:

PROPOSITION 2.140.    (i) $Z \cap Q_1 \cong Z_2$;

                                   (ii) *No two involutions of $Z$ are conjugate in $G$*;

                                   (iii) $m(S) = 4$; *and*

                                   (iv) $R_2$ *is not cyclic or quaternion.*

We restrict ourselves to a few comments. Observe that if (i) fails, then necessarily $S = Q_1 \times Q_2$ with $Q_1 \cong Z_2 \times Z_2$. Since $S$ is nonabelian by Proposition 2.78(ii), $Q_2$ is therefore nonabelian and, in particular, noncyclic. Furthermore, by Proposition 2.88(v), $Q_2$ is not dihedral, quaternion, or quasi-dihedral. Hence $\Omega_1(Q_2) \cong Z_2 \times Z_2$ by (2.60). We derive a contradiction in this case by studying $N_G(A)$, where $A = Z_1 \times \Omega_1(Q_2)$.

If (ii) fails, we force $S \cong (Z_2 \times Z_2) \int Z_2$, in which case Proposition 2.88(iii) is contradicted. Similarly, if (iii) fails, we force $S \cong Z_2 \times D_{2^n}$, in which case Proposition 2.87(v) is contradicted.

Finally using (iii), (iv) follows easily.

The remaining case is the most difficult. First of all, the subgroup $H = C_{Q_1} Q_1 R_2$ plays a key role. Note that as $Q_1 \lhd S$, $H$ is $S$-invariant and it follows at once that $Q_1 R_2 \in \mathrm{Syl}_2(H)$. Setting $\tilde{H} = H/Q_1$, then as $Q_1 \cap R_2 = 1$, we see that $\tilde{R}_2 \cong R_2$ with $\tilde{R}_2 \in \mathrm{Syl}_2(\tilde{H})$. But $R_2$ has sectional 2-rank $\leqslant 2$, so the structure of $\tilde{H}$ is determined from Proposition 2.95. In particular, we obtain the following result.

PROPOSITION 2.141. *One of the following holds:*

  (i) $t$ *is isolated in $H$; or*

  (ii) $R_2$ *is dihedral, quasi-dihedral, or homocyclic abelian of rank* 2.
      *Furthermore, if* (ii) *holds, then $R_2$ centralizes $Q_1$.*

Notice that elimination of the isolated case gives the exact structure of $Q_1 R_2$ and as $Q_1 R_2$ has index at most 2 in $S$, $S$ will be almost pinned down.

In fact, our aim is to prove:

(1) $t$ is not isolated in $H$; and
(2) $S = R_1 \times R_2$, where $Q_1 \leqslant R_1$ and $R_1$ is either dihedral        (2.63)
or quasi-dihedral.

Once this is achieved, Proposition 2.88(v) will again yield a contradiction and Theorem 2.128 will at last be proved.

One would imagine that with so much information available, verification of (2.63) would be a straightforward matter. However, this is far from the case. One must first analyze the $G$-fusion of $E_{16}$-subgroups of $S$ and determine the consequences of the existence of two such nonconjugate subgroups. After that, a fusion argument eliminates the case $L_1 \cong L_3(3)$, so that $Q_1$ is dihedral and $L_1 \cong L_2(q_1)$, $q_1$ odd, or $A_7$. At this point, one argues *for a suitable choice* of $t$ that $t$ is $G$-conjugate to an involution $y$ of $Q_1 R_2 - \langle t \rangle$. Now taking $\langle t, y \rangle \leqslant A \leqslant Q_1 R_2$ with $A \cong E_{16}$ and setting $N = N_G(A)$ and $\tilde{N} = N/O(N)$, one shows that $\tilde{N}$ is a $\{2, 3\}$-group with elementary 3-subgroup of order *nine*. This last conclusion is critical for showing that $t$ is not isolated in $H$ and establishing the direct product decomposition of $S$.

Finally we come to case VI, which because of the even greater indeterminancy involved, is the most difficult. The final result is as follows:

THEOREM 2.142.  *If case VI holds, then a Sylow 2-subgroup $S$ of $G$ is of type $L_4(q)$, $q$ odd, $PSp_4(q)$, $q$ odd, $L_5(q)$, $q \equiv -1 \pmod 4$, $G_2(q)$, $q$ odd, $L_3(4)$, $A_8, A_{10}, J_2$, Mc, or Ly.*

Since cases II, III, IV, and V do not hold, Propositions 2.92 and 2.93 now yield:

PROPOSITION 2.143.  *If $t$ is an involution of $G$ such that $C = C_t$ is nonsolvable, then $t$ is not 2-central and if we set $\overline{C} = C/O(C)$ and $\overline{L} = L(\overline{C})$, then $\overline{L} \cong L_2(q)$, $q$ odd, $L_3(3)$, $U_3(3)$, $A_7$, or $M_{11}$.*

Note that the possibilities $\overline{L} \cong SL_2(q)$, $q$ odd, or $\hat{A}_7$, or $L_3(q)$, $U_3(q)$, $q$ odd, $q \geqslant 5$, are excluded, for in these cases it would follow by essentially the same argument as in Lemma 2.130 and Proposition 2.138 that the centralizer of some 2-central involution $z$ of $G$ is nonsolvable and hence non-2-constrained (as case II fails), whence $L(C_z/O(C_z)) \neq 1$, contrary to the fact that case V fails.

We choose $t$ as in the proposition so that a Sylow 2-subgroup $R$ of $C$

has as large an order as possible, and we assume without loss that $R \leqslant S$. Thus $R < S$. Furthermore, as before, we let $Q_1, Q_2$ be the pre-images in $R$ of $\bar{R} \cap \bar{L}$ and $C_{\bar{R}}(\bar{L})$, so that again $Q_1$ is either dihedral, quasi-dihedral of order 16, or wreathed of order 32, and $Q_2$ has sectional 2-rank $\leqslant 2$. Also $Q_1$ and $Q_2$ are again normal in $R$. We fix this notation.

The first step of the proof is to obtain a restriction on the structure of $Q_2$. This major reduction leads to three basic subcases, which must then be individually analyzed.

PROPOSITION 2.144. *One of the following holds*:

(1) $Q_2 = \langle t \rangle$ *and* $\bar{L} \cong L_2(q), A_7, L_3(3),$ *or* $M_{11}$;
(2) $Q_2 \cong Z_2 \times Z_2,$ *and* $\bar{L} \cong L_2(q), q \equiv 3,5 \pmod 8$;
(3) *Either* $Q_2 = \langle t \rangle$ *or* $Q_2 \cong Z_4,$ *and* $\bar{L} \cong U_3(3)$; *or*
(4) $S$ *is of type* $L_4(r), r$ *odd,* $PSp_4(r), r$ *odd, or* $L_5(r),$
$\quad r \equiv -1 \pmod 4$.

Indeed, suppose (1), (2), and (3) fail. First, using the maximality of $R$, we argue that if we take $y \in N_S(R) - R$ with $y^2 \in R$, then

$$Q_2 Q_2^y = Q_2 \times Q_2^y. \tag{2.64}$$

This places a very strong restriction on $Q_2$, since it forces $Q_2^y$ and hence $Q_2$ to be isomorphic to a subgroup of $R/Q_2$, a group which can be identified with a subgroup of $\mathrm{Aut}(\bar{L})$.

For example, if $\bar{L} \cong L_2(q), q \equiv 3,5 \pmod 8$, then $R/Q_2$ and hence also $Q_2$ is isomorphic to a subgroup of $D_8$. Since (2) fails, we thus have $Q_2 \cong Z_4$ or $D_8$. The first possibility is easily eliminated, whence $R = Q_2 \times Q_2^y \cong D_8 \times D_8$. This in turn leads to the conclusion $S = R\langle y \rangle \cong D_8 \wr Z_2$, so $S$ is of type $L_4(r), r \equiv 3,5 \pmod 8$ and (4) holds. We can therefore assume that $\bar{L}$ is not of this form.

We argue next that also

$$Q_1 Q_1^y = Q_1 \times Q_1^y. \tag{2.65}$$

In particular, this rules out $\bar{L} \cong U_3(3)$, for then $R$ would have sectional rank 6—contradiction. Furthermore, considering the centralizer of $\bar{Q}_1$ in $\mathrm{Aut}(L)$, one easily obtains that $Q_1 Q_2$ has index at most 2 in $C_R(Q_1)Q_1$, so that by (2.65):

$$Q_2 \text{ is isomorphic to a subgroup of } Q_1 \text{ of index at most 2.} \tag{2.66}$$

If $R = Q_1 \times Q_2$, one argues that $S = R\langle y \rangle$, so that $S \cong D \int Z_2$, where $D$ is either dihedral of order at least 8 $\left(\text{as } \bar{L} \not\cong L_2(q), q \equiv 3,5(\text{mod } 8)\right)$ or quasi-dihedral of order 16. It follows that $S$ is of type $L_4(r)$, $r \equiv 1,7(\text{mod } 8)$ or $L_5(r)$, $r \equiv -1(\text{mod } 4)$ and again (4) holds.

Hence we can also assume that $R > Q_1 \times Q_2$. Since $\text{Aut}(M_{11}) = \text{Inn}(M_{11})$, this eliminates $\bar{L} \cong M_{11}$. Furthermore, if $\bar{L} \cong L_3(3)$, we find that the sectional 2-rank hypothesis is violated, so this case is also excluded. Thus we are left only with the possibilities $\bar{L} \cong L_2(r)$, $r \equiv 1,7(\text{mod } 8)$, or $A_7$, in which cases $Q_1 \cong D_{2^n}$, $n \geqslant 3$.

This is the most difficult case. First a lengthy fusion argument shows that either some element $u$ of $R$ induces a nontrivial field automorphism on $\bar{L}(\cong L_2(q))$ or a transposition on $\bar{L}(\cong A_7)$. We choose $u$ of minimal order. If $R = Q_1 Q_2 \langle u \rangle$, it is shown that $S = R\langle y \rangle \cong D_{2^n} \int Z_2$ and again (4) holds. Hence we can assume that $R > Q_1 Q_2 \langle u \rangle$, which in turn rules out the case $\bar{L} \cong A_7$. We also show that $Q_2 \langle u \rangle$ is dihedral or quasi-dihedral of order $2^m$, where either $m = n$ or $n + 1$.

In this remaining case, we argue that $S = R\langle y, u, x \rangle$, where $x^2 \in \langle t \rangle$ and $\bar{L}\langle \bar{t}, \bar{x} \rangle / \langle \bar{t} \rangle \cong PGL_2(q)$, thus obtaining a rather tight hold on the structure of $S$. However, to pin $S$ down precisely, we must distinguish the cases $m = n$ and $m = n + 1$. In the first case a lengthy fusion analysis forces $S$ to be of type $PSp_4(r)$, $r \equiv 1,7(\text{mod } 8)$ and again (4) holds. On the other hand, if $m = n + 1$, a similar analysis shows ultimately that $S$ must have sectional rank $\geqslant 5$, giving a final contradiction.

We are left therefore to deal with the first three possibilities of Proposition 2.144. We first reduce to the $L_2(q)$, $A_7$ case.

PROPOSITION 2.145. *If* $\bar{L} \cong L_3(3)$, $M_{11}$, *or* $U_3(3)$, *then* $S$ *is of type* $G_2(r)$, $r \equiv 1,7(\text{mod } 8)$, $J_2$, *or* $Ly$.

In these cases either $Q_2 = \langle t \rangle$ or $\bar{L} \cong U_3(3)$ and $Q_2 \cong Z_4$, by Proposition 2.144. Also $Q_1$ is either quasi-dihedral of order 16 or wreathed of order 32. We let $y$ be as before and also fix a normal four-subgroup $U$ of $S$.

We first treat the case $R = Q_1 Q_2$. We consider $E_8 \cong A \leqslant R$ (with $A \lhd R$ in the wreathed case), and by analyzing the structure of $N_G(A)$, we argue first that

$$|S| \geqslant 4 |R|. \tag{2.67}$$

Hence if we set $T = C_S(U)$, then $|T| > |R|$ (as $|S:T| \leqslant 2$). In particular, $t \notin U$.

If $Q_1$ is quasi-dihedral, then $R$ has no normal $E_8$-subgroups. In particular, $t \notin T$ (otherwise $U\langle t \rangle$ would be such a normal subgroup). Similarly it follows that $S$ has no normal $E_8$-subgroups (otherwise if $E$ were such a subgroup $C_E(t)\langle t \rangle$ would be a normal $E_8$-subgroup of $R$). Thus $S = T\langle t \rangle$ and so by Thompson's transfer lemma, $t$ is $G$-conjugate to an involution of $T$. On the basis of this information, one argues that the involutions of $U$ are $G$-conjugate, so the proposition follows from MacWilliams's theorem (Theorem 2.80).

On the other hand, if $Q_1$ is wreathed, the goal of the analysis is to show that $S = R\langle y \rangle$, contrary to (2.65). The argument involves a delicate two-stage pushing-up. First, $Q_1$ contains a unique $Q_8$-subgroup $V_1$; and setting $V = V_1 Z(R)$ and $H = N_G(V)$, it is shown that $Y = R\langle y \rangle \in \mathrm{Syl}_2(H)$ and that $C_C(V) \lhd H$. Then if $W$ denotes the pre-image of $Z(H/C_C(V))$ in $Y$, we argue that $V$ char $W$ and $W$ char $N_S(Y)$, whence $S = Y = R\langle y \rangle$.

Thus we can assume that $R > Q_1 Q_2$, which again excludes $\bar{L} \cong M_{11}$. If $\bar{L} \cong L_3(3)$, a fusion analysis forces $R - Q_1 Q_2$ to contain an involution $y$, whence $R = Q_1 Q_2 \langle y \rangle$. Then an argument very similar to that of the preceding paragraph yields that $S = R\langle y \rangle$, whence $|S| = 2^7$. Finally it is shown, on the one hand, that $S$ must contain a subgroup (of index 2) isomorphic to $D_8 \times D_8$; and, on the other, that $S$ cannot contain such a subgroup.

Finally, if $\bar{L} \cong U_3(3)$, our sectional 2-rank hypothesis is shown to force $Q_2 = \langle t \rangle$ and $R = Q_1 \langle u \rangle$, where $u^2 = t$. Now we again argue that $G$ has sectional 2-rank $\geqslant 5$—contradiction.

Next we reduce to the $Q_2 = \langle t \rangle$ case.

PROPOSITION 2.146. *If $\bar{L} \cong L_2(q)$, $q \equiv 3,5 \pmod 8$ and $Q_2 \cong Z_2 \times Z_2$, then $S$ is of type $L_3(4)$, $A_8$, $J_2$, or Mc.*

Thus $Q_1 \cong Z_2 \times Z_2$ and $Q_1 Q_2 \cong E_{16}$. Again there are two cases according as $R = Q_1 Q_2$ or $R > Q_1 Q_2$, in both of which it is shown that $R$ possesses a unique $E_{16}$-subgroup $A$ (of index at most 2). We set $N = N_G(A)$, $\tilde{N} = N/O(N)$, and analyze the structure of $\tilde{N}$. If $R = Q_1 Q_2 = A$, it is shown that either $S = R\langle y \rangle \cong (Z_2 \times Z_2) \wr Z_2$ [contrary to Proposition 2.88(iii)] or $\tilde{N} \cong A_4$ with a Sylow 2-subgroup $T$ of $N$ of type $L_3(4)$ or $A_8$ (according as $\tilde{N}$ splits or does not split over $\tilde{A}$). Without loss $T \leqslant S$. Since the proposition holds if $T = S$, we can assume the contrary.

If $T$ splits over $A$, we argue that $|S| = 2^7$ and that MacWilliams's assumptions are satisfied, so that $S$ is of type $J_2$ by her theorem. In the contrary case, we show that $S$ is a uniquely determined split extension of $Z_4 \times Z_4$ by $Z_2 \times Z_2$ and then derive a fusion contradiction.

Thus again we can assume that $R > Q_1 Q_2$. This time it is shown that either $S \cong E_{16} \cdot Z_4$ [contrary to Proposition 2.88(iii)] or $S$ is of type $A_8$ (in which case the proposition holds) or else $\tilde{N} \cong \Sigma_4$ (so we can assume that this last case holds). Again there are two possibilities according as the pre-image $T$ of $O_2(\tilde{N})$ in $N$ is of type $L_3(4)$ or $A_8$. Without loss, $S \cap N \in \mathrm{Syl}_2(N)$. In both cases we argue that $S \leqslant N$, so that $|S| = 2^7$. If $T$ is of type $L_3(4)$, it is immediate that $S$ is of type $Mc$, while if $T$ is of type $A_8$, we reach a fusion contradiction. Thus the proposition holds in these cases as well.

In view of Propositions 2.144, 2.145, and 2.146, we are left to establish Theorem 2.142 in the following case:

$$\bar{L} \cong L_2(q), \ q \text{ odd, or } A_7, \text{ and } Q_2 = \langle t \rangle. \tag{2.68}$$

Even this last case must be split into two parts. We first prove the following result:

PROPOSITION 2.147. *If $R/\langle t \rangle$ is not dihedral or quasi-dihedral, then $S$ is of type $PSp_4(q)$, $q \equiv 1,7 \pmod 8$, or $A_{10}$.*

We assume false and let $\tilde{R} = R/Q_1\langle t \rangle$. Then $\tilde{R} \neq 1$, so if $\bar{L} \cong A_7$, then $\bar{C} \cong \Sigma_7$. The situation is more complex when $\bar{L} \cong L_2(q)$. We can identify $\tilde{R}$ with a subgroup of $\mathrm{Aut}(\bar{L})/\mathrm{Inn}(\bar{L})$, which has Sylow 2-subgroups of the form $T \times F$, where $T \cong Z_2$ is the image of $PGL_2(q)$ and $F$ is cyclic, inducing a group of field automorphisms on $\bar{L}$. Since the proposition is assumed false, we easily check that either $|\tilde{R}| \geqslant 4$ or $\tilde{R}$ corresponds to $\Omega_1(F)$. However, in the first case there are three possibilities: namely, for some nontrivial subgroup $F_1$ of $F$, either (a) $\tilde{R}$ corresponds to $F_1$, (b) $\tilde{R}$ corresponds to $T \times F_1$, or (c) $\tilde{R}$ is cyclic (of order at least 4) and corresponds to a "diagonal" subgroup of $T \times F$.

A delicate fusion analysis is required to eliminate this unpleasant last possibility. Once this case is ruled out, it follows that $R$ contains an element $u$ with $u^2 = 1$ or $t$ such that either $\bar{y}$ induces a field automorphism on $\bar{L}(\cong L_2(q))$ or a transposition on $\bar{L}(\cong A_7)$. Also if $\bar{C}$ contains a $PGL_2(q)$ subgroup, there is $x \in R$ with $x^2 = 1$ or $t$ such that $\bar{L}\langle \bar{t}, \bar{x} \rangle / \langle \bar{t} \rangle \cong PGL(q)$. In the contrary case, we put $x = 1$. (In particular, $x = 1$ if $\bar{L} \cong A_7$.) Again by a fusion analysis, we force

$$u^2 = 1, \text{ and so } u \text{ is an involution.} \tag{2.69}$$

Also from the structure of $\text{Aut}\big(L_2(q)\big)$ and $\text{Aut}(A_7)$, we have

$$C_{\bar{L}}(\bar{u}) \cong PGL_2(q^{1/2}) \text{ or } \Sigma_5, \tag{2.70}$$

where $\bar{L} \cong A_7$ in the latter case. In particular, either $C_{\bar{L}}(\bar{u})$ is nonsolvable or $\bar{L} \cong L_2(9)$.

As we shall see in the next section, this situation is an important one for the general analysis of centralizers of involutions in groups of component type and must be treated without the sectional 2-rank assumption under which we are operating here. In the present case we proceed as follows:

First, suppose $x = 1$ and $\bar{L} \not\cong L_2(9)$, so that $C_{\bar{L}}(\bar{u})$ is nonsolvable. We easily argue that for $v = u$ or $ut$, $R \leqslant C_v$ and $R \notin \text{Syl}_2(C_v)$. But $C_v$ is nonsolvable [as $C_{\bar{L}}(\bar{u}) = C_{\bar{L}}(\bar{v})$], so our maximal choice of $R$ is contradicted. Hence we are left with the following two possibilities:

$$\bar{L} \cong L_2(q) \text{ and either } x \neq 1 \text{ or } q = 9. \tag{2.71}$$

We argue next that $u$ is not a square in $R$. Indeed, in the contrary case, $q \neq 9$ and so $C_u$ is nonsolvable. Moreover, $C_u$ can be shown to have a Sylow 2-subgroup of at least the order of $R$, so again by the maximality of $R$, equality must hold and then $C_u$ must have the same general form as $C$. But it is easily seen that this is incompatible with the structure of $C_R(u)$. This immediately implies that $R$ has the following form:

$$R = Q_1\langle x, u, t \rangle \text{ and } Q_1\langle t, u \rangle = Q_1 \times \langle t, u \rangle. \tag{2.72}$$

We next prove:

> The element $y$ can be chosen to normalize an
> elementary subgroup $A$ of $Q_1\langle t, u \rangle$ of order 16. $\tag{2.73}$

If $x \neq 1$, all $E_{16}$-subgroups of $Q_1\langle t, u \rangle$ are conjugate in $R$ and (2.73) follows easily. Thus we can assume $x = 1$ and $\bar{L} \cong L_2(9)$ (whence $Q_1 \cong D_8$). If $y$ leaves invariant a four-subgroup of $Q_1$, we argue that $C_u$ is nonsolvable and has Sylow 2-subgroups of order exceeding $R$, giving the same contradiction as before. In the contrary case, if $A_i, i = 1, 2$, denote the two $E_{16}$-subgroups of $R$, we reach a contradiction by comparing the structures of $N_G(A_1)$ and $N_G(A_2)$, using the fact that $\big(N_G(A_1)\big)^y = N_G(A_2)$.

Now let $A$ and $y$ be as in (2.73) and set $N = N_G(A)$ and $N^* = N/O(N)$. Since $N$ is solvable (as case II does not hold), $N^*$ is, in fact, a $\{2,3\}$-group. To complete the proof, each possible structure for $N$ must be systematically analyzed, with the aid of the sectional 2-rank hypothesis. We comment here only on the one case which leads to specific possibilities for $S$: namely, $N^* \cong \Sigma_4$.

We let $Y$ be a Sylow 2-subgroup of the pre-image of $O_2(N^*)$, so that $|Y| = 2^6$. If $|Z(Y)| = 2$, it is not difficult to show that $Y$ must be of type $A_8$, in which case $Y$ contains a unique normal subgroup $Q \cong Q_8 * Q_8$. Since the centralizer of every 2-central involution of $G$ is solvable, we see that the hypotheses of Proposition 2.110 [including (c)] are satisfied by $N_G(Q)$. Hence we conclude from that result that either $S$ is of type $A_{10}$ (in which case $N$ contains a Sylow 2-subgroup of $G$) or else $S = (S_1 * S_2)\langle w \rangle$, where $S_1, S_2$ are either dihedral or quasi-dihedral of order 16 and interchanged by an involution $w$. If $S_1$ is dihedral, then $S$ is of type $PSp_4(q)$, $q \equiv 1,7 \pmod 8$, while if $S_1$ is quasi-dihedral, Proposition 2.88(ii) yields a contradiction.

On the other hand, if $|Z(Y)| \geqslant 4$, we force $N$ to contain a Sylow 2-subgroup of $G$ (whence $|S| = 2^7$) and then derive a fusion contradiction, using Thompson's transfer lemma.

We are thus left with the final possibility that $R/\langle t \rangle$ is dihedral or quasi-dihedral. It is here that we require a pushing-up sequence of unrestricted length in order to pin down $S$, so that, in fact, the order of $S$ is unrelated to that of $R$. [Of course, after pinning down $S$ and applying Theorem 2.4, it will then follow that $S$ is necessarily of type $G_2(q)$, $q \equiv 3,5 \pmod 8$ (and hence of type $M_{12}$ and order $2^6$), $M_{12}$ being the only simple group in $\mathscr{F}$ satisfying the hypotheses of case VI.]

We first abstract the pushing-up portion of the argument. It arises in the following context: for some $E_8$-subgroup $A$ of $Q_1 \times \langle t \rangle$, if we set $N = N_G(A)$ and $\bar{N} = N/O(N)$, then $\bar{N}$ has the following structure:

(a) $\bar{N} = \bar{B}\bar{F}$, where
(b) $\bar{B} \cong Z_4 \times Z_4$,
(c) $\bar{F}$ is either cyclic of order 6 or nonabelian of order     (2.74)
    12, but $\bar{F} \not\cong A_4$, and
(d) $\bar{B} \lhd \bar{N}$ and $C_{\bar{F}}(\bar{B}) = 1$.

By (d), $\bar{F}$ is isomorphic to a subgroup of $\mathrm{Aut}(\bar{B}) = \mathrm{Aut}(Z_4 \times Z_4)$, while by (c), $O_3(\bar{F}) \cong Z_3$ and $\bar{F} \cong Z_3 \times Z_2, \Sigma_3 \times Z_2$, or $Z_3 \cdot Z_4$. Moreover, in all cases, $O_2(\bar{N}) = \bar{B}\langle \bar{x} \rangle$, where $\bar{x}$ is an involution inverting $\bar{B}$.

Harada analyzes (2.74) in complete generality and proves the following result:

THEOREM 2.149. *Let $X$ be a fusion simple group with an elementary subgroup $A \cong E_8$ such that if $N = N_X(A)$ and $\bar{N} = N/O(N)$, then $\bar{N}$ satisfies*

(2.73). *Under these conditions, $\bar{F} \cong \Sigma_3 \times Z_2$ and a Sylow 2-subgroup of $X$ is of type $G_2(q)$, $q$ odd, or Ly.*

Using Proposition 2.88, we immediately obtain Theorem 2.149 in the following case:

LEMMA 2.150. *If $N$ contains a Sylow 2-subgroup $T$ of $X$, then $\bar{F} \cong \Sigma_3 \times Z_2$ and $T$ is of type $G_2(q)$, $q \equiv 3,5 \pmod 8$.*

Thus in proving the theorem we can assume that $N$ does not contain a Sylow 2-subgroup of $X$, so that we can begin the pushing-up process. The first stage leads to three possible "branchings." We put $A = A_1$, $N = N_1$, and we let $A_2$ be a Sylow 2-subgroup of the pre-image of $O_2(N)$ in $X$ and set $N_2 = N_X(A_2)$, $\bar{N}_2 = N_2/O(N_2)$. Harada first proves the following result:

LEMMA 2.151. *We have $\bar{N}_2 = \bar{B}_2 \bar{F}_2$, where*

(i) *$\bar{B}_2$ is either of type $Z_8 \times Z_8$, $L_3(4)$, or $U_3(4)$;*
(ii) *$\bar{F}_2 \cong \bar{F}$; and*
(iii) *$\bar{B}_2 \lhd \bar{N}_2$ and $C_{\bar{F}_2}(\bar{B}_2) = 1$.*

We next treat the $L_3(4)$, $U_3(4)$ cases.

LEMMA 2.152. *If $\bar{B}_2$ is of type $L_3(4)$ or $U_3(4)$, then $\bar{F} \cong \Sigma_3 \times Z_2$ and a Sylow 2-subgroup of $X$ is of type Ly.*

Indeed, it is shown then that $N_2$ contains a Sylow 2-subgroup $T$ of $X$ and that $T$ is of type $L_3(4)^f$, $U_3(4)$, $L_3(4)^f$, $Ly$ (with $\bar{F} \cong \Sigma_3 \times Z_2$), $U_3(4)^f$, or $U_3(4)^{f'}$. All but the $Ly$ case are excluded by Proposition 2.88 as $X$ is fusion simple.

Hence Harada is left to analyze the case that $\bar{B}_2 \cong Z_8 \times Z_8$. It is here that the pushing-up sequence has arbitrary length. We place the argument in an inductive setting. Thus we assume that there exists a 2-subgroup $A_n$ of $X$, $n \geq 2$, such that if we set $N_n = N_X(A_n)$ and $\bar{N}_n = N_n/O(N_n)$, then

(a) *$\bar{N}_n = \bar{B}_n \bar{F}_n$, where*
(b) *$\bar{B}_n \cong Z_{2^{n+1}} \times Z_{2^{n+1}}$,*                                    (2.75)
(c) *$\bar{F}_n \cong \bar{F}$, and*
(d) *$\bar{B}_n \lhd \bar{N}_n$ and $C_{\bar{F}_n}(\bar{B}_n) = 1$.*

We then let $A_{n+1}$ be a Sylow 2-subgroup of the pre-image of $O_2(\overline{N}_n)$ in $N_n$ and set $N_{n+1} = N_X(A_{n+1})$ and $\overline{N}_{n+1} = N_{n+1}/O(N_{n+1})$ and prove:

LEMMA 2.153. *One of the following holds*:

(i) $N_n$ *contains a Sylow 2-subgroup of* $X$; *or*
(ii) $N_{n+1}$ *satisfies* (2.74) *with* $n + 1$ *in place of* $n$.

In view of the lemma, it follows that for $m$ sufficiently large, $N_m$ contains a Sylow 2-subgroup $T$ of $X$. Moreover, $T$ is of the form $B_m \cdot F$, where $B_m \cong Z_{2^m} \times Z_{2^m}$ and $F$ is isomorphic to a Sylow 2-subgroup of $\overline{F}$ (whence $F \cong Z_2, Z_4$, or $Z_2 \times Z_2$). The first two cases for $F$ are excluded by Proposition 2.88, and in the last case $T$ is isomorphic to a Sylow 2-subgroup of $G_2(q)$ for $q \equiv \pm 1 \pmod{2^{n+1}}$. Thus we have the following result:

PROPOSITION 2.154. *If* $\overline{B}_2 \cong Z_8 \times Z_8$, *then* $\overline{F} \cong \Sigma_3 \times Z_2$ *and a Sylow 2-subgroup of* $X$ *is of type* $G_2(q)$, $q \equiv \pm 1 \pmod{2^n}$ *for some* $n \geqslant 3$.

This completes the proof of Theorem 2.149.
Now we return to $G$ and prove finally:

PROPOSITION 2.155. *If* $R/\langle t \rangle$ *is dihedral or quasi-dihedral, then* $S$ *is of type* $G_2(q)$, $q$ *odd*, $A_8, A_{10}$, *or* $Ly$.

The first step of the proof is to show that the element $y$ of $N_S(R)$ can be chosen to normalize an $E_8$-subgroup $A$ of $Q_1 \times \langle t \rangle$. This is easily seen to hold for the initial $y$ if $R$ does not split over $\langle t \rangle$ or if $R/\langle t \rangle$ is either a four-group or quasi-dihedral. In the remaining case, a typical fusion argument yields that either the desired conclusion holds or else $S = R\langle y \rangle$ with $S$ containing a cyclic normal subgroup of index 4. However, this last possibility is excluded by Proposition 2.88.

For such a choice of $y$, set $N = N_G(A)$ and $\overline{N} = N/O(N)$. Harada argues that

(a) $\overline{N} = \overline{B}\overline{F}$, where
(b) either $\overline{B} \cong E_{16}$ or $Z_4 \times Z_4$; and                              (2.76)
(c) (2.73)(c) and (d) hold.

Hence if $\overline{B} \cong Z_4 \times Z_4$, Theorem 2.149 applies and yields that $S$ is either of type $G_2(q)$, $q$ odd, or $Ly$ (with $\overline{F} \cong \Sigma_3 \times Z_2$).

Thus we can assume that $\bar{B} \cong E_{16}$. If $R$ does not split over $\langle t \rangle$ or if $R \cong E_8$, we argue easily that $N$ contains a Sylow 2-subgroup $S$ of $G$ and that $\Omega_1(S) \cong (Z_2 \times Z_2) \wr Z_2$ with $\Omega_1(S)$ correspondingly of index 2 or 1 in $S$. In either case, Proposition 2.88 contradicts the simplicity of $G$. Hence $R = R_1 \times \langle t \rangle$ with $|R_1| \geqslant 8$, which immediately yields

$$\bar{F} \cong \Sigma_3 \times Z_2. \tag{2.77}$$

It follows easily now that a Sylow 2-subgroup $T$ of $N$ is of type $A_8$. We can assume therefore that $T \notin \mathrm{Syl}_2(G)$. We let $B$ be the unique $E_{16}$-subgroup of $T$ and set $H = N_G(B)$. Harada analyzes the structure of $H/O(H)$, arguing first that a Sylow 2-subgroup of $H$ is of type $A_{10}$ and then that it is a Sylow 2-subgroup of $G$. Thus $S$ is of type $A_{10}$ and so the proposition holds in this case as well.

Theorem 2.142 is therefore proved in all cases. This completes the discussion of the sectional 2-rank $\leqslant 4$ theorem.

## 2.8. Other Low 2-Rank Problems

We shall now describe the further low 2-rank problems that have been used in the course of the classification of simple groups. The term "low 2-rank" here should be interpreted loosely, for in most cases the initial assumptions are placed on only a *portion* of a group $G$—e.g., the existence of a *single* 2-local subgroup (in particular, the centralizer of an involution) of low 2-rank with a specified structure—but no further restriction is put on the 2-rank of $G$ itself.

The first of these, due to Harada [I: 162], gives a basic application of the sectional 2-rank $\leqslant 4$ theorem.

THEOREM 2.155. *If $G$ is a group which contains an elementary subgroup $A \cong E_8$ such that $A$ is a Sylow 2-subgroup of $C_A$, then $G$ has sectional 2-rank at most 4 (and hence $G$ is a K-group).*

We have stated the theorem in this form, but it is actually a result about 2-groups. Indeed, if $A \leqslant S \in \mathrm{Syl}_2(G)$, then

$$A = C_S(A). \tag{2.78}$$

Using this condition alone, Harada forces $S$ to have sectional rank at most 4. In view of MacWilliams's theorem, he can assume that $S$ contains an elementary normal subgroup $E \cong E_8$. If $A$ can be taken as $E$, the desired conclusion is immediate as then $S/A \leqslant \mathrm{Aut}(A) \cong L_3(2)$; so he can assume that $E \neq A$.

As one would expect, the possible structures of $N_S(A)$ are important for the analysis. Critical to the argument are the following two lemmas, the second of which depends upon the first.

LEMMA 2.157. *If $S$ contains a subgroup $Q \cong Q_8 * Q_8$ such that $C_S(Q) \leqslant Q$, then $S$ has sectional 2-rank $\leqslant 4$.*

LEMMA 2.158. *If $S$ contains a subgroup $Q = (Z_2 \times Z_2) \wr Z_2$ such that $C_S(Z(Q)) = Q$, then $S$ has sectional 2-rank $\leqslant 4$.*

Harada has also established two results concerning groups $G$ which possess a self-centralizing elementary subgroup $A$ of order 16. The first of these, involving conditions on the structure of $N_G(A)$, has been considerably extended by Stroth (however, by arguments requiring elaborate 2-local analysis), and the full result will be stated below (Theorem 2.167). We state Harada's second result now, as its proof is similar to that of Theorem 2.156, again forcing a Sylow 2-subgroup of the given group $G$ to have sectional rank $\leqslant 4$.

THEOREM 2.159. *Let $G$ be a group, $x$ an involution of $G$, and $T$ a Sylow 2-subgroup of $C_x$, and assume the following conditions:*

(a) *$T = D \times B$, where $D$ is dihedral or quasi-dihedral and $B$ is a four-group; and*

(b) *$T$ is a Sylow 2-subgroup of $C_b$ for each $b \in B^{\#}$.*

*Then $G$ has sectional 2-rank $\leqslant 4$ (and hence is a K-group).*

(Of course, if we take $A \leqslant T$ with $A \cong E_{16}$, then $x \in A$ and it follows that $A$ is a Sylow 2-subgroup of $C_A$.)

The theorem is used by Foote in his work on groups in which the centralizer of some involution has a 2-component $L$ with $L/O(L) \cong L_2(q)$, $q$ odd, or $A_7$ (see Proposition 4.56 below).

Our next result concerns the following situation which arises as a minimal configuration in the proof of the $B$-theorem:

For some $t \in \mathscr{I}(G)$, if we set $C = C_t$, $\bar{C} = C/O(C)$, and $\bar{L} = L(\bar{C})$, then

$$
\begin{aligned}
&\text{(a)} \quad \bar{L} \cong L_2(q), \ q \text{ odd, or } A_7; \text{ and} \\
&\text{(b)} \quad C_{\bar{C}}(\bar{L}) \text{ has cyclic Sylow 2-subgroups.}
\end{aligned}
\qquad (2.79)
$$

Here we are interested in an exceptional case of the analysis of (2.79), which we shall briefly explain. We can, of course, assume that $G$ has sectional 2-rank $\geqslant 5$. It is immediate then from the structure of $\mathrm{Aut}\big(L_2(q)\big)$ and $\mathrm{Aut}(A_7)$ that

$$t \text{ is not a 2-central involution.} \tag{2.80}$$

Under the sectional 2-rank $\leqslant 4$ hypothesis, (2.79) and (2.80) were analyzed as part of case VI; and the exceptional case here relates to an extension of that argument when $\bar{L}$ has dihedral subgroups of *order* 8. To motivate our interest in this special case, we make a few further reductions. First, as in the corresponding sectional 2-rank $\leqslant 4$ analysis, we easily reduce to the following situation:

$$C_{\bar{C}}(\bar{L}) = \langle \bar{t} \rangle. \tag{2.81}$$

Furthermore, by applying Harada's theorem above, we obtain the following result:

PROPOSITION 2.160. *Let $G$ be a group containing an involution $t$ such that $C = C_t$ satisfies (2.79) and (2.81). If $C/\langle t \rangle$ has dihedral or quasi-dihedral Sylow 2-subgroups, then $G$ has sectional 2-rank $\leqslant 4$ (and hence $G$ is a K-group).*

Indeed, let $R \in \mathrm{Syl}_2(C)$ and suppose that $R/\langle t \rangle$ is dihedral or quasi-dihedral. Since $\bar{L}\langle \bar{t} \rangle = \bar{L} \times \langle \bar{t} \rangle$, $m(R) \geqslant 3$. Choose $A \cong E_8$ in $R$ with $t \in A$. We shall argue that $A \in \mathrm{Syl}_2(C_A)$, in which case the proposition will follow from Theorem 2.156.

Let $T \in \mathrm{Syl}_2(C_A)$, so that $A \leqslant T$. Since $t \in A$, $T \leqslant C$. Replacing $T$ by a suitable conjugate in $C_A$, we can assume without loss that $T \leqslant R$. Set $\tilde{C} = C/\langle t \rangle$, so that $\tilde{A} \cong Z_2 \times Z_2$ and $\tilde{T} \leqslant C_{\tilde{R}}(\tilde{A})$. But by assumption, $\tilde{R}$ is dihedral or quasi-dihedral, so $C_{\tilde{R}}(\tilde{A}) = \tilde{A}$. Thus $\tilde{T} = \tilde{A}$ and $T = A$, as required.

We can therefore assume that $C/\langle t \rangle$ does not have dihedral or quasi-dihedral Sylow 2 subgroups. Now it follows as in the sectional 2-rank $\leqslant 4$ case that $C$ contains an element $u$ such that either

$$\begin{array}{ll}
\text{(a)} \ \ \bar{L} \cong L_2(q), \ q \equiv 1,7 (\mathrm{mod}\ 8), \text{ with } u \text{ inducing a field} \\
\quad \text{automorphism of order 2 on } \bar{L}; \text{ or} \\
\text{(b)} \ \ \bar{L} \cong A_7 \text{ with } \bar{u} \text{ acting like a transposition on } \bar{L}.
\end{array} \tag{2.82}$$

Likewise the usual fusion argument allows us to make the following additional assumption:

$$\text{We can take } u \text{ to be an involution.} \tag{2.83}$$

Furthermore, if we set $\bar{K} = L\big(C_{\bar{L}}(\bar{u})\big)$, then as before, we have

(a) If $\bar{L} \cong L_2(q)$, $q > 9$, then $\bar{K} \cong L_2(q^{1/2})$;
(b) If $\bar{L} \cong L_2(9)$, then $\bar{K} = 1$ [as $C_L(\bar{u}) \cong PGL_2(3) \cong \Sigma_4$ is   (2.84)
     solvable]; and
(c) If $\bar{L} \cong A_7$, then $\bar{K} \cong A_5$.

Apart from the solvable case $(b)$, a general line of argument is now available for dealing with the problem: namely, if we set $K = L_{2'}\big(C_L(u)\big)$ (in which case $K$ maps on $\bar{K}$) and $\bar{C}_u = C_u/O(C_u)$, $\bar{L}_u = L(\bar{C}_u)$, then by $L$-balance [I, 4.73], we conclude that

$$\bar{K} \text{ is a 2-component of } C_{\bar{L}_u}(\bar{t}). \tag{2.85}$$

We see then that if we are in a suitable inductive setup, we shall be able to determine the possibilities for $\bar{L}_u$ and then "play off" the relationship between the two groups $C$ and $C_u$. This has indeed been successfully carried out by Harris [59] and will be fully described in Section 4.6. It is certainly possible that the $L_2(9)$ case is amenable to similar treatment, for Aschbacher has established a form of $L$-balance for "solvable components" [2] (here of type $A_4$), which although somewhat weaker than the customary $L$-balance, is nevertheless an effective replacement for it in many situations. However, this approach to the $L_2(9)$ problem has not been attempted; rather the analysis has followed the 2-local pushing-up method à la the sectional 2-rank $\leqslant 4$ theorem, the goal of which is as there to pin down the structure of a Sylow 2-subgroup of $G$.

It is this residual $L_2(9)$ case in which we are interested here. However, the critical condition is that $\bar{L}$ has *Sylow 2-subgroups of order* 8, so that, in fact, the argument applies if either $\bar{L} \cong A_7$ or $\bar{L} \cong L_2(q)$, $q \equiv 7,9 \pmod{16}$.

The problem at hand has received two similar, but independent treatments: one by Harris and Solomon [57, 62], the other by Fritz [33]. However, both proofs consist, in effect, of reductions to the classification of groups with Sylow 2-subgroups of order $\leqslant 2^{10}$. Of the two analyses, Fritz's comes closer to pinning down the possible Sylow 2-subgroups of $G$, and so we shall follow his argument.

As we have noted in the Introduction of the book, a minimal counterexample $G$ to the $B$-Theorem need not be simple, but satisfies the weaker condition $F^*(G)$ simple, so we must state Fritz's result for such groups. [In his paper, Fritz assumes that $G$ is simple, but his proof is easily adapted with essentially no change to the $F^*(G)$ simple case.]

THEOREM 2.161. *Let $G$ be a group with $F^*(G)$ simple of sectional 2-rank $\geqslant 5$ and assume that $G$ contains an involution $t$ such that if we set $C = C_t$, $\bar{C} = C/O(C)$, and $\bar{L} = L(\bar{C})$, then $\bar{L} \cong L_2(q)$, $q \equiv 7,9 \pmod{16}$ or $A_7$ and $C_{\bar{C}}(\bar{L}) = \langle \bar{t} \rangle$. Then one of the following holds*:

(i) $\bar{L} \cong L_2(9)$ *and* $G = F^*(G) \cong HS$;
(ii) $\bar{L} \cong L_2(9)$, $G = F^*(G)\langle t \rangle$ *with* $t \notin F^*(G)$, *and* $F^*(G) \cong PSp_4(4)$, $L_5(2)$, *or* $U_5(2)$; *or*
(iii) $\bar{L} \cong A_7$, $G = F^*(G)\langle t \rangle$ *with* $t \notin F^*(G)$, *and* $F^*(G) \cong He$.

Specifically Fritz proves the following result: Either $G$ is fusion simple with $S$ of type $HS$ or else $G$ possesses a normal subgroup of index 2 and $S_o = S \cap G_o$ has one of the following structures:

(a) $S_o$ is of type $PSp_4(4)$, $L_5(2)$, or $U_5(2)$; or
(b) $S_o$ has order $\leqslant 2^{10}$ and has one of approximately        (2.86)
     four well-specified structures, for which there exists
     no known simple group with any such Sylow 2-subgroups.

For example, one possibility for $S_0$ in (b) is $S_0 = E_1 E_2 \langle v \rangle$, where $E_1 \cong E_2 \cong E_{32}$ with $E_1$, $E_2$ normalizing each other and $E_1 \cap E_2 \cong E_8$, and $v$ is an involution interchanging $E_1$ and $E_2$ (additional details concerning the structure of $S_0$ in (b) appear below). Furthermore, in the published paper, Fritz does not quite reach (a), but is content to invoke the $\leqslant 2^{10}$ theorem to complete the proof [as he is if (b) holds]. However, he points out that from the information he has derived, the precise structure of $S_0$ in each of the three cases of (a) is easily obtained with slight additional work. [We also note that a careful reading of the Harris–Solomon proof would yield a similar (but somewhat longer) list of possibilities for $S$ and $S_0$.]

If $S$ is of type $HS$ or $S_0$ satisfies (2.86)(a), then correspondingly $S$ or $S_0$ is of type $\mathscr{F}$ and so $F^*(G)$ is determined from Theorem 2.1. On the other hand, if (2.86)(b) holds, one must either invoke the $\leqslant 2^{10}$ theorem to complete Theorem 2.160 or else give an independent proof that these cases lead to a contradiction. On the basis of the fusion and 2-local information obtained by Fritz in reaching (b), the latter procedure should not be very difficult; however, as of this writing, it has not been carried out.

We shall describe Fritz's proof very briefly. Again let $R \in \mathrm{Syl}_2(C)$ with $\langle t, u \rangle \leqslant R$ and let $Q_1$ be the pre-image in $R$ of $\bar{R} \cap \bar{L}$, so that $Q_1 \cong D_8$. Also set $R_1 = Q_1\langle t, u \rangle$, so that

$$R_1 = Q_1 \times \langle t, u \rangle \quad \text{and} \quad |R : R_1| \leqslant 2. \tag{2.87}$$

Thus we also have

$$R_1 = A_1 A_2, \text{ where } A_1 \cong A_2 \cong E_{16}. \tag{2.88}$$

Furthermore, if $|R : R_1| = 2$, there is $x \in R$ with $x^2 = 1$ or $t$ such that $Q_1\langle x \rangle / \langle x^2 \rangle \cong D_{16}$. If $R = R_1$, we again set $x = 1$ for completeness.

In the present situation it is no longer possible to establish the analog of (2.72), so that the cases in which $A_1$ and $A_2$ are conjugate or nonconjugate in $G$ require detailed separate analyses.

We set $N_i = N_G(A_i)$ and $N_i^* = N_i/O(N_i)$, $i = 1,2$.

Fritz first proves:

LEMMA 2.162. *If $x = 1$, then one of the following holds*:

(i) $A_1$ *and* $A_2$ *are not conjugate in $G$; or*
(ii) $A_1$ *and* $A_2$ *are conjugate in* $N_G(R)$ *and* $N_i^* \cong \Sigma_4$, $i = 1,2$.

Assuming false, a delicate pushing-up argument involving arbitrarily many steps pins down a Sylow 2-subgroup of $G$, which then turns out to have sectional 2-rank 4, contradiction.

Next by pushing-up an appropriate 2-subgroup of $N_1$, Fritz proves the following further result:

LEMMA 2.163. *If Lemma 2.162(ii) holds, then $G$ has a normal subgroup $G_0$ of index 2 with Sylow 2-subgroup $S_0$ of type (2.86)(b) and order $\leqslant 2^8$.*

Since $|S_0| \leqslant 2^8$, the structure of $G_0$ is determined from the $\leqslant 2^{10}$ theorem (Theorem 2.166 below) and we check that there is no group $G$ containing an involution with a centralizer of the form of $C$. Hence Lemma 2.162(ii) does not hold.

Combined with further fusion analysis, this in turn yields:

LEMMA 2.164.   (i) $N_i^* \cong \Sigma_4 \times Z_2$, $i = 1$ and 2; and

(ii) $A_1$ and $A_2$ are conjugate in $G$ if and only if they are conjugate in $R$ (equivalently if and only if $x \neq 1$).

Now let $T_1 \in \mathrm{Syl}_2(N_G(R_1))$, with $R \leqslant T_1$. Using Lemma 2.164, Fritz determines a subgroup $T$ of $T_1$ of order 32 whose structure dominates the subsequent analysis. $T$ has one of the following forms:

$$
\begin{array}{ll}
\text{(a)} & T \cong E_{32}; \\
\text{(b)} & T \cong Z_4 \times Z_4 \times Z_2; \\
\text{(c)} & T \cong Q_8 * Q_8 \text{ with } \langle u \rangle = Z(T); \text{ or} \\
\text{(d)} & Z(T) \cong E_8.
\end{array}
\qquad (2.89)
$$

Each of these cases requires separate treatment, involving the pushing-up of appropriate 2-subgroups of $T_1$. The final result is as follows:

PROPOSITION 2.165. *According as* (2.89)(a), (b), (c), *or* (d) *holds, we have*

(i) *G has a normal subgroup $G_0$ of index 2 with Sylow 2-subgroups of type $PSp_4(4)$;*

(ii) *G has a normal subgroup $G_0$ of index 2 with Sylow 2-subgroups of type (2.86)(b);*

(iii) *G has Sylow 2-subgroups of type HS; or*

(iv) *G has a normal subgroup $G_0$ of index 2 with Sylow 2-subgroups of type $L_5(2)$ or $U_5(2)$.*

As already remarked, in his paper Fritz does not pin down the exact structure of $S \cap G_0$ in (i) and (iv), but it can be done with some additional work.

Finally if $G$ has a normal subgroup $G_0$ of index 2, Proposition 2.165 implies that a Sylow 2-subgroup of $G$ has order $\leqslant 2^{10}$ and so the structure of $G$ is determined from the classification of such groups. In the contrary case, Proposition 2.165(iii) must hold and so $G \cong HS$ by Theorem 2.51. In either case we obtain one of the alternatives of Theorem 2.161.

The above outline should make clear the further analysis which would be required to make Theorem 2.161 completely independent of the $\leqslant 2^{10}$ theorem itself, which we state in the following way:

THEOREM 2.166. *If $G$ is a simple group of sectional 2-rank $\geqslant 5$ with*

*Sylow 2-subgroups of order $\leqslant 2^{10}$ and class $\geqslant 3$, then $G \cong L_6(q)$, $U_6(q)$, $PSp_6(q)$, $P\Omega_7(q)$, $q \equiv 3, 5 \pmod 8$, $L_5(2)$, $U_5(2)$, $A_n$, $12 \leqslant n \leqslant 15$, He, or .3.*

As previously indicated, the pattern of proof follows that of the sectional 2-rank $\leqslant 4$ analysis. Thus first comes the solvable 2-local case, which is again obtained from the classification of such groups of characteristic 2 type. Hence the analysis again begins with the 2-constrained 2-local case, which was carried out by Beisiegel [8]. This reduces us to the case in which the centralizer $C$ of some involution of $G$ is not 2-constrained. In [33], Fritz covers, in particular, the cases that $\bar{C} = C/O(C)$ has a component $\bar{L}$ with either dihedral or quasi-dihedral Sylow 2-subgroups [whence $\bar{L} \cong L_2(q)$, $L_3(q)$, $U_3(q)$, for suitable odd $q$ or $A_7$ or $M_{11}$] such that $C_{\bar{C}}(\bar{L})$ has cyclic Sylow 2-subgroups. The remaining cases were treated by Fritz [33], Nah [88], and Stingl [115].

Likewise the theorem requires Sylow 2-group characterizations of those listed groups which are not in $\mathscr{F}$: Foote [32], Harris [I:164], Solomon [112], and Yamaki [140].

Next we mention Stroth's extension of Harada's work on self-centralizing $E_{16}$-subgroups. As we have already indicated, it is a long and difficult argument, involving the same techniques as in previous low 2-rank problems, eventually pinning down a Sylow 2-subgroup of the group $G$ under investigation and then invoking appropriate Sylow 2-group characterization theorems. Furthermore, G. Mason has used the result in the course of his analysis of quasi-thin groups of characteristic 2 type [83], although it is clear that this was a convenience rather than a necessity for him, since he had derived considerable additional 2-local information by that point in his argument.

Again we limit ourselves to a statement of the theorem.

THEOREM 2.167. *If $G$ is a simple group of sectional 2-rank $\geqslant 5$ which contains an elementary subgroup $A \cong E_{16}$ such that $A$ is a Sylow 2-subgroup of $C_A$ and $N_G(A)$ is nonsolvable, then $G \cong L_4(q)$, $q \equiv 1 \pmod 8$, $U_4(q)$, $q \equiv -1 \pmod 8$, $L_6(q)$, $q \equiv -1 \pmod 4$, $U_6(q)$, $q \equiv 1 \pmod 4$, $L_5(2)$, $A_n$, $16 \leqslant n \leqslant 19$, $M_{24}$, He, HS, or .3.*

The corresponding result when $N_G(A)$ is solvable appears to be more difficult, even under the assumption that $N_G(A)/C_A$ has order divisible by 3. Note that such a result would yield Theorem 2.161 as a corollary, for in the notation of (2.87), if $E_{16} \cong A \leqslant R_1$, it is indeed the case that $A \in \mathrm{Syl}_2(C_A)$

(as $C_A \leqslant C_t = C$) and that $|N_C(A)/C_C(A)|$ is divisible by 3 [as $N_{\bar{L}}(\bar{A} \cap \bar{L}) \cong \Sigma_4$].

The classification of simple groups of component type depends upon a number of other low 2-rank centralizer of involution problems. However, those that remain all fall within the domain of the standard component theorem, which will be fully outlined in Chapter 5. It will therefore be better to defer discussion of these low 2-rank problems.

In conclusion, we mention two other interesting results of this general nature which were not actually used in the classification theorem itself. First, there is M. Hall's determination of all simple groups of order at most 1,000,000 [50]. Using primarily character theory and Sylow's theorem, Hall was able to force the possible orders of such simple groups together with the possible structures of their Sylow 2-subgroups and then invoke appropriate classification theorems. Of course, the result can now be reduced fairly easily to the $\leqslant 2^{10}$ theorem, for if $|G| \leqslant 1,000,000$ and $G$ has a Sylow 2-subgroup $S$ of order at least $2^{11}$, then

$$|G : S| < 500;$$

in particular, $|G : S|$ is a product of at most five (not necessarily distinct) odd primes, so there are very few cases left to consider.

Finally we mention Stroth's classification of simple groups of 2-rank 3, which he obtained as a corollary of a more general result classifying simple groups in which two distinct Sylow 2-subgroups necessarily intersect in a group of rank $\leqslant 2$ [110]. O'Nan independently treated the special case of simple groups of 2-rank 3 in which every 2-local subgroup is 2-constrained [91].

THEOREM 2.168. *If $G$ is a simple group of 2-rank 3, then $G \cong G_2(q)$ or* $^3D_4(q)$ *$q$ odd,* $^2G_2(3^n)$, *$n$ odd, $n > 1$, $L_2(8)$, $Sz(8)$, $U_3(8)$, $J_1$, $M_{12}$, or ON.*

# 3

# Centralizers of Involutions in
# Simple Groups

The solution of the needed low 2-rank problems brought simple group theory to a new plateau. We were at last in a position to strike out towards higher ground and study centralizers of involutions in groups of arbitrary 2-rank. As already pointed out in the Introduction, the focus for these investigations was the fundamental *B-theorem*:

> For any finite group $X$ with $O(X) = 1$ and any involution
> $t$ of $X$, if we set $C = C_X(t)$ and $\bar{C} = C/O(C)$, then
> $\overline{L(C)} = L(\bar{C})$.

The significance of the *B*-theorem is that it removes, once and for all, the obstruction caused by the cores of centralizers of involutions, for it implies that each component of $\bar{C}$ arises as an image of a component of $C$.

In this chapter, we shall give an historical overview of the existing proof of the *B*-theorem, and in particular, shall present a complete proof in an important special case. We shall also show that the general theorem is closely linked to another fundamental result: the classification of *locally unbalanced* groups. (Recall that a group $X$ is locally unbalanced if $O(C_X(t)) \neq 1$ for some $t \in \mathscr{I}(X)$ [I, 4.55.]) Its statement is as follows:

UNBALANCED GROUP THEOREM. *Let $G$ be a group with $F^*(G)$ simple. If $O(C_t) \neq 1$ for some involution $t$ of $G$, then one of the following holds:*

(i) *$F^*(G)$ is of Lie type of odd characteristic;*
(ii) *$F^*(G) \cong A_n$, $n$ odd, $n \geqslant 7$; or*
(iii) *$F^*(G) \cong L_3(4)$ or He.*

Note that the assumption $F^*(G)$ simple rather than $G$ itself simple allows $t$ to induce either an outer or inner automorphism on $G$. [In fact, $t$ is outer when $F^*(G) \cong L_3(4)$ or $He$.]

The bulk of the chapter will, however, be devoted to an outline of two basic theorems of Aschbacher, which play a central role in the proof of both the $B$-theorem and the unbalanced group theorem—his *component* theorem [I: 11] and his *classical involution* theorem [I: 13]. Even though the primary application of the component theorem occurs after the $B$-theorem is proved, where it is used to reduce the study of groups of component type to the solution of *standard form* problems, it has similar consequences for proper sections of a minimal counterexample to the $B$-theorem inasmuch as such sections have the $B$-property.

The classical involution theorem, on the other hand, gives a characterization of the groups of Lie type over $GF(q)$, $q$ odd, in terms of their generation by $SL_2(q)$-subgroups, that provides an effective tool for studying locally unbalanced groups. Indeed, in a number of situations one is able to force the group $G$ under investigation to possess a "classical involution" and then invoke Aschbacher's theorem to conclude that $F^*(G)$ is of Lie type of odd characteristic.

DEFINITION.   An involution $t$ of a group $G$ is *classical* if $C_t$ possesses a subnormal subgroup $L$ with quaternion Sylow 2-subgroups such that $t \in L$.

The structure of the image $\bar{L}$ of $L$ in $\bar{C}_t = C_t/O(C_t)$ is, of course, given by the dihedral Sylow 2-group classification theorem (Theorem 1.47) as $L/\langle \bar{t} \rangle$ has dihedral Sylow 2-subgroups. Indeed, that result implies that $\bar{L}$ contains a normal subgroup $\bar{L}_0$ with $\bar{L}/\bar{L}_0$ cyclic (of order not divisible by 4) such that

$$\bar{L}_0 \cong SL_2(q), q \text{ odd}, \hat{A}_7, \text{ or } Q_{2^n} \text{ for some } n.$$

The critical fact here is that $\bar{t}$ is the unique involution of $\bar{L}_0$ (as well as of $\bar{L}$). Note also that if $\bar{L}_0 \cong SL_2(q)$, $q > 3$, or $\hat{A}_7$, then $\bar{L}_0$ is quasisimple, in which case the subnormality of $\bar{L}$ implies that $\bar{L}_0$ is a component fo $\bar{C}_t$.

Aschbacher's remarkable theorem, in effect, characterizes the groups of Lie type of odd characteristic from the existence of a *single* classical involution. [The precise result (Theorem 3.64 below) includes a few other groups as well.]

## 3.1. The $B$-Theorem; General Considerations

The $B$-property is by no means an obvious one, for in general there is very little connection between the layer of a group $X$ and the layer of $X/O(X)$. Indeed, if we set $\bar{X} = X/O(X)$, the only assertion we can make in general is

$$\overline{L(X)} \text{ is a product of components of } L(\bar{X}), \tag{3.1}$$

an immediate consequence of the definitions.

We should like to illustrate by an example how great the discrepancy between $\overline{L(X)}$ and $L(\bar{X})$ may be. Let $Y$ be the direct product of $n$ (nonabelian) simple groups, $n$ an arbitrary positive integer, and represent $Y$ faithfully on a vector space over $GF(p)$, $p$ any odd prime (any such representation will do). Form the semidirect product of $V$ by $Y$ (see [I, Section 1.4]). Thus $X = VY$ with $V \lhd X$, $V \cap Y = 1$, and conjugation of $V$ by the element $y \in Y$ is determined by the action of $y$ on $V$ in the given representation. Since $p$ is odd, $V \leqslant O(X)$ and as $X/V \cong Y$ is the direct product of simple groups, it follows that $V = O(X)$. Moreover, as $Y$ acts faithfully on $V$ and $V$ is abelian, also $V = C_X(V)$, so $O(X) = C_X(O(X))$. On the other hand, $[L(X), O(X)] \leqslant L(X) \cap O(X)$ and is normal in $L(X)$. Since it has odd order and hence is solvable, the last inclusion forces $[L(X), O(X)] \leqslant Z(L(X))$, whence $[L(X), O(X), L(X)] = 1$. But $L(X)$ is perfect, so by the three-subgroups lemma [I, 4.19], $[L(X), O(X)] = 1$, whence $L(X)$ centralizes $O(X)$. Since $C_X(O(X)) = O(X)$, we thus conclude that $L(X) = 1$. Therefore $X$ has a trivial layer in this example. On the other hand, $\bar{X} = X/O(X) = X/V = \bar{Y} \cong Y$, and consequently $L(\bar{X}) \cong Y$, the direct product of arbitrarily many simple groups.

We next reformulate the $B$-property in terms of 2-layers [recall that a 2-*component* of a group $X$ is by definition the unique subnormal subgroup of $X$ which is minimal subject to mapping on a component of $X/O_{2'}(X) = X/O(X)$ [I, 4.71] and the 2-*layer* $L_{2'}(X)$ is the product of all 2-components of $X$, with $L_{2'}(X) = 1$ if no such 2 components exist (equivalently if $L(X/O(X)) = 1$)]. In particular, it is immediate from the definition that

$$L_{2'}(X) \text{ maps onto } L(X/O(X)). \tag{3.2}$$

Observe now that if $K$ is a 2-component of $X$ and we set $\bar{X} = X/O(X)$, then $\bar{K}$, being a component of $\bar{X}$, is quasisimple, so $\bar{K} = [\bar{K}, \bar{K}]$. Hence $[K, K]$ also maps on $\bar{K}$; moreover, as $[K, K]$ char $K$, also $[K, K]$ is subnormal in $X$. Hence $K = [K, K]$ by the minimality of $K$, and so $K$ is perfect.

The following lemma gives a condition for a 2-component of $X$ to be a component.

LEMMA 3.1. *A 2-component $K$ of the group $X$ is a component if and only if $K$ is quasisimple; and this is the case if and only if $[K, O(K)] = 1$.*

Indeed, if $K$ is a component of $X$, it is certainly quasisimple. Conversely, if $K$ is quasisimple, then as 2-components are subnormal, $K$ is a component of $X$, so the first assertion is clear. Likewise if $K$ is quasisimple, then $O(K) \leqslant Z(K)$, so $[K, O(K)] = 1$. Conversely, if $[K, O(K)] = 1$, then $O(K) \leqslant Z(K)$, so if $H$ denotes the pre-image of $Z(K/O(K))$ in $K$, we have $[K, H] \leqslant O(K) \leqslant Z(K)$ and as $K$ is perfect, the three-subgroups lemma implies that $K$ centralizes $H$. Thus $H \leqslant Z(K)$ and as $K/H$ is simple, it follows that $K$ is quasisimple.

As an immediate corollary, we have the following result:

PROPOSITION 3.2. $L(X)$ *maps on* $L(X/O(X))$ *if and only if* $L_{2'}(X) = L(X)$.

Now we can restate the $B$-theorem in terms of 2-layers. We give both the core-free and general versions.

$B$-THEOREM (*Core-Free Form*). *If $X$ is a group with $O(X) = 1$, then for every involution $t$ of $X$,*

$$L_{2'}(C_X(t)) = L(C_X(t)).$$

$B$-THEOREM (*General Form*). *For any group $X$ and any involution $t$ of $X$,*

$$[L_{2'}(C_X(t)), O(C_X(t))] \leqslant O(X).$$

Passing to $X/O(X)$ and using Lemma 3.1, one easily shows that the two forms of the theorem are equivalent.

For convenience, if the appropriate form of the theorem holds in a given group $X$ (for every involution $t$ of $X$), we say that $X$ has the *B-property*.

We would now like to describe the general shape of a minimal counterexample.

THEOREM 3.3. *If G is a minimal counterexample to the B-theorem, then $F^*(G)$ is simple of index at most 2 in G. Moreover, every proper section of G has the B-property.*

The second assertion is clear from the definition of a minimal counterexample. In proving the first, we need a general property of 2-components (which follows easily from the definitions together with the three-subgroups lemma).

LEMMA 3.4. *For any group X, we have*

(i) *Every 2-component of X is normal in $L_{2'}(X)$; and*
(ii) *If K, L are distinct 2-components of X, then $[K, L] \leqslant O(K) \cap O(L)$. In particular, if either K or L are components of X, then $[K, L] = 1$.*

Now let $G$ be a minimal counterexample to the $B$-theorem. If $G/O(G)$ has the $B$-property, it is immediate that $G$ does as well, which is not the case. Hence $O(G) = 1$ and so $L = L(G) = L_{2'}(G)$. Likewise as $G$ is a counterexample, there is $t \in \mathcal{I}(G)$ such that $L_{2'}(C_t) \neq L(C_t)$. Hence $C_t$ possesses a 2-component $K$ which is not quasisimple.

We now apply $L$-balance [I, 4.73] to obtain that $L_{2'}(C_t) \leqslant L_{2'}(G) = L$. Hence $K$ is a subnormal perfect subgroup of $C_L(t)$ with $K/O(K)$ quasisimple and so $K$ is a 2-component of $C_L(t)$. Thus the $B$-theorem fails in $L\langle t\rangle$ (as $K$ is not quasisimple), so $G = L\langle t\rangle$ by the minimality of $G$.

Suppose $t$ either centralizes or does not leave invariant a component $J$ of $L$. Then either $J_0 = L\bigl(C_{JJ^t}(t)\bigr)$ is $J$ or $J_0$ is a "diagonal" of $JJ^t$ and hence a nontrivial homomorphic image of $J$. In either case, $J_0$ is quasisimple and subnormal in $C_t$, so $J_0$ is a component of $C_t$. Since $K$ is not quasisimple, $K \neq J_0$, so $K$ centralizes $J_0$ by Lemma 3.4(ii). In particular, $K$ centralizes $J$ if $J_0 = J$. In the contrary case, $K$ leaves $JJ^t$ invariant and hence leaves both $J$ and $J^t$ invariant (as $K$, being perfect, has no normal subgroups of index 2). Since $J_0$ projects onto $J/Z(J)$, it follows that $K$ centralizes $J$ in this case as well. Similarly, $K$ centralizes $J^t$.

But then $K \leqslant H = C_G(JJ^t)\langle t\rangle$, whence $K$ is a nonquasisimple 2-component of $C_H(t)$, forcing $H = G$ by the minimality of $G$. Thus $J \leqslant H$ and hence $J$ centralizes itself, which is absurd as $J$ is quasisimple. We conclude that $t$ centralizes no component of $L$ and that $t$ leaves invariant every component of $L$.

Next set $\bar{G} = G/Z(L)$, so that $\bar{t}$ is an involution. We check that $\bar{K}$ is a 2-

component of $C_{\overline{G}}(\overline{t})$ and again by the three-subgroups lemma, $\overline{K}$ is not quasisimple, so again by the minimality of $G$, we must have $\overline{G} = G$ and $Z(L) = 1$. Thus $L$ is the direct product of simple groups. Furthermore, the preceding argument shows that $K$ centralizes no component of $L$.

Finally we argue that $L$ is simple, which will complete the proof. Suppose false, in which case $L$ is the direct product of simple groups $L_1, L_2, ..., L_n$, $n \geqslant 2$. Let $K_i$ be the projection of $K$ on $L_i$, $1 \leqslant i \leqslant n$ (since $K \leqslant L$, every $x \in K$ can be expressed uniquely in the form $x_1 x_2 \cdots x_n$ with $x_i \in L_i$, and by definition $K_i = \{x_i \mid x \in K\}$). In particular, each $K_i$ is clearly a homomorphic image of $K$, and as $K$ does not centralize any $L_i$, each $K_i \neq 1$. Thus each $K_i$ is perfect with $K_i/O(K_i)$ quasisimple. But $t$ leaves each $L_i$ invariant and so centralizes $K_i$, whence $K_i$ is subnormal in $C_{L_i}(t)$. Since $L_i \langle t \rangle < G$ (as $n \geqslant 2$), it follows that each $K_i$ is quasisimple. But then each $K_i$ is a component of $C_t$ and is distinct from $K$, so $K$ centralizes each $K_i$ by Lemma 3.4(ii). Hence, by definition of the $K_i$, $K$ centralizes itself, a contradiction as $K$ is perfect.

The $B$-conjecture was, of course, motivated by the structure of the centralizers of involutions in the known simple groups (cf. [I, 4.241–4.247]). In fact, combined with the argument of Theorem 3.3, we have the following theorem:

THEOREM 3.5. *If $G$ is a K-group, then $G$ has the B-property.*

Indeed, repeating the proof of Theorem 3.3, a minimal counterexample $G$ to the theorem again satisfies $F^*(G)$ simple. But then as $F^*(G)$ is a $K$-group, [I, 4.244] implies that $G$ has the $B$-property.

Combined with the general form of Aschbacher's theorem [I, 4.28] and the sectional 2-rank $\leqslant 4$ theorem, we obtain as a corollary:

THEOREM 3.6. *If $G$ has a proper 2-generated core or if $G$ has sectional 2-rank $\leqslant 4$, then $G$ has the B-property.*

This in turn yields the following further theorem:

THEOREM 3.7. *If $G$ is a balanced group, then $G$ has the B-property.*

Indeed, we can assume that $G$ is core-free of sectional 2-rank $\geqslant 5$, in which case $O(C_t) = 1$ for every $t \in \mathscr{I}(G)$ by [I, 4.54]. However, it is immediate now from the definition that $L_{2'}(C_t) = L(C_t)$ when $O(C_t) = 1$, so $G$ has the $B$-property in this case.

Thus we have the following result:

THEOREM 3.8. *If $G$ is a minimal counterexample to the B-theorem, then we have*

   (i) *$G$ is an unbalanced group;*
   (ii) *Some element of $\mathscr{L}(G)$ is locally unbalanced; and*
   (iii) *$O(C_t) \neq 1$ for some $t \in \mathscr{I}(G)$.*

First, (i) follows from Theorem 3.7. Clearly (i) implies (iii). Furthermore, (i) and [I, 4.58] imply the crucial conclusion (ii).

We conclude with one further property of a minimal counterexample, related to $L$-balance.

PROPOSITION 3.9. *Let $G$ be a group with $F^*(G)$ simple in which every proper section has the B-property and let $x$, $y$ be commuting involutions of $G$. Let $J$ be a 2-component of $L_{2'}(L_{2'}(C_x) \cap C_y)$ and set $\bar{C}_y = C_y/O(C_y)$. Then we have*

   (i) *$\bar{J}$ is a component of $C_{\bar{C}_y}(\bar{x})$; and*
   (ii) *If $D$ is a $J$-invariant subgroup of $C_x \cap C_y$ of odd order, then $[J, D] \leqslant O(C_y)$.*

Indeed, if $K$ denotes the normal closure of $J$ in $L_{2'}(C_y)$, then by [I, 4.79], $K$ is $x$-invariant and $J$ is a 2-component of $C_K(x)$, whence $\bar{J}$ is a 2-component of $C_{\bar{K}}(\bar{x})$. But as $F^*(G)$ is simple, $\bar{K}\langle \bar{x}\rangle$ is a proper section of $G$ and so has the $B$-property. Hence $\bar{J}$ is quasisimple and so is a component of $C_{\bar{K}}(\bar{x})$. Since $C_{\bar{K}}(\bar{x})$ is subnormal in $C_{\bar{C}_y}(\bar{x})$, (i) follows.

Finally if $D$ is as in (ii), then $\bar{E} = [\bar{J}, \bar{D}] \leqslant \bar{L} = L\big(C_{\bar{C}_y}(\bar{x})\big)$ as $\bar{J} \leqslant \bar{L}$ [by (i)] and $\bar{L} \lhd C_{\bar{C}_y}(\bar{x})$. In fact, $\bar{J} \lhd \bar{L}$ by (i). Since $\bar{E}$ is $\bar{J}$-invariant, it follows that $[\bar{J}, \bar{E}]$ is normal in $\bar{J}$ and of odd order, which, as usual, implies that $[\bar{J}, \bar{E}] = 1$. Hence $[\bar{J}, \bar{D}, \bar{D}] = 1$ and as $\bar{J}$ is perfect, [I, 4.8] now yields that $[\bar{J}, \bar{D}] = 1$. Thus $[J, D] \leqslant O(C_y)$, so (ii) also holds.

## 3.2. A Special Case

Walter and I were the first to investigate the $B$-theorem. We knew [Theorem 3.8(ii)] that locally unbalanced elements of $\mathscr{L}(G)$ were the principal obstruction to its verification. Furthermore, as the typical such locally unbalanced group is a group of Lie type of *odd* characteristic, we first focused on the case in which *every* element of $\mathscr{L}(G)$ is of this type and thus imposed suitable generational and local balance conditions on the elements of $\mathscr{L}(G)$ which were known to hold in the groups of Lie type of odd characteristic (apart from some small exceptions). Eventually (under suitable connectivity conditions), we produced an essentially formal proof of the $B$-theorem for such groups, which could be viewed as an extension of the balanced, connected group argument of [I, 4.54] to the case of 2-balanced groups.

We should like to present in detail a particular case of our analysis, since it provides an excellent illustration of the signalizer functor method and at the same time points up the importance of balance and generation for the study of centralizers of involutions in simple groups. We need two preliminary terms, related to connectivity and generation.

DEFINITION 3.10. A group $X$ will be called *strongly* 3-*connected* provided the following two conditions hold:

(a) If $A, B$ are any two elementary $E_8$-subgroups of $X$, then there exists a sequence $A = A_1, A_2, ..., A_n = B$ of elementary $E_8$-subgroups of $X$ such that $A_i$ centralizes $A_{i+1}$ for all $i$, $1 \leqslant i \leqslant n - 1$; and

(b) Every four-subgroup of $X$ is contained in an $E_8$-subgroup of $X$.

Condition (a) alone defines 3-connectivity. Note also that (a) and (b) together imply that $X$ is necessarily connected.

Furthermore, if $X$ has 2-rank $\geqslant 4$, (a) implies:

(c) Every $E_8$-subgroup of $X$ is contained in an $E_{16}$-subgroup.

Indeed, if $A$ is such an $E_8$-subgroup, then as $m(X) \geqslant 4$, (a) implies that $A$ centralizes and $E_8$-subgroup $A^*$ of $X$ with $A^* \neq A$. But then $AA^*$ is elementary of rank $\geqslant 4$ and (c) follows.

The following, easily verified result shows that these conditions are satisfied by all "suitably large" 2-groups.

LEMMA 3.11. *If $X$ is a 2-group which contains an elementary normal subgroup of order* $\geqslant 32$, *then $X$ is strongly 3-connected.*

DEFINITION 3.12. A group $X$ is said to be *layer* generated provided whenever a noncyclic elementary 2-group $A$ of rank $\leqslant 4$ acts on $K$, we have

$$X = \langle L\big(C_X(B)\big) \mid B \leqslant A, |A : B| \leqslant 2 \rangle.$$

Note that this is a stronger requirement than *hyperplane* generation, which is defined by the condition

$$X = \langle C_X(B) \mid B \leqslant A, |A:B| \leqslant 2 \rangle.$$

Apart from certain families of low Lie rank, the groups of Lie type of odd characteristic are layer generated. However, the remaining known simple groups are not layer generated. Thus layer generation is an odd characteristic phenomenon. Likewise we note that, except for $L_2(q)$ and $SL_2(q)$, $q$ odd, groups of Lie type of odd characteristic are locally 2-balanced [I, 4.61].

We shall prove the following result:

THEOREM 3.13. *Let $G$ be a group of 2-rank $\geqslant 4$ with $F^*(G)$ simple which satisfies the following conditions*:

(a) *A Sylow 2-subgroup of $G$ is strongly 3-connected*;

(b) *Every element of $\mathscr{L}(G)$ is layer generated and locally 2-balanced*; and

(c) *Every proper section of $G$ has the B-property*.

*Then $G$ has the B-property.*

In particular, these conditions are satisfied if every element of $\mathscr{L}(G)$ is of Lie type of odd characteristic (excluding certain small groups), and if, in addition, $G$ has a suitably large Sylow 2-subgroup.

Under these assumptions, it is possible to establish two basic properties of the "extended 2-layer" $L_2^*(C_x)$, $x \in \mathscr{I}(G)$, which we shall refer to as $L^*$-*balance* and $L^*$-*generation* which will allow us to mimic portions of the proof of [I, 4.35 and 4.54] for balanced, connected groups. Recall that by definition [I, 4.80], for any group $X$,

$$L_{2'}^*(X) = L_{2'}(X) \, O(X) \, O\big(C_X(T)\big), \tag{3.2}$$

where $T \in \mathrm{Syl}_2\big(L_{2'}(X) \, O_{2'2}(X)\big)$. (By Sylow's theorem, the definition is independent of the choice of $T$.)

The following properties are immediate.

LEMMA 3.14. *For any group $X$, we have*

(i) $L_{2'}^*(X) \lhd X$;

(ii) $L_{2'}^*(X) = L_{2'}(X)F$, *where F is of odd order and is invariant under a Sylow 2-subgroup of X; and*

(iii) *If K is a 2-component of X, then* $K \lhd L_{2'}^*(X)$.

The importance of the extended 2-layer $L_{2'}^*(X)$ rests on the following basic "core balance" property of arbitrary finite groups $X$ [I, 4.82]:

$$\text{For any 2-subgroup } R \text{ of } X, \; O\big(C_X(R)\big) \leqslant L_{2'}^*(X). \tag{3.3}$$

For brevity, we shall write $L_x$ and $L_x^*$ for $L_{2'}(C_x)$ and $L_{2'}^*(C_x)$, respectively, for any involution $x$ of $G$.

The following proposition incorporates the definition of $L^*$-balance.

PROPOSITION 3.15. *If $x, y$ are any two commuting involutions of $G$, then*

$$L_{2'}^*(L_x^* \cap C_y) \leqslant L_y^*.$$

Indeed, using (3.2) and Lemma 3.14, it is not difficult to show that the left side has the form $KF$, where $K = L_{2'}(L_x \cap C_y)$ and $F \leqslant O\big(C_{C_x}(R)\big)$ for some 2-subgroup $R$ of $C_x \cap C_y$ containing $\langle x, y \rangle$. But by $L$-balance [I, 4.73] $K \leqslant L_y \leqslant L_y^*$. Furthermore, as $x \in R$, $F \leqslant O(C_R)$, so as $y \in R$, $F \leqslant O\big(C_{C_y}(R)\big)$. Therefore $F \leqslant L_y^*$ by (3.3) and so $KF \leqslant L_y^*$, as asserted.

The next result incorporates the definition of $L^*$-generation.

PROPOSITION 3.16. *For any involution $x$ of $G$ and any noncyclic elementary 2-subgroup $A$ of $C_x$ of 2-rank $\leqslant 4$, we have*

$$L_x^* = \big\langle L_{2'}^*\big(C_{L_x}(B)\big) \mid B \leqslant A, |A:B| \leqslant 2 \big\rangle.$$

Indeed, denote the right side by $Y$. Now $O(C_x)$ is generated by its subgroup $C_{O(C_x)}(B)$ by [I, 4.13], while by (3.3) each $C_{O(C_x)}(B) \leqslant L_{2'}^*(C_B)$. Since also $O(C_x) \leqslant L_x^*$, it follows easily that each $C_{O(C_x)}(B) \leqslant L_{2'}^*\big(C_{L_x}(B)\big)$ and consequently $O(C_x) \leqslant Y$. Hence if we set $\bar{C}_x = C_x/O(C_x)$, it will suffice to show that $\bar{L}_x^* = \bar{Y}$ and hence that $\bar{L}_x^* \leqslant \bar{Y}$.

But the "$L^*$-functor" behaves well under homomorphic images by groups of odd order, so that if we set $\bar{L}^* = L_{2'}^*(\bar{C}_x)$, we have

$$\bar{L}_x^* = \bar{L}^* \quad \text{and} \quad \bar{Y} = \big\langle L_{2'}^*\big(C_{\bar{L}^*}(\bar{B})\big) \mid \bar{B} \leqslant \bar{A}, |\bar{A}:\bar{B}| \leqslant 2 \big\rangle. \tag{3.4}$$

Now by hypothesis each element of $\mathcal{L}(G)$ is layer generated and hence so is each component of $\bar{L} = L(\bar{C}_x)$. But $\bar{A}$ acts as a group of permutations of the components of $\bar{L}$, and it can be shown by a rather straightforward argument that $\bar{L}$ itself is layer generated:

$$\bar{L} = \langle L\big(C_{\bar{L}}(\bar{B})\big) \mid \bar{B} \leqslant \bar{A}, |\bar{A} : \bar{B}| \leqslant 2 \rangle. \tag{3.5}$$

It is immediate therefore from (3.4) and (3.5) that $\bar{L} \leqslant \bar{Y}$.

Finally if we let $\bar{T}$ be an $\bar{A}$-invariant Sylow 2-subgroup of $\bar{L}O_2(\bar{C}_x)$, (3.2) implies that $\bar{L}^* = \bar{L}\bar{F}$, where $\bar{F} = O\big(C_{\bar{C}_x}(\bar{T})\big)$. In particular, $\bar{F}$ is $\bar{A}$-invariant and so it, too, is generated by its subgroups $C_{\bar{F}}(\bar{B})$. Hence, to complete the proof, it remains only that each $C_{\bar{F}}(\bar{B}) \leqslant L_{2'}^*\big(C_{\bar{L}^*}(\bar{B})\big)$. But this conclusion is not difficult to establish with the aid of Glauberman's automorphism theorem [I, 4.102].

Now we are ready to prove Theorem 3.13. In view of Theorem 3.5 we can assume that $F^*(G)$ is not a $K$-group. The first step of the argument is directly modeled after that for balanced, connected groups [I, 4.35]. There for any four-subgroup $U$ of $G$, we set $\theta(G; U) = \langle O(C_u) \mid u \in U^\# \rangle$, and argued first that $\theta(G; U) = \theta(G; V)$ for any two four-subgroups $U, V$ of $S$, $S \in \mathrm{Syl}_2(G)$, and then that $\Gamma_{S,2}(G) \leqslant N_G\big(\theta(G; U)\big)$, which as $G$ was core-free, allowed us to conclude that $\theta(G; U)) = 1$. Here we obtain an analogous conclusion with $\theta(G; U)$ replaced by

$$\theta^*(G; U) = \langle L_u^* \mid u \in U^\# \rangle. \tag{3.6}$$

PROPOSITION 3.17. *For any four-subgroup $U$ of $G$, either $\theta^*(G; U) = 1$ or $\theta^*(G; U) \geqslant F^*(G)$.*

Indeed, let $S \in \mathrm{Syl}_2(G)$. We can assume $U \leqslant S$ and we argue first that

$$\theta^*(G; U) = \theta^*(G; V) \tag{3.7}$$

for any four-subgroup $V$ of $S$.

Since $S$ is connected, it suffices to treat the case in which $V$ centralizes $U$. By symmetry, we need only prove that $\theta^*(G; U) \leqslant \theta^*(G; V)$, and this will follow provided for each $u \in U^\#$ we have

$$L_u^* \leqslant \theta^*(G; V). \tag{3.8}$$

But by $L^*$-generation, we have

$$L_u^* = \langle L_{2'}^*(L_u^* \cap C_v) \mid v \in V^\# \rangle, \tag{3.9}$$

while by $L^*$-balance, each term on the right lies in the corresponding $L_v^*$, so $L_u^* \leqslant \langle L_v^* \mid v \in V^\# \rangle = \theta^*(G; V)$, as required.

On the other hand, it is immediate from the definition that for each four-subgroup $U$ of $G$ and each $g \in G$, we have

$$\left(\theta^*(G; U)\right)^g = \theta^*(G; U^g). \tag{3.10}$$

Hence if we set $N = N_G\left(\theta^*(G; U)\right)$, it follows by essentially the same argument as in $[\mathrm{I}, 4.35]$ that

$$\Gamma_{S,2}(G) \leqslant N. \tag{3.11}$$

But as $F^*(G)$ is not a $K$-group, $G$ does not have a proper 2-generated core by Aschbacher's theorem $[\mathrm{I}, 4.28]$, so $\Gamma_{S,2}(G) = N = G$. It follows that $\theta^*(G; U)$ is normal in $G$ and so, as $F^*(G)$ is simple, one of the alternatives of the proposition must hold.

We now use this result to prove an analog in 2-balanced groups of "core-killing" in balanced groups.

PROPOSITION 3.18. *We have* $\Delta_G(U) = 1$ *for every four-subgroup* $U$ *of* $G$.

(Recall that by definition $\Delta_G(U) = \bigcap_{u \in U^\#} O(C_u)$ $[\mathrm{I}, 4.59]$).

Suppose false for some $U$. Without loss $U \leqslant S$. Our goal will be to contradict the previous proposition. First, as $G$ is strongly 3-connected of 2-rank $\geqslant 4$, $U \leqslant A$ for some $E_{16}$-subgroup $A$ of $G$. Since each element of $\mathscr{L}(G)$ is locally 2-balanced, $[\mathrm{I}, 4.64]$ implies that $G$ is 2-balanced. Hence if for $a \in A^\#$, we define

$$\theta(C_a) = \langle C_a \cap \Delta_G(E) \mid Z_2 \times Z_2 \cong E \leqslant A \rangle, \tag{3.12}$$

then $\theta$ is an $A$-signalizer functor on $G$ by $[\mathrm{I}, 4.65]$ and so by the signalizer functor theorem $[\mathrm{I}, 4.38]$, its completion $\theta(G; A) = \langle \theta(C_a) \mid a \in A^\# \rangle$ is of odd order. Since $\Delta_G(U) \neq 1$, it follows that $\theta(G; A) \neq 1$.

Set $M = N_G\left(\theta(G; A)\right)$. Since $\theta(G) = 1$ and $O(G; A)$ is nontrivial of odd order, we have

$$O(M) \neq 1 \quad \text{and} \quad M < G. \tag{3.13}$$

Again without loss we can assume $A \leqslant S$. Since $G$ is 2-balanced and $S$

is 3-connected, it follows by essentially the same argument as in [I, 4.35] that

$$\Gamma_{S,3}(G) \leqslant M.$$

If a classification of groups with a proper 3-generated core existed, we would be done at this point, without recourse to Proposition 3.17. However, as no such classification theorem has been established, we proceed as follows: For $u \in U^\#$, $A \leqslant C_u$ and so by $L^*$-generation,

$$L_u^* = \langle L_2^*\big(C_{L_u^*}(B)\big) \mid B \leqslant A, |A:B| \leqslant 2 \rangle. \qquad (3.14)$$

But each $B \cong E_8$ or $E_{16}$ (as $A \cong E_{16}$) and so each $N_G(B) \leqslant \Gamma_{S,3}(G) \leqslant M$. Hence $L_u^* \leqslant M$ by (3.14) for each $u \in U^\#$ and so

$$\theta^*(G; U) \leqslant M. \qquad (3.15)$$

On the other hand, as $O(M) \neq 1$ and $F^*(G)$ is simple, it is immediate that $F^*(G) \not\leqslant M$. Hence by (3.15), $\theta^*(G; U) \not\geqslant F^*(G)$ and so by the previous proposition $\theta^*(G; U) = 1$. However, this is a contradiction as $1 \neq O(C_u) \leqslant L_u^* \leqslant \theta^*(G; U)$ for $u \in U^\#$.

   *If* $O(C_x) = 1$ for every $x \in \mathcal{I}(G)$ (as is the case in balanced, connected groups), it is obvious that $G$ has the $B$-property [cf. Theorem 3.8(i)]. However, to reach the same conclusion from the weaker condition $\Delta_G(U) = 1$ for every four-subgroup $U$ of $G$ requires a further argument, which depends, in particular, on the fact that every proper section of $G$ has the $B$-property.

   Suppose then that $G$ does not have the $B$-property, so that $L_x = L_{2'}(C_x) \neq L(C_x)$ for some $x \in \mathcal{I}(G)$. Then $L_x$ does not centralize $O(L_x)$ by Lemma 3.1. Now $x \in A$ for some $E_{16}$-subgroup $A$ of $G$. Setting $\bar{C}_x = C_x/O(C_x)$, (3.5) gives

$$\bar{L}_x = \langle L\big(C_{\bar{L}_x}(\bar{B})\big) \mid \bar{B} \leqslant \bar{A}, |\bar{A}:\bar{B}| \leqslant 2 \rangle. \qquad (3.16)$$

However, as $L_x$ does not centralize $O(L_x)$, it follows at once from (3.16) that $K = L_{2'}\big(C_{L_x}(B)\big)$ does not centralize $O(L_x)$ for some $E_8 \cong B \leqslant A$ [as $K$ maps onto $L\big(C_{\bar{L}_x}(\bar{B})\big)$]. But also $O(L_x)$ is generated by its subgroups $C_{O(L_x)}(D)$ as $D$ ranges over the four-subgroups of $B$ by [I, 4.13]. Since $K$ centralizes $B$, it centralizes each such $D$ and hence leaves each $C_{O(L_x)}(D)$ invariant. Since $K$ does not centralize $O(L_x)$, we conclude that

$$F = [C_{O(L_x)}(D), K] \neq 1 \qquad (3.17)$$

for some four-subgroup $D$ of $B$. But as $K$ is perfect, [I, 4.8] implies that

$$F = [F, K].\tag{3.18}$$

We shall argue that $F \leqslant \varDelta_G(D)$, which as $F \neq 1$, will imply that $\varDelta_G(D) \neq 1$, thus contradicting Proposition 3.18. By definition of $\varDelta_G(D)$, it will suffice to show that $F \leqslant O(C_d)$ for each $d \in D^\#$.

For $d \in D^\#$, set $K_d = L_{2'}(C_{L_x}(d))$. Then it is immediate by $L$-balance that $K \leqslant K_d$. Hence if we set $F_d = [O(L_x), K_d]$, it follows that $F \leqslant F_d$. However, as $F^*(G)$ is simple and every proper section of $G$ has the $B$-property, Proposition 3.9(ii) implies that

$$[F_d, K_d] \leqslant O(C_d),\tag{3.19}$$

whence $F = [F, K] \leqslant O(C_d)$, as required. This completes the proof of Theorem 3.13.

Let us continue our analysis of $G$. We have seen in Proposition 3.17 that $\theta^*(G; U)$ is either trivial or contains $F^*(G)$ for any four-subgroup $U$ of $G$; and that argument applies whether or not $G$ has the $B$-property. We can therefore ask, now that we know that $G$ does, in fact, have the $B$-property, whether this condition on $\theta^*(G; U)$ has any further consequences for the structure of the centralizers of involutions in $G$. Indeed, it does!

THEOREM 3.19. *Let $G$ be a group of component type satisfying the hypothesis of Theorem* 3.13. *Among all $x \in \mathscr{I}(G)$, choose $x$ so that a component $L$ of $C_x$ has largest possible order. Then $C_G(L)$ has 2-rank 1 and hence has cyclic or quaternion Sylow 2-subgroups.*

Note that as $G$ has the $B$-property, $L_{2'}(C_y) = L(C_y)$ for any $y \in \mathscr{I}(G)$, and so as $G$ is of component type, it follows that $L(C_y) \neq 1$ for some $y \in \mathscr{I}(G)$. Hence such a component $K$ exists.

The proof is a direct consequence of the following general result.

PROPOSITION 3.20. *Let $G$ be a group of component type with the $B$-property and let $K$ be a component of $C_a$, $a \in \mathscr{I}(G)$, which satisfies the following conditions:*

(a) *$|K|$ is maximal among the components of centralizers of involutions of $G$;*

(b) *Subject to (a), the 2-rank of $C_{C_a}(K)$ is as large as possible; and*

(c) *$K$ has 2-rank $\geqslant 2$.*

*If $B$ is an elementary 2-subgroup of $C_{C_a}(K)$ of maximal rank, then for any $b \in B^\#$, $K$ is a component of $C_b$.*

Indeed, by $L$-balance, $K \leqslant L_b = L(C_b)$ (as $G$ has the $B$-property); and, moreover, if $J$ denotes the normal closure of $K$ in $L_b$, then $J$ is either a single $a$-invariant component of $L_b$ or the product of two such components interchanged by $b$, and in either case $K$ is a component of $L(C_J(a))$. In particular, $K \leqslant J$.

If $J$ is a single component, maximality of $|K|$ forces $K = J$, so $K$ is a component of $C_b$ in this case. Thus we can assume that $J$ is the product of two components $J_1, J_2$ interchanged by $a$, in which case $K$ is a "diagonal" of $J$ and hence isomorphic to a homomorphic image of $J_1$. Again maximality of $|K|$ implies that $|K| = |J_1|$ whence $K \cong J_1$. Thus $J_1, b$ satisfy condition (a) in place of $K, a$.

We shall argue now that $m(C_{C_b}(J_1)) > m(B)$, thus contradicting (b). Clearly we can evaluate the left side in $\bar{C}_y = C_y/O(C_y)$. For simplicity of notation, we replace $\bar{C}_y$ by $C_y$ [thus assuming, in effect, that $O(C_y) = 1$]. Then $J_1 \cap J_2$ is a 2-group invariant under $a$. But $K$ is determined as the image of the map $x \mapsto xx^a$, $x \in J_1$, and as $K \cong J_1$, it follows that $xx^a \neq 1$ for all $1 \neq x \in J_1$. This forces $J_1 \cap J_2 = 1$, otherwise we could take $x$ to be an involution of $J_1 \cap J_2$ centralized by $a$ and get $xx^a = 1$. We thus conclude that

$$J = J_1 \times J_2. \tag{3.20}$$

By the maximality of $B$, we have $a \in B$. Furthermore, $B$ induces a group of permutations of the components of $C_b$ and $B$ centralizes $K$, which immediately implies that $B$ leaves $J$ invariant and $B = E\langle a \rangle$, where $E$ leaves both $J_1$ and $J_2$ invariant. But then $E$ centralizes the projection of $K$ on each $J_i$ and as $K \cong J_1 \cong J_2$, it follows that $E$ centralizes $J_i$ for both $i = 1, 2$. In particular, $J_1$ centralizes $J_2 E$.

Observe next that $J_2 \cap E$ is centralized by $a$ as $E\langle a \rangle = B$ is abelian. But $J_2^a = J_1$ and $J_1 \cap J_2 = 1$ by (3.20), forcing $J_2 \cap E = 1$. It follows therefore that

$$J_2 E = J_2 \times E. \tag{3.21}$$

Thus as $m(J_2) = m(K) \geqslant 2$ and $m(E) = m(B) - 1$, we conclude that

$$m(C_{C_b}(J_1)) \geqslant m(J_2 E) = m(J_2) + m(E) \geqslant 2 + (m(B) - 1) > m(B), \tag{3.22}$$

giving the desired contradiction.

Now suppose the theorem is false, in which case $m(C_G(L)) = m(C_L) \geqslant 2$. Since $x \in C_L$, it follows that $m(C_{C_x}(L)) \geqslant 2$. Now choose an involution $a$ and a component $K$ of $C_a$ to satisfy conditions (a) and (b) of the previous proposition. Maximality of $|L|$ forces $|L| = |K|$ and so as $m(C_{C_x}(L)) \geqslant 2$, (b) implies that $m(C_{C_a}(K)) \geqslant 2$.

If $m(K) = 1$, then by Theorem 1.62, $K = K/O(K) \cong SL_2(q)$, $q$ odd, or $\hat{A}_7$. However, it is immediate then that $\bar{K}$ is not layer generated, contrary to the fact that $\bar{K}$, being an element of $\mathscr{L}(G)$, is layer generated by hypothesis. Thus $m(K) \geqslant 2$ and so the previous proposition is applicable.

Let $B$ be an elementary 2-subgroup of $C_{C_a}(K)$ of maximal rank, so that $B$ is noncyclic, and let $U$ be a four-subgroup of $B$. Then by the proposition, $K$ is a component of $L_u^*$ for each $u \in U^\#$, and consequently by Lemma 3.14(iii), $K$ is normal in each $L_u^*$. Hence

$$K \lhd \langle L_u^* \mid u \in U^\# \rangle = \theta^*(G; U). \tag{3.23}$$

In particular, $\theta^*(G; U) \neq 1$, so by Proposition 3.17, $\theta^*(G; U) \geqslant F^*(G)$ and therefore $F^*(G)$ normalizes $K$. Since $F^*(G)$ is simple, this forces $K = F^*(G)$. It follows that $C_G(K) = C_K \leqslant K$, whence $C_K = 1$ (as $K$ is simple), contrary to the fact that $K$ centralizes $a$.

In effect, Theorem 3.19 expresses a fundamental property of the groups of Lie type of odd characteristic, which Walter and I referred to as *standard form*. We also called $K$ a *standard component* of $C_a$ (or of $G$).

Observe that if we set $H = C_{C_a}(K)$, then $m(H) = 1$ (as $H \leqslant C_K$) and so the structure of $H/O(H)$ is well determined from Theorem 1.62. Since $m(K) \geqslant 2$, $K$ is clearly the unique component of $C_a$ of its isomorphism type, so $K \lhd C_a$ and hence also $H \lhd C_a$. Setting $\tilde{C}_a = C_a/H$, it follows that $\tilde{C}_a \leqslant \text{Aut}(\tilde{K})$ with $\tilde{K} \cong K/Z(K)$, so the structure of $\tilde{C}_a$ is also well determined. Thus the structure of $C_a$ itself is very precisely pinned down. In fact, at this point, we can justifiably assert that $C_a$ "closely resembles" the centralizer of an involution in some "known" simple group.

We see then that the entire thrust of the $B$-property and standard form is to reduce the determination of simple groups (of component type) to the solution of certain well-specified centralizer of involution problems.

Walter and I were unable to push this line of argument much beyond the case covered by Theorems 3.13 and 3.19. In the published version of these results [I, 144], we assumed only the weaker condition of "core-layer" generation (a group $X$ is *core-layer* generated if

$$X = \langle O(C_X(B)) \, L(C_X(B)) \mid B \leqslant A, |A : B| \geqslant 2 \rangle$$

for any elementary 2-subgroup $A$ of rank $\leqslant 4$ which acts on $X$). The advantage of core-layer generation is that it covers most of those groups of Lie type of odd characteristic which are not already layer generated.

With some additional assumptions, it is also possible to extend the arguments to the case in which only *some* of the elements of $\mathscr{L}(G)$ satisfy the hypotheses of Theorem 3.13. Even though elements of $\mathscr{L}(G)$ of Lie type of odd characteristic constitute the primary obstruction to the $B$-theorem and despite the basic simplicity of the proofs of Theorems 3.13 and 3.19, Walter and I were never able to extend the arguments to the general situation. (However, Walter has stuck with this approach, and it is the starting point for his further work on the $B$-theorem.)

## 3.3. Some History

The $B$-theorem has been such a central part of our thinking about simple groups that it will be useful to describe the intricate sequence of investigations which led to its ultimate verification.

The first general result was Aschbacher's fundamental "component theorem" concerning groups of component type which *have* the $B$-property. In this theorem Aschbacher determines the correct general formulation of the standard form theorem of the previous section (Theorem 3.19).

His result is expressed in terms of a certain natural ordering of the components of centralizers of involutions. To deal with 2-components of the centralizers of involutions in groups which do not necessarily have the $B$-property, it is preferable to use a slight modification of his ordering, which applies equally well in his theorem.

For clarity we repeat the general description of *pump-ups* of centralizers of 2-components of centralizers of involutions. (We have already used the result in the previous section.)

PROPOSITION 3.21. *For any group $G$, Let $L$ be a 2-component of $C_x$ for $x \in \mathscr{I}(G)$, let $y$ be an involution of $C_x$, let $K$ be a 2-component of $L_{2'}(C_{LL_y}(y))$, and let $K^* = \langle K^{L_y} \rangle$ be the normal closure of $K$ in $L_y$. Then we have*

(i)  $K^*$ *is either a single $x$-invariant 2-component of $C_y$ or the product of two 2-components of $C_y$ interchanged by $x$; and*
(ii)  $K$ *is a 2-component of $C_{K^*}(x)$.*

The proposition follows at once from [I, 4.77] (the $K$-group assumption of that result is not required for the prime 2). Note also that either $L = LL^y$ or $L$ and $L^y$ are distinct 2-components of $C_x$ interchanged by $y$.

We are here interested in the special case that $y$ centralizes $L/O(L)$.

DEFINITION 3.22. If $y$ centralizes $L/O(L)$, we write $L < K_1$ for any 2-component $K_1$ of $K^*$. Moreover, we extend $<$ to a partial ordering $\ll$: namely, we write $L \ll L^*$ provided there exists a sequence $L = L_1, L_2, ..., L_n = L^*$ of 2-components of centralizers of involutions of $G$ such that for each $i$, $1 \leqslant i \leqslant n - 1$, we have $L_i < L_{i+1}$.

DEFINITION 3.23. A 2-component $L$ of the centralizer of an involution of $G$ will be called *maximal* if whenever $L \ll L^*$, then necessarily

$$L/O(L) \cong L^*/O(L^*). \tag{3.24}$$

In other words, modulo cores, maximal 2-components do not increase properly under any sequence of pump-ups. We denote by $\mathscr{L}^*(G)$ the set of $L/O(L)$ such that $L$ is a maximal 2-component. Thus $\mathscr{L}^*(G) \subseteq \mathscr{L}(G)$.

Note that if we choose $L$ among 2-components of centralizers of involutions of $G$ with $L/O(L)$ of largest possible order, it is immediate that $L$ is a maximal 2-component. Thus if $\mathscr{L}(G) \neq \varnothing$, then $\mathscr{L}^*(G) \neq \varnothing$.

Aschbacher's component theorem is an assertion (under the assumption that $G$ has the $B$-property) concerning the embedding in $G$ of suitable maximal components $L$. Apart from a single exceptional case in which $L$ has 2-rank 1 [whence by Theorem 1.62, $L/O(L) \cong SL_2(q)$, $q$ odd, or $\hat{A}_7$], he proves that $L$ must have the following properties, which form the basis of the general definition of standard component:

DEFINITION 3.24. A quasisimple subgroup $L$ of a group $X$ is called a *standard component* of $X$ (or, for brevity, is *standard* in $X$) provided the following conditions are satisfied:

(1) $H = C_X(L)$ has even order;
(2) $H$ is tightly embedded in $X$ [i.e., $|H \cap H^x|$ is odd for $x \in X - N_X(H)$];
(3) $L$ commutes with none of its $X$-conjugates; and
(4) $N_X(L) = N_X(H)$.

Observe that these conditions are automatically satisfied when $m(L) \geqslant 2$ and $m(H) = 1$, so that the component $L$ of Theorem 3.19 is, in fact, standard in Aschbacher's sense. Here then is the proper general concept of standard form. We shall give a detailed outline of Aschbacher's theorem in the next section.

The crucial condition in the definition of standard component is that of a tightly embedded subgroup. With the proof of the component theorem, Aschbacher introduced this fundamental new idea into finite group theory. The notion made such a deep impression on Thompson that it inspired him to attack the $B$-theorem in full generality.

A minimal counterexample $G$ clearly contains an involution $x$ whose centralizer possesses a *nonquasisimple* 2-component. Thompson denoted the product of such 2-components by $B(C_x)$ (the "bad" 2-components) and chose $x$ so that $B(C_x)$ was suitably maximal. Following Bender, he now attempted to analyze the structure of a maximal subgroup $M$ of $G$ containing $C_x$. As usual in the Bender approach, there are two cases to consider according as $F^*(M)$ is or is not a $p$-group for some odd prime $p$, both of which under this degree of generality are extremely complex. Our interest here is in the $p$-group case, for it soon reduced to another fundamental classification problem, this time related to tightly embedded subgroups.

As we have seen, the basic difficulty with the Bender method in the $p$-group case arises from failure of $p$-stability in $M$, so that Glauberman's $ZJ$-theorem [I, 4.111] is not applicable. In the present situation, Thompson was able to force $M$ to contain a 2-component $J$ which he could identify by means of the quadratic pair classification theorems [I, 4.120–4.123] (since the quadratic pair theorem is not completely settled for $p = 3$, there were certain residual problems in that case): $\bar{J} = J/O(J)$ is of Lie type of characteristic $p$ [excluding the groups $L_2(p^n)$ and, when $p = 3$, $^2G_2(3^n)$]. Now it is known that, these exceptions aside, the groups in Chev($p$) are generated in a canonical way by their $SL_2(p^n)$ subgroups; in particular, they possess a tightly embedded such subgroup (apart from the orthogonal groups, where one must take $SL_2(p^n) * SL_2(p^n)$ to obtain a tightly embedded subgroup). Thompson's aim was to use this fundamental property to produce an involution $t$ of $J$ and a 2-component $I$ of $L_{2'}(C_J(t))$ with $t \in I$ and $I/O(I) \cong SL_2(p^n)$ (a "solvable" 2-component in the case $p^n = 3$) such that $I$ was *tightly embedded* in $M$. Considering the pump-up of $I$ in $C_t$, this would lead to a subgroup $H$ of $C_t$ with $t \in H$ and $H/O(H) \cong SL_2(p^m)$ with the property that $H$ was tightly embedded in $G$. Thus, in effect, his idea was to reduce this phase of the problem to the classification of groups $G$ which possess a *tightly embedded subgroup having quaternion Sylow 2-subgroups*.

Thompson now turned his attention to the latter problem, which he hoped to show characterized the groups of Lie type of odd characteristic, and made significant inroads. In particular, he realized the significance of a certain "torus" associated with such a group $G$, which Aschbacher later came to call the "Thompson subgroup."

Unfortunately at about the same time, Thompson's elaborate reduction to the tightly embedded quaternion problem was encountering seemingly intractible difficulties (in particular, to verification of the tight embedding of $I$ in $M$), which led him to put aside his investigations.

Shortly thereafter, Aschbacher took up the problem of groups containing a tightly embedded subgroup with quaternion Sylow 2-subgroups, but divorced from the $B$-conjecture per se, beginning with Thompson's torus and pushing on with a geometric graph-theoretic approach.

At the same time, during Thompson's extended effort on the $B$-theorem and thereafter, Walter was continuing his attempt to obtain an inductive characterization of the groups of Lie type of odd characteristic in terms of conditions on the structure of the 2-components of centralizers of involutions. (We have described early portions of that effort in the previous section.) In the process of this analysis, he developed some further basic ideas concerning the role of the groups $SL_2(p^n)$ for such a characterization. In particular, his ideas meshed very neatly with the notion of tightly embedded subgroups with quaternion Sylow 2-subgroups and he attempted to combine them in an analysis of groups which possess a tightly embedded subgroup with *cyclic* Sylow 2-subgroups of order at least 4. (Note that in any group $X$, a subgroup of order 2 is automatically tightly embedded, so the restriction on the order here is necessary to make the problem meaningful.) In addition, he realized that the desired inductive characterization itself was somehow related to the classification of locally unbalanced groups.

All these ideas crystalized at the Sapporo, Japan group theory conference in September, 1974. There Walter explained his approaches both to the tightly embedded cyclic Sylow 2-group problem and to the inductive characterization of the groups of Lie type of odd characteristic to Aschbacher and Thompson. Out of these conversations, the full implications of these ideas gradually emerged. It was soon realized that, under the assumption the group $G$ under investigation does not possess a tightly embedded subgroup with quaternion Sylow 2-subgroups, the tightly embedded cyclic Sylow 2-group problem leads in a natural way to the existence of a 2-component $L$ in the centralizer of some involution with $L/O(L) \cong \mathrm{Spin}_7(p^n)$, $p$ odd [by definition $\mathrm{Spin}_m(p^n)$ is the 2-fold cover $P\Omega_m(p^n)/Z_2$ of the orthogonal group $P\Omega_m(p^n)$] and that this Spin group is linked to a certain line of argument concerning components of Lie type of odd characteristic.

But probably more significant was the realization that using results on balanced groups one could reformulate the $B$-theorem itself in terms of the classification of locally unbalanced groups in a manner more amenable to an inductive analysis. Indeed, one can prove the following result:

PROPOSITION 3.25. *If $G$ is a minimal counterexample to the $B$-theorem, then there exists a pair of commuting involutions $x$, $y$ of $G$ and a 2-component $L$ of $C_x$ such that*

(i) *$D = O(C_x) \cap C_y$ and $y$ both leave $L$ invariant;*
(ii) *If we set $\bar{C}_x = C_x/O(C_x)$, then $\bar{D}$ does not centralize $\bar{L}$; and*
(iii) *If we set $\bar{H} = \bar{L}\bar{D}\langle \bar{y} \rangle$ and $\tilde{H} = \bar{H}/C_{\bar{H}}(\bar{L})$, then $\tilde{L}$ is simple, $\tilde{H} \leqslant \mathrm{Aut}(\tilde{L})$, $\tilde{y}$ is an involution, and $1 \neq \tilde{D} \leqslant O(C_{\tilde{H}}(\tilde{y}))$.*

Indeed, by Theorem 3.8(i), $G$ is unbalanced, so by definition $D = O(C_y) \cap C_x \not\leqslant O(C_x)$ for some pair of commuting involutions $x$, $y$ of $G$. But then $\bar{D}$ does not centralize $F^*(\bar{C}_x)$. However, $\bar{D} \leqslant O(C_{\bar{C}_x}(\bar{y}))$, and it follows directly from the $(A \times B)$-lemma [I, 4.9] that $\bar{D}$ centralizes $O_2(\bar{C}_x)$, so $\bar{D}$ does not centralize $L(\bar{C}_x)$. On the other hand, by core balance [I, 4.82] $D \leqslant L_x^*$ and so $D$ leaves each 2-component of $C_x$ invariant. It now follows easily that $\bar{D}$ centralizes every component of $\bar{L}_x = L(\bar{C}_x)$ either moved by $\bar{y}$ or centralized by $\bar{y}$. Thus $\bar{D}$ does not centralize $\bar{L}$ for some 2-component $L$ of $C_x$ left invariant by $y$, so (i) and (ii) hold. Moreover, $\bar{y}$ does not centralize $\tilde{L}$.

Since $\bar{D}\langle \bar{y} \rangle = \bar{D} \times \langle \bar{y} \rangle$, $\tilde{H}$ is a group and as $\bar{L}$ is quasisimple, $\tilde{L}$ is simple and $\tilde{H} \leqslant \mathrm{Aut}(\tilde{L})$. Furthermore, as $\bar{y}$ does not centralize $\tilde{L}$, $\tilde{y}$ is an involution. In addition, $\bar{D} \leqslant O(C_{\bar{H}}(\bar{y}))$, which by general considerations implies that $\tilde{D} \leqslant O(C_{\tilde{H}}(\tilde{y}))$. Since $\tilde{D} \neq 1$, (iii) therefore also holds.

We refer to the triple $(x, y, L)$ as an *unbalancing triple* for $G$.

The key conclusion is (iii), which, in particular, asserts that $\tilde{L}$ is *locally unbalanced*. Of course, as $O(C_y) \neq 1$, $G$ itself is also locally unbalanced, so, in effect, the situation is inductive. Hence with the proper formulation, one should be able to determine inductively the possibilities for $\tilde{L}$. This indicates the importance for the $B$-theorem of obtaining a classification of all locally unbalanced groups. The list of such simple $K$-groups is given in [I, 4.245] and forms the basis for the natural conjecture stated in the introduction of the chapter as the unbalanced group theorem, which for brevity we refer to as the *U-theorem*.

Since the $B$-theorem holds for $K$-groups and since a minimal counterexample $G$ to the $B$-theorem is locally unbalanced with $F^*(G)$ simple, we have the following result:

PROPOSITION 3.26. *The U-theorem implies the B-theorem.*

Observe next that if $G$ is a counterexample to the $U$-theorem, then as with the $B$-theorem, $G$ is not balanced, otherwise [I, 4.54] would imply that $O(C_x) = 1$ for every $x \in \mathscr{I}(G)$, in which case $G$ is obviously locally balanced–contradiction. Thus the proof of Proposition 3.25 applies equally well to a counterexample to the $U$-theorem, and so we conclude:

THEOREM 3.27. *If $G$ is a minimal counterexample to both the B-theorem and U-theorem, then*

   (i)   *$F^*(G)$ is simple and not a K-group;*
   (ii)  *Every proper section of G has the B-property;*
   (iii) *G possesses an unbalancing triple; and*
   (iv)  *If $(x, y, L)$ is an unbalancing triple and we set $\bar{L} = L/O(L)$, then either $\bar{L}/Z(\bar{L}) \in \mathrm{Chev}(p)$, $p$ odd, or $\bar{L}/Z(\bar{L}) \cong A_n$, $n$ odd, $L_3(4)$, or He.*

Sapporo turned out to be a watershed for the analysis of groups of component type. Aschbacher was soon led to the final formulation of his fundamental "classical involution theorem," characterizing all groups of Lie type of odd characteristic (including orthogonal groups) in terms of their $SL_2(q)$-subgroups and covering the classification of groups having a tightly embedded subgroup with quaternion Sylow 2-subgroups as a special case (his theorem will be outlined in Section 3.5–3.8 below). Using Proposition 3.25, Aschbacher's theorem, and an analysis of the $\mathrm{Spin}_7$ configuration, Thompson shortly thereafter produced his beautiful "Notes on the $B$-conjecture" [I:291], which provided a major reduction in the $B$- and $U$-theorems (Theorem 3.30 below). Likewise, using Aschbacher's theorem and analyzing the resulting $\mathrm{Spin}_7$ configuration, Walter gave an inductive characterization of the groups of Lie type of characteristic $p \geqslant 5$ in groups which have the $B$-property [119]. [Extension to the case $p = 3$ is considerably more involved, in part due to the solvability of the group $SL_2(3)$.]

Thompson's result represented the first of the several basic steps which together provide a complete proof of the $B$- and $U$-theorems. For clarity, we list these here as Theorems A, B, C, etc.

THEOREM A. *Let $G$ be a minimal counterexample to the B-theorem and the U-theorem and let $(x, y, L)$ be an unbalancing triple in G. If $L/O(L)$ is of Lie type of odd characteristic $p$, then $L/O(L) \cong L_2(p^n)$ for some n.*

In addition to the classical involution theorem, Thompson's argument depends both on Solomon's prior "intrinsic" $\hat{A}_n$ 2-component theorem [I : 264] (theorem B below) and on certain properties of groups of Lie type of odd characteristic. These latter properties, related to the existence of $SL_2$ subgroups satisfying special conditions (and stated in [I, 4.246 and 4.247]) were actually not verified until shortly after Thompson first put forth his proof (by Burgoyne [I : 51]). In the end, Thompson's "Notes" remained unpublished; but his argument is included in its entirety in Burgoyne's paper.

The Sapporo developments formed the basis for all subsequent work on the $B$- and $U$-theorems, focusing attention on those possibilities for $L/O(L)$ in an unbalancing triple $(x, y, L)$ not covered by Thompson's reduction. However, the key to his argument lay in a special property of the groups of Lie type of odd characteristic which has no analog for the remaining groups. Because of this property, Thompson was able to work entirely within the set of unbalancing triples of $G$ and so could identify every critical 2-component encountered in his analysis by means of Theorem 3.25(iv).

But, in the general situation, one may well lose control of the "unbalancing" condition. Indeed, if $(x, y, L)$ is an unbalancing triple, $x^* \in \mathscr{I}(C_x)$, $K$ is a 2-component of $L_{2'}(C_L(x^*))$ and $K$ pumps-up in $L_{x^*}$ to a single 2-component $L^*$, there is no guarantee of the existence of an involution $y^*$ in $C_{x^*}$ for which $(x^*, y^*, L^*)$ is an unbalancing triple. How then will we identify $L^*/O(L^*)$, for surely our analysis will require its determination?

Here is the information preserved by the pumping-up process:

$$\text{In } \bar{C}_{x^*} = C_{x^*}/O(C_{x^*}), \bar{K} \text{ is a component of } C_{\bar{L}^*}(\bar{x}^*). \qquad (3.25)$$

[This follows from Proposition 3.9(i) as every proper section of $G$ has the $B$-property.] Furthermore, as $L/O(L)$ is a covering group of one of the listed locally unbalanced groups, the possibilities for $\bar{K}$ are determined from the structure of the centralizers of involutions acting on the known group $L/O(L)$.

If $G$ had been taken to be a minimal counterexample to the classification of _all_ simple groups, then $\bar{L}^*$ would be a $K$-group, and the possibilities for $\bar{L}^*$ would be determined by $\bar{K}$. However, as our aim is, as we have said, to establish the $B$- and $U$-theorems as independent results, there is no _a priori_ reason for $\bar{L}^*$ to be a $K$-group. On the contrary, the problem here is to establish this conclusion solely from the existence of the known component $\bar{K}$.

To carry this out in practice, it is clear that one must free oneself from the unbalancing triple situation and simply begin with a 2-component $L$ of $C_x$ for some involution $x$ of $G$ with $L/O(L)$ of some specified isomorphism type and on the basis of this information alone determine the possibilities for $F^*(G)$. In other words, one requires a theorem which asserts that $F^*(G)$ is determined from the existence of an element of $\mathscr{L}(G)$ which is a $K$-group of specified type.

In the course of the analysis of such a theorem, one will still face the problem of determining $\bar{L}^*$ in (3.25) from $\bar{K}$; but if the desired theorem has been formulated in sufficiently broad terms (i.e., if its statement allows the elements of $\mathscr{L}(G)$ to range over a suitably broad collection of quasisimple $K$-groups), then the structure of $\bar{L}^*$ will be determined by induction.

This is precisely the procedure which Solomon followed in the $\hat{A}_n$ situation (and likewise Walter in his inductive characterizations of the groups of Lie type of odd characteristic). If $L/O(L) \cong \hat{A}_n$, then $\bar{K} \cong K/O(K) \cong \hat{A}_m$ for some $m \leqslant n$, and Solomon was able to conclude by induction that either $\bar{L}^* \cong \hat{A}_r$, $r \geqslant m$, or else $\bar{L}^* \cong Mc$ or $Ly$ (with correspondingly $m = 8$ or $11$).

Ultimately, as we shall see, the inductive analysis in such 2-component problems reduces to the solution of appropriate *standard form* problems [specified by the allowed possibilities for the elements of $\mathscr{L}(G)$ in the given problem]. In the $\hat{A}_n$ situation, it is not difficult to show that these in turn reduce to [I, 2.32, 2.33]: the classification of groups having $\hat{A}_n$ as exact centralizer of an involution. Therefore, when Solomon began his work on 2-components of $\hat{A}_n$ type, the $\hat{A}_n$ standard form problem had already essentially been solved.

On the other hand, if we consider the case $L/O(L) \cong A_n$ [whence $\bar{K} \cong A_m$, $m \leqslant n$, in (3.25)], we have not yet described even a partial solution of the corresponding $A_n$ standard form problem. Hence this problem must be studied before we can hope to determine all groups in which the centralizer of some involution has a 2-component of type $A_n$.

Thus the discussion clearly indicates that a complete proof of the $B$- and $U$-theorems ultimately depends on the solution of certain standard form problems related to the possibilities for $L/O(L)$ in an unbalancing triple $(x, y, L)$. The precise list will be stated in Section 4.3. However, we note that a certain "closure" process is implicitly at work here. For example, beginning with $\hat{A}_n$ type 2-components, Solomon is led in the course of his analysis to consider 2-components of type $Mc$ and $Ly$, and he must also be able to identify *their* possible pump-ups to complete the argument. To carry this out, he is then forced to solve the corresponding $Mc$ and $Ly$ standard form problems. Thus, starting with a given family of 2-components, other quasisimple groups get "picked up" along the way and must be added to the list to obtain a theorem amenable to an inductive analysis. This closure condition will therefore be reflected in the exact list of standard form problems required for the $B$- and $U$-theorems.

For brevity, we shall refer here to the solution of this specified list of standard form problems as *Hypothesis S* (see Section 4.3 for the exact statement).

Even though some of these problems were solved either prior to or within the framework of the $B$- and $U$-theorems, for the sake of clarity we prefer to defer all discussion of standard form problems until Chapter 5, where we shall treat them systematically. In particular, for consistency we now state Solomon's $\hat{A}_n$ 2-component theorem only under Hypothesis $S$. We need a preliminary definition.

DEFINITION 3.28. Let $t \in \mathscr{I}(G)$ and let $J$ be a 2-component of $C_t$. Then $J$ is said to be *intrinsic* if $t \in J$.

Note that if we set $\bar{C}_t = C_t/O(C_t)$, the given condition is equivalent to the assertion $\bar{t} \in Z(\bar{J})$; in particular, $\bar{J}$ is not simple. Also, in the special case $\bar{J} \cong SL_2(q)$, $q$ odd, $\langle \bar{t} \rangle = Z(\bar{J})$, so that $t$ is then a classical involution.

THEOREM B. *Assume Hypothesis $S$, and let $G$ be a group with $F^*(G)$ simple. If for some involution $t$ of $G$, $C_t$ contains an intrinsic 2-component $J$ with $J/O(J) \cong \hat{A}_n$, $n \geqslant 8$, then $F^*(G) \cong Mc$ or $Ly$ (and correspondingly $n = 8$ or $11$).*

The reason for the restriction $n \geqslant 8$ is that for $n = 5, 6$, or $7$, $J$ has quaternion Sylow 2-subgroups, in which case $F^*(G)$ is already determined by Aschbacher's classical involution theorem.

Furthermore, as remarked above, Theorem B is used in Thompson's proof of Theorem A and, in addition, Solomon establishes Theorem B without the assumption of Hypothesis $S$.

Solomon's argument is based on Goldschmidt-type signalizer functors, which enable him to show that the hypotheses of either Aschbacher's 2-*component fusion* theorem, a key subsidiary result of his component theorem (Theorem 3.39 below), or his strong embedding criterion [I, 4.31] are satisfied. In addition, the classification of groups with Sylow 2-subgroups of type $Mc$ or $Ly$ (Theorems 2.44 and 2.45) enter at one point in the argument.

All this having been said, let us return to the $B$- and $U$-theorems. Having settled the $\hat{A}_n$ 2-component problem, it was natural for Solomon to take up the case of unbalancing triples $(x, y, L)$ with $L/O(L) \cong A_n$, $n$ odd. The case $n = 7$ presents special difficulties and is best treated along with the case $L/O(L) \cong L_2(q)$, $q$ odd, inasmuch as $A_7$ and $L_2(q)$ have dihedral Sylow 2-subgroups. Since also $A_5 \cong L_2(5)$, Solomon thus restricted himself to the case $n \geqslant 9$ [I : 266].

Again his arguments involve signalizer functors (so also do those of the classical involution theorem). Thus, after the Sapporo conference, it can be asserted that the signalizer functor method replaced the Bender method as the principal underlying tool for investigating centralizers of involutions in simple groups.

For simplicity, we state only the consequences of Solomon's $A_n$ 2-component results for the $B$- and $U$-theorems. Indeed, combined with Theorems A and B, they imply the following:

THEOREM C. *Assume Hypothesis S. If $(x, y, L)$ is an unbalancing triple in a minimal counterexample $G$ to the B-theorem and U-theorem, then $L/O(L) \cong L_2(q)$, q odd, $A_7$, $L_3(4)$, He, or a covering group of $L_3(4)$.*

To establish Theorem C, Solomon was forced to extend a certain line of argument he had developed for maximal 2-components of type $\hat{A}_n$ to the corresponding $A_n$ case. However, he soon realized that the required analysis had little to do with alternating 2-components per se, but instead could be formulated as a completely general result about maximal 2-components. This led to his fundamental *maximal 2-component* theorem [I : 265] (Theorem 4.9 below), which asserts under suitable conditions that the centralizer of any involution of $G$ possesses at most *one* maximal 2-component of a given isomorphism type (mod core). In effect, the proof amounts to a reduction, with the aid of signalizer functor theory, to Aschbacher's 2-component fusion theorem.

Not only was the maximal 2-component theorem basic for Solomon's contributions to the $B$- and $U$-theorems, but it was also used crucially by Foote [I : 104] in his analysis of maximal 2-components of type $L_2(q)$ or $A_7$, one of the later steps in the proof of the $B$- and $U$-theorems.

The residual cases left by Theorem C are closely intertwined, as the following facts show:

(a) The centralizer of a field automorphism of $L_3(4)$ of period 2 is $L_2(7)$ $(\cong L_3(2))$.

(b) *He* admits an outer automorphism of period 2 whose centralizer is a covering group of $\Sigma_7$ by $Z_3$.

(c) *He* contains an involution whose centralizer is a covering group of $L_3(4)$ by $Z_2 \times Z_2$.

It should also be noted that the 2-part of the Schur multiplier of $L_3(4)$ is $Z_4 \times Z_4$ [I, table 4.1], so $L_3(4)$ possesses several nonisomorphic covering groups.

Gilman and Solomon [34], on the one hand, and Griess and Solomon [48], on the other, investigated some aspect of the remaining cases of the $B$- and $U$-conjecture. As in the alternating group case, they were forced into consideration of 2-components which were not part of any unbalancing triple.

In analyzing this possibility, Gilman and Solomon were led to introduce the notion of a *maximal* unbalancing triple $(x, y, L)$ and to determine conditions that forced $L$ to be maximal in the sense of Definition 3.23. Although the notion is directly connected to the pumping-up process, it is possible to define maximal unbalancing triples without considering this ordering, using certain of their characteristic properties:

DEFINITION 3.29. An unbalancing triple $(x, y, L)$ of $G$ is said to be *maximal* provided the following two conditions hold for any unbalancing triple $(x^*, y^*, L^*)$ for which $x^* \in \mathscr{I}(C_x)$, $x^*$ centralizes $L/O(L)$, and $L^*$ is a 2-component of the normal closure of $L_{2'}\big(C_L(x^*)\big)$ in $L_{x^*}$:

(a) $L/O(L) \cong L^*/O(L^*)$; and
(b) If $S \in \mathrm{Syl}_2\big(N_{C_x}(L)\big)$, and $x^*$ is an involution of $Z(S)$ centralizing $L/O(L)$, then $S \in \mathrm{Syl}_2\big(N_{C_{x^*}}(L^*)\big)$.

In particular, condition (a) implies that $L$ does not pump-up properly within other unbalancing triples; likewise condition (b) implies that a Sylow 2-subgroup of the normalizer of $L$ does not "pump-up" properly.

We emphasize that if $(x, y, L)$ is a maximal unbalancing tiple, $L$ need not be a maximal 2-component of $G$ in the sense of Definition 3.23, for, in general, the unbalancing property may be lost under the pumping-up process.

Ultimately, using a combination of fusion analysis and signalizer functor theory, their combined arguments yielded the following result:

THEOREM D. *Assume Hypothesis S. If G is a minimal counterexample to the B-theorem and U-theorem, then G possesses a maximal 2-component L with $L/O(L) \cong L_2(q)$, q odd, or $A_7$.*

Thus at this point the $B$- and $U$-theorems were reduced to the independent problem of determining all groups in which the centralizer of some involution has a maximal 2-component of $L_2(q)$ or $A_7$ type. Although Foote's precise result involves an inductive assumption on proper sections, we can avoid making it explicit by limiting ourselves to its consequences for the $B$- and $U$-theorems. Again the argument involves a combination of signalizer functor theory and fusion analysis (plus, as we have said, Solomon's maximal 2-component theorem).

THEOREM E. *Assume Hypothesis S. If G is a minimal counterexample to the B-theorem and U-theorem, then G possesses a 2-component L such that*

(i) $L/O(L) \cong L_2(q)$, q odd, or $A_7$; and
(ii) $C_G\big(L/O(L)\big)$ has 2-rank 1.

Using the Fritz–Harris–Solomon theorem (Theorem 2.160) Aschbacher's tightly embedded quaternion subgroup theorem (Theorem 3.65 below), and Harada's self-centralizing $E_8$ theorem (Theorem 2.156) together with a short fusion argument, it is not difficult to reduce the problem one step further to the following bedrock situation:

THEOREM F. *Assume Hypothesis S. If G is a minimal counterexample to the B-theorem and U-theorem, then G possesses a 2-component L such that*

(i) $L/O(L) \cong L_2(q), q \equiv \pm 1 \pmod{16}$;
(ii) $C_G\big(L/O(L)\big)$ has Sylow 2-subgroups of order 2; and
(iii) $N_G(L)$ contains an involution u which induces a nontrivial field automorphism on $L/O(L)$.

One would think that surely by this point we must be very close to completion of the proofs of the two theorems (modulo Hypothesis $S$), but this turns out to be slightly deceptive. Indeed, in order to reach a fusion-theoretic contradiction from these conditions, it is necessary to identify the pump-up $L^*$ in $C_u$ of $K = L_{2'}\big(C_L(u)\big)$. [Note that as $q \equiv \pm 1 \pmod{16}$, $C_L(u)$ is nonsolvable, so that $K \neq 1$; more precisely, $K/O(K) \cong L_2(q^{1/2})$.] Since the possibilities for $L^*$ include many groups of Lie type of odd characteristic, in order to identify $L^*$ one is unfortunately first forced to establish a broad inductive characterization of the groups of Lie type of odd characteristic.

It was Harris who studied this minimal situation, first deriving the required inductive characterization theorem and then using it to obtain the desired fusion contradiction. Thus, finally his results yield our elusive objective.

MAIN REDUCTION THEOREM. *The B-theorem and U-theorem both hold under Hypothesis S.*

As we have stated above, the precise formulation of Hypothesis $S$ will be given in Section 4.3 and an outline of its verification in Chapter 5.

We add a few words about Walter's approach to the $B$- and $U$- theorems, which places greater emphasis on the inductive characterization of the groups of Lie type of odd characteristic [126, 128, 129]. By making this characterization explicit at the beginning of his analysis, he is able to incorporate the work of Gilman and Solomon, Griess and Solomon, Foote, and Harris within a single framework. However, it is a very long and intricate argument; so even though it has the advantage of a uniform treatment, it unfortunately does not provide a major simplification of their contributions to the proof of these theorems.

There in brief is the tortuous path that has been followed in pursuit of the fundamental $B$-property of the centralizers of involutions in finite groups. A detailed outline of the main reduction theorem will be given in Chapter 4.

Finally it should perhaps be noted that both the 2-rank $\leqslant 2$ and sectional 2-rank $\leqslant 4$ theorems involve solutions of several standard form problems, so that, theoretically at least, one could organize their proofs as reductions to an enlarged Hypothesis $S$, verification of which could likewise be deferred to Chapter 5. From this point of view, the noncharacteristic 2 theorem itself would become an elaborate reduction to the soluion of standard form problems. Even though this approach has the advantage of consistency, it would force a rather cumbersome exposition as well as tend to distort the historical perspective. The observation itself is worth making, but it seems definitely preferable to have these two theorems at one's disposal before undertaking a general study of centralizers of involutions in arbitrary groups of component type.

Perhaps one other historical comment should be added. By the time of the Sapporo conference (however, in no way specifically connected with it), there also developed the feeling that any "explicit configuration" arising in a general classification problem "can be handled." It was this overriding confidence that sustained most of us throughout the extended effort each such general classification theorem entailed. The excitement tended to arise instead from the very inductive process that reduced each problem to a specific set of minimal configurations. Elimination of the latter was viewed as somehow more routine, and no matter how long and painful their analyses, there was a sense of inevitable ultimate success. Of course, in the few instances in which a configuration led, not to a contradiction, but to a new simple group, excitement was quickly rekindled. (There were false hopes as well—the light of day transforming the dream of many a new simple group into just one more impossible configuration.)

## 3.4. Aschbacher's Component Theorem

Shortly after Aschbacher established the component theorem, Gilman [I:107] reorganized the proof somewhat (at the same time extending it to certain types of "2-constrained" components). Gilman's treatment, focusing as it does on "nonembedded" components, allows for a slightly more conceptual exposition and so we shall follow it here.

Whereas Aschbacher's formulation is in terms of arbitrary elements of $\mathscr{L}^*(G)$, Gilman limits himself to such elements which are maximal in a suitable further ordering. (This ordering is similar to that used by Walter and me in the course of the proof of Theorem 3.19.)

DEFINITION 3.30. For any group $G$, if $x, y \in \mathscr{I}(G)$ and $L, L^*$ are 2-components of $C_x$, $C_y$, respectively, with $L/O(L^*)$ both in $\mathscr{L}^*(G)$, we shall write $L <^* L^*$ provided the following conditions hold:

(a) $L/O(L) \cong L^*/O(L^*)$;
(b) $m(C_{C_y}(L^*/O(L^*))) \geqslant m(C_{C_x}(L/O(L)))$; and
(c) If equality holds in (b), then a Sylow 2-subgroup of $C_{C_y}(L^*/O(L^*))$ has order at least as great as one of $C_{C_x}(L/O(L))$.

2-components maximal in this ordering will be called *strongly maximal*.

Clearly, for every $K \in \mathscr{L}^*(G)$, there exists a strongly maximal 2-component $L$ with $L/O(L) \cong K$.

In this terminology Aschbacher's component theorem assumes the following form:

THEOREM 3.31. *Let $G$ be a group of component type with the B-property in which $F^*(G)$ is quasisimple. Then one of the following holds for any strongly maximal component*:

(i) $L$ is standard in $G$; or

(ii) $L$ has quaternion Sylow 2-subgroups and $G$ possesses a classical involution.

Aschbacher's actual result is somewhat stronger: he obtains the same conclusions for any $L$ for which $L/O(L) \in \mathscr{L}^*(G)$ and, in addition, establishes a more precise result when $m(L) = 1$. Moreover, subsequently Foote [31] classified all groups satisfying these latter conditions. However, in view of Aschbacher's classical involution theorem, the present version is entirely sufficient for all applications.

Gilman's proof divides into two parts, based on the idea of a nonembedded component. Although the definition is easily extended to arbitrary groups, we limit it to our group $G$, which by assumption has the $B$-property.

DEFINITION 3.32. A component $L$ of $C_x$, $x \in \mathscr{I}(G)$, will be called *nonembedded* in $G$ if for every $y \in \mathscr{I}(C_L)$, $L$ is a component of $C_y$.

Gilman proves the following two results:

PROPOSITION 3.33. *Every strongly maximal component is nonembedded in $G$.*

PROPOSITION 3.34. *If $L$ is a nonembedded component of $G$, then either*

(i) $L$ is standard; or

(ii) $m(L) = 1$ and the unique involution of $L$ is classical.

Thus together the two propositions will yield the component theorem.

The proof of Proposition 3.33 is rather technical, but basically straightforward (and similar to that of Proposition 3.20). Assuming false, the idea is to contradict the strong maximality of $L$.

By assumption $L$ is a component of $C_x$ for some $x \in \mathscr{I}(G)$ with $x$ and $L$ satisfying the conditions of strong maximality. Let $T \in \mathrm{Syl}_2(C_L)$ with $C_T(x) \in \mathrm{Syl}_2(C_{C_x}(L))$. Since we are arguing by contradiction, $L$ is not a component of $C_y$ for some $y \in \mathscr{I}(C_L)$, and we choose $y$ so that $R = C_T(y)$ has largest possible order.

If $y \in \mathscr{I}(Z(T))$, then, of course, $y$ centralizes $x$. In the contrary case, $L$ is a component of $C_z$ for every $z \in \mathscr{I}(Z(T))$ and so $x \in Z(T)$ by definition of strong maximality. Thus $y$ centralizes $x$ in this case as well. Since $L/O(L) \in \mathscr{L}^*(G)$, it follows therefore from Proposition 3.21 that $\langle L^{L_y} \rangle = K \times K^x$ with $K \cong L$. Note also that $R$ leaves $KK^x$ invariant as it centralizes the diagonal $L$ of $KK^x$. Thus a subgroup $Q$ of index at most 2 in $R$ leaves $K$ and $K^x$ invariant and so $Q$ centralizes both $K$ and $K^x$. Thus

$$K \text{ centralizes } K^x Q \text{ and } [K^x, Q] = 1. \tag{3.26}$$

One argues now on the basis of (3.26) that $K$ is greater in the ordering $<^*$ than $L$, contrary to the strong maximality of $L$.

Proposition 3.34 is more elaborate. However, we can quickly dispose of two important subcases.

LEMMA 3.35. *Proposition* 3.34 *holds if either* $m(L) = 1$ *or* $L$ *commutes with none of its G-conjugates.*

Indeed, if $m(L) = 1$ and $t$ is the unique involution of $L$ [by Theorem 1.62, $L/O(L) \cong SL_2(q)$, $q$ odd, or $\hat{A}_7$, and $\langle t \rangle = Z(L)$], then $t$ centralizes $L$ and as $L$ is nonembedded, $L$ is a component of $C_t$. Hence by the definition, $t$ is a classical involution.

On the other hand, if $L$ commutes with none of its $G$-conjugates, we argue that $L$ is standard in $G$. Set $H = C_L$. Since $L$ is a component of $C_x$ for some $x \in \mathscr{I}(G)$, $x \in H$ and so $|H|$ is even. Since $C_H \leqslant C_x$, $L$ is a component of $C_H$. But $N_G(H)$ normalizes $L(C_H)$ [as $L(C_H)$ char $C_H \lhd N_G(H)$], and as $L$ commutes with none of its $G$-conjugates, it follows that $N_G(H)$ leaves $L$ invariant, whence $N_G(H) \leqslant N_G(L)$. Since $N_G(L)$ normalizes $C_L = H$, we conclude that $N_G(H) = N_G(L)$.

It thus remains to show that $H$ is tightly embedded in $G$. Suppose then that $|H \cap H^g|$ is even for $g \in G$ and let $y$ be an involution of $H \cap H^g$. Then $y$ centralizes both $L$ and $L^g$ and as each is nonembedded, it follows that each is a component of $C_y$. But then as $L$ commutes with none of its $G$-conjugates, we must have $L = L^g$, whence $g \in N_G(L) = N_G(H)$. Hence by the definition, $H$ is tightly embedded in $G$ and therefore $L$ is standard, as claimed.

Thus, in proving Proposition 3.34, we may assume

$$m(L) \geqslant 2 \text{ and } L \text{ commutes with some } G\text{-conjugate.} \tag{3.27}$$

Set $L = L_1$ and let $L_2, L_3, ..., L_r$ be the distinct $G$-conjugates of $L$ that commute with $L$. Each $L_i$ is nonembedded in $G$ as it is $G$-conjugate to $L$, $2 \leqslant i \leqslant r$. Hence by $L$-balance each is a component of $C_L$. Since the $L_i$ are distinct, it follows that $[L_i, L_j] = 1$ for $2 \leqslant i$, $j \leqslant r$, $i \neq j$. Since also $[L, L_j] = 1$ for $2 \leqslant j \leqslant r$, we thus conclude:

LEMMA 3.36. $L_1 L_2 \cdots L_r$ *is a semisimple group.*

We now set $M = N_G(L_1 L_2 \cdots L_r)$ and note that as $F^*(G)$ is quasisimple by hypothesis, $M < G$. We next establish three basic properties of $M$.

LEMMA 3.37. *Let* $t \in \mathscr{I}(G)$ *and* $g \in G$. *Then we have*

(i) *If* $t$ *centralizes* $L_i$ *and* $t^g$ *centralizes* $L_j$ *for some* $i, j$, $1 \leqslant i, j \leqslant r$, *then* $g \in M$;

(ii) *If* $t$ *centralizes* $L_i$ *and* $t^g \in M$, *then one of the following holds:*
   (1) $r = 2$;
   (2) $t^g$ *leaves each* $L_j$ *invariant,* $1 \leqslant j \leqslant r$; *or*
   (3) $g \in M$.

(iii) *If* $L_i^g \leqslant M$ *for some* $i$, $1 \leqslant i \leqslant r$, *then* $g \in M$.

PROOF. If $t$ and $t^g$ are as in (i), then $L_i^g$ and $L_j$ are components of $C_{t^g}$ (as each in nonembedded), so either $L_j = L^g$ or $L_j$ centralizes $L^g$. Since $L_j$ and $L^g$ are $G$-conjugate, it follows from the definition of $L_1, L_2, ..., L_r$ that

$$L_j = L_k^g \text{ for some } k, \ 1 \leqslant k \leqslant r. \tag{3.28}$$

Set $K = L_1 L_2 \cdots L_r$. By $L$-balance and nonembeddedness, it is immediate that $K$ is a product of components of $N_G(L_j)$. Similarly $K^g$ is a product of components of $N_G(L_k^g)$ and hence of $N_G(L_j)$. Hence, distinct components of $KK^g$ centralize each other. However, as $K$ includes every $G$-conjugate of $L$ that centralizes $L$, this forces $K^g = K$, whence $g \in N_G(K) = M$, proving (i).

Note that as an immediate corollary of (i), we obtain

$$C_y \leqslant M \text{ for every } y \in \mathscr{I}(C_{L_i}) \text{ and all } i, \ 1 \leqslant i \leqslant r. \tag{3.29}$$

Suppose in (ii) that $r \geqslant 3$ and that $t^g$ interchanges, say, $L_1$ and $L_2$. Set $I = L\big(C_{L_1 L_2}(t^g)\big)$, so that $I$ is a diagonal of $L_1 L_2$, and let $u \in \mathscr{I}(I)$. Then $u$ centralizes $L_3$. We shall argue that $u^{g^{-1}}$ centralizes $L_k$ for some $k$, $1 \leqslant k \leqslant r$, in which case $g^{-1}$ and hence $g$ is in $M$ by (i), and this will prove (ii).

Now as $t^g \in M$, $t^g$ leaves $L_3 L_4 \cdots L_r$ invariant, so $t^g$ centralizes some involution $y$ of $L_3 L_4 \cdots L_r$. By $L$-balance and nonembeddedness, $L_1, L_2$ are components of $C_y$. Hence again by Proposition 3.21, the normal closure $J$ of $I$ in $L_{tg}$ is either a single $y$-invariant component or the product of two components interchanged by $y$. On the other hand, $t$ centralizes some $L_i$ by hypothesis, so $L_i^g$ is a component of $L_{tg}$. If $L_i^g \not\leqslant J$, then $L_i^g$ centralizes $J$ and hence $I$. Since $u \in I$, $u^{g^{-1}}$ thus centralizes $L_i$, as required. On the other hand, if $L_i^g$ is a component of $J$, then so is $L_i^{gy}$. Since $L_i^g$ and $L_i^{gy}$ are either equal or centralize each other, it follows that $J = L_i^g L_i^{gy}$ is a product of one or two components of $K^g$. Since $r \geqslant 3$, $J$ and hence $I$ therefore centralizes some component $L_k^g$ of $K$. Thus $u^{g^{-1}}$ centralizes $L_k$ and again we are done. Hence (ii) also holds.

Finally assume (iii) is false, so that $L_i^g \leqslant M$ for some $g \in G - M$ and some $i$, say $i = 1$. Thus $L^g \leqslant M$. Consider first the case that $L^g$ does not leave $L$ invariant. Since $L^g$ is perfect, clearly then $r \geqslant 3$. Let $y \in \mathscr{I}(L^g)$. Then $t = y^{g^{-1}} \in L$ centralizes $L_2$ and $y = t^g \in M$. Since $r \geqslant 3$ and $g \notin M$, it follows from (ii) that $y$ leaves $L$ invariant. Hence $y$ centralizes some involution $u$ of $L$. Furthermore, as $y$ centralizes $L_2^g$, $C_y \leqslant M^g$ by (3.29), applied to $M^g$. Thus

$$u \in M^g. \tag{3.30}$$

Now let $v$ be any element of $L^g$ which does not leave $L$ invariant. We argue that $v$ centralizes $u$, so assume false and set $z = [v, u]$. Then $z = v^{-1}u^{-1}vu = u^v u$ (as $u$ is an involution) and $L^v$ centralizes $L$, so $z$ is an involution of $LL^v$. In particular, as $r \geqslant 3$, $z$ centralizes some $L_j$. On the other hand, $z = v^{-1}v^u \in L^g L^{gu}$ and as $u \in M^g$, $L^g$ and $L^{gu}$ are each components of $K^g$, so $z$ centralizes some $L_k^g$ and consequently $z^{g^{-1}}$ centralizes $L_k$. But then $g^{-1}$ and hence $g$ is in $M$ by (i), which is not the case. Thus $v$ centralizes $u$, as asserted.

Since this conclusion holds for all $v \in L^g$ which do not leave $L$ invariant and since $L^g$ is perfect, it follows that $L^g$ centralizes $u$, whence $u^{g^{-1}}$ centralizes $L$. But $u \in L$ centralizes $L_2$, so another application of (i) yields that $g \in M$—again a contradiction.

Therefore $L^g$ leaves $L$ invariant. Let $R \in \mathrm{Syl}_2(L^g)$. Then $R$ centralizes some involution $a$ of $L$ (as $R$ normalizes a Sylow 2-subgroup of $L$). Again by (3.29), $C_R \leqslant M^g$, so $a \in M^g$. Thus $a$ normalizes $K^g$ and centralizes the Sylow 2-subgroup $R$ of $L^g$, so $a$ must leave $L^g$ invariant. Furthermore, $a$ does not centralize $L^g$, otherwise it follows once again from (i) that $g \in M$, so $a$ does not centralize $L^g/Z(L^g)$. However, $[a, L^g] \lhd L^g$, and as $L^g/Z(L^g)$ is simple, this forces $[a, L^g] = L^g$. But $a \in L$ and $L^g$ normalizes $L$, so $L^g = [a, L^g] \leqslant L$. Thus $g$ normalizes $L$. Furthermore, as noted above, $K$ is a product of components of $N_G(L)$. Hence by definition of $K$, $K \lhd N_G(L)$, so $g$ normalizes $K$, whence $g \in M$—contradiction. Thus (iii) also holds and the lemma is proved.

Since $L_i$ and $L_j$ are $G$-conjugate for all $i, j$, $1 \leqslant i, j \leqslant r$, Lemma 3.37(iii) implies that they are, in fact $M$-conjugate. Hence as a corollary, we have the following result:

LEMMA 3.38. *M permutes* $L_1, L_2, ..., L_r$ *transitively under conjugation.*

To complete the proof of the component theorem, Aschbacher must show that the conclusions of Lemmas 3.37 and 3.38 lead to a contradiction. The ensuing analysis turns out to be completely general and Aschbacher places it in a context independent of the component theorem. In fact, the argument holds equally well under the weaker assumption that $K/O(K)$ (rather than $K$ itself) is semisimple and hence when the $L_i$ are assumed only to be 2-components; it is therefore applicable to the study of centralizers of involutions in groups which do not necessarily have the $B$-property.

Since the conclusions of Lemma 3.37 are essentially fusion-theoretic, we refer to the following result as Aschbacher's 2-*component fusion* theorem.

THEOREM 3.39. *Let G be a group with* $F^*(G)$ *quasisimple and let K be a subgroup of G such that* $K = L_1 L_2 \cdots L_r$, *where each* $L_i$ *is a 2-component of K, and* $M = N_G(K)$ *permutes the* $L_i$ *transitively under conjugation,* $1 \leqslant i \leqslant r$. *Let* $g \in G$ *and* $t \in \mathscr{I}(G)$ *and assume that each of the following three conditions implies that* $g \in M$:

(a) *t centralizes* $L_i/O(L_i)$ *and* $t^g$ *centralizes* $L_j/O(L_j)$ *for some* $i, j$, $1 \leqslant i, j \leqslant r$;
(b) *t centralizes* $L_i/O(L_i)$ *and* $t^g$ *does not leave* $L_j$ *invariant for some* $i, j$, $1 \leqslant i, j \leqslant r$; *or*
(c) $L_i \leqslant O(K)(M^g \cap L_i)$ *for some* $i$, $1 \leqslant i \leqslant r$.

*Under these conditions, either* $r = 1$ *or* $m(L_i) = 1$, $1 \leqslant i \leqslant r$.

Aschbacher's theorem is actually considerably stronger, for he reaches the same conclusion without assuming the transitive action of $M$ on $L_1, L_2, ..., L_r$ and without condition (b). However, the present form of the theorem is sufficient to yield the component theorem, for by Lemmas 3.37 and 3.38, Theorem 3.39 is applicable with $K = L_1 L_2 \cdots L_r$, where the $L_i$ are as in Lemma 3.37, $1 \leqslant i \leqslant r$. Hence either $r = 1$ or $m(L_i) = 1$, contrary to (3.27).

On the other hand, later applications require the result without the assumption of transitivity. Aschbacher's proof in this more general form depends upon some important properties of tightly embedded subgroups (which play a key role in many applications) and will be discussed at the end of the section.

It should also be mentioned that a particular case of the 2-component fusion theorem was obtained by Powell and Thwaites [101]. Aschbacher's proof was obtained independently of their result, but Gilman incorporates some of their ideas in his argument.

The thrust of the proof of Theorem 3.39 is to show that the conditions of Aschbacher's strong embedding criterion [I, 4.31] are satified. Under the present hypotheses, this reduces to verification of the following single condition:

PROPOSITION 3.40. *Theorem* 3.39 *holds if the following condition is satisfied for some* $z \in \mathcal{I}(L_i)$ *and some* $i$, $1 \leqslant i \leqslant r$:

(*) *Let* $y \in L_j$ *for some* $j$, $1 \leqslant j \leqslant r$, *with* $y$ *centralizing* $z$ *and* $|y| \leqslant 2$. *If* $(yz)^g \in M$ *for* $g \in G$, *then necessarily* $g \in M$.

Indeed, as usual, assumption (a) of Theorem 3.39 implies that $C_z \leqslant M$. Hence if $z^x \in C_z$ for $x \in G$, then by (*) (with $y = 1$), we have $x \in M$. Thus $M$ controls the fusion of $z$ in $C_z$. Furthermore, as $x \in M$, $z^x \in L_j$ for some $j$, $1 \leqslant j \leqslant r$. Hence if $z^x \neq z$ and $z^x$ centralizes $z$, (*) implies that $M$ also controls the fusion of $zz^x$ in $M$. Hence the hypotheses of Aschbacher's strong embedding criterion are indeed satisfied.

In [I, 4.31], we have for simplicity stated his result only in the case that $G$ is simple; however, the result holds more generally. Under the present assumption $F^*(G)$ quasisimple, it yields that either $z \in Z(F^*(G))$ or else the normal closure $N = \langle z^G \rangle$ of $z$ in $G$ possesses a strongly embedded subgroup. We argue that Theorem 3.39 holds in either case.

In the first case, as $C_z \leqslant M$, $F^*(G) \leqslant M$. Since $K \triangleleft M$ and $K$ is a product of 2-components, this immediately forces $F^*(G) = K$, so $K$ is quasisimple and $r = 1$, which is one of the alternatives of Theorem 3.39. Hence we can assume that $N$ has a strongly embedded subgroup. If $z \in O_{2'2}(N)$, then as $N \triangleleft G$, $z \in Z(F^*(G))$ and again $r = 1$, so we can also assume that $z \notin O_{2'2}(N)$. But then $m(N) \geqslant 2$ by the Brauer–Suzuki theorem [I, 4.88] and so by (the more general form of) Bender's strong embedding theorem [I, 4.24], $F^*(N) = F^*(G) \cong L_2(2^n)$, $Sz(2^n)$, or $U_3(2^n)$. It follows at once now that the centralizer of every involution of $G$ is solvable, which again forces $r = 1$ [otherwise involutions of $L_1$ would centralize $L_2/O(L_2)$

and so would have nonsolvable centralizers]. Thus Theorem 2.39 is indeed reduced to verification of (*).

The proof itself is divided into two distinct cases according as $r \geqslant 3$ or $r = 2$, the latter being considerably more difficult. We assume throughout that $m(L_i) > 1$, $1 \leqslant i \leqslant r$.

PROPOSITION 3.41. *Theorem* 3.39 *holds if* $r \geqslant 3$.

We can suppose that condition (*) of Proposition 3.40 fails for suitable $z, y, g, i$, and $j$. Since $zy$ centralizes $L_k/O(L_k)$ for some $k$, $1 \leqslant k \leqslant r$, assumption (b) of Theorem 3.39 implies that $(zy)^g$ leaves each $L_m$ invariant, $1 \leqslant m \leqslant r$, while assumption (a) implies that $C_{(zy)^g} \leqslant M^g$. Then $(zy)^g$ centralizes an involution $u$ of $L_i$ and $u \in M^g$. Furthermore, as $u \in L_i$, $u^g \in L_i^g$. We see then that Proposition 3.40 fails with $M^g$ in place of $M$ and $u^g, g^{-1}$ in the roles of $zy$ and $g$, respectively. Therefore the situation is symmetric in $M$ and $M^g$.

This symmetry provides considerable tension, which eventually enables one to force the structure of a Sylow 2-subgroup of $L_i$. Note first that the above argument shows that for each $i$, $N_{M^g}(L_i)$ contains $(zy)^g$ and hence has even order. Therefore by symmetry $N_M(L_i^g)$ does as well.

Furthermore, using assumption (a) of Theorem 3.39, it is immediate that

$$|L_i L_j \cap L_m^g L_n^g| \text{ is odd for all } 1 \leqslant i, j, m, n \leqslant r. \qquad (3.31)$$

One now studies the action of a Sylow 2-subgroup $T$ of $N_{L_m^g L_n^g}(L_i)$ on $L_i$. By what we have shown, we have

$$T \neq 1 \quad \text{and} \quad T \cap L_i = 1. \qquad (3.32)$$

Using the fusion assumptions in the usual way, one next proves the following result:

LEMMA 3.42. *Set* $J_i = L_i \cap M^g$. *Then we have*

   (i)  $C_{L_i}(t) \leqslant J_i$ *for all* $t \in T^{\#}$;
   (ii) $[T, J_i] \leqslant O(J_i)$;
   (iii) $J_i$ *does not cover* $L_i/O(L_i)$; *and*
   (iv) $T$ *acts faithfully on* $L_i/O(L_i)$.

These conditions [including (3.32)] place severe restrictions on $T$.

Indeed, whenever a 2-group $T$ acts on an arbitrary group $L_i$ in accordance with these conditions, one can prove:

LEMMA 3.43. (i) $T$ *is abelian; and*
(ii) $T$ *is isomorphic to a subgroup of* $J_i$.

Consider this situation for a moment. On the one hand, $T$ contains the product of Sylow 2-subgroups $T_k$ of $N_{L^g}(L_i)$ for *both* $k = m$ and $n$; on the other hand, $T$ is isomorphic to a subgroup of a Sylow 2-subgroup of $L_i \cap M^g$. Using the symmetry between $M$ and $M^g$ and together with the initial assumption $m(L_i) > 1$, there turns out to be only one way this can occur:

LEMMA 3.44. *Let* $R_i \in \mathrm{Syl}_2(L_i \cap M^g)$, $1 \leqslant i \leqslant r$, *set* $K = L_1 L_2 \cdots L_r$ *and* $\bar{K} = K/O(K)$. *Then interchanging* $M$ *and* $M^g$, *if necessary, we have*

(i) $\bar{R}_i \leqslant Z(\bar{L}_i)$, $1 \leqslant i \leqslant r$; *and*
(ii) $\bar{R}_1 = \bar{R}_2 = \cdots = \bar{R}_r$.

In particular, no $t \in T^{\#}$ centralizes any 2-element of $\bar{L}_i - Z(\bar{L}_i)$, a very restrictive condition. Without loss $T$ normalizes $R_i$, and we expand $R_i$ to a $T$-invariant Sylow 2-subgroup $Q_i$ of $L_i$. The given action of $T$ on $Q_i$ now forces the structure of $Q_i/R_i$.

LEMMA 3.45. *Set* $\tilde{Q}_i = Q_i/R_i$. *Then* $\tilde{Q}_i$ *has class at most 2 and if equality holds, then either*

(i) $|\tilde{Q}_i/Z(\tilde{Q}_i)| \leqslant 4$; *or*
(ii) $|\tilde{Q}_i| = 2^{2d+1}$, *where* $d = m(T_m) \geqslant 2$.

Thus $L_i^* = \bar{L}_i/Z(\bar{L}_i)$ has Sylow 2-subgroup $Q_i^*$ of class at most 2 (as $Q_i^*$ is a homomorphic image of $\tilde{Q}_i$). Also $\bar{L}_i$ is nonsimple as $1 \neq \bar{R}_i \leqslant Z(\bar{L}_i)$. We now invoke the classification of groups with abelian or class 2 Sylow 2-subgroups [I, 4.129]. If $Q_i^*$ is abelian, then using [I, Table 4.1], it follows that the only possibility is $\bar{L}_i \cong SL_2(q)$ for some $q \equiv 3, 5 \pmod 8$, in which case $m(L_i) = 1$, contrary to assumption. On the other hand, if $Q_i^*$ has class 2, [I, 4.129] implies that $|Q_i^*/Z(Q^*)| > 4$ and $|Q_i^*| \geqslant 2^6$, so Lemma 3.45(ii) must hold with $d \geqslant 3$. But $T_m$ induces a group of outer automorphism of $\bar{L}_i$, so by the structure of $\mathrm{Aut}(L_i^*)$ [I, 4.237], $d = m(T_m) \leqslant 2$—contradiction. This completes the outline of Proposition 3.41.

Two aspects of the preceding argument do not carry over, in general, to the case $r = 2$. First, if $a \in \mathcal{I}(G)$ centralizes $L_i/O(L_i)$, $i = 1$ or 2, there may exist $x \in G$ such that $a^x$ interchanges $L_1$ and $L_2$ under conjugation, a possibility excluded by assumption (b) of Theorem 3.39 when $r \geqslant 3$. Secondly, the conclusion $\bar{R}_i \leqslant Z(\bar{L}_i)$ in Lemma 3.44 depended crucially on the fact that our 2-group $T$ contained Sylow 2-subgroups of $N_{L_k^g}(L_i)$ for *two* distinct values of $k$, $1 \leqslant k \leqslant r$, which will no longer be the case.

*If* no such $a^x$ exists interchanging $L_1$ and $L_2$, then the conclusions of Lemmas 3.42 and 3.43 again hold, provided we take $T$ to be a Sylow 2-subgroup of $N_{L_2^g}(L_1)$ (note then that $T$ centralizes $\bar{L}_1 = L_1^g/O(L_1^g)$, a necessary condition for the analysis). Let $R$ be a $T$-invariant Sylow 2-subgroup of $L_1 \cap M^g$. If, in addition, $\bar{R} \leqslant Z(\bar{L}_1)$, then the preceding argument goes through essentially unchanged: however, as just remarked, we must consider the contrary possibility as well. In that case, we conclude from the symmetry between $M$ and $M^g$ that $R \cong T$.

Thus if we expand $R$ to a $T$-invariant Sylow 2-subgroup $Q$ of $L_1$, our task is to determine the structure of $Q$ from the following conditions:

$$\begin{aligned} &(1) \;\; C_Q(t) = R \text{ for every } t \in T^\#; \\ &(2) \;\; T \text{ is abelian; and} \qquad\qquad\qquad\qquad\qquad (3.33) \\ &(3) \;\; T \cong R. \end{aligned}$$

It is interesting to note that O'Nan encountered the identical situation of (3.33) in his work on doubly transitive permutation groups [I: 229], which he then analyzed as an abstract problem. Of course, Aschbacher is forced to deal with the problem, too, but does so within the context of the component theorem itself. The result is as follows:

PROPOSITION 3.46. *If $T$, $Q$, and $R$ are nontrivial 2-groups satisfying the conditions of (3.33), then one of the following holds*:

  (i) *$T$ is not elementary abelian, $|Q:R| = 2$, and every element of $Q - R$ inverts $R$;*
 (ii) *$|T| = 2$, and $Q$ is cyclic, dihedral, or quaternion; or*
(iii) *$T$ is elementary abelian of order at least 4 and either*
     (1) *$R \leqslant Z(Q)$, $Q/R$ is elementary abelian, and $|Q/R| \leqslant |R|$; or*
     (2) *If we set $Q_0 = N_Q(TR)$, then $|Q:Q_0| = 2$ and $|Q_0| = |R|^2$.*

Thus, in essence, the proposition forces the structure of the Sylow 2-subgroup $Q$ of $L_1$. Each possibility must now be analyzed in turn in conjunction with the given fusion hypothesis. The argument is quite delicate and eventually $Q$ is forced to be quaternion, whence $m(L_1) = 1$—contradiction.

However, this last analysis was predicated on the assumption that whenever $a^x$ centralizes $L_i/O(L_i)$ and $a^x \in M$, $a \in \mathcal{I}(G)$, $x \in G$, then necessarily $a^x$ leaves $L_1$ and $L_2$ invariant. In general, if $S$ is a Sylow 2-subgroup of $L_2^g \cap M$ containing $T$, then $|S:T| \leqslant 2$ and elements of $S - T$ interchange $L_1$ and $L_2$. Moreover, $S - T$ may even contain involutions. The preceding argument must therefore be extended to the case $S > T$. In effect, this comes down to establishing an appropriate extension of Proposition 3.46 under the present fusion conditions.

This completes the discussion of Aschbacher's component theorem per se. We turn now to the general properties of tightly embedded subgroups and the sharpened form of his 2-component fusion theorem, which he obtained in the course of his analysis of groups with a standard component.

The first result is a direct consequence of his strong embedding criterion [I, 4.28]. We limit the statement to the case $F^*(G)$ simple.

THEOREM 3.47. *Let $G$ be a group with $F^*(G)$ simple and $H$ a tightly embedded subgroup of $G$. Set $N = N_G(H)$. If $H^g \cap N$ has odd order for each $g \in G - N$, then either $H \lhd G$ or $F^*(G)$ has a strongly embedded subgroup.*

Indeed, it is immediate from the definition of tight embedding that $C_u \leqslant N$ for every $u \in \mathcal{I}(H)$. Hence if we fix $z \in \mathcal{I}(H)$, then $C_z \leqslant N$. Furthermore, by the assumption of the theorem, $z^g \in N$ if and only if $g \in N$. In particular, if $z^g \in C_z$ then $g \in N$ (as $C_z \leqslant N$). Since $H \lhd N$, it follows that $z^g \in H$. Thus if $z^g \neq z$, then $zz^g \in \mathcal{I}(H)$ and we conclude that likewise $C_{zz^g} \leqslant N$. Hence the assumptions of [I, 4.31] are satisfied and the theorem follows.

Thus, in practice, one can always assume that there exists an element $g \in G - N$ such that $X = H^g \cap N$ has even order. Let $T \in \mathrm{Syl}_2(X)$ and note that the tight embedding of $H$ implies that

$$T \cap H = 1. \tag{3.34}$$

Furthermore, if $R \in \mathrm{Syl}_2(N_H(T))$, then again by tight embedding $R \leqslant N^g$ and if we let $Q \in \mathrm{Syl}_2(N_H(R))$, then $Q \leqslant X$. Hence, there is a symmetry between $H, N$ and $H^g, N^g$ and we see that for a suitable choice of $H$ or $H^g$, we can assume without loss that

$$|T| \geqslant |R|. \tag{3.35}$$

Finally, observe that as $H$ is tightly embedded in $G$, $X$ is tightly embedded in $HX$. Aschbacher analyzes this situation independently of the ambient group $G$ to obtain structural properties of $T$ and its embedding in $HT$. His precise result is as follows [clearly there will be no loss in assuming $O(H) = 1$]:

THEOREM 3.48. *Assume the group $X$ acts on the group $H$ with $O(H) = 1$ in such a way that $X$ is tightly embedded in $HX$. Let $T \in \mathrm{Syl}_2(X)$, let $R \in \mathrm{Syl}_2(N_H(T))$, and expand $TR$ to $S \in \mathrm{Syl}_2(HX)$. If $T \cap H = 1$ and $|T| \geqslant |R|$, then $T \cong R$ and one of the following holds:*

(i) *$T$ is cyclic, $S \cap H$ is cyclic, quaternion, or dihedral, and $S$ is either dihedral, quasi-dihedral, or wreathed;*
(ii) *$L(H) \cong L_2(2^n)$ and $L(H)T \cong L_2(2^n) \times E_{2^n}$ for some $n \geqslant 2$;*
(iii) *$T \leqslant O_2(HT)$ and if $Y$ denotes the normal closure of $T$ in $HT$, (whence $Y \leqslant O_2(H)$), then either*
   (1) *$Y = T$ (whence $T \lhd HT$);*
   (2) *$T$ is abelian and $Y = T \times T^x$ for some $x \in N_S(T)$. Moreover, if $m(S/Y) \geqslant 2$, then $T$ is elementary; or*
   (3) *$T$ is elementary, $Y$ is special of order $|T|^3$ with $T \cong Z(Y)$ and $Y$ is of index 2 in $S$.*

We hope this discussion has given a reasonable picture of Aschbacher's component theorem as well as some feeling for his related results on fusion and tight embedding. The theorem is fundamental, for in essence it asserts that in a (simple) group of component type with the $B$-property, the centralizer of some involution "closely resembles" the centralizer of an involution in some *known* (simple) group, the notion of "standard" constituting the formal expression of this resemblance.

## 3.5. The Groups of Lie Type of Odd Characteristic

Aschbacher's remarkable classical involution theorem gives a complete classification of all simple groups $G$ in which the centralizer of some involution possesses a suitably embedded $SL_2(q)$-subgroup $J$, $q = p^m$, $p$ an odd prime. It asserts that such a group $G$ is necessarily isomorphic to a group of Lie type of characteristic $p$ [apart from the groups $L_2(p^r)$ and $^2G_2(3^r)$, which contain no $SL_2(q)$-subgroups at all] with the single additional possibility $G \cong M_{11}$.

The group $G$ obviously acts as a (transitive) permutation group on the set $\Omega$ of $G$-conjugates of $J$ and $G$ is generated by the elements of $\Omega$. Thus Aschbacher's task is to recover the full Lie subgroup structure of $G$ from the existence of the *single* $SL_2(q)$-subgroup $J$ and the nature of the action of $G$ on $\Omega$.

Clearly, in order to grasp Aschbacher's proof, one must first understand how the groups of Lie type are built up from their $SL_2(q)$-subgroups and, in addition, how they can be identified from this description; for it is this picture of the groups of Lie type which underlies Aschbacher's entire argument. In this section, we summarize these basic properties, occasionally interpolating some comments on the classical involution theorem.

Recall that in [I, Section 2.1] we have given a general description of a finite Chevalley group $G(q)$ over $GF(q)$, $q = p^n$, $p$ a prime. (Our initial discussion will apply for all primes $p$.) That description included the definition of a root system $\Sigma$ of $G(q)$, the division of $\Sigma$ into disjoint subsets $\Sigma^+$, $\Sigma^-$ of positive and negative roots, the associated Dynkin diagram $\Delta$ of $\Sigma$, as well as the generation of $G(q)$ by root subgroups $\chi_\alpha = \langle X_\alpha(t) \mid t \in GF(q) \rangle$.

In particular, each $\chi_\alpha$ is elementary abelian of order $q$, isomorphic to the additive group of the underlying field $GF(q)$, and invariant under a Cartan subgroup $H$ of $G(q)$. Moreover, one has that

$$Q^+ = \langle \chi_\alpha \mid \alpha \in \Sigma^+ \rangle, \qquad Q^- = \langle \chi_\alpha \mid \alpha \in \Sigma^- \rangle \tag{3.36}$$

are Sylow $p$-subgroups of $G(q)$, so that

$$G(q) = \langle Q^+, Q^- \rangle. \tag{3.37}$$

Furthermore, the elements of $\Sigma$ are vectors in a suitable $r$-dimensional Euclidean vector space $E$ associated with $G(q)$ [where $r$ is the Lie rank of $G(q)$ = the rank of its Weyl group $W$ as a Coxeter group]; as such, each element of $\Sigma$ has a length. In particular, this is the case for the roots $\alpha_1, \alpha_2, ..., \alpha_r$ of a simple root system $\pi$ (the roots in $\pi$ form a basis for $E$). Moreover, if the $\alpha_i$, $1 \leqslant i \leqslant r$, are indexed by successive nodes of $\Delta$, then the weight partitioning of $\Delta$ [I, 2.8] preserves root length—i.e., either $\Delta$ is of type $A_n$, $D_n$, $E_6$, $E_7$, or $E_8$ and all roots have the same length or else $\Delta$ is of type $B_n$, $C_n$, $G_2$, or $F_4$ and the roots are of one of two lengths, the *long* roots corresponding to the nodes to the left of the partition, the *short* roots to its right. (In the former case one calls all the $\alpha_i$ long roots.)

We note also that every nonzero root of $\Sigma$ is a member of some simple

**Table 3.1.** Dynkin Diagrams and Root Subgroup Structures of Twisted Groups of Lie Type

| $G(q)$ | Type of $\varDelta$ | Long root | Short root |
|---|---|---|---|
| ${}^2A_{2n}(q)$ | $B_n$ | Type $U_3(q)$ | $E_{q^2}$ |
| ${}^2A_{2n-1}(q)$ | $C_n$ | $E_q$ | $E_{q^2}$ |
| ${}^2D_n(q)$ | $B_{n-1}$ | $E_q$ | $E_{q^2}$ |
| ${}^2E_6(q)$ | $F_4$ | $E_q$ | $E_{q^2}$ |
| ${}^3D_4(q)$ | $G_2$ | $E_q$ | $E_{q^3}$ |

root system, so that every $\alpha \neq 0$ of $\Sigma$ is either short or long. More precisely, the Weyl group $W$ of $G(q)$ transitively permutes the roots of a given type and likewise its pre-image in $G(q)$ transitively permutes the corresponding root subgroups under conjugation.

We have also indicated there that an entirely analogous description holds for the Steinberg variations $G(q)$ of the Chevalley groups [and likewise for the Suzuki and Ree groups, but we omit these latter three families from the present discussion as well as the groups $L_2(q)$]. Hence each of these twisted groups possesses a root system $\Sigma$ and an associated Dynkin diagram $\varDelta$. Every nonzero root of $\Sigma$ is either long or short and the Weyl group transitively permutes the roots of a given type as well as their corresponding root subgroups. Likewise, $G(q)$ is generated by its root subgroups. However, it is now no longer necessarily the case that root subgroups are elementary abelian of order $q$.

Table 3.1 lists for each such $G(q)$ the type of its Dynkin diagram and the structure of the long- and short-root subgroups of a simple root system.

In particular, except for long-root subgroups of even-dimensional unitary groups, all root subgroups are elementary abelian of order $q$, $q^2$, or $q^3$.

In [I, 2.12] we have stated the fundamental Chevalley commutator formula, which gives a precise expression for the commutator $[X_\alpha(t), X_\beta(u)]$, $t$, $u \in GF(q)$, for any two linearly independent roots $\alpha, \beta$ of a given root system $\Sigma$ of an arbitrary finite Chevalley group. Likewise this formula has been extended by Steinberg to the twisted groups.

Combining his result with Chevalley's, we can express the commutator formula in the following general form. Let $G(q)$ be an arbitrary finite Chevalley or Steinberg group, $q = p^n$, $\Sigma$ a root system of $G(q)$, $\Sigma^+, \Sigma^-$ the subsets of $\Sigma$ of positive and negative roots, and for each $\alpha \in \Sigma$, let $Q_\alpha$ denote the corresponding root subgroup of $G(q)$. Then we have the following result:

**Table 3.2.** Comparison of Lie and
Classical Notation

| $G(q)$ | Classical name |
|---|---|
| $A_n(q)$ | $L_{n+1}(q)$ |
| $^2A_n(q)$ | $U_{n+1}(q)$ |
| $B_n(q)$ | $P\Omega_{2n+1}(q)$ |
| $C_n(q)$ | $PSp_{2n}(q)$ |
| $D_n(q)$ | $P\Omega_{2n}^+(q)$ |
| $^2D_n(q)$ | $P\Omega_{2n}^-(q)$ |

THEOREM 3.49. *The following conditions hold*:

(i) $[Q_\alpha, Q_\beta] \leqslant \prod_{i,j>0, i\alpha+j\beta \in \Sigma} Q_{i\alpha+j\beta}$ *for any two linearly independent roots* $\alpha, \beta \in \Sigma$;

(ii) $Q = \langle Q_\alpha \mid \alpha \in \Sigma^+ \rangle$ *is a Sylow p-subgroup of* $G(q)$; *and*

(iii) $G(q) = \langle Q_\alpha \mid \alpha \in \Sigma \rangle$.

To avoid later confusion, it will be well to recall once again the relationship between the Lie notation and the more classical designation of the linear, unitary, symplectic, and orthogonal groups. We limit ourselves to the case in which $G(q)$ is simple. The comparison is given in Table 3.2.

Now we begin to describe the generation of $G(q)$ by its $SL_2(q)$-subgroups. Fix a simple root system $\pi = \{\alpha_1, \alpha_2, ..., \alpha_r\}$ of $\Sigma$ with $\pi \subseteq \Sigma^+$. For $1 \leqslant i \leqslant r$, set

$$Q_i = Q_{\alpha_i}, \qquad Q_{-i} = Q_{-\alpha_i}, \tag{3.38}$$

and put

$$U_i = Z(Q_i), \qquad U_{-i} = Z(Q_{-i}). \tag{3.39}$$

Using Table 3.1 and the corresponding results in the Chevalley case, one easily proves:

PROPOSITION 3.50. *The following conditions hold for any* $i$, $1 \leqslant i \leqslant r$:

(i) $Q_i \cong Q_{-i}$ *(whence* $U_i \cong U_{-i}$*)*;

(ii) $U_i$ *is elementary abelian; and*

(iii) *One of the following holds*:

(1) $U_i = Q_i \cong E_q$;

(2) $G(q)/Z\big(G(q)\big) \cong {}^2A_{2n}(q)$; or

(3) $G(q)/Z\big(G(q)\big) \cong B_n(q),\ {}^2A_{2n-1}(q),\ {}^2D_n(q),\ {}^2E_6(q),\ or\ {}^3D_4(q)$, and $\alpha_i$ is a short root.

For each $i$, $1 \leqslant i \leqslant r$, we now set

$$J_i = \langle U_i, U_{-i} \rangle \quad \text{and} \quad K_i = \langle Q_i, Q_{-i} \rangle. \tag{3.40}$$

The following result gives the structure of $J_i$ and $K_i$.

PROPOSITION 3.51. *The following conditions hold*:

(i) *If $U_i \cong E_q$, then $J_i \cong SL_2(q)$*;

(ii) *If $\alpha_i$ is a long root, then $J_i \cong SL_2(q)$; and*

(iii) *If all $\alpha_i$ have the same length, then $J_i = K_i$.*

For completeness, we list in Table 3.3 the structure of $J_i$ and $K_i$ when either of them is not isomorphic to $SL_2(q)$.

The groups $K_i$ and their $G(q)$-conjugates are called *fundamental* subgroups; *long* or *short* fundamental subgroups according as $\alpha_i$ is a long or short root.

The proposition shows that the situation is somewhat simpler when $\alpha_i$ is a long root. Moreover, if the $\alpha_i$ are all of the same length, then $J_i = K_i$, so that $J_i$ itself is a fundamental subgroup.

Although some of the subsequent results hold for all primes $p$, we assume in Propositions 3.52–3.54 that $p$ is *odd*. In this case, if $J_i \cong SL_2(q)$, then $J_i$ contains a unique involution $z_i$ and $\langle z_i \rangle = Z(J_i)$.

**Table 3.3.** The Exceptional Structures of $J_i$ and $K_i$

| $G(q)$ | $\alpha_i$ | $J_i$ | $K_i$ |
|---|---|---|---|
| ${}^2A_{2n}(q)$ | Long | $SL_2(q)$ | $U_3(q)$ |
| ${}^2A_{2n}(q)$ | Short | $L_2(q^2)$ | $L_2(q^2)$ |
| $B_n(q)$, $q$ odd | Short | $L_2(q)$ | $L_2(q)$ |
| ${}^2A_{2n-1}(q)$ | Short | $L_2(q^2)$ | $L_2(q^2)$ |
| ${}^2D_n(q)$ | Short | $L_2(q^2)$ | $L_2(q^2)$ |
| ${}^2E_6(q)$ | Short | $SL_2(q^2)$ | $SL_2(q^2)$ |
| ${}^3D_4(q)$ | Short | $SL_2(q^3)$ | $SL_2(q^3)$ |

We are interested in the embedding of $J_i$ in $G(q)$ and the precise form of the generation of $G(q)$ by conjugates of $J_i$.

We let $\alpha_i$ be a long root, we set $J = J_i$ [so that $J \cong SL_2(q)$], $\langle z \rangle = Z(J)$, $U \in \mathrm{Syl}_2(J)$, and put $G = G(q)$. Then with this notation, we have the following two basic results, which motivate both the exact statement and the strategy of the proof of the classical involution theorem.

PROPOSITION 3.52. *The following conditions hold*:

(i) $O^{p'}(N_G(J)) = JX$, where $J$ centralizes $X$ and $X$ is the Levi factor of the parabolic subgoup $N_G(U)$;
(ii) $X$ is generated by G-conjugates of $J$ except in the case of $\Omega_7(q)$ or $\Omega_8^-(q)$;
(iii) *Apart from the two exceptions of* (ii), $JX$ *is the subgroup generated by the G-conjugates of* $J$ *contained in* $C_G(z)$;
(iv) *If either* $G$ *is not an orthogonal group or* $G$ *is a universal orthogonal group* $\Omega_n^\pm(q)$, *then* $N_G(J) = C_G(z)$; *and*
(v) *The possible structures of* $X$ *are given in Table* 3.4 *below*.

[In the exceptional cases of $\Omega_7(q)$ and $\Omega_8^-(q)$, the G-conjugates of $J$ contained in $X$ generate only the $SL_2(q)$ factor of $X$.]

As a corollary of Proposition 3.52(iv), we have the following basic result:

PROPOSITION 3.53. *If either* $G$ *is not an orthogonal group or* $G$ *is a universal orthogonal group, then* $J$ *is tightly embedded in* $G$.

Indeed, if $J^g \cap J$ has even order for some $g \in G$, then $z^g = z$ as $z$ is the *unique* involution of $J$, so $g \in C_G(z)$ and consequently $J^g = J$ by Proposition 3.52(iv). Thus by definition, $J$ is tightly embedded in $G$.

Since $J$ has quaternion Sylow 2-subgroups, we see that most of the

**Table 3.4.** The Levi Factors of $N_G(U)$

| $G(q)$ | $X$ | $G(q)$ | $X$ |
|--------|-----|--------|-----|
| $L_n(q)$ | $SL_{n-2}(q)$ | $F_4(q)$ | $Sp_6(q)$ |
| $PSp_n(q)$ | $Sp_{n-2}(q)$ | $^2E_6(q)$ | $SU_6(q)$ |
| $U_n(q)$ | $SU_{n-2}(q)$ | $E_6(q)$ | $SL_6(q)/Z_d$, |
| $P\Omega_n^\pm(q)$ | $SL_2(q) * \Omega_{n-4}^\pm(q)$ | | where $d = (q-1, 3)$ |
| $G_2(q)$ | $SL_2(q)$ | $E_7(q)$ | $\Omega_{12}(q)$ |
| $^3D_4(q)$ | $SL_2(q^3)$ | $E_8(q)$ | $E_7(q)$ |

groups of Lie type of odd characteristic possess a tightly embedded subgroup with quaternion Sylow 2-subgroups.

However, when $G$ is an orthogonal group $P\Omega_n^{\pm}(q)$, the normal closure of $J$ in $C_G(z)$ has the form

$$JJ^y \cong SL_2(q) * SL_2(q) \tag{3.41}$$

for some $y \in C_G(z)$. In particular, $z \in J \cap J^y$ and $y \in G - N_G(J)$, so $J$ is definitely not tightly embedded in this case.

Thus a classification of (simple) groups with tightly embedded subgroups having quaternion Sylow 2-subgroups will not cover the orthogonal groups $P\Omega_n^{\pm}(q)$. On the other hand, the second property of the $G$-conjugates of $J$ encompasses all the groups of Lie type of odd characteristic including $P\Omega_n^{\pm}(q)$ [apart, of course, from the groups $L_2(q)$ and $^2G_2(3^m)$].

PROPOSITION 3.54. *The following two conditions hold for any two distinct G-conjugates $J_1$ and $J_2$ of $J$*:

(i) $J_1 \cap J_2$ *has Sylow 2-subgroups of order* $\leqslant 2$, *and if equality holds, then $J_1$ centralizes $J_2$; and*

(ii) *If $y$ is a 2-element of $J_1$ of order $\geqslant 4$ which centralizes $Z(J_2)$, then $y$ normalizes $J_2$.*

It is this last proposition which leads to the general form of the classical involution theorem (see the statement of Theorem 3.64 below, which is formulated in sufficiently broad terms to allow for an inductive proof). The key to Aschbacher's argument is the fact that the subgroup $X$ of Table 3.4 is (apart from the orthogonal case) a quasisimple group of Lie type of odd characteristic and $X$ *is generated by $X$-conjugates of $J$*.

Of course, this is a property of the groups of Lie type themselves, whereas Aschbacher is forced to work with a minimal counterexample $G$ to his proposed theorem and there is no *a priori* reason for $C_J = C_G(J)$ to contain a subgroup of the given form of $X$. But this is his precise strategy: namely, to reduce to the case that $X^* = F^*(C_J)$ is quasisimple and satisfies the same hypothesis as $G$, thus indeed enabling him to identify $X^*$ by induction as one of the groups $X$ of Table 3.4.

The reduction itself is, hardly surprising, very complex. On the one hand, there are a large number of low Lie rank cases, requiring their own individual analyses, which must be disposed of first. (Their identification is made primarily by invoking the sectional 2-rank $\leqslant 4$ theorem.) Then, to carry through the induction, Aschbacher is forced to consider a general situation leading to 2-constrained solutions for $F^*(G)$ [again including some small exceptional cases, as well as a generic solution related to the alternating groups, the latter being identified through a presentation of $F^*(G)$ by generators and relations]. Finally Aschbacher encounters a number of special problems due to the exceptional nature of the orthogonal groups indicated above. However, it should be added that his overall proof is organized in such a way that these various exceptional cases arise as natural subconfigurations. This will be more fully detailed in the next three sections.

Once Aschbacher has constructed the required subgroup $X^*$, it remains for him to determine the isomorphism type of $G$ from that of $X^*$. For the groups of Lie type themselves, apart from certain low Lie rank cases, if $g$ is an element of $G(q)$ such that $J^g \leqslant X$, one has the basic generation:

$$G(q) = \langle X, X^g \rangle. \tag{3.42}$$

In the case $G(q) \cong P\Omega_n^{\pm}(q)$, one replaces $X$ $\left( \cong SL_2(q) * \Omega_{n-4}^{\pm}(q) \right)$ by its component $X_1 \cong \Omega_{n-4}^{\pm}(q)$.

In view of (3.42), Aschbacher's aim is therefore to prove that

$$F^*(G) = \langle X^*, X^{*g} \rangle \tag{3.43}$$

for any $g \in G$ such that $J^g \leqslant X^*$ [where again $X^* = F^*(C_J)$]; and then to determine the isomorphism type of $F^*(G)$ from the interrelationships between the subgroups $X^*$ and $X^{*g}$.

In effect, this last step amounts to verification of the Steinberg presentation for $G$ in terms of generators and relations. However, there exist certain reformulations of the Steinberg relations, due to Curtis [I: 74] and Phan [97], which, in practice, are easier to check; and except in the case of the symplectic groups, which Aschbacher finds easier to identify by a geometric characterization due to Buekenhout and Shult [12], he uses one form or another of the Curtis–Phan relations to determine the isomorphism type of $G$. [However, as subsequent work of Seitz on the groups of Lie type of characteristic 2 indicates [106], it is very likely that by making a judicious choice of the element $g$ in (3.43) in each case, the identification of $G$ can be obtained by a more uniform procedure.]

The principal result here is due to Curtis, which asserts that the group $G(q)$ is essentially characterized by its root subgroups $Q_\alpha$ together with the commutator relations of Theorem 3.49, for all $q$, even or odd. To state his theorem precisely, for each pair of integers $i, j$, $1 \leqslant i, j \leqslant r$ (not necessarily distinct), let $\Sigma_{ij}$ be the set of those $\alpha \in \Sigma$ which are linear combinations of $\alpha_i$ and $\alpha_j$. We can identify $\Sigma$ with the union of the sets $\Sigma_{ij}$.

THEOREM 3.55. *Let* $\tilde{G}$ *be a group generated by a collection of subgroups* $\tilde{Q}_\alpha$, $\alpha \in \Sigma$, *and suppose the following two conditions hold:*

(a) *For each* $\alpha \in \Sigma$, *there exists an isomorphism* $\Theta_\alpha$ *of* $\tilde{Q}_\alpha$ *on* $Q_\alpha$; *and*
(b) *If* $\Theta$ *denotes the union of the* $\Theta_\alpha$'s, $\alpha \in \Sigma$, *then for each pair* $i, j$, $1 \leqslant i, j \leqslant r$, *and each pair of linearly independent roots* $\alpha, \beta \in \Sigma_{ij}$, $\Theta$ *preserves the commutator relations of Theorem* 3.49.

*Under these conditions,* $\Theta$ *extends to a homomorphism of* $\tilde{G}$ *onto* $G(q)$ *whose kernel is in the center of* $\tilde{G}$.

In other words, under the given assumptions $\tilde{G}$ is isomorphic to a covering group of $G(q)$.

It is this theorem which Aschbacher uses to identify most orthogonal groups as well as the groups $F_4(q)$, $^2E_6(q)$.

For the groups $A_n(q)$, $D_n(q)$, $E_6(q)$, $E_7(q)$, $E_8(q)$, which have roots all of the same length, Curtis's theorem leads to a very nice characterization in terms of their fundamental $SL_2(q)$-subgroups $J_i$, $1 \leqslant i \leqslant r$. To state it, we first introduce some notation due to Phan.

Let $\Delta$ be a Dynkin diagram. Thus $\Delta$ is of type $A_n$, $B_n$, $C_n$, $D_n$, $G_2$, $F_4$, $E_6$, $E_7$, or $E_8$. We take $\Delta$ to be indexed from 1 to $r$. Let $G$ be an arbitrary group containing a collection of subgroups $J_i \cong SL_2(q)$, indexed by $\Delta$. Let $Y$ be either the group $L_3(q)$ or $U_3(q)$.

DEFINITION 3.56. We say that the set of $J_i$, $i \in \Delta$, is a *Y-generating system of type* $\Delta$ for $G$ provided

(1) $G = \langle J_i \mid i \in \Delta \rangle$;
(2) $J_i$ centralizes $J_j$ if $i, j$ are not connected by an edge of $\Delta$, $i \neq j$;
(3) $\langle J_i, J_j \rangle$ is isomorphic to a covering group of $Y$ if $i, j$ are connected by an edge of $\Delta$; and
(4) $Z(J_i)$ centralizes $Z(J_j)$ for all $i, j \in \Delta$.

In this terminology, we have the following basic characterization theorem of Curtis, which Aschbacher uses to identify the groups $A_n(q)$, $D_n(q)$, $E_6(q)$, $E_7(q)$, $E_8(q)$.

THEOREM 3.57. *Let* $\{J_i\}$ *be an* $L_3(q)$-*generating system of type* $A_n$, $D_n$,

$E_6$, $E_7$, or $E_8$ for the simple group $G$. Then there exists an isomorphism of $G$ and $G(q) = A_n(q)$, $D_n(q)$, $E_6(q)$, $E_7(q)$, or $E_8(q)$, respectively, under which the groups $J_i$ correspond to fundamental subgroups of $G(q)$.

Phan has extended this result to obtain an analogous characterization of the unitary groups. However, when $q = 3$, it requires an extra condition, which is expressed in terms of the following definition: a *chain* $(i, j, k)$ in $\Delta$ is a set of three nodes $i, j, k \in \Delta$ such that both $i, j$ and $j, k$ are connected by edges.

THEOREM 3.58.   Let $\{J_i\}$ be a $U_3(q)$-generating system of type $A_{n-1}$ of the simple group $G$, $q$ odd. If $q = 3$, assume in addition that $\langle J_{i-1}, J_i, J_{i+1} \rangle$ is isomorphic to a covering group of $U_4(3)$ for each chain $(i-1, i, i+1)$ in $\Delta$. Then there is an isomorphism of $G$ and $U_n(q)$ under which the groups $J_i$ correspond to fundamental subgroups of $U_n(q)$.

Aschbacher uses Phan's theorem to identify the unitary groups. Similarly he uses the following additional result of Phan to identify the groups $D_4(q) \cong P\Omega_8^+(q)$.

THEOREM 3.59.   If $\{J_i\}$ is a $U_3(q)$-generating system of type $D_4$ of the simple group $G$, $q$ odd, $q > 3$, then there is an isomorphism of $G$ and $D_4(q)$ under which the groups $J_i$ correspond to fundamental subgroups of $D_4(q)$.

Finally, we describe the Buekenhout–Shult geometric characterization of the symplectic groups in terms of their action on the so-called "Shult space" constructed from the set of $G(q)$-conjugates of a Sylow 2-subgroup $U = Q$ of $J$.

DEFINITION 3.60.   A *Shult space* $S$ is a set of points $\mathscr{P}$ together with a set of distinguished subsets $\mathscr{L}$ of $\mathscr{P}$ called lines, each of cardinality $\geqslant 3$, satisfying the following condition:

For any line $L$ of $\mathscr{L}$ and any point $P$ of $\mathscr{P}$ not on $L$, if $\{P, Q\}$ is contained in a line of $\mathscr{L}$ for at least two distinct points $Q$ of $L$, then $\{P, Q\}$ is contained in a line of $\mathscr{L}$ for every point $Q$ of $L$.

Now assume $G(q)/Z\big(G(q)\big) \cong PSp_n(q)$, $q = p^m$, $p$ a prime, set $G = G(q)$, let $J = K$ be a fundamental subgroup of $G$, and let $Q \in \mathrm{Syl}_p(J)$. Furthermore, set

$$\mathscr{P} = \{Q^g \mid g \in G\}.$$

For $P \in \mathscr{P}$, let

$$\Omega(P) = \{J^g \mid J^g \text{ centralizes } P, \ g \in G\},$$

$$L(P) = \{P\} \cup \{P^y \mid y \in O_p\langle\Omega(P)\rangle\},$$

and finally set

$$\mathscr{L} = \{L(P) \mid P \in \mathscr{P}\}.$$

PROPOSITION 3.61. *If* $S = S(G)$ *denotes the graph whose points are the elements of* $\mathscr{P}$ *and whose lines are the elements of* $\mathscr{L}$, *then* $S$ *is a Shult graph.*

There is a direct connection between the incidence relation of $S$ and properties of the $G$-conjugates of $J$ in the above situation.

PROPOSITION 3.62. *If* $P_1, P_2$ *are distinct points of* $\mathscr{P}$, *then one of the following holds:*

(i) $\langle P_1, P_2 \rangle = J^g$ *for some* $g \in G$; *or*
(ii) $P_i \leqslant J^{g_i}$, $i = 1, 2$, *with* $[J_1, J_2] = 1$ *for suitable elements* $g_1, g_2 \in G$.

This proposition shows that $S$ can be reconstructed from the set $\Omega$ of $G$-conjugates of $J$ with two vertices $J_1, J_2 \in \Omega$ connected by an edge if and only if $[J_1, J_2] = 1$.

In particular, for any Shult graph $S$ and any point $P$ of $S$, let $S(P)$ be the set of points $R$ of $S$ "adjacent" to $P$ (i.e., for which $\{P, R\}$ is contained in some line of $S$). Then in the case $S = S(G)$ above, if $Q = Q_1$ and $Q_2$ are distinct Sylow $p$-subgroups of $J$, we see that $\langle S(Q_1) \cap S(Q_2)\rangle$ is precisely the subgroup $X$ in the symplectic row of Table 3.4. Thus we have

$$\langle S(Q_1) \cap S(Q_2)\rangle \cong Sp_{n-2}(q). \tag{3.44}$$

The symplectic case of the Buekenhout–Shult theorem is, in essence, a converse to these assertions. (Their general result includes a similar characterization of the orthogonal groups.) We state it in the following form, which suffices for Aschbacher's identification of $F^*(G)$ as a symplectic group. [Here $\mathrm{Aut}(S)$ denotes the group of one–one line-preserving transformations of the points of $S$.]

THEOREM 3.63. *Let S be a Shult graph such that for any two nonadjacent points $P_1, P_2$ of S, the subgraph $S(P_1) \cap S(P_2)$ is isomorphic to the Shult graph of $PSp_m(q)$ for some fixed m and prime power q. Then*

$$\mathrm{Aut}(S) \cong \mathrm{Aut}\big(PSp_n(q)\big), \text{ where } n = m + 2.$$

In effect, their proof amounts to the assertion that the graph $S$ is *uniquely* determined from the given conditions.

Aschbacher points out that it is possible to establish analogous geometric characterizations of all the groups of Lie type, but that in many cases the identification of $F^*(G)$ by this approach requires more effort than using the Curtis–Phan relations.

## 3.6. Aschbacher's Classical Involution Theorem

The possibility of a classification of groups containing a tightly embedded subgroup with quaternion Sylow 2-subgroups was first suggested by Thompson in his American Mathematical Society Colloquium lectures at the January 1974 meeting of the American Mathematical Society in San Francisco, as an outgrowth of his proposed attack on the $B$-conjecture. As we have already described in Section 3.3, conversations between Aschbacher, Thompson, and Walter at the Sapporo conference in 1974 led to its final formulation, encompassing both the quaternion tightly embedded subgroup and subnormal $SL_2(q)$ problems.

Here is its general statement.

THEOREM 3.64. *Let G be a group with $F^*(G)$ simple such that for some involution z of G, $C_z$ possesses a subnormal subgroup J satisfying the following conditions*:

(1) *J has quaternion Sylow 2-subgroups and z is the unique involution of J.*
(2) *If u is any element of $C_z$, then either u normalizes J or $[J, J^u] \leqslant O(J) \cap O(J^u)$ ($\leqslant O(C_z)$); and*
(3) *If y is a 2-element of J of order $\geqslant 4$ and $y^g$ centralizes z for $g \in G$, then $y^g$ normalizes J.*

*Under these conditions, either $F^*(G)$ is a group of Lie type of odd characteristic, $M_{11}$, $M_{12}$, $PSp_6(2)$, or $D_4(2)$.*

Conditions (2) and (3) are very similar to Proposition 3.54(i) and (ii). Indeed, the principal distinction is that in the prior result one has $J \cong SL_2(q)$, $q$ odd, whereas in the present theorem $J$ is an *arbitrary* group with quaternion Sylow 2-subgroups, in which case it follows from the Brauer–Suzuki theorem [I, 4.88] and the classification of groups with dihedral Sylow 2-subgroups (Theorem 1.47) that either

(1) $J/O(J) \cong SL_2(q)$, $q$ odd, $\hat{A}_7$, or a quaternion group; or
(2) $J/O(J)$ contains a subgroup of index 2 isomorphic to $SL_2(q)$.          (3.45)

(Cf. Theorem 1.62.)

[Note that for all odd $q$, the group $K = SL_2(q)$ possesses a unique extension by an element $t$ of order 4 such that $\langle t^2 \rangle = Z(K)$ with $K\langle t \rangle$ having quaternion Sylow 2-subgroups (the group $K\langle t \rangle / \langle t^2 \rangle$ is isomorphic to $PGL_2(q)$). In (2), one has $J/O(J) \cong K\langle t \rangle$.]

In particular, it is the existence of the core $O(J)$ which forces one to impose only the weaker relation $[J, J^u] \leqslant O(J) \cap O(J^u)$ in condition (2) rather than the stronger condition $[J, J^u] = 1$, which holds in the groups of Lie type themselves. Moreover, a central portion of Aschbacher's proof is concerned with removal of the obstruction $O(J)$, which is accomplished with the aid of signalizer functor theory.

The tightly embedded quaternion Sylow 2-group theorem and the standard formulation of the classical involution theorem are each corollaries of Theorem 3.64. Indeed, if $J$ satisfies condition (1) of Theorem 3.64 and, in addition, is tightly embedded in $G$, it is immediate that $J$ satisfies (2) and (3). Hence we have the following theorem:

THEOREM 3.65. *If $G$ is a group with $F^*(G)$ simple which contains a tightly embedded subgroup with quaternion Sylow 2-subgroups, then $F^*(G)$ is either a group of Lie type of odd characteristic, $M_{11}$, or $M_{12}$.*

On the other hand, derivation of the classical involution theorem from Theorem 3.64 depends on the following result which Aschbacher establishes in a short, separate paper [2] dealing with intrinsic 2-components of type $SL_2(q)$, $q$ odd, or $\hat{A}_7$ (including the solvable case $q = 3$).

PROPOSITION 3.66. *Let $J, K$ be distinct intrinsic 2-components of the group $G$ with $J/O(J) \cong K/O(K) \cong SL_2(q)$, $q$ odd, or $\hat{A}_7$, and let $z(J), z(K)$ be the unique involutions of $J, K$, respectively. Then we have*

(i) *If y is a 2-element of K of order $\geqslant 4$ which centralizes $z(J)$, then y normalizes J; and*

(ii) *If $z(J)$ normalizes K, then $z(K)$ normalizes J.*

Aschbacher's proof does not require $J/O(J)$ and $K/O(K)$ to be isomorphic, but it is the only case we require here. Moreover, only part (i) is needed for the classical involution theorem. However, as (ii) is critical for the proofs of both (i) and Theorem 3.64, we have included its statement here.

Theorem 3.64 and Proposition 3.66(i) together immediately yield the following result:

THEOREM 3.67 (*Classical involution theorem*). *If G is a group with $F^*(G)$ simple such that for some involution z of G, $C_z$ possesses a subnormal subgroup J with $J/O(J) \cong SL_2(q)$, q odd, or $\hat{A}_7$ and $z \in J$, then $F^*(G)$ is either a group of Lie type of odd characteristic or $M_{11}$.*

Because of its importance, we outline the proof of Proposition 3.66. It is an elementary, but very pretty argument, involving a combination of *L*-balance (including Aschbacher's solvable 2-component extension), 2-fusion, and specific properties of the groups $SL_2(q)$ and $\hat{A}_7$. For clarity of exposition, we assume

$$O(C_u) = 1 \text{ for every } u \in \mathscr{I}(G). \tag{3.46}$$

This simplifies the notation, but has no effect on the basic argument. Note also that because of the symmetry between $J$ and $K$ in the proposition, all assertions will hold for any pair of distinct intrinsic components isomorphic to $J$.

We let $t, z$ be the involutions of $J, K$, respectively. The first observation to be made is the following:

LEMMA 3.68. *If t centralizes a quaternion subgroup of K, then J centralizes K.*

Indeed, in this case, it follows from the structure of $K$ that $K_0 = C_K(t) \cong SL_2(q_0)$ for suitable $q_0$ or $\hat{A}_7$ and that $z \in K_0$. Hence by *L*-balance [and (3.46)], $K_0 \leqslant L(C_t)$. But by hypothesis $J$ is a component of $C_t$, which immediately forces $z \in K_0$ to centralize $J$. Hence again by *L*-balance $J \leqslant L(C_z)$ and likewise $K$ is a component of $C_z$. Since $J$ and $K$ are distinct, we conclude that $J$ centralizes $K$, as asserted.

Similarly one proves:

LEMMA 3.69. *If z leaves J invariant and $K^t \neq K$, then the following conditions hold*:

  (i)  $C_{KK^t}(t) = \langle z \rangle \times L$, *where $L \cong L_2(q)$ or $A_7$*;
  (ii) *L centralizes J; and*
  (iii) *z centralizes no element of J of order* 4.

Indeed, our assumptions imply that $KK^t \cong SL_2(q) * SL_2(q)$ or $\hat{A}_7 * \hat{A}_7$ and (i) follows at once. Since $J$ is a $z$-invariant component of $C_t$ and $L \leqslant L(C_t)$ by $L$-balance, $L$ must centralize $J$ and (ii) holds. Furthermore, if $z$ centralizes $a$ of order 4 in $J$, then $a^2 = t$ (as $K$ has quaternion Sylow 2-subgroups) and $a \in C_z$. But $a$ centralizes $L$ by (ii). Since $a$ induces a permutation of the components of $C_z$ and $L$ is a diagonal of $KK^t$, it follows that $a$ leaves $KK^t$ invariant, whence $a^2 = t$ leaves *both* $K$ and $K^t$ invariant, contrary to the assumption $K^t \neq K$. Thus (iii) holds as well.

Now we can prove the following:

LEMMA 3.70. *Proposition* 3.66(ii) *holds*.

Indeed, suppose $z$ normalizes $J$, but $K^t \neq K$. Since $t \in J$, $J$ does not centralize $K$, so by Lemma 3.68, $z$ does not centralize $J$. Hence by the structure of $J$,

$$z \text{ is } J\text{-conjugate to } zt. \tag{3.47}$$

Let $L$ be as in the previous lemma and let $u$ be an involution of $L$. Then $u$ centralizes $J$ and so by (3.47)

$$uz \text{ is } J\text{-conjugate to } uzt. \tag{3.48}$$

On the other hand, by the structure of $KK^t$ and the embedding of $L$ in $KK^t$, we have

$$\begin{array}{l} (1) \ uzt \text{ is } KK^t\text{-conjugate to } t; \text{ and} \\ (2) \ u \text{ is } KK^t\text{-conjugate to } uz. \end{array} \tag{3.49}$$

Combined with (3.48), this yields:

$$u \text{ is } G\text{-conjugate to } t. \tag{3.50}$$

Hence $C_u$ contains an intrinsic component $I \cong J$. Now from the embedding of $L$, $u$ centralizes an element of order 4 in $K$. We conclude therefore from the previous lemma with $I$, $u$, $K$, $z$ in the roles of $K$, $z$, $J$, $t$, respectively, that $z$ normalizes $I$. Furthermore, as $u$ does not centralize $K$, neither does $I$, so by Lemma 3.68, $z$ does not centralize $I$ and consequently

$$z^a = zu \text{ for some } a \in I. \tag{3.51}$$

But by $(3.49)(2)$, $u^b = uz$ for some $b \in KK^t$. Hence if we set $U = \langle u, z \rangle$ $(\cong Z_2 \times Z_2)$, then as $\langle a, b \rangle \leqslant N_G(U)$, it follows that

$$N_G(U)/C_G(U) \cong \Sigma_3. \tag{3.52}$$

Finally for any involution $x$ of $G$, let $X_x$ denote the product of all intrinsic components $X$ of $C_x$ with $X \cong J$ (set $X_x = 1$ if no such components exist).

Then we have $U \leqslant KK^t \leqslant X_z$, so by (3.52), $z \in U \leqslant X_u$. Since, as noted above, $z$ does not centralize $J$, we conclude that $X_u$ does not centralize $J$. However, as $u$ centralizes $J$, Lemma 3.68 implies that each component of $X_u$ and hence $X_u$ itself centralizes $J$, a contradiction.

Our discussion of Proposition 3.66(i) will be sketchier; its proof is similar to the above argument, but slightly more delicate. We argue by contradiction, whence clearly $J$ does not centralize $K$.

Let $y$ be as in the proposition, so that $y \in J - \langle t \rangle$ centralizes $z$ and $K^y \neq K$. Since $t \in \langle y \rangle$, $z \in C_t$. But $y \in J - \langle t \rangle$ and $z$ induces a permutation of the components of $C_t$, thus forcing $z$ to leave $J$ invariant. Hence by Proposition 3.66(ii), we conclude:

LEMMA 3.71. *z normalizes $J$ and $t$ normalizes $K$.*

In particular, $z$ centralizes $t$. Since $J$ does not centralize $K$, Lemma 3.68 implies that $z \neq t$, so $V = \langle z, t \rangle \cong Z_2 \times Z_2$. Furthermore, again by Lemma 3.68 and the structure of $K$, either $t$ induces an inner automorphism on $K$ or else $K\langle t \rangle$ has quasi-dihedral Sylow 2-subgroups, with a similar conclusion holding for $J$ and $z$. Using this information, a fusion argument now yields the following result:

LEMMA 3.72. (i) $N_G(V)/C_G(V) \cong \Sigma_3$; *and*
(ii) *$t$ induces an inner automorphism on $K$.*

Let $Q$ be a $t$-invariant Sylow 2-subgroup of $K$ such that $C_Q(t) \in$ $\mathrm{Syl}_2(C_K(t))$. Since $t$ induces an inner automorphism on $K$ and $Q$ is quaternion, $C_Q(t)$ is cyclic of index 2 in $Q$ and so $Q = C_Q(t)\langle d \rangle$ for some element $d \in Q - C_Q(t)$ of order 4. Also $d$ is inverted by an element $e$ of order 4 in $C_Q(t)$. Finally set $x = ee^y$. Since $e^2 = z$ and $y$ centralizes $z$, $x$ is an involution of $KK^y - \langle z \rangle$.

Arguing in the same spirit as above, Aschbacher next establishes the following facts:

LEMMA 3.73.   (i) $J^x \neq J$;

                (ii) $tz$ *leaves invariant every component of* $X_x$; *and*

                (iii) $t \in KK^y$.

But all noncentral involutions of $KK^y$ are conjugate in $KK^y$; hence, in particular, $x$ and $t$ are conjugate. However, $x$ centralizes a subgroup of $K$ of order 4 and so by Lemma 3.69(iii), $z$ leaves invariant every component of $X_x$. Combined with Lemma 3.73(ii), this yields:

LEMMA 3.74. $t$ *leaves invariant every component of* $X_x$.

Finally as $x$ and $t$ are conjugate, $X_x$ is nonempty. But now if $I$ is a component of $X_x$, $t$ leaves $I$ invariant, so by Proposition 3.66(ii), $x$ leaves invariant every component of $X_t$, contrary to the fact that $J \leqslant X_t$ and $J^x \neq J$ by Lemma 3.73(i). This completes the discussion of Proposition 3.66.

We return now to Theorem 3.64. Setting $J_0 = O^{2'}(J)$ (the subgroup of $J$ generated by its 2-elements), then except in the case $J/O(J) \cong SL_2(3)$, one has $J_0/O(J_0) \cong J/O(J)$. Moreover, the conditions of the theorem hold with $J_0$ in place of $J$. Since $J_0 = O^{2'}(J_0)$, we can therefore assume:

$$\text{Either } J = O^{2'}(J) \text{ or } J/O(J) \cong SL_2(3). \tag{3.53}$$

Furthermore, when $J/O(J) \cong SL_2(3)$, we can similarly reduce to the case in which $J/O^{2'}(J)$ is a cyclic 3-group.

In order to carry out an inductive argument, Aschbacher is forced to consider a somewhat more general situation than that of Theorem 3.64, which he refers to as *Hypothesis* $\Omega$.

HYPOTHESIS $\Omega$.   $G$ *is a group with* $O(G) = 1$ *and* $\Omega$ *is a G-invariant collection of subgroups of $G$ such that for each choice of distinct* $J, K \in \Omega$, *we have*

(1) $J$ has quaternion Sylow 2-subgroups and contains a unique involution, denoted by $Z(J)$;

(2) Either $J = O^{2'}(J)$ or $J/O(J) \cong SL_2(3)$, and $J/O^{2'}(J)$ is a cyclic 3-group;

(3) $J/O(J) \cong K/O(K)$;

(4) Sylow 2-subgroups of $J \cap K$ have order $\leqslant 2$, and if equality holds, then $[J, K] \leqslant O(J) \cap O(K)$ (i.e., $J$ and $K$ centralize each other mod cores); and

(5) If $y$ is a 2-element of $K$ of order $\geqslant 4$ that centralizes $Z(J)$, then $y$ normalizes $K$.


Theorem 3.64 corresponds essentially to the case of Hypothesis $\Omega$ in which

$$
\begin{array}{ll}
(1) \ \ G = \langle \Omega \rangle; & \\
(2) \ \ G \text{ is transitive on } \Omega; \text{ and} & (3.54) \\
(3) \ \ F^*(G) \text{ is simple.} &
\end{array}
$$


Note that the subnormality condition in Theorem 3.64 is a consequence of Hypothesis $\Omega(4)$:


LEMMA 3.75. *If $J \in \Omega$, then $J$ is subnormal in $C_{z(J)}$, where $z(J)$ is the unique involution of $J$.*

Indeed, setting $C = C_{z(J)}$, then for any $y \in C$, $z(J) \in J \cap J^y$, so by $\Omega(4)$ $[J, J^y] \leqslant O(J)$ and consequently $J^y$ normalizes $J$. Thus $J$ is normal in $H = \langle J^y \mid y \in C \rangle$. But $H \lhd C$, so by definition $J$ is subnormal in $C$.

However, for the induction it is necessary for Aschbacher to completely classify groups $G$ satisfying Hypothesis $\Omega$ and only (3.54)(1). In particular, in order to determine inductively the structure of the centralizers of certain involutions he must allow the possibility that $O_2(G) \neq 1$ and/or $G$ is intransitive on $\Omega$.

It is a long and difficult argument. Indeed, to establish his main result, Aschbacher proves *seven* auxiliary theorems and for most of the proof considers a minimal counterexample to all *eight* results (i.e., to both the principal classification and seven supporting theorems). We shall state (all but one of) these results, for they will provide a preliminary picture of Aschbacher's global strategy. Throughout $G$ will denote a group satisfying Hypothesis $\Omega$.

Central to the entire analysis is a fundamental graph $\mathscr{D}$ which Aschbacher associates with $\Omega$. Since several of the auxiliary statements explicitly involve $\mathscr{D}$, it is best to define it first. Its definition depends upon the following preliminary result.

PROPOSITION 3.76. *If* $J, K$ *are distinct members of* $\Omega$ *and a Sylow* 2-*subgroup of* $K$ *normalizes* $J$, *then a Sylow* 2-*subgroup of* $J$ *normalizes* $K$.

Indeed, let $Q, R$ be Sylow 2-subgroups of $J, K$, respectively, and suppose that $R$ normalizes $J$. Replacing $Q$ by a suitable $J$-conjugate, we can assume that $R$ normalizes $Q$. Set $\langle z \rangle = Z(R)$, so that $\langle z \rangle = Z(K)$ and put $Q_1 = C_Q(z)$. We must show that $Q_1$ normalizes $K$ and that $Q_1 = Q$.

Now $R$ normalizes some subgroup $Q_0$ of $Q$ of order 4 as $R$ normalizes $Q$. Since $R$ is nonabelian, $z$ centralizes $Q_0$, so $Q_0 \leqslant Q_1$. Since $Q$ is quaternion, it follows that $Q_1$ is either cyclic of order $\geqslant 4$ or quaternion. Hence by Hypothesis $\Omega(5)$, $Q_1$ normalizes $K$. Thus the lemma holds if $Q_1 = Q$, we can assume that $Q_1 < Q$.

Clearly $[Q_1, R] \leqslant J \cap K$ and so by Hypothesis $\Omega(4)$, $[Q, R]$ has odd order. But $R = C_R(z)$ normalizes $Q_1$ and hence $[Q_1, R] \leqslant Q_1$ is a 2-group, forcing $[Q_1, R] = 1$. Thus $R$ centralizes $Q_1$.

Finally as $Q_1 < Q$, $N_{RQ}(RQ_1) > RQ_1$ by $[I, 1.10]$ and so has the form $RQ_2$ with $Q_2 > Q_1$. But $Q \cong R$, so $Q_1$ is isomorphic to a *proper* subgroup of $Q$. Since $R$ is quaternion and $Q_1$ centralizes $R$, it is immediate that $\langle z \rangle$ char $RQ_1$, so $Q_2$ normalizes $\langle z \rangle$. Thus $Q_2 \leqslant C_Q(z_1) = Q_1$, contrary to the fact that $Q_2 > Q_1$.

Now we can define $\mathscr{D}$.

DEFINITION 3.77. *Let* $\mathscr{D}$ *be the graph with vertex set* $\Omega$ *and two distinct vertices* $J, K$ *joined by an edge if some Sylow* 2-*subgroup of* $K$ *normalizes* $J$.

In view of Proposition 3.76, $\mathscr{D}$ is well-defined, the vertices $J, K$ being joined by an edge if and only if a Sylow 2-subgroup $R(J)$ of $J$ *and* a Sylow 2-subgroup $R(K)$ of $K$ normalize $K$ and $J$, respectively. Moreover, for appropriate choice of $R(J)$ and $R(K)$, one then has

$$[R(J), R(K)] = 1. \tag{3.55}$$

Furthermore, as $R(J)$ normalizes $K$ and $R(J)$ and $R(K)$ are quaternion, it is immediate that the involuion of $R(J)$ centralizes $K/O(K)$. Hence if $z(J)$ denotes the unique involution of $J$, then

$$z(J) \text{ centralizes } K/O(K). \tag{3.56}$$

Clearly the group $G$ acts by conjugation on $\mathscr{D}$. In the "generic" case, $\mathscr{D}$ is connected, $G$ acts transitively on $\mathscr{D}$, and $G$ is generated by the elements of $\Omega$. A large portion of the argument deals with the reduction to this situation and the auxiliary results form an integral part of that reduction.

For all but the final result, which deals with the structure of $G$ in the intransitive case, we shall assume

$$\begin{array}{ll} (1) & G = \langle \Omega \rangle; \text{ and} \\ (2) & G \text{ is transitive on } \mathscr{D}. \end{array} \tag{3.57}$$

Aschbacher's main theorem is as follows:

THEOREM 3.78. *Either* $z(J) \in O_2(G)$ *for* $J \in \Omega$ *or* $F^*(G)$ *is quasisimple and isomorphic to one of the groups listed in Theorem* 3.64.

His first auxiliary result is a complete determination of the possibilities for $G$ when $z(J) \in O_2(G)$, which includes a number of small 2-constrained groups. For simplicity, at this time we state the precise conclusions only in the connected case.

Recall first from [I, p. 262] the definition of the natural module $V$ for $A_n$. According as $n$ is even or odd, $V$ is an elementary abelian 2-group of rank $n - 2$ or $n - 1$. We denote the semidirect product of $V$ by $A_n$ as $(A/E)_n$.

PROPOSITION 3.79. *If* $z(J) \in O_2(G)$ *for every* $J \in \Omega$, *then we have*

(i) *Either* $J = G$ *or* $J/O(J)$ *is a 2-group; and*
(ii) *If* $\mathscr{D}$ *is connected and contains an edge, then* $G/Z(G) \cong (A/E)_n$.

Note that if $\mathscr{D}$ is connected and contains no edges, then $\mathscr{D}$ consists of the single point $J$, in which case we have the trivial possibility $J = G$.

In particular, (i) shows that the case $z(J) \in O_2(G)$ arises only in connection with the tightly embedded quaternion Sylow 2-group problem. Hence as far as the classical involution theorem itself is concerned, if $G$ satisfies (3.57) and $\mathscr{D}$ has more than one vertex, then $F^*(G)$ is necessarily quasisimple.

In the next three results, we assume in addition to (3.57):

$$F^*(G) \text{ is simple.} \tag{3.58}$$

The first two concern the nonconnected case.

PROPOSITION 3.80. *If* $\mathscr{D}$ *contains no edges, then*

$$F^*(G) \cong L_2(q), L_3(q), U_3(q), G_2(q), \text{ or } {}^3D_4(q), q \text{ odd.}$$

[Although $L_2(q)$ contains no quaternion subgroups, when $q$ is a square its extension $L_2(q)^*$ by the product of the graph automorphism and a field automorphism of period 2 has quasi-dihedral Sylow 2-subgroups and hence contains a quaternion subgroup. Moreover, $L_2(q)^*$ satisfies Hypothesis $\Omega$ and the conditions of the proposition.]

PROPOSITION 3.81. *If* $\mathscr{D}$ *possesses an edge, but is disconnected, then*

$$F^*(G) \cong L_4(q), U_4(q), PSp_4(q), P\Omega_7(q), P\Omega_8^{\pm}(q), q \text{ odd, or } PSp_6(2)$$

We note the isomorphisms:

$$PSp_4(q) \cong P\Omega_5(q), L_4(q) \cong P\Omega_6^+(q), U_4(q) \cong P\Omega_6^-(q), \tag{3.59}$$

so that one can rephrase the conclusion to read:

$$F^*(G) \cong P\Omega_n^{\pm}(q), 5 \leqslant n \leqslant 8, q \text{ odd,} \qquad \text{or } PSp_6(2). \tag{3.60}$$

The next result singles out arbitrary dimensional orthogonal groups, whose exceptional character among the groups of Lie type has been stressed in the previous section.

PROPOSITION 3.82. *If there exist two distinct elements of* $\Omega$ *containing the same involution, then*

$$F^*(G) \cong P\Omega_n^{\pm}(q), q \text{ odd, } n \geqslant 5, PSp_6(2), \text{ or } D_4(2).$$

Now we describe the result which Aschbacher proves to eliminate the problem of cores. Its proof ultimately depends, of course, on the construction of suitable signalizer functors on $G$. However, the key object of attention is defined with respect to *four*-subgroups.

DEFINITION 3.83. For any group $X$ and four-subgroup $B$ of $X$, set

$$\alpha_X(B) = \langle [O(C_X(b)), B] \mid b \in B^{\#} \rangle \Delta_X(B)$$

[where, as usual, $\Delta_X(B) = \bigcap_{b \in B^{\#}} O(C_X(b))$].

PROPOSITION 3.84. *If $J$ and $K$ are distinct elements of $\Omega$ that are connected by an edge and $B$ is a four-subgroup of $JK$, then $\alpha_G(B) = 1$.*

In the terminology of [I, 4.68], this is essentially the assertion that $G$ is *strongly locally* 2-*balanced* with respect to $B$.

Finally we consider the intransitive case. To avoid technicalities, we combine Aschbacher's two results concerning this case and, in addition, state it in somewhat simplified form.

PROPOSITION 3.85. *Assume $G = \langle \Omega \rangle$, let $\Omega_1, \Omega_2, ..., \Omega_r$ be the distinct transitive constituents of $\Omega$ under the action of $G$, and set $G_i = \langle \Omega_i \rangle$, $1 \leqslant i \leqslant r$. Then $G_i$ centralizes $G_j$ for all $i \neq j$, $1 \leqslant i, j \leqslant r$.*

(In particular, every vertex of $\Omega_i$ is connected by an edge to every vertex of $\Omega_j$ for $i \neq j$.)

Thus the proposition reduces the general case to the transitive situation. Hence Aschbacher shows that the generic group $G$ satisfying Hypothesis $\Omega$ with $G = \langle \Omega \rangle$ and $J/O(J)$ not a 2-group is a central product of groups of Lie type of odd characteristic.

## 3.7. The Generic Nonorthogonal Case

In this section we outline Aschbacher's proof of the general nonorthogonal case of the classical involution theorem, where the induction works most smoothly. To simplify the analysis, we shall impose more stringent conditions on $\mathscr{D}$ than does Aschbacher. However, these assumptions preserve the general features of his argument.

Throughout we assume

G is a minimal counterexample to Theorem 3.78 and
Propositions 3.79–3.85; and subject to this condition, $|\Omega|$ is
minimal.                                                                                (3.61)

To state the result precisely, we need some preliminary terminology. First, for any $J \in \Omega$, let $\Omega(J)$ denote the set of $K \in \Omega$, distinct from $J$, which are connected to $J$ by an edge. Furthermore, define $\mathscr{D}(J)$ to be the subgraph of $\mathscr{D}$ whose vertex set is $\Omega(J)$ and put $X(J) = \langle K \mid K \in \Omega(J) \rangle$. $\Omega(J)$, $\mathscr{D}(J)$, and $X(J)$ play essential roles throughout the analysis. Also, as in the previous section, for any $J \in \Omega$, denote the unique involution of $J$ by $z(J)$.

In view of Proposition 3.82, the nonorthogonal case is characterized by the following definition.

DEFINITION 3.85. We say that $\mathscr{D}$ is *multiplicity free* if $z(J) \neq z(K)$ for any pair of distinct elements $J, K \in \Omega$.

Note that if $J, K$ are connected by an edge, then $[J, K] \leqslant O(J) \cap O(K)$ and consequently $z(J)$ centralizes $z(K)$. Hence, if $\mathscr{D}$ is multiplicity free, it then follows that

$$\langle z(J), z(K) \rangle \cong Z_2 \times Z_2.$$                              (3.62)

DEFINITION 3.86. Assume $\mathscr{D}$ is multiplicity free. We call $\mathscr{D}$ *strongly connected* if $\mathscr{D}$ is connected and for any $J \in \Omega$ and any $K, L \in \Omega(J)$,

$$|\Omega(J) \cap \Omega(K) \cap \Omega(L)| \geqslant 3.$$

In particular, if $K, L \in \Omega(J)$, there is $I \in \Omega(J)$ with $I, K$ and $I, L$ each joined by an edge, so $K$ is connected to $L$ and consequently $\Omega(J)$ is connected.

Here is the general result we shall discuss.

THEOREM 3.87. *Assume the following conditions*:

(a) $\mathscr{D}$ *is multiplicity free and strongly connected*;
(b) *If* $J \in \Omega$, *then* $X(J)$ *acts transitively on* $\Omega(J)$; *and*
(c) $J/O(J)$ *is not a* 2-*group*.

*Then* $F^*(G) \cong A_n(q)$, ${}^2A_n(q)$, $B_n(q)$, $E_6(q)$, ${}^2E_6(q)$, $E_7(q)$, *or* $E_8(q)$ *for suitable n and odd q*.

For the proof, we fix $J \in \Omega$ and first prove the following result:

PROPOSITION 3.88. *If g is an element of G such that $J^g \in \Omega(J)$, then*

$$G = \langle X(J), X(J)^g \rangle.$$

Aschbacher establishes the proposition without conditions (b) and (c) and with connectivity in place of strong connectivity in (a), in a short, separate paper involving some purely graph-theoretic results [3]. However, under our hypotheses, its proof is quite direct.

Set $X = X(J) = \langle \Omega(J) \rangle$ and put $G^* = \langle X, X^g \rangle$. It will suffice to prove that $G^*$ contains every element of $\Omega$, for then $G^* \geqslant \langle \Omega \rangle = G$, whence $G^* = G$, as asserted.

We argue that for any $K \in \Omega$, there is an element $u \in G^*$ such that

$$K = J^u \text{ or } J^{gu}. \tag{3.63}$$

Since $\langle J, J^g \rangle \leqslant \langle X, X^g \rangle = G^*$, it will then follow that $K \leqslant G^*$, as required.

Suppose false for some $K \in \Omega$. By assumption, $\mathscr{D}$ is connected. We choose $K$ so that the connecting chain

$$J = J_1, J_2, \ldots, J_r = K$$

has minimal length. Clearly $K \neq J$, so $r > 1$ and by the minimality of $r$, (3.63) holds for $L = J_{r-1}$ and appropriate $u \in G^*$. But $L$ and $K$ are joined by an edge and hence so are $L^{u^{-1}} = J$ or $J^g$ and $K^{u^{-1}}$. Thus $K^{u^{-1}} \in \Omega(J)$ or $\Omega(J^g)$. However, by assumption $X$ is transitive on $\Omega(J)$, whence also $X^g$ is transitive on $\Omega(J^g)$. Since $J^g \in \Omega(J)$ and $J \in \Omega(J^g)$, it follows that for some $x \in X$ or $X^g$, we have correspondingly

$$K^{u^{-1}x} = J^g \text{ or } J. \tag{3.64}$$

Thus $K = J^{gv}$ or $J^v$, where $v = ux^{-1} \in G^*$, contrary to our choice of $K$.

We turn next to the question of cores. The main result is as follows:

PROPOSITION 3.89. *For any $K \in \Omega(J)$, we have*

$$\alpha_G(\langle z(J), z(K) \rangle) = 1.$$

We shall outline the argument. It will be convenient to introduce the following terminology.

DEFINITION 3.90. A *triangle* $\mathscr{E}$ of $\mathscr{D}$ is a set of three vertices $J_1, J_2, J_3$ of $\Omega$, each pair connected by an edge. The subgroup $A(\mathscr{E}) = \langle z(J_i) \mid 1 \leqslant i \leqslant 3 \rangle$ will be called the *axis* of $\mathscr{E}$. By (3.62), $A(\mathscr{E})$ is elementary abelian of order 4 or 8. Correspondingly we say that $\mathscr{E}$ has *rank* 2 or 3.

The key signalizer functor result is the following:

LEMMA 3.91. *Let $\mathscr{E}$ be a triangle of rank 3 with axis $A$ and fix a four-subgroup $B$ of $A$. If for any $a \in A^{\#}$, we set*

$$\theta_B(a) = [O(C_a), B](O(C_B) \cap C_a),$$

*then $\theta_B$ is an $A$-signalizer functor on $G$.*

This is obtained as a consequence of [I, 4.69], so Aschbacher must show that the strong local 2-balance assumption of that proposition is satisfied. The term *strong local 2-balance* was introduced in [I] solely for expository purposes. However, the precise conditions which are required for the proposition are the following:

(1) For any $a \in A^{\#}$, if we set $\overline{C}_a = C_a / O(C_a)\langle a \rangle$, then $\overline{C}_a$ contains at most one $A$-invariant component on which $\overline{A}$ acts faithfully; and
(2) If such a component $\tilde{L}$ exists and we set $\tilde{H} = N_{\overline{C}}(\tilde{L})/C_{\overline{C}_a}(\tilde{L})$ (whence $\tilde{L}$ is simple and $\tilde{A} \cong Z_2 \times Z_2$), then $a_{\tilde{H}}(\tilde{A}) = 1$.

$$(3.65)$$

We remark that the proof that $\theta_B$ is an $A$-signalizer functor under the conditions of (3.65) involves some careful, but elementary, commutator calculations. Underlying it are the following two preliminary facts, which are a direct consequence of (3.65) and are derived in much the same way as [I, 4.58 and 4.64]. For any $a \in A^{\#}$, we have

(1) $[O(C_{a'}) \cap C_a, B] \leqslant O(X_a)$ for any $a' \in A^{\#}$.
(2) $\Delta_G(B) \cap C_a \leqslant O(C_a)$.

$$(3.66)$$

Hence, to complete the discussion of the lemma, it remains to verify (3.65), which is an assertion about the structure of $C_a$ for $a \in A^{\#}$. Determination of that structure depends crucially on the minimal choice of $G$, which allows Aschbacher to apply the various results in Section 3.6 to the group $C_a^* = C_a/O(C_a)$. Note that as $O_2(C_a^*) \neq 1$, this also explains why he was forced to allow groups $G$ with $O_2(G) \neq 1$ in his general theorem.

Let $J_1, J_2, J_3$ be the vertices of $\mathscr{E}$, so that $J_i/O(J_i) \cong J/O(J)$, $1 \leqslant i \leqslant 3$. Also set $z_i = Z(J_i)$, $1 \leqslant i \leqslant 3$. Since $\mathscr{E}$ is a triangle and $A = \langle z_1, z_2, z_3 \rangle$, each $H_i = C_{J_i}(A)$ covers $J_i/O(J_i)$. According as $J/O(J)$ is nonsolvable or solvable, set $I_i = O^{2'}(H_i)$ or $O^{2'}(H_i)D_i$, where $D_i$ is a cyclic 3-subgroup of $H_i$ not centralizing $H_i/O(H_i)$. Then for each $i$, $1 \leqslant i \leqslant 3$, we have

> (1) $I_i/O(I_i) \cong J/O(J)$;
> (2) Either $I_i = O^{2'}(I_i)$ or $I_i/O(I_i) \cong SL_2(3)$ and $I_i/O^{2'}(I_i)$        (3.67)
>     is a cyclic 3-group; and
> (3) $z_i \in I_i$.

If we now set $C_a^* = C_a/O(C_a)$ and let $\Omega^*$ be the collection of $C_a^*$-conjugates of $I_i^*$, $1 \leqslant i \leqslant 3$, then as $O(C_a^*) = 1$, we see that Hypothesis $\Omega$ holds with $C_a^*$ and $\Omega^*$ in the roles of $G$ and $\Omega$ [and $z_i = z(I_i)$, $1 \leqslant i \leqslant 3$]. Since $F^*(G)$ is simple, $C_a < G$, so we can indeed apply the results in Section 3.6 to $C_a^*$.

Observe next that as $\Omega^*$ is $C_a^*$-invariant, $G^* = \langle \Omega^* \rangle$ is a normal subgroup of $C_a^*$. In particular, $O(G^*) = 1$. Proposition 3.85 is therefore applicable to the subgroups $G_j^* = \langle \Omega_j^* \rangle$ of $G^*$, where $\Omega_j^*$, $1 \leqslant j \leqslant r$, denote the distinct orbits of $\Omega^*$ under the action of $G^*$. We thus conclude that

$$G^* = G_1^* G_2^* \cdots G_r^* \quad \text{and} \quad [G_j^*, G_k^*] = 1 \quad \text{for all } j \neq k. \quad (3.68)$$

But by Theorem 3.78 and Proposition 3.79, $L_j^* = F^*(G_j^*)$ is quasisimple [or $SL_2(3)$]. Since $G^* \lhd C_a^*$, (3.68) therefore implies that each $L_j^*$ is a component (possibly solvable) of $C_a^*$.

Now we can determine the embedding of $A$ in $C_a$ and establish (3.65). Indeed, by the definition of $G^*$, each $I_i$ is contained in a unique $G_j^*$ ($j$ depending on $i$). Moreover, from the exact structure of $G_j^*$ (Theorem 3.78), it follows that $z_i^* \in L_j^*$. Hence if we set $\bar{C}_a = C_a^*/\langle a^* \rangle$ $(\cong C_a/O(C_a)\langle a \rangle)$, we conclude for $1 \leqslant i \leqslant 3$ that

$$\bar{z}_i \text{ is contained in a unique component } \bar{L}_j \text{ of } \bar{C}_a, \quad (3.69)$$

(where $\bar{L}_j$ is the image of $L_j^*$ in $\bar{C}_a$). Moreover, $\bar{z}_i$ centralizes every component of $\bar{C}_a$ except possibly the component $\bar{L}_j$ in which it lies. In particular, $\bar{A} = \langle \bar{z}_i \mid 1 \leqslant i \leqslant 3 \rangle$ leaves invariant each component of $\bar{C}_a$.

Finally as $\bar{A} \cong Z_2 \times Z_2$, $\bar{A}$ is, in fact, generated by exactly two of the $\bar{z}_i$, say $\bar{z}_1, \bar{z}_2$. But then if $\bar{z}_1, \bar{z}_2$ are contained in distinct components of $\bar{C}_a$, $\bar{A}$ does not act faithfully on any component of $\bar{C}_a$ and (3.65) holds. In the contrary case, $\bar{A} = \langle \bar{z}_1, \bar{z}_2 \rangle \leqslant \bar{L}$ for some component $\bar{L} = \bar{L}_j$ of $\bar{C}_a$, $\bar{L}$ is quasisimple, and $\bar{L}$ is the unique component of $\bar{C}_a$ on which $\bar{A}$ acts faithfully. But now by Proposition 3.84, applied to $G_j^*$, and/or by properties of the groups of Lie type of odd characteristic, it follows that $\alpha_{\tilde{H}}(\tilde{A}) = 1$, where $\tilde{H} = N_{\bar{C}_a}(\tilde{L})/C_{\bar{C}_a}(\tilde{L})$, so (3.65) holds in this case as well.

As an immediate corollary, Aschbacher obtains the following result:

LEMMA 3.92. *Under the assumptions of Lemma 3.91, $\alpha_G(B)$ has odd order.*

Indeed, by the rank 3 signalizer functor theorem [I, 4.38] the completion $\theta_B(G; A) = \langle \theta_B(a) \mid a \in A^\# \rangle$ of $\theta_B$ has odd order. But as $\Delta_G(B) \leqslant O(C_B)$, $[O(C_b), B] \Delta_G(B) \leqslant \theta_B(b)$ for each $b \in B^\#$ and consequently

$$\alpha_G(B) = \langle [O(C_b), B] \mid b \in B^\# \rangle \Delta_G(B) \leqslant \theta_B(G; A); \qquad (3.70)$$

so $\alpha_G(B)$ has odd order, as asserted.

By commutator calculations similar to those used in proving that $\theta_B$ is an $A$-signalizer functor, Aschbacher also obtains the following critical equality:

LEMMA 3.93. *Under the assumptions of Lemma 3.91,*

$$\alpha_G(B) = \alpha_G(B')$$

*for any two four-subgroups $B, B'$ of $A$.*

We now bring our assumption of strong connectivity into play to prove the following further result:

LEMMA 3.94. *If $K, L$ are any two distinct elements of $\Omega(J)$, then*

$$\alpha_G(\langle z(J), z(K) \rangle) = \alpha_G(\langle z(J), z(L) \rangle) \text{ and is of odd order.}$$

Indeed, by strong connectivity $\Omega(J) \cap \Omega(K) \cap \Omega(L)$ contains three distinct elements $I_1, I_2, I_3$ of $\Omega$. Since $\mathscr{D}$ is multiplicity free, $z(I_i) = z(J) z(K)$

for at most *one* value of $i$ and likewise $z(I_i) = z(J) z(L)$ for at most *one* value of $i$, $1 \leqslant i \leqslant 3$. Hence for some $i$, if we set $I = I_i$, we have

$$z(I) \neq z(J) z(K) \quad \text{and} \quad z(I) \neq z(J) z(L). \tag{3.71}$$

But then as $I \in \Omega(J) \cap \Omega(K) \cap \Omega(L)$, it follows that

$$\mathscr{E}_K = \{I, J, K\} \quad \text{and} \quad \mathscr{E}_L = \{I, J, L\} \tag{3.72}$$

are each triangles of rank 3. Hence Lemmas 3.92 and 3.93 apply to their respective axes $A_K$ and $A_L$. But then

$$\begin{array}{l} (1) \ \alpha_G(\langle z(I), z(J)\rangle) \text{ has odd order; and} \\ (2) \ \alpha_G(\langle z(J), z(K)\rangle) = \alpha_G(\langle z(I), z(J)\rangle) = \alpha_G(\langle z(J), z(L)\rangle), \end{array} \tag{3.73}$$

which together yield the desired conclusions.

Similarly we can prove:

LEMMA 3.95. *For any* $K \in \Omega(J)$, $X(J)$ *normalizes* $\alpha_G(\langle z(J), z(K)\rangle)$.

Indeed, by definition $X(J) = \langle \Omega(J)\rangle$, so it will suffice to prove that each $L \in \Omega(J)$ normalizes $\alpha_G(\langle z(J), z(K)\rangle)$.

Suppose $L \neq K$ and let the notation be as in the previous lemma. Also set $B_K = \langle z(J), z(K)\rangle$, $B_L = \langle z(J), z(L)\rangle$, and $Y = \alpha_G(B_K)$. By Lemmas 3.93 and 3.94, we have for any four-subgroup $B$ of $A_L$:

$$Y = \alpha_G(B_K) = \alpha_G(B_L) = \alpha_G(B). \tag{3.74}$$

But it is immediate from the definition of $\alpha$ that for any four-subgroup $B$ of $G$,

$$N_G(B) \leqslant N_G(\alpha_G(B)). \tag{3.75}$$

Hence by (3.74), (3.75) and the definition of $\Gamma$, it follows that

$$\Gamma_{A_L, 2}(G) \leqslant N_G(Y). \tag{3.76}$$

But as $A_L \cong E_8$ acts on $O(C_{z(L)})$, $O(L) \leqslant O(C_{z(L)}) \leqslant \Gamma_{A_L, 2}(G)$ by [I, 4.13], so $O(L)$ normalizes $Y$. On the other hand, we have already seen in

Lemma 3.91 that $C_{A_L}$ covers $L/O(L)$, so $\Gamma_{A_L,2}(G)$ covers $L/O(L)$ and hence $L$ normalizes $Y$.

Similarly $K$ normalizes $\alpha_G(B_K) = Y$. We therefore conclude that every element of $\Omega(J)$ (including $K$) normalizes $Y$, as required.

We can now quickly complete the proof of Proposition 3.89. Suppose false for some $K \in \Omega(J)$, so that $Y = \alpha_G(\langle z(J), z(K) \rangle) \neq 1$. Since $G$ is transitive on $\Omega$, there is $g \in G$ such that $J^g \in \Omega(J)$. But by the hypothesis of Theorem 3.87, also $X(J)$ is transitive on $\Omega(J)$, so there is $x \in X(J)$ such that $J^{gx} = K$. Replacing $g$ by $gx$, we can therefore suppose to begin with that $J^g = K$.

By the previous lemma, $X(J)$ normalizes $Y$. We claim that also $X(J)^g$ normalizes $Y$. Indeed, $X(J)^g = \langle \Omega(J) \rangle^g = \langle \Omega(J^g) \rangle = \langle \Omega(K) \rangle$. But as $J$ and $K$ are connected by an edge, $J \in \Omega(K)$, so by the previous lemma with $J$ and $K$ interchanged, $X(J)^g$ normalizes $\alpha_G(\langle z(K), z(J) \rangle) = Y$, as claimed.

We thus conclude that $\langle X(J), X(J)^g \rangle$ normalizes $Y$, whence by Proposition 3.88, $G$ normalizes $Y$. But as $J, K$ are vertices of a triangle of rank 3, Lemma 3.92 implies that $Y$ has odd order. Thus $Y \leqslant O(G)$, contrary to the fact that $O(G) = 1$, while $Y \neq 1$ by assumption.

As a corollary, we have the following additional result:

PROPOSITION 3.96. $z(J)$ centralizes $X(J)$.

Indeed, as $X(J) = \langle \Omega(J) \rangle$, we need only show that $z(J)$ centralizes each $K \in \Omega(J)$. But by (3.56), $C_K(z(J))$ covers $K/O(K)$, so it will suffice to prove that $z(J)$ centralizes $O(K)$. However, by Lemma 3.75, $K$ is subnormal in $C_{z(K)}$ and hence $O(K) \leqslant O(C_{z(K)})$, so we need only only show that $z(J)$ centralizes $O(C_{z(K)})$. But

$$[O(C_{z(K)}), z(J)] \leqslant \alpha_G(\langle z(J), z(K) \rangle) = 1 \tag{3.77}$$

by Proposition 3.89, so the desired assertion holds.

This last result together with the minimality of $G$ enables us to determine the structure of $X(J)$.

PROPOSITION 3.97. $X(J)$ is quasisimple and isomorphic to a group of Lie type of odd characteristic.

Indeed, set $X = X(J)$, $\bar{X} = X/O(X)$, and for $L \in \Omega(J)$, let $\bar{\Omega}$ be the set of $\bar{X}$-conjugates of $\bar{L}$ and $\bar{\mathcal{D}}$ the graph induced on $\bar{\Omega}$ by $\Omega(J)$. Clearly the conditions of Theorem 3.87 carry over to $\bar{X}$ in its action on $\bar{\Omega}$, so $\bar{X}$ is transitive on $\bar{\Omega}$, $\bar{X} = \langle \bar{\Omega} \rangle$, $\mathscr{D}$ is connected, and $\bar{L}/O(\bar{L})$ is not a 2-group. Hence by the minimality of $G$, Theorem 3.78 and Proposition 3.79 hold for $\bar{X}$. Under the present assumptions, $\bar{X}$ is necessarily quasisimple of Lie type of odd characteristic.

It therefore remains only to show that $X$ is quasisimple. But by the previous proposition, $X \leqslant C = C_{z(J)}$, and once again as in Lemma 3.91, it follows that $X^*$ is a component of $C^* = C/O(C)$, whence $O(X) \leqslant O(C)$. Hence for any $K \in \Omega(J)$, $[O(X), z(K)] \leqslant \alpha_G(\langle z(J), z(K) \rangle) = 1$, again by Proposition 3.89. Thus $z(K)$ centralizes $O(X)$. Furthermore, again by Theorem 3.78 and Proposition 3.79, we have $\overline{z(K)} \notin O_2(\bar{X})$ for some $K \in \Omega(J)$. Since $\bar{X}/Z(\bar{X})$ is simple, this in turn implies that $C_x(O(X))$ covers $\bar{X}$, so, in particular, $C_x(O(X))$ covers $K/O(K)$. But either $K = O^{2'}(K)$ or $K/O(K) \cong SL_2(3)$ and $K/O^{2'}(K)$ is a cyclic 3-group. In either case we conclude that $K$ centralizes $O(X)$, whence $X = \langle K \mid K \in \Omega(J) \rangle$ centralizes $O(X)$. Thus $O(X) \leqslant Z(X)$ and it follows that $X$ is quasisimple, as required.

Because of our restricted assumptions on $\mathscr{D}(J)$, certain low rank possibilities for $X(J)$ are excluded here, as well as orthogonal groups. Moreover, as $G = \langle X(J), X(J)^g \rangle$ for $J^g \in \Omega(J)$ by Proposition 3.88, we are thus very close to determining the exact structure of $G$; it remains only to identify the latter group by means of an appropriate recognition theorem. Indeed, for each choice of $X(J)/Z(X(J))$, we can anticipate the result of this identification by comparison with Table 3.4.

To simplify the discussion, we shall limit ourselves to the linear and unitary cases, where the analysis is most transparent. Following Aschbacher, we set

$$SL_n(q) = SL_n^+(q) \quad \text{and} \quad SU_n(q) = SL_n^-(q), \tag{3.78}$$

with corresponding designations of $L_n(q)$, $U_n(q)$. Thus for our final two results, we assume

$$X(J)/Z(X(J)) \cong L_m^\varepsilon(q), \qquad \varepsilon = \pm 1. \tag{3.79}$$

To carry out the identification, Aschbacher must first determine the structure of one further subgroup of $G$. [The analogous result is also needed for the other possibilities for $X(J)$.]

Let $K \in \Omega(J)$, set $a = z(J) z(K)$, and let $\Omega(C_a)$ be the subset of $\Omega$ contained in $C_a$. It is the subgroup $Y = \langle \Omega(C_a) \rangle$ whose structure is required. To this end, set $Y(J) = \langle J^{C_a} \rangle$ and $Y(X) = L(C_x(z(K)))$. Aschbacher proves:

PROPOSITION 3.98. *The following conditions hold*:

(i) $Y(J) \cong SL_4^\varepsilon(q)$ *and* $K \leqslant Y(J)$;
(ii) $Y(X) \cong SL_{m-2}^\varepsilon(q)$; *and*
(iii) $Y = Y(J) Y(X)$ *with* $[Y(J), Y(X)] = 1$.

The proof is essentially inductive. Some care must be taken for low values of $m$ (including use of Proposition 3.82, characterizing the orthogonal groups). Here we limit the discussion to large values of $m$.

By the structure of $X$, $Y(X) \cong SL_{m-2}^\varepsilon(q)$, $Y(X)$ centralizes $K$, and $Y(X)$ is generated by conjugates of $J$ contained in $C_X(z(K))$. In particular, (ii) holds. But also $J$ centralizes $X$ and hence $Y(X)$ (by Proposition 3.96), so $a$ centralizes $Y(X)$ and consequently $Y(X) \leqslant Y$.

Next choose $L \in \Omega(L) \cap \Omega(K)$, so that $L \leqslant Y(X)$ and $z(L)$ centralizes both $J$ and $K$. Since $G$ is transitive on $\Omega$, we have $L = J^h$ for some $h \in G$. Then $X^h$ centralizes $z(L)$ (again by Proposition 3.96), $X^h = \langle \Omega(L) \rangle$ and $J, K \in \Omega(L)$. But now if $Y_0(J)$ denotes the normal closure of $J$ in $C_{X^h}(a)$, it follows by induction that $Y_0(J) \cong SL_4^\varepsilon(q)$ and that $K \leqslant Y_0(J)$. Moreover, clearly $Y_0(J) \leqslant Y(J) \leqslant Y$.

Arguing now as in Lemma 3.91 and using our quasisimplicity results, it follows that $Y_0(J)$ and $Y(X)$ lie in components $Y_1, Y_2$, respectively, of $C_a$. In particular, $L \leqslant Y_2$ and $K \leqslant Y_1$. Now by its definition, $Y_0(J)$ is a component of $C_{C_a}(z(L))$; and similarly $Y(X)$ is a component of $C_{C_a}(z(K))$. If $Y_1 \neq Y_2$, $z(L)$ centralizes $Y_1$ and $z(J)$ centralizes $Y_2$, in which case we conclude by $L$-balance that $Y_0(J) = Y_1$ and $Y(X) = Y_2$ are components of $C_a$. Thus $Y_0(J) = Y(J)$, proving (i), and $[Y(J), Y(X)] = 1$. Likewise it is clear that $Y = \langle \Omega(C_a) \rangle = Y(J) Y(X)$, so (iii) holds as well. Hence the proposition holds in this case.

On the other hand, if $Y_1 = Y_2$, then $a = z(J) z(K) \in Y_0(J) \leqslant Y_1$, so $Y_1$ is an intrinsic component. Furthermore, by induction $Y_1$ is of Lie type of odd characteristic. In addition, $Y_0(J)$ is a component of $C_{Y_1}(z(L))$ and $Y(X)$ is a component of $C_{Y_1}(z(K))$. However, it is easy to check that for $m$ sufficiently large, there is no group of Lie type of odd characteristic satisfying these conditions [the key restriction is that the central involution $a$ of $Y_1$ is contained in $Y_0(J)$].

Finally Aschbacher proves the following result:

PROPOSITION 3.99. *If* $X(J)/Z(X(J)) \cong SL_m^\varepsilon(q)$, *then* $G/Z(G) \cong L_n^\varepsilon(q)$, *where* $n = m + 2$.

Again we restrict ourselves to large values of $m$. We shall show that the

Curtis–Phan relations for $L_{m+2}(q)$, $U_{m+2}(q)$, respectively, hold in $G$ (Theorems 3.57 and 3.58). Thus we shall prove that $G$ possesses an $L_3^\varepsilon(q)$-generating system, indexed by the set $\{1, 2,..., m+1\}$.

We let $J$, $K$, $L$, and $h$ be as above. Some care must be taken to index the canonical generators properly. We first set

$$J_1 = J, \ J_3 = K, \text{ and } J_{m+1} = L. \tag{3.80}$$

The previous proposition gives the structure of the groups $Y(J)$, $Y(X)$, and $Y$. In addition, we set $Y_1(X) = L(C_{Y(X)}(z(L)))$. Since $Y(X) \cong SL_{m-2}^\varepsilon(q)$, it follows that

$$Y_1(X) \cong SL_{m-4}^\varepsilon(q). \tag{3.81}$$

Furthermore, we have

(1) $(Y(J), Y_1(X)) = 1$; and
(2) $\langle Y(J), Y_1(X) \rangle \leqslant X^h$. $\qquad(3.82)$

Since $Y(J) \cong SL_4^\varepsilon(q)$, $Y_1(X) \cong SL_{m-4}^\varepsilon(q)$, and $X^h/Z(X^h) \cong X/Z(X) \cong L_m^\varepsilon(q)$, each of these groups possesses an $L_3^\varepsilon(q)$-generating system. Moreover, in view of (3.82), we can choose the generating system for $X^h$ to include the canonical generators of both $Y(J)$ and $Y_1(X)$. Thus there exist elements $J_2$ and $J_i$, $4 \leqslant i \leqslant m-1$, of $\Omega$ contained in $X^h$ such that:

(1) $\{J_i \mid 1 \leqslant i \leqslant 3\}$ is an $L_3^\varepsilon(q)$-generating system for $Y(J)$;
(2) $\{J_i \mid 5 \leqslant i \leqslant m-1\}$ is an $L_3^\varepsilon(q)$-generating system for $Y_1(X)$; and
(3) $\{J_i \mid 1 \leqslant i \leqslant m-1\}$ is an $L_3^\varepsilon(q)$-generating system for $X^h$. $\qquad(3.83)$

In particular, by (3), we have for $1 \leqslant i, j \leqslant m-1$,

(1) $\langle J_i, J_{i+1} \rangle$ is isomorphic to a covering group of $L_3^\varepsilon(q)$.
(2) $[J_i, J_j] = 1$ if $j \neq i, i+1$. $\qquad(3.84)$

[Likewise the additional $U_4(3)$ condition holds when $q = 3$ and $\varepsilon = -1$.]

Similarly $\langle J_3, J_{m+1} \rangle = \langle K, L \rangle$ centralizes $Y_1(X)$, $\langle Y_1(X), J_{m+1} \rangle = Y(X)$, and each of these groups is contained in $X$. Hence there exists an element $J_m$ of $\Omega$ contained in $X$ such that

$$\{J_i \mid 3 \leqslant i \leqslant m+1\} \text{ is an } L_3^\varepsilon(q)\text{-generating system for } X. \tag{3.85}$$

But $J$ centralizes $X$, which contains $\langle J_m, J_{m+1} \rangle$, so we also have

$$[J_1, J_m] = [J_1, J_{m+1}] = 1. \tag{3.86}$$

Hence by (3.84)(3), (3.85), and (3.86), the set $\{J_i \mid 1 \leqslant i \leqslant m+1\}$ is an $L_3^\varepsilon(q)$-generating system for the subgroup $G^*$ of $G$ which they generate, so by the Curtis–Phan theorems $G^*$ is quasisimple and

$$G^*/Z(G^*) \cong L_{m+2}^\varepsilon(G). \tag{3.87}$$

On the other hand, by (3.84)(3) and (3.85), $\langle X, X^h \rangle \leqslant G^*$. But as $L = J^h \in \Omega(J)$, Proposition 3.88 implies that $G = \langle X, X^h \rangle$. We conclude that $G = G^*$ and the proposition is proved.

This completes our discussion of the generic nonorthogonal case of the classical involution theorem.

## 3.8. Special Configurations

As remarked above, it is a long and involuted path that Aschbacher must traverse to reach the "high rank" cases of the preceding section. Indeed, his paper contains 31 sections and it is not until Section 29 that the complete reduction to the generic case is achieved. Almost all of the first 28 sections (apart from some preliminary material and the description of the groups of Lie type) deal with various types of degenerate configurations. We can therefore do little more here than indicate some of the highlights of his reduction.

Since the graph $\mathscr{D}$ is defined in terms of the set $R(G)$ of Sylow 2-subgroups of the elements of $\Omega$, it is clear that this set will play an important role in the proof. In Thompson's earlier investigations he used the set $R(G)$ to prove the existence of a "Weyl group" of a suitable "torus" of $G$. This Weyl group $\mu(G)$, which Aschbacher appropriately calls the *Thompson subgroup* of $G$, gives him considerable control in analyzing configurations which lead to groups of low 2-rank.

The definition of $\mu(G)$ is somewhat involved. One first considers the set $U(G)$ of cyclic subgroups of index 2 contained in elements of $R(G)$.

If $R \in R(G)$ has order $\geqslant 16$, then $R$ contains a *unique* such cyclic subgroup, but if $|R| = 8$, $R$ contains *three* such subgroups. For this reason, the definition of $\mu(G)$ and some of its properties are easier to state (and to establish) when the elements of $R(G)$ have order $\geqslant 16$, so *we shall make this assumption* throughout the discussion.

Using reasoning similar to that of Proposition 3.66, Aschbacher (and Thompson as well) prove the following result:

**PROPOSITION 3.100.** *For any* $S \in \text{Syl}_2(G)$, *the set* $U(S)$ *of elements of* $U(G)$ *contained in* $S$ *generate an abelian normal subgroup* $T(S)$ *of* $S$.

[When the elements of $R(G)$ have order 8, $T(S)$ is defined in a more roundabout fashion. It is again abelian and normal in $S$, but it is not uniquely determined by $S$ nor does it contain every element of $U(S)$.]

Since the ultimate objective is to show that $G$ is a group of Lie type of odd characteristic, one should visualize $T(S)$ as the 2-part of a suitable torus of $G$ (not necessarily of a Cartan subgroup).

Next we consider the set $R(S)$ of elements of $R(G)$ contained in $S$. Since $T(S) \lhd S$, $\langle R(S) \rangle$ normalizes $T(S)$. Finally we set

$$\mu(G; S) = \langle R(S) \rangle T(S) O(C_{T(S)}) / T(S) O(C_{T(S)}). \tag{3.88}$$

**DEFINITION 3.101.** $\mu(G; S)$ is called the *Thompson* subgroup of $G$.

$\mu(G; S)$ is a suitable section of the normalizer of $T(S)$ and so by analogy with the Weyl group of a group of Lie type, it can be viewed as the Weyl group of the given torus $T(S)$. Furthermore, as $\mu(G; S)$ is determined up to conjugacy by $S$, we write, as usual in denoting Weyl groups, $\mu(G)$ for $\mu(G; S)$. [When the elements of $R(G)$ have order 8, even though $T(S)$ is not uniquely determined, it turns out that the group $\langle R(S) \rangle T(S) O(C_{T(S)} / T(S) O(C_{T(S)})$ is, up to isomorphism, independent of the choice of $T(S)$, so one can speak of *the* Thompson subgroup of $G$ in this case as well.]

Set $N = N_G(T(S))$ and $\bar{N} = N / T(S) O(C_{T(S)})$, so that $\mu(G:S) = \langle \overline{R(S)} \rangle$ is a subgroup of $\bar{N}$. Furthermore, if $R \in R(S)$, then $R$ contains a unique subgroup $U$ of index 2 in $U(S)$ [as we are assuming that elements of $R(G)$ have order $\geqslant 16$] and $U \leqslant T(S)$ by Proposition 3.100. On the other hand, $R \not\leqslant T(S)$ as $T(S)$ is abelian, so $|\bar{R}| = 2$. Thus $\overline{R(S)}$ is a collection of subgroups of order 2 and so $\mu(G:S)$ is generated in a natural way by the corresponding set of involutions.

Aschbacher first pins down the precise structure of $\mu(G)$, again using arguments similar to those of Proposition 3.66.

PROPOSITION 3.102. (i) $\mu(G)$ *is the central product of dihedral groups of order* 12 *with a group generated by* 3-*transpositions*; *and*
(ii) *If G is transitive on* $\Omega$, *then either* $\mu(G)$ *is a central product of dihedral groups of order* 12 *or* $\mu(G)$ *is a group generated by* 3-*transpositions.*

We next describe some of the ways in which small Thompson subgroups arise in the analysis. First, Aschbacher proves the following two results.

PROPOSITION 3.103. *If* $\mathscr{D}$ *has no edges and G is transitive on* $\Omega$, *then* $\mu(G) \cong Z_2$ *or* $\Sigma_3$.

PROPOSITION 3.104. *If for* $J \in \Omega$, $\Omega(J)$ *consists of a single element, then* $\mu(G) \cong Z_2 \times Z_2$ *or* $\Sigma_4$.

Furthermore, in the course of the proof of his main theorem, Aschbacher brings into play a second graph $\mathscr{D}^*$ containing $\mathscr{D}$ as a subgraph, and small rank cases also emerge in comparing the connectedness of $\mathscr{D}$ and $\mathscr{D}^*$.

DEFINITION 3.105. Let $\mathscr{D}^*$ be the graph with vertex set $\Omega$ and two distinct vertices $J, K$ joined by an edge if and only if an element of $U(G)$ contained in $K$ normalizes $J$.

Again Aschbacher shows that the condition is symmetric in $J$ and $K$, so that $\mathscr{D}^*$ is well-defined. Furthermore, if $J, K \in \Omega$ are connected by an edge in $\mathscr{D}$, in which case a Sylow 2-subgroup of $K$ normalizes $J$, then clearly an element of $U(G)$ contained in $K$ normalizes $J$, so $J, K$ are also joined by an edge in $\mathscr{D}^*$. Thus indeed $\mathscr{D}$ is a subgraph of $\mathscr{D}^*$.
Aschbacher proves the following result:

PROPOSITION 3.106. *Either every connected component of* $\mathscr{D}$ *is a connected component of* $\mathscr{D}^*$ *or else* $\mu(G) \cong \Sigma_3, \Sigma_4$ *or* $\Sigma_3/Q_8 * Q_8$.

In the last case, the extension splits and $\mu(G)$ is isomorphic to the Weyl group of the Dynkin diagram $D_4$.
We describe another situation that gives rise to small Thompson subgroups. If $S \in \mathrm{Syl}_2(G)$, we know that $\mu(G) = \mu(G; S)$ is generated by the involutions $\overline{R(S)}$, we can therefore consider the *commuting graph* associated

with $\mu(G)$ whose vertices are the elements of $\overline{R(S)}$ with two vertices connected by an edge if and only if they commute.

Aschbacher proves:

PROPOSITION 3.107. *If the commuting graph of $\mu(G)$ is disconnected, then $\mu(G) \cong \Sigma_3, \Sigma_4$, or $\Sigma_3/Q_8 * Q_8$.*

Finally the case $\mu(G) \cong D_{12}$, the dihedral group of order 12 ($\cong \Sigma_3 \times Z_2$), is a further small Thompson group, but the precise conditions under which it arises are somewhat technical and so we shall not state them.

Table 3.5 lists the nonsolvable possibilities for $G$ in each of these small cases (including 2-constrained solutions); separate analyses are required to reach each of these solutions and together they account for a considerable portion of the total proof. [Apart from these exceptional cases, the only other solutions requiring independent treatment are the orthogonal groups $P\Omega_n^\pm(q)$ and the 2-constrained groups $(A/E)_n$ for all $n$.]

For simplicity of notation, we set $L_n^+(q) \cong L_n(q)$ and $L_n^-(q) \cong U_n(q)$ and put $\pi = \pm 1$ according as $q \equiv \pi \pmod 4$. Also $L_2(q^2)*$ will denote the extension of $L_2(q^2)$ by a group of order 2 having quasi-dihedral Sylow 2-subgroups (it exists whenever $q$ is a square and is induced by the product of a graph and a field automorphism of period 2), and $L_3^\pm(q)*$ will denote the extension of $L_3^\pm(q)$ by a graph automorphism of period 2. Recall also from (3.59) that $P\Omega_5(q) \cong PSp_4(q)$, $P\Omega_6^+(q) \cong L_4(q)$, and $P\Omega_6^-(q) \cong U_4(q)$ as well as the definition of $(A/E)_n$ preceding Proposition 3.79.

Note that $M_{12}$, $G_2(q)$, $U_3(3)*$, and $L_3(2)/E_8$ appear in the table *twice*. These groups [and likewise the family $^3D_4(q)$] are generated by two distinct conjugacy classes of subgroups that satisfy Hypothesis $\Omega$. [In the case of $^3D_4(q)$, the resulting two Thompson groups turn out to be isomorphic (to

**Table 3.5.** Solutions for $G$ with Small Thompson Subgroups

| $\mu(G)$ | $G$ |
|---|---|
| $Z_2$ | $M_{11}, L_3^{-\pi}(q), L_2(q^2)*$ |
| $\Sigma_3$ | $M_{12}, L_3^\pi(q), L_3^\pm(q)*, G_2(q), {}^3D_4(q), L_3(2)/E_8$ |
| $Z_2 \times Z_2$ | $P\Omega_5(q), P\Omega_6^{-\pi}(q), L_5^{-\pi}(q), (A/E)_5$ |
| $D_{12}$ | $M_{12}, G_2(q), U_3(3)*, L_3(2)/E_8$ |
| $\Sigma_4$ | $P\Omega_6^\pi(q), P\Omega_7(q), P\Omega_8^-(q), L_3^\pi(q), Sp_6(2),$ |
|  | $(A/E)_6, (A/E)_7, L_3(2)/E_{64}$ |
| $\Sigma_3/Q_8 * Q_8$ | $D_4(2), P\Omega_8^+(q), (A/E)_8, (A/E)_9$ |

$\Sigma_3$).] In the cases leading to these solutions, Aschbacher first determines all possibilities for $G$ having a $\Sigma_3$ Thompson subgroup. As a result, when he comes to deal with the $D_{12}$ case, since he is working throughout with a minimal counterexample, his argument consists, in effect, of a reduction to the previously treated $\Sigma_3$ case.

As in the previous section, we assume (3.61) throughout, so that $G$ is a minimal counterexample to Theorem 3.78 and Propositions 3.79–3.85, and subject to this, $|\Omega|$ is minimal.

Before systematically analyzing these small cases, Aschbacher proves the following result:

PROPOSITION 3.108. *G acts transitively on* $\Omega$, $G = \langle \Omega \rangle$, *and* $F^*(G)$ *is simple.*

In order not to disrupt the present discussion, we defer our comments on the proof of the proposition until the end of the section. (Note that the 2-constrained solutions of Table 3.5 will arise in the course of its proof.)

We can do little more here than explain how Aschbacher ultimately identifies each possible small solution with $F^*(G)$ simple (as well as the generic orthogonal solution), as the various cases of the analysis are very delicate, involving detailed fusion analysis, and depend heavily on the minimality of $G$ to identify specific proper sections of $G$ [in particular, to identify $X(J)$ for $J \in \Omega$, which under the present conditions includes some 2-contrained possibilities].

Thus we assume

$$G \text{ is transitive on } \Omega \text{ and } F^*(G) \text{ is simple.} \qquad (3.89)$$

First, if $\mathcal{D}$ has no edges, then by Proposition 3.103 and (3.89), $\mu(G) \cong Z_2$ or $\Sigma_3$. In the first case, Aschbacher argues that for $J \in \Omega$, $N_G(J)$ contains a quasi-dihedral subgroup $P$ with $P \cap J \in \mathrm{Syl}_2(J)$ and that $P$ is strongly closed with respect to $G$ in a Sylow 2-subgroup $S$ of $G$. Now Goldschmidt's general strong closure theorem [I, 4.156] and the classification of groups with quasi-dihedral Sylow 2-subgroups (Theorem 1.63) yields [as $F^*(G)$ is simple and $G = \langle \Omega \rangle$] that

$$\begin{aligned} &G \cong L_2(q^2)^*, q \text{ odd}, L_3(q), q \equiv -1 \pmod 4, U_3(q), \\ &q \equiv 1 \pmod 4, \text{ or } M_{11}. \end{aligned} \qquad (3.90)$$

$(\Omega_1(P)$ is dihedral and also strongly closed in $S$ with respect to $G$; we note that this particular case of Goldschmidt's theorem was treated by J. Hall [49].)

On the other hand, if $\mu(G) \cong \Sigma_3$, let $z = z(J)$ and let $X$ be the subgroup of $C_z$ generated by the set of all elements of $U(G) \cup R(G)$ contained in $C_z$. Setting $\bar{X} = X/O(X)$, Aschbacher argues that $X$ contains a Sylow 2-subgroup of $G$ and that either $O_2(\bar{X}) \cong Q_8 * Q_8$ with $X/O_2(X) \cong \Sigma_3$ and $S$ of type $M_{12}$ or else $\bar{X}$ contains a normal subgroup of index 2 isomorphic to $SL_2(q_1) * SL_2(q_2)$ for suitable odd $q_1, q_2$ with $S$ of type $G_2(q_1)$. Hence by the classification of groups with such type Sylow 2-subgroups (Theorem 2.4):

$$G \cong G_2(q), \text{ or } {}^3D_4(q), q \text{ odd, or } M_{12}. \tag{3.91}$$

Thus Aschbacher can assume henceforth that

$$\mathscr{D} \text{ possesses and edge.} \tag{3.92}$$

He focuses next on the *orthogonal* case—i.e., the case in which $\mathscr{D}$ is not multiplicity free, so that for $J \in \Omega$, $\Omega(J)$ contains an element $K$ with $z(J) = Z(K)$. The argument is very elaborate and covers several sections of the paper. The generic case arises when $\mathscr{D}$ is connected and its analysis is similar in spirit to the generic nonorthogonal case of the previous section. In particular, Aschbacher argues first that the standard functor $\alpha$ on four-groups is trivial. Moreover, if $\Omega_0(J)$ denotes the set of $K \in \Omega(J)$ such that $z(K) = z(J)$ and $X_0(J) = \langle \Omega_0(J) \rangle$, he shows, using the minimality of $G$, that

$$\begin{aligned}
&(1) \;\; X_0(J)/Z\big(X_0(J)\big) \cong P\Omega_n^\varepsilon(q), \text{ where } \varepsilon = \pm 1, \\
&\quad\;\; n \geqslant 5, \text{ and } q \text{ is odd; and} \\
&(2) \;\; G \cong \langle X_0(J), X_0(J)^g \rangle \text{ for any } g \in G \\
&\quad\;\; \text{such that } J^g \in \Omega_0(J).
\end{aligned} \tag{3.93}$$

To establish the Curtis root subgroup relations, Aschbacher needs the structure of one additional subgroup: namely, if $K \in \Omega_0(J)$, $y$ is an involution of $JK - z(J)$, and $Y$ denotes the subgroup of $C_y$ generated by the elements of $R(G)$ contained in $C_y$, he proves that

$$Y/Z(Y) \cong P\Omega_{n+2}^{\varepsilon\pi}(q). \tag{3.94}$$

Finally on the basis of (3.93) and (3.94), he argues that the conditions of Theorem 3.55 are satisfied, thus proving the following:

PROPOSITION 3.109. *If $\mathscr{D}$ is connected and not multiplicity free, then* $G \cong P\Omega_{n+4}^{\varepsilon}(q)$ *[where $\varepsilon$, $n$, and $q$ are as in (3.93)].*

We can therefore assume (through Proposition 3.113) that

$$\mathscr{D} \text{ is disconnected and not multiplicity free.} \tag{3.95}$$

In this case, Aschbacher first proves:

PROPOSITION 3.110. *For any $J \in \Omega$, $\Omega(J) = \Omega_0(J)$.*

Assuming false and considering a connected component $\mathscr{D}_1$ of $\mathscr{D}$, Aschbacher ultimately derives a contradiction by arguing that $G_1 = \langle \mathscr{D}_1 \rangle$ is normal in $G$. The proof is based on the following previously derived normality criterion for $G_1$ under the present conditions:

> Let $S \in \mathrm{Syl}_2(G)$ with $S \cap J \in \mathrm{Syl}_2(J)$. If for each $K \in \Omega(J)$ and each involution $y \in JK \cap T(S)$, we have $C_y \leqslant N_G(G_1)$, then $G_1$ is normal in $G$. $\tag{3.96}$

To verify (3.96), Aschbacher again brings the four-group functor $\alpha$ into play and divides the argument into two cases, according as $\alpha$ is nontrivial or trivial.

Aschbacher next proves the following result:

PROPOSITION 3.111. *For any $J \in \Omega$, either $\Omega(J)$ consists of a single element or $G \cong P\Omega_8^+(q)$ or $D_4(2)$.*

Assuming $\Omega(J)$ has more than one element, he first shows that $\mu(G) \cong \Sigma_3/Q_8 * Q_8$. The analysis in this case is, apart from two exceptional subcases, quite similar to that of the connected case above. If $K$ is any element of $\Omega(J)$, he argues first that $\alpha(A) = 1$ for every four-subgroup $A$ of $JK$. Next, if $y$ is an involution of $JK - z(J)$ and $y$ again denotes the subgroup generated by the elements of $R(G)$ contained in $C_y$, he shows, using the minimality of $G$, that one of the following holds:

> (1) $Y/\langle y \rangle \cong P\Omega_6^+(q)$ and $G = \langle \Omega(J), Y \rangle$; or
> (2) $Y/\langle y \rangle \cong L_3(2)/E_{64}$ and $J \cong Q_8$. $\tag{3.97}$

First consider the $P\Omega_6^+(q)$ case. If $q > 3$, he uses the given information to show that $G$ possesses an $L_3(q)$- or $U_3(q)$-generating system of type $D_4$ according as $q \equiv +1$ or $-1$ (mod 4) and then concludes from the Curtis–Phan Theorems (Theorems 3.57 and 3.63) that $G \cong P\Omega_8^+(q)$.

However, the case $q = 3$ requires special treatment. Here Aschbacher works with a graph constructed from the set of $G$-conjugates of a Sylow 3-subgroup of $J$, which he shows to be a Shult graph and identifies from the general form of the Buckenhout–Shult theorem [12] to be the graph of $P\Omega_6^+(3)$. Now he is able to prove that $G$ satisfies the root subgroup conditions of Theorem 3.55 and thus conclude that $G \cong P\Omega_8^+(3)$.

Suppose then that $Y/\langle y \rangle \cong L_3(2)/E_{64}$. Expand $J$ to $S \in \mathrm{Syl}_2(G)$ and let $N$ be the pre-image of $\mu(G; S)$ in $N_G(T(S))$. Aschbacher first proves that

$$H = \langle N, Y \rangle \cong (A/E)_8. \tag{3.98}$$

Next, using the minimality of $G$, and information on the fusion of $z(J)$, he produces a subgroup $L$ of $G$ isomorphic to a parabolic subgroup of $D_4(2)$ of the following shape:

$$L \cong L_3(2)/Q, \text{ where } Q \text{ has class 2 and order } 2^9$$
with the extension split. $\tag{3.99}$

Finally, using the subgroups $H$ and $L$, he produces eight conjugates of $z(J)$, indexed as $t_{\pm i}$, $1 \leqslant i \leqslant 4$, and argues that the subgroups $\langle t_{\pm i} \rangle$, $1 \leqslant i \leqslant 4$, satisfy the Curtis relations for $D_4(2)$ and generate $G$, so $G \cong D_4(2)$ by Theorem 3.55, thus completing the proof.

In view of the proposition, we can assume henceforth that $\Omega(J)$ consists of a single element $K$. Since $\Omega(J) = \Omega_0(J)$, we have $z(J) = z(K)$ and consequently

$$JK/O(JK) \cong SL_2(q) * SL_2(q), \hat{A}_7 * \hat{A}_7, \text{ or } Q_{2^n} * Q_{2^n}. \tag{3.100}$$

Furthermore, by Proposition 3.104, $U(G) \cong Z_2 \times Z_2$ or $\Sigma_4$. Aschbacher treats these cases separately, and, in addition, splits the analysis in the $\Sigma_4$ case (which we discuss first) into two parts.

Let $S \in \mathrm{Syl}_2(G)$ and set $N = N_G(T(S))$. Aschbacher first proves the following result:

PROPOSITION 3.112. *If $\mu(G) \cong \Sigma_4$ and $Z(N) \cap T(S) \neq 1$, then we have* $G \cong P\Omega_7(q)$, $P\Omega_8^-(q)$, *or* $Sp_6(2)$.

Taking an appropriate involution $t \in Z(N) \cap T(S)$ and considering the subgroup $X$ generated by all elements of $R(G)$ contained in $C_t$, Aschbacher concludes from the minimality of $G$ that one of the following holds (here $Y$ is as in the preceding proposition):

$$
\begin{aligned}
&(1) \ \ Y/\langle y \rangle\, O(Y) \cong P\Omega_5(q) \text{ or } P\Omega_6^{-\pi}(q) \text{ and} \\
&\qquad X/\langle t \rangle\, O(X) \cong P\Omega_6^{\pi}(q); \text{ or} \\
&(2) \ \ Y/O(Y) \cong (A/E)_5 \text{ and } X/O(X) \cong (A/E)_6.
\end{aligned}
\tag{3.101}
$$

[Initially, there is a third possibility for $X$ and $Y$, which Aschbacher eliminates by forcing $S$ to contain a nontrivial strongly closed abelian subgroup and then using Goldschmidt's theorem to reach a contradiction.]

In the first case, Aschbacher again proves that $\alpha(A) = 1$ for every four-subgroup $A$ of $JK$ and then easily concludes that

$$
\begin{aligned}
&(1) \ \ Y/\langle y \rangle \cong P\Omega_5(q) \text{ or } P\Omega_6^{-\pi}(q) \text{ and } X/\langle t \rangle \cong P\Omega_6^{\pi}(q); \text{ and} \\
&(2) \ \ G = \langle X, Y \rangle.
\end{aligned}
\tag{3.102}
$$

If $q \equiv 1 \pmod 4$, he now argues that either Curtis's conditions of Theorem 3.55 or 3.57 or Phan's conditions of Theorem 3.58 are satisfied, whence according to the two possibilities for $Y$, $G \cong P\Omega_7(q)$ or $P\Omega_8^-(q)$. On the other hand, when $q \equiv -1 \pmod 4$, he once again employs the graph-theoretic approach used to identify $P\Omega_8^+(3)$, to show that $G$ satisfies the root subgroup conditions of Theorem 3.55. Hence $G \cong P\Omega_7(q)$ or $P\Omega_8^-(q)$ in this case as well.

The analysis in the second case of (3.101) is quite delicate. In particular, two separate subcases require proof that the four-group functor $\alpha$ is trivial. Set $z = z(J)$, $\bar{C}_z = \bar{C}_z/O(C)$, and let $\bar{Z}$ be the normal closure of $\bar{t}$ in $\bar{C}_z$. Aschbacher proves that $\bar{Z}$ is elementary abelian and if $Z$ denotes the pre-image of $\bar{Z}$ in $S$, then $Z$ normalizes $Y$ and $ZY/O_{2'2}(Y) \cong \Sigma_6$, which implies that $P = ZY \cap S$ is of type $Sp_6(2)$ [with $P \in \mathrm{Syl}_2(ZY)$]. If $A$ denotes the unique elementary abelian subgroup of $P$ of order 64, he then goes on to show that $G$ contains an elementary subgroup $A \cong E_{64}$ with $N_G(A)/C_A \cong L_3(2)$ and then proves that $P = S$, in which case Solomon's theorem (Theorem 2.72) implies that $G \cong Sp_6(2)$.

Aschbacher next proves:

PROPOSITION 3.113. *If $\mu(G) \cong \Sigma_4$ and $Z(N) \cap T(S) = 1$, then we have* $G \cong P\Omega_6^{\pi}(q)$.

Let $P$ be a Sylow 2-subgroup of the pre-image of $\mu(G; S)$ in $N$ with $P \leqslant S$. Aschbacher separates the cases that a Sylow 2-subgroup $R$ of $J$ has order 8 or exceeding 8. In the first case, he argues easily that $P$ is of type $\hat{A}_8$, and then goes on to prove by a very delicate fusion analysis that $S = P$. The possibilities for $F^*(G)$ are then given by the sectional 2-rank $\leqslant 4$ theorem and it follows at once that $G \cong P\Omega_6^\tau(q)$ $[\cong L_4(q),\ q \equiv 1 \pmod 4$, or $U_4(q)$, $q \equiv -1 \pmod 4)]$. On the other hand, if $|R| > 8$, he first "kills cores" in the usual fashion and then argues for a suitable involution $t$ of $G$, with $X$ again denoting the subgroup generated by the elements of $R(G)$ contained in $C_t$:

> (1) $X/Z(X) \cong L_3(q)$ or $U_3(q)$ according as $q \equiv +1$
>     or $-1 \pmod 4$; and
> (2) $G = \langle J, X \rangle$. $\hspace{3cm}$ (3.103)

Finally he argues that the Curtis–Phan relations of Theorems 3.57 or 3.58 are satisfied and concludes that correspondingly $G \cong L_4(q)$ or $U_4(q)$, whence $G \cong P\Omega_6^\tau(q)$ in this case as well.

To complete the orthogonal case, Aschbacher proves finally:

PROPOSITION 3.114. *If* $\mu(G) \cong Z_2 \times Z_2$, *then* $G \cong P\Omega_6^{-\tau}(q)$ *or* $P\Omega_5(q)$.

Let $J \in \Omega$, $\{K\} = \Omega(J)$, $Q \in \mathrm{Syl}_2(JK)$ and expand $Q$ to $S \in \mathrm{Syl}_2(G)$. The condition on $\mu(G)$ implies that $Q \cap J$ and $Q \cap K$ are the only elements of $R(G)$ contained in $S$, whence $Q \lhd S$. Set $A = C_S(Q)$, so that $z = z(J) \in A$. This time let $X$ be the subgroup of $C_A$ generated by the elements of $R(G)$ contained in $C_A$. Aschbacher first proves

> (1) Either $A = \langle z \rangle$ or $A \cong Z_2 \times Z_2$; and
> (2) Either $X/Z^*(X) \cong P\Omega_5(q)\ (\cong PSp_4(q))$ $\hspace{1.5cm}$ (3.104)
>     or $X/O(X) \cong (A/E)_5$.

The next step is to force $A = \langle z \rangle$, which is achieved by a delicate fusion analysis. In particular, transfer is used to eliminate certain configurations, for which $G$ is shown to have a normal subgroup of index 2, contrary to the minimality of $G$. Thus

$$A = \langle z \rangle. \hspace{3cm} (3.105)$$

A further transfer argument is used to show that no element of $S$ induces a nontrivial field automorphism on $K$, which in turn yields

$$S/Q \cong Z_2 \quad \text{or} \quad Z_2 \times Z_2. \tag{3.106}$$

At this point, Aschbacher is able to pin down the precise structure of $S$: namely, correspondingly $S$ is of type $P\Omega_5(q)$ or $P\Omega_6^{-\pi}(q)$ [i.e., of type $PSp_4(q)$, $L_4(q)$, or $U_4(q)$]. The possibilities for $G$ are therefore determined by Theorems 2.21 and 2.45, and Aschbacher concludes at once in the present case that the only possibilities are $G \cong P\Omega_5(q)$ or $P\Omega_6^{-\pi}(q)$.

This completes the outline of the nonmultiplicity free case, and we therefore assume for the balance of the discussion of the low-rank cases that

$$\mathscr{D} \text{ is multiplicity free.} \tag{3.107}$$

Aschbacher again divides the analysis into two parts.

PROPOSITION 3.115. *If* $\mu(G) \cong Z_2 \times Z_2$, *then we have* $G \cong L_5(q)$, $q \equiv -1 \pmod 4$, *or* $U_5(q)$, $q \equiv 1 \pmod 4$.

Let $J \in \Omega$ and let $K \in \Omega(J)$, so that $\langle z(J), z(K) \rangle \cong Z_2 \times Z_2$ as $\mathscr{D}$ is multiplicity free. Let $y$ be the involution $z(J) z(K)$ and again let $Y$ be the subgroup of $C_y$ generated by the elements of $R(G)$ contained in $C_y$.

Aschbacher first proves that one of the following holds:

$$
\begin{aligned}
&(1) \ \ X(J)/O(X(J)) \cong L_3(q), q \equiv -1 \pmod 4 \text{ or } U_3(q), \\
&\quad\quad q \equiv 1 \pmod 4, \text{ and } Y/O(Y)\langle y \rangle \cong P\Omega_6^{-\pi}(q); \text{ or} \\
&(2) \ \ X(J)/O(X(J)) \cong M_{11} \text{ or } L_2(q^2)^* \text{ and} \\
&\quad\quad \langle z(J), z(K) \rangle \leqslant Z^*(C_y).
\end{aligned}
\tag{3.108}
$$

Next he argues that $G = \langle J, K, \Omega(J), \Omega(K) \rangle$, which in turn implies:

$$\mathscr{D} \text{ is connected.} \tag{3.109}$$

Now by a combination of fusion analysis and transfer, together with the connectivity of $\mathscr{D}$, he shows that (3.108)(1) must hold and that the four-group functor $\alpha$ is trivial, whence

$$X(J) \cong L_3(q) \text{ or } U_3(q) \text{ and } Y/\langle y \rangle \cong P\Omega_6^{-\pi}(q). \tag{3.110}$$

With this information, he quickly produces an $L_3(q)$- or $U_3(q)$-generating system for $G$ of type $A_4$ and thus concludes that $G \cong L_5(q)$ or $U_5(q)$, as asserted.

Finally Aschbacher proves the following result:

PROPOSITION 3.116. *We have $\mu(G) \not\cong \Sigma_4$ or $\Sigma_3/Q_8 * Q_8$.*

Let $J \in \Omega$ and let $K \in \Omega(J)$. Since $\mathscr{D}$ is multiplicity free, $z(J) \neq z(K)$. Set $y = z(J) z(K)$. By analyzing the $G$-fusion of $y$, Aschbacher first proves

$$X(J) \neq KO\big(X(J)\big). \tag{3.111}$$

Now let $R \in \mathrm{Syl}_2(J)$, expand $R$ to $S \in \mathrm{Syl}_2(G)$, and let $N$ be the pre-image of $\mu(G; S)$ in $N_G\big(T(S)\big)$.

If $\mu(G) \cong \Sigma_4$, then by (3.111) and the minimality of $G$, $X(J)/O\big(X(J)\big) \cong M_{11}$, $L_2(q^2)^*$, $L_3(q)$, or $U_3(q)$ for suitable odd $q$. But now if $P \in R(G)$ is contained in $N$, then using the structure of $X(J)$, Aschbacher argues that $z(P)$ must centralize $R$. However, by the structure of $N$, there exists such an element $P$ for which $z(P)$ does not centralize $R$—contradiction.

On the other hand, if $\mu(G) \cong \Sigma_3/Q_8 * Q_8$, he lets $Z$ be the subgroup of $T(S)$ generated by all conjugates of $z(J)$ contained in $T(S)$ and sets $A = Z \cap Z(N)$. He knows from his earlier analysis of $\mu(G)$ that in the present case $A \cong Z_2 \times Z_2$ with $y \in A$. His goal is to prove that $A$ is strongly closed in $S$ with respect to $G$, thus contradicting Goldschmidt's theorem. As usual, the argument is fusion–theoretic. It depends critically on the structure of the subgroup $Y$ of $C_y$ generated by the elements of $R(G)$ contained in $C_y$, which because of (3.111) (and the minimality of $G$) has one of the following forms: $Y/Z^*(Y) \cong P\Omega_8^+(q)$, $Y/O(Y) \cong (A/E)_8$, or $O(Y) T(S) \lhd Y$.

This completes our outline of Aschbacher's analysis of the small Thompson subgroup cases. We turn now to Proposition 3.108, which we have been assuming throughout the discussion. Again the proof splits into two cases, according as $O_2(G)$ is trivial or nontrivial.

PROPOSITION 3.117. *Proposition 3.108 holds if $O_2(G) = 1$.*

Since $O(G) = 1$, we have $F^*(G) = L(G)$. Suppose $F^*(G)$ is not simple. Then $G = \langle \Omega \rangle$ by the minimality of $G$. By analyzing the $G$-fusion of the involutions of the elements of $\Omega$, Aschbacher argues for any $J \in \Omega$ that $J$ leaves invariant each component of $L(G)$ and that $z(J)$ is contained in a component $L$ of $L(G)$ with $L \lhd G$. (Since $G = \langle \Omega \rangle$ and $L \lhd G$, $L$ is thus independent of the choice of $J$.) This in turn implies that $J$ centralizes every component of $L(G)$ other than $L$. But then $G = \langle \Omega \rangle$ does as well. Since $C_G(G) = 1$ under our present assumptions, this forces $L = L(G)$, so $F^*(G) = L$ is simple. In particular, $G \leqslant \mathrm{Aut}(L)$.

But now if either $G$ is not transitive on $\Omega$ or if $G \neq \langle \Omega \rangle$, then $F^*(G)$ is determined by the minimality of $G$ and it is easy to see that $G$ is not a counterexample to any of the specified results of Section 3.6.

Likewise, the proof in the case $O_2(G) \neq 1$ splits into two parts. Aschbacher first establishes the following reduction.

PROPOSITION 3.118. *If $O_2(G) \neq 1$, then we have*

  (i) $z(J) \in O_2(G)$ *for each* $J \in \Omega$;
 (ii) $G = \langle \Omega \rangle$; *and*
(iii) *One of the following holds*:
    (1) *$G$ is transitive on $\Omega$; or*
    (2) $\mu(G) \cong D_{12}$, *$G$ has two orbits $\Omega_1$ and $\Omega_2$ on $\Omega$, and for any $J \in \Omega_1$, there is $K \in \Omega_2$ such that $K \in \Omega(J)$ and $z(K) = z(J)$.*

[(iii) is the analog of the $F^*(G)$ simple nonorthogonal and orthogonal cases.]

First of all, if $z(J) \notin O_2(G)$ for every $J \in \Omega$ and we set $\bar{G} = G/O_2(G)$, then Hypothesis $\Omega$ holds for $\bar{G}$ with $\bar{\Omega} = \{\bar{J} \mid J \in \Omega\}$. Hence the structure of $\bar{G}$ is determined by the minimality of $G$. However, it is not difficult to show that the elements of $R(G)$ necessarily centralize $O_2(G)$. Combining this fact with the known structure of $\bar{G}$, Aschbacher argues that $G$ itself is not a counterexample. Now (i) and (ii) follow easily from the minimality of $G$.

Therefore $z(J) \in O_2(G)$ for some $J \in \Omega$. The next step is to show that this holds for *every* $J \in \Omega$, thus proving (i). To establish this, Aschbacher invokes a portion of the following lemma, which plays an important role throughout the analysis of the 2-constrained case.

LEMMA 3.119. *Suppose $z(J) \in O_2(G)$ for some $J \in \Omega$ and assume $K \in \Omega - \{J\}$. Then we have*

  (i) *$J$ is a 2-group*;
 (ii) *If $\Omega \supset \{J\} \cup \Omega(J)$, then $J \cap O_2(G)$ is cyclic*;
(iii) *If $K$ does not centralize $J$, then $z(K) \in O_2(G)$*;
(iv) *If $K$ lies in the same $G$-orbit as $J$, then $\big(z(K), z(J), z(J)\big) = 1$*;

(v) *If $z(K)$ normalizes $J$, then $J$ and $K$ are connected by an edge in $\mathscr{D}*$; and*

(vi) *If $z(K)$ does not normalize $J$, then $J \cap O_2(G) = \langle z(J) \rangle$ and either $|J| = 8$, or $|J| = 16$ and $z(K)$ centralizes $z(J)$.*

Next by the minimality of $G$, we have $G = \langle \Omega \rangle$. Finally to establish Proposition 3.118(iii), Aschbacher considers the distinct orbits $\Omega_i$ of $G$ on $\Omega$ and the groups $G_i = \langle \Omega_i \rangle$, $1 \leqslant i \leqslant m$. Assuming $m > 1$, each $G_i$ is determined by the minimality of $G$. Since $z(J) \in O_2(G)$ for every $J \in \Omega$, each $G_i$ is, in fact, a 2-constrained group (the exact list of possibilities will be given in the next proposition). Aschbacher now attempts to prove that $G_i$ centralizes $G_j$ for all $i \neq j$, in which case $G$ is not a counterexample. However, there turns out to be an exceptional case in which for some pair $i, j$, say $i = 1$, $j = 2$, the Thompson subgroup of the group $\langle G_1, G_2 \rangle$ is isomorphic to $D_{12}$. If $G = \langle G_1, G_2 \rangle$, Aschbacher argues that (iii)(2) holds. On the other hand, if $m \geqslant 3$, he shows that $\langle G_1, G_2 \rangle$ centralizes $G_k$ for all $k \geqslant 3$. Since the structure of $\langle G_3, G_4, ..., G_m \rangle$ is determined by the minimality of $G$, it follows once again that $G$ is not a counterexample, so all parts of the proposition hold.

The above-mentioned exceptional case emerges from the analysis of yet another subgroup of $G$, which plays a critical role throughout the proof of the classical involution theorem, but which because of its technical nature we have avoided making explicit in the discussion. Here is its definition:

For each $J \in \Omega$, let $A(J)$ be the set of $K \in \Omega - \{J\}$ such that

(1) $K \notin \Omega(J)$, but $J$ and $K$ are connected by an edge in $\mathscr{D}*$; and

(2) $J$ and $K$ are not conjugate by an element of $\langle N_J(z(K)), N_K(z(J)) \rangle$.

Thus, more precisely, Aschbacher shows that (iii)(1) or (iii)(2) holds according as $A(J)$ is empty for all $J \in \Omega$ or nonempty for some $J \in \Omega$. In fact, these two possibilities are treated separately throughout the entire analysis. Moreover, as Proposition 3.118(iii)(2) indicates, it is the nonempty case that corresponds to Thompson subgroups isomorphic to $D_{12}$.

Finally Aschbacher considers the situation of Proposition 3.118, establishing the following more complete version of Proposition 3.79.

PROPOSITION 3.120. *Let the assumptions be as in Proposition 3.118, let $J \in \Omega$, and let $S \in \text{Syl}_2(G)$. Then one of the following holds:*

(i) $J \underline{\lhd} G$;

(ii) $T(S) \lhd G$, $G/T(S) \cong \mu(G)$, and either $\mu(G) \cong D_{12}$ or $\mu(G)$ is a group generated by 3-transpositions;

(iii) $G \cong L_3(2)/E_8$ and the extension does not split;

(iv) $G/Z(G) \cong L_3(2)/E_{64}$ and the extension splits;

(v) $G/Z(G) \cong L_3(2)/Q$, where $Q$ is a specified 2-group of order $2^9$ and class 2. Moreover $G/Z(G)$ is isomorphic to a parabolic subgroup of $D_4(2)$; or

(vi) $G/Z(G) \cong (A/E)_n$ for some $n \geqslant 5$.

[The solutions arising from Proposition 3.118(iii)(2) are $G/T(S) \cong D_{12}$ and $L_3(2)/E_8$. However, the latter is also a possibility when Proposition 3.118(iii)(1) holds.]

We can assume that (i) does not hold, so that $z(J)$ is not isolated in a Sylow 2-subgroup $S$ of $G$ for any $J \in \Omega$. Also it follows easily from the minimality of $G$ that $Z(G) = 1$.

Consider first the case of Proposition 3.118(iii)(2) and let $J, K$ be as in its statement. Using Lemma 3.119, Aschbacher argues first that the graph $\mathscr{D}^*$ is connected and then that the normal closure $A$ of $z(J)$ in $G$ is elementary abelian and contained in $JK$. Since $z(J)$ is not isolated, $A > \langle z(J) \rangle$ and so $|A| = 4$ or 8.

In the first case, it follows easily that $T(S) \lhd G$ and $G/T(S) \cong D_{12}$. On the other hand, if $|A| = 8$, then using his knowledge of the $G$-conjugates of $z(J)$ and the fact that $G = \langle \Omega \rangle$, Aschbacher argues that $A = C_A$ and that $G/A \cong L_3(2)$ with $S$ of type $M_{12}$. The latter implies that the extension does not split. Thus the proposition holds in either case.

We can therefore assume henceforth that $G = \langle \Omega \rangle$ with $G$ transitive on $\Omega$ [Proposition 3.118(iii)(1)]. As usual, Aschbacher divides the analysis into two subcases, according as $\mathscr{D}$ is or is not connected. Suppose first that $\mathscr{D}$ is connected. Let $\Omega_1(J)$ be a connected component of $\Omega(J)$ of maximal length under the action of $X(J)$ and set $X_1(J) = \langle \Omega_1(J) \rangle$. As in earlier connected cases, Aschbacher proves that if $J^g \in \Omega_1(J)$ for $g \in G$, then

$$G = \langle X_1(J), X_1(J)^g \rangle. \qquad (3.112)$$

By the minimality of $G$, $X_1(J)$ has one of the forms listed in Proposition 3.120. As $\Omega$ is connected, (i) and (ii) are excluded. On the other hand, if (iii) or (v) holds for $X_1(J)$, Aschbacher argues that $X_1(J)^g$ centralizes $X_1(J)$, whence $X_1(J) \lhd G$, which is not the case. Thus

$$X_1(J)/Z(X_1(J)) \cong (A/E)_n \quad \text{for some } n \geqslant 5. \qquad (3.113)$$

Now set $A_1 = O_2(X_1(J))$ and $A = A_1 A_1^g$. Aschbacher argues next that

(1) $A$ is the normal closure of $z(J)$ in $G$.
    In particular, $A \lhd G$; and
(2) $A$ is elementary abelian of order $2^{n+3}$ or $2^{n+2}$     (3.114)
    according as $n$ is odd or even.

Next, using the standard presentation for the alternating group and the fact that $\bar{G} = G/A = \langle \overline{X_1(J)}, \overline{X_1(J)^g} \rangle$, Aschbacher proves

$$G/A \cong A_{n+4}. \tag{3.115}$$

The final step is to determine the isomorphism type of $G$ itself. Because of the slight variation in the structure of $(A/E)_m$ according as $m$ is odd or even, Aschbacher must treat the two cases separately. [When $n$ is odd, the triviality of $Z(G)$ is used to rule out one possible configuration.] Ultimately he shows in either case that

$$G \cong (A/E)_{n+4}. \tag{3.116}$$

Hence for the balance of the discussion, we can assume that

$$\mathscr{D} \text{ is not connected.} \tag{3.117}$$

Thus we are left small Thompson subgroups in the case that $G$ is 2-constrained.

Aschbacher first proves:

$$\text{If } \mu(G) \cong Z_2 \times Z_2, \text{ then } G \cong (A/E)_5. \tag{3.118}$$

Indeed, if $K \in \Omega(J)$, he considers the permutation representation of $G$ on the set $\bar{\Omega}$ of all $G$-conjugates of $JK$. Let $H$ be the kernel of the representation and set $\bar{G} = G/H$. Using a previously established property of the embedding of $J$ in a Sylow 2-subgroup $S$ of $G$ [namely, if $|J \cap S| > 2$, then $J \cap T(S) \in U(S)$], he now concludes that $\bar{J}$ acts semiregularly on $\Omega$, in which case Shult's theorem [I: 251] (see also [I, 3.39]) yields the possibilities for $\bar{G}$. Under the present assumptions, $\bar{G} \cong L_2(4)$ ($\cong A_5$) or $U_3(4)$. Aschbacher argues to a contradiction in the latter case, and in the former he shows that $H = O_2(G) \cong E_{16}$ with $\bar{G}$ acting nontransitively on $H^*$, which in turn implies that $G \cong (A/E)_5$, as asserted.

Using (3.118), Aschbacher next argues that

$$\mathscr{D}^* \text{ is connected.} \qquad (3.119)$$

Since $\mathscr{D}$ is not connected, Proposition 3.106 now yields

$$\mu(G) \cong \Sigma_3, \Sigma_4, \text{ or } \Sigma_3/Q_8 * Q_8. \qquad (3.120)$$

Aschbacher shows next that $z(K) = z(J)$ for every $K \in \Omega(J)$, which implies that $z(J)$ is a 2-central involution and hence that $z(J) \in O_2(G)$. Thus if $A$ is the normal closure of $z(J)$ in $G$, it follows that

$$A \text{ is an elementary abelian subgroup of } O_2(G). \qquad (3.121)$$

Next let $B$ be the subgroup of $A$ generated by the set of $z(J)^g$ such that either $J^g = J$ or $J^g$ is connected to $J$ by an edge in $\mathscr{D}^*$. Aschbacher proves

$$|B| \geqslant 8. \qquad (3.122)$$

He next considers the case that $A = B$ and proves:

If $A = B$, then $G \cong L_3(2)/E_8, L_3(2)/E_{2^6}$, or a parabolic subgroup of $D_4(2)$ and correspondingly $\mu(G) \cong \Sigma_3, \Sigma_4$, or $\Sigma_3/Q_8 * Q_8$. $\qquad (3.123)$

Under this assumption, Aschbacher argues first that $|J| = |A| = 8$, that $Q = C_A$, and that $G$ acts transitively on $A^\#$. Since $J \cap Q = z(J)$, it follows that $G/Q \cong L_3(2)$. Next he argues that $A$ is the unique nontrivial normal subgroup of $Q$ contained in $N_Q(J)$.

Now he treats the three possibilities for $\mu(G)$ separately. If $\mu(G) \cong \Sigma_3$, the last condition of the previous paragraph forces $Q = A$ and the isomorphism type of $S$ is that of $M_{12}$, which together imply that $G \cong L_3(2)/E_8$ with the extension nonsplit.

On the other hand, if $\mu(G) \cong \Sigma_4$, he first forces $Q/A \cong E_8$ and then $Q \cong E_{64}$, whence $G \cong L_3(2)/E_{2^6}$. Finally, using information on the fusion of involutions, he argues that the extension necessarily splits.

Finally if $\mu(G) \cong \Sigma_3/Q_8 * Q_8$, he argues first that $\bar{Q} = Q/A$ has order $2^6$ and that $G/Q$ has two noncentral chief factors on $\bar{Q}$ (each elementary of order 8), each of which as an irreducible $L_3(2)$-module is the dual of the $L_3(2)$-module $A$. Let $Q_1$ be a minimal normal subgroup of $\bar{G} = G/A$

contained in $\bar{Q}$ (so that $\bar{Q}_1 \cong E_8$) and $Q_1$ its pre-image in $Q$. Aschbacher next applies induction to the group $G_1 = Q_1 J$ and concludes that $G_1 \cong L_3(2)/E_{2^6}$ with the extension split. It follows that $G$ splits over $Q$ and that $Q = Q_1 Q_2$, where $A \leqslant Q_2 \cong E_{2^6}$ and $Q_2 \lhd G$. Finally he shows that $Q$ has class 2 with a uniquely determined structure, and one checks that $G$ is, in fact, isomorphic to a parabolic subgroup of $D_4(2)$.

In particular, Proposition 3.119 holds when $A = B$, so it remains to treat the case

$$A > B. \tag{3.124}$$

This condition immediately implies

$$\mu(G) \ntrianglelefteq \Sigma_3. \tag{3.125}$$

Aschbacher next proves

$$\begin{array}{l} \text{If } \Omega(J) \text{ consists of a single element } K, \\ \text{then } G \cong (A/E)_6 \text{ or } (A/E)_7. \end{array} \tag{3.126}$$

The two possibilities for $G$ arise according as $B \leqslant JK$ or $B \nleqslant JK$. In the first case, he argues that $|B| = 8$, $|A| = 16$, and determines enough about the action of $G = \langle \Omega \rangle$ on $A$ to conclude first that $G/A \cong A_6$ and then that $S$ is of type $U_4(3)$, which together force $G = (A/E)_6$.

On the other hand, if $B \nleqslant JK$ and $N$ denotes the pre-image of $\mu(G; S)$ in $N_G(T(S))$, Aschbacher first argues that $\langle y \rangle = Z(N) \cong Z_2$ and that if $Y$ denotes the subgroup of $C_y$ generated by the elements of $\Omega$ contained in $C_y$, then $Y/\langle y \rangle \cong (A/E)_6$. He then shows that $y$ has eactly 7 $G$-conjugates in $A$, that $Q = A \cong E_{64}$, and that $G/A \cong A_7$ with $G/A$ acting naturally on $A$, whence $G \cong (A/E)_7$.

Using (3.120), (3.125), and (3.126), it remains to treat the case

$$\mu(G) \cong \Sigma_3/Q_8 * Q_8. \tag{3.127}$$

In this case, Aschbacher proves

$$G \cong (A/E)_8. \tag{3.128}$$

Let $N$ be as above, and this time let $y$ be a noncentral involution of $JK$, where $K \in \Omega(J)$, and define $Y$ as above. Aschbacher argues that $y \in A$,

$A \cong E_{64}$, $Y/\langle y \rangle \cong (A/E)_6$, and if $P \in \mathrm{Syl}_2\big(C_N(y)\big)$, then $PY$ is isomorphic to the stabilizer of a suitable vector in $(A/E)_8$. An analysis of fusion now yields that $G$ acts as a rank 3 permutation group on the conjugates of $y$ in $A$ with orbits of length 1, 12, and 15. With this information, Aschbacher concludes that $G/Q \cong A_8$. Finally he argues that $Q = A$ and that $G/A$ splits over $A$, whence $G \cong (A/E)_8$, as asserted.

This completes the outline of the reduction to the case that $G = \langle \Omega \rangle$, $G$ is transitive on $\Omega$, and $F^*(G)$ is simple. However, we note that in the course of establishing this reduction as well as treating both the orthogonal and small Thompson subgroup cases, Aschbacher explicitly verifies the various assertions of Propositions 3.79–3.85, and is thus left with the task of identifying $G$ under the above assumptions. Moreover, in the previous section we have indicated in detail how he carries this out in the special case that $\mathscr{D}$ is strongly connected.

We hope that, taken together, the discussion of Sections 3.6, 3.7, and 3.8 gives the reader a good picture of Aschbacher's global strategy for proving the classical involution theorem. At the least, this discussion vividly reveals his spectacular mathematical powers!

# 4

# The *B*-Theorem and
# Locally Unbalanced Groups

In Section 3.3, we have given an overview of the proofs of the *B*- and *U*-theorems. In particular, we have noted that, in effect, these proofs amount to reductions to the solution of certain standard form problems (Hypothesis *S*). In the present chapter we expand this sketch to a full outline.

However, as Thompson's initial reduction as well as Solomon's maximal 2-component theorem do not require Hypothesis *S*, we prefer to discuss these results before making Hypothesis *S* explicit in Section 4.3. After that, we shall describe the work of Foote, Gilman, Griess, Harris, and Solomon, which together complete the reduction of the *B*- and *U*-theorems to Hypothesis *S*.

## 4.1. Unbalancing Triples of Odd Characteristic

As we have already remarked, Thompson's reduction of the *B*- and *U*-theorems depends on Solomon's intrinsic $\hat{A}_n$ 2-component theorem (Theorem 4.24 below). However, because of its special significance for the *B*- and *U*-theorems, we consider it first.

Solomon's proof is, of course, independent of Thompson's reduction. However, it does require Hypothesis *S*, so to that extent (only) Thompson's result does as well.

THEOREM 4.1. *Assume Hypothesis S and let G be a minimal counterexample to the B-theorem and U-theorem. If $(x, y, L)$ is an unbalancing triple in G with $L/O(L)$ of Lie type of characteristic p, p odd, then $L/O(L) \cong L_2(p^n)$ for some n.*

Assume the theorem false and set $D = O(C_y) \cap C_x$ and $\bar{C}_x = C_x/O(C_x)$. Since $(x, y, L)$ is an unbalancing triple, $\langle D, y \rangle$ leaves $L$ invariant and $\bar{D}$ does not centralize $\bar{L}$. Moreover, if we set $\hat{H} = \bar{L}\langle \bar{D}, \bar{y} \rangle$ and $\tilde{H} = \hat{H}/C_{\hat{H}}(\bar{L})$, then $\tilde{y}$ is an involution, $1 \neq \tilde{D} \leqslant O(C_{\tilde{H}}(\tilde{y}))$, $\tilde{L}$ is simple, and $\tilde{H} \leqslant \text{Aut}(\tilde{L})$ (Proposition 3.25). In particular, $\bar{L}$ is locally unbalanced. Since $\bar{L}$ is of Lie type of odd characteristic, but $\bar{L} \not\cong L_2(p^n)$, we can therefore apply [I, 4.246] to conclude:

LEMMA 4.2. *There exists an involution $z \in L$ and a 2-component $J$ of $L_{2'}(C_L(z))$ such that*

(i) $\bar{J} \cong SL_2(q)$ *for some odd $q$;*
(ii) $\langle \bar{z} \rangle = Z(\bar{J})$;
(iii) $\langle D, y \rangle$ *normalizes $J$; and*
(iv) $\bar{D}$ *does not centralize $\bar{J}$.*

In particular (ii) implies that $\bar{z}$ is a *classical* involution of $\bar{L}$.

Since $y$ normalizes $J$, $\bar{y}$ centralizes $\bar{z}$, so we can assume that $z$ is chosen to centralize $y$. Thus $z$ acts on $D$ and as $\bar{z} \in Z(\bar{J})$ with $D$ normalizing $J$, it follows that $[D, z] \leqslant O(J)$, whence $[\bar{D}, \bar{z}] = 1$. Since $D = [D, z]C_D(z)$ by Lemma 1.51, we conclude that $E = C_D(z)$ does not centralize $J$. It is this fact which allows Thompson to stay within the set of unbalancing triples, for as $E \leqslant D \leqslant O(C_y)$, we have

$$E \leqslant O(C_y) \cap C_z. \tag{4.1}$$

We now shift our attention to $C_z$. By Proposition 3.21, the normal closure $K$ of $J$ in $L_z$ is either a single $x$-invariant 2-component or a product of two 2-components interchanged by $x$, and in either case $J$ is a 2-component of $L_{2'}(C_K(x))$. Set $\bar{C}_z = C_z/O(C_z)$ and observe that $\bar{z} \in \bar{J} \leqslant \bar{K}$ with $K$ centralizing $\bar{z}$, so $\bar{z} \in Z(\bar{K})$. In particular, if $\bar{K}$ is a single component, then $\bar{K}$ is intrinsic.

Using the classical involution theorem, we first eliminate certain $SL_2$ possibilities:

LEMMA 4.3. *$\bar{K}$ is not isomorphic to $SL_2(r)$ or $SL_2(r) * SL_2(r)$, $r$ odd.*

Indeed, in the contrary case, $Z(\bar{K}) \cong Z_2$ and so $\langle \bar{z} \rangle = Z(\bar{K})$, whence $\langle \bar{z} \rangle = Z(\bar{K}_1)$ for any component $\bar{K}_1$ of $\bar{K}$. Since $\bar{K}_1 \cong SL_2(r)$, $z$ is therefore a classical involution (of $G$) and so $F^*(G)$ is a K-group by Aschbacher's theorem, contrary to Proposition 3.26(i).

We also have the following result:

LEMMA 4.4. *If* $K_1$ *is a 2-component of* $K$, *then* $(z, y, K_1)$ *is an unbalancing triple.*

Indeed, $E$ leaves each 2-component of $L_z$ invariant by $L^*$-balance and either $K = K_1$ or $K = K_1 K_1^x$. Since $\bar{J} \leqslant \bar{K}$ and $\bar{E}$ does not centralize $\bar{J}$, it follows in either case that $E$ and hence $O(C_y) \cap C_z$ does not centralize $\bar{K}_1$. Thus it suffices to show that $y$ leaves $K_1$ invariant.

Since $y$ leaves $J$ invariant, and induces a permutation of the 2-components of $L_z$, $y$ leaves $K$ invariant. In particular, $y$ leaves $K_1$ invariant if $K = K_1$. In the contrary case, if $y$ interchanges $K_1$ and $K_1^x$, then as $E \leqslant O(C_y)$ and $E$ leaves $C_K(y)$ invariant, we check easily that $\bar{E}$ centralizes $C_{\bar{K}}(\bar{y})$ and then that $\bar{E}$ centralizes $\bar{K}$, whence $\bar{E}$ centralizes $\bar{J}$—contradiction. Thus $y$ leaves $K_1$ invariant in this case as well, and so $(z, y, K_1)$ is an unbalancing triple, as asserted.

Now we can identify $\bar{K}_1$.

LEMMA 4.5. $\bar{K}_1$ *is of Lie type of odd characteristic.*

Indeed, if $\bar{K}$ is a product of two components, then $\bar{J} \cong SL_2(q)$ is a diagonal of $\bar{K}$, whence $\bar{J}$ is a homomorphic image of $\bar{K}$, in which case $\bar{K}_1 \cong \bar{J} \cong SL_2(q)$ and the lemma holds. Thus we can assume that $K_1 = K$.

Since $(z, y, K)$ is an unbalancing triple by the previous lemma, either $\bar{K}$ is of Lie type of odd characteristic, as desired, or else $\bar{K}/Z(\bar{K}) \cong A_n$, $n$ odd, $L_3(4)$, or $He$ by Proposition 3.27(iv). However, as $\bar{z} \in Z(\bar{K})$, the last possibility is excluded here as $He$ has trivial Schur multiplier by [I, Table 4.2]. On the other hand, if $\bar{K} \cong L_3(4)$, then necessarily $\bar{J} \cong L_2(7)$ $\left(\cong L_3(2)\right)$ and $\bar{x}$ induces a field automorphism of $\bar{K}$. Furthermore, $\bar{E} \leqslant O\left(C_{\bar{C}_z}(\bar{y})\right)$ and $\bar{E}$ normalizes, but does not centralize $\bar{J}$. However, it is not difficult to show that $\text{Aut}\left(L_3(4)\right)$ does not possess a subgroup with the properties of $\bar{E}$.

Thus it remains to eliminate the case $\bar{K}/Z(\bar{K}) \cong A_n$. Since $\bar{K}$ is intrinsic, we must have $\bar{K} \cong \hat{A}_n$. But now if $n \leqslant 7$, $K$ has quaternion Sylow 2-subgroups and so $F^*(G)$ is a $K$-group, again by the classical involution theorem. On the other hand, if $n \geqslant 8$, $F^*(G)$ is again a $K$-group by Solomon's intrinsic $\hat{A}_n$ 2-component theorem (Theorem 4.24 below). In either case, Proposition 3.27(i) is contradicted.

Now, in the case $K = K_1$, we bring into play the critical conclusion that $\bar{K}$ is intrinsic with $\bar{K}$ not isomorphic to $SL_2(r)$. Indeed, we can then invoke another general property of groups of Lie type of odd characteristic to conclude that $\bar{J}$ is a diagonal of an $SL_2(q) \times SL_2(q)$-subgroup of $\bar{K}$. The precise result is as follows:

LEMMA 4.6. $C_K(y)$ contains a four-subgroup $Z$ with the following properties:

(i)   There exist two 2-components $J_1, J_2$, in $L_2(C_K(Z))$ such that $\bar{J}_1 \cong \bar{J}_2 \cong SL_2(q)$ and $\bar{J}_1\bar{J}_2 = \bar{J}_1 \times \bar{J}_2$;
(ii)  $Z \cap J_i = \langle z_i \rangle \cong Z_2, i = 1, 2$ and $z = z_1 z_2$;
(iii) $\langle E, y \rangle$ normalizes $J_i$ for both $i = 1$ and 2; and
(iv)  $\bar{E}$ does not centralize $\bar{J}_i$ for both $i = 1$ and 2.

If $K = K_1$, then as $\bar{z} \in z(\bar{J}) \cap Z(\bar{K})$ with $\bar{J}$ a component of $L(C_{\bar{K}}(\bar{x}))$ and $\bar{J} \cong SL_2(q)$ and as $\bar{E} \leqslant O(C_{\bar{C}_z}(\bar{y}))$ normalizes, but does not centralize $\bar{J}$, the existence of the required $Z$ follows readily from the above-mentioned property of the groups of Lie type of odd characteristic.

In the contrary case, $\bar{K} = \bar{K}_1 \times \bar{K}_1^{\bar{x}} \cong SL_2(q) \times SL_2(q)$ by Lemma 4.3. Also, as we have shown in Lemma 4.4, $y$ leaves both $K_1$ and $K_1^x$ invariant, whence $\bar{y}$ centralizes $Z(\bar{K}) \cong Z_2 \times Z_2$. Then $C_K(y)$ contains a four-subgroup $Z$ with $z \in Z$ which maps on $Z(\bar{K})$ and $L_2(C_K(Z)) = J_1 J_2$ with $\bar{J}_1 = \bar{K}_1$ and $\bar{J}_2 = \bar{K}_1^{\bar{x}}$; and it is immediate that $Z$ and $J_1, J_2$ have the required properties.

As with $J$ above, it follows now that $F = C_E(Z)$ does not centralize $\bar{J}_i$, and so we can repeat the above argument with $J_i, z_i, z$ in the roles of $J, z, x$, respectively, for both $i = 1$ and 2. However, this time as $J_1 J_2 \leqslant C_{z_i}$ and $z_i \in J_i$, our conclusions are much sharper. The argument requires us to consider only one value of $i$, say $i = 1$. We let $N$ denote the normal closure of $J_1$ in $L_{z_1}$ and set $\bar{C}_{z_1} = C_{z_1}/O(C_{z_1})$. We shall prove:

LEMMA 4.7. The following conditions hold:

(i)   $N$ is a single $z$-invariant 2-component of $C_{z_i}$;
(ii)  $(z, y, N)$ is an unbalancing triple;
(iii) $\bar{N}$ is of Lie type of odd characteristic with $\bar{N} \not\cong SL_2(r)$;
(iv)  $J_2 \leqslant N$; and
(v)   $\bar{z}$ does not centralize $\bar{N}$.

First, if (i) fails, then as usual $N$ is the product of two 2-components interchanged by $z$; in particular, $z \notin L_{z_i}$. But by $L$-balance, $J_1 J_2 \leqslant L_{z_i}$, so $z \in Z \leqslant L_{z_1}$. Thus (i) holds. Now repeating the proofs of Lemmas 4.3, 4.4, and 4.5, we conclude that (ii) and (iii) also hold.

By Proposition 3.21, $J_1$ is a 2-component of $L_2\big(C_N(z)\big)$. But then if $\bar{z}$ centralized $\bar{N}$, it would follow that $\bar{J}_1 = \bar{N}$, in which case $\bar{N} \cong SL_2(q)$, contrary to (iii). Thus $\bar{z}$ does not centralize $\bar{N}$ and so (v) also holds.

Finally, again by Proposition 3.21, if $M$ denotes the normal closure of $J_2$ in $L_{z_1}$, then $M$ is either a single $z$-invariant 2-component of $C_{z_1}$, or the product of two 2-components of $C_{z_1}$ interchanged by $z$. Hence, either $\bar{M} = \bar{N}$ or $\bar{M}$ centralizes $\bar{N}$. However, in the latter case, $\bar{J}_2 \leqslant \bar{M}$ centralizes $\bar{N}$ and hence so does $\bar{z}_2 \in \bar{J}_2$. Since obviously $\bar{z}_1$ centralizes $\bar{N}$, it follows that $\bar{z} = \bar{z}_1 \bar{z}_2$ does as well, contrary to (v). Thus $\bar{M} = \bar{N}$ and hence $J_2 \leqslant N$, proving (iv).

We can now pin down $\bar{N}$ precisely.

PROPOSITION 4.8. *We have* $\bar{N} \cong Spin_7(q)$.

Indeed, as $J_1$ and $J_2$ are 2-components of $L_2\big(C_K(Z)\big)$ and $Z = \langle z_1, z \rangle$, they are each 2-components of $L_2\big(C_N(z)\big)$, again by Proposition 3.21. Since $\bar{N}$, being a $K$-group, has the $B$-property, it follows that $\bar{J}_1$ and $\bar{J}_2$ are each components of $L\big(C_{\bar{K}}(\bar{z})\big)$. In particular, $\bar{z}_1 \in \bar{J}_1\bar{J}_2 \leqslant \bar{N}$, so $\bar{z}_1 \in Z(\bar{N})$. Thus $\bar{N}$ is intrinsic.

Setting $\tilde{N} = \bar{N}/\langle \bar{z}_1 \rangle$, we see that $\tilde{J}_1 \cong L_2(q)$, $\tilde{J}_2 \cong SL_2(q)$ [whence $\tilde{J}_1\tilde{J}_2 \cong L_2(q) \times SL_2(q)$], and that $\tilde{J}_1, \tilde{J}_2$ are components of $C_{\tilde{N}}(\tilde{z})$. Since $\tilde{N}$ is of Lie type of odd characteristic, it follows now, with the aid of [I, 4.247], that there is only one possibility for $\tilde{N}$: namely, $\tilde{N} \cong P\Omega_7(q)$. Thus $\bar{N}$ is a covering group of $P\Omega_7(q)$ by $Z_2$ and so by [I, Table 4.1], $\bar{N} \cong Spin_7(q)$, as asserted.

Now we can reach a final contradiction at once by playing off the structure $\mathrm{Aut}\big(P\Omega_7(q)\big)$ against the action of $F = C_E(Z)$ on $\bar{N}$. Indeed, $\bar{F}$ leaves $\bar{N}$ invariant, again by $L^*$-balance, and $\bar{y}$ leaves $\bar{N}$ invariant as it leaves $\bar{J}_1\bar{J}_2 \leqslant \bar{N}$ invariant and induces a permutation of the components of $\bar{L}_{z_1}$. Hence if we set $\bar{H} = \bar{N}\bar{F}\langle \bar{y} \rangle$ and $\tilde{H} = \bar{H}/C_{\bar{H}}(\bar{N})$, then, as usual, $\tilde{H} \leqslant \mathrm{Aut}(\tilde{N})$, $\tilde{N} \cong P\Omega_7(q)$, $\tilde{y}$ is an involution, and $\tilde{F} \leqslant O\big(C_{\tilde{H}}(\tilde{y})\big)$ with $\tilde{F}$ acting nontrivially on both $\tilde{J}_1$ and $\tilde{J}_2$. However, one can easily show that $\mathrm{Aut}\big(P\Omega_7(q)\big)$ does not possess an involution with the properties of $\tilde{y}$ and $\tilde{F}$. This completes the proof of Theorem 4.1.

## 4.2. Maximal 2-Components

Underlying the preceding argument is the basic property of $L$-balance, which allows one to compare 2-components in the centralizers of distinct involutions. However, this point is partially submerged in the analysis because of the dominant role in the proof of the groups of Lie type of odd characteristic. In contrast, Solomon's maximal 2-component theorem, which we discuss next, will illustrate the power of $L$-balance in a very general context. In addition, it provides an excellent application of Aschbacher's component fusion theorem (Theorem 3.39). For these reasons we present the argument in some detail.

To state the theorem, we require the following terminology: Let $G$ be a group of component type and let $\mathscr{H}$ denote an isomorphism type of elements of $\mathscr{L}^*(G)$. Then for each $t \in \mathscr{I}(G)$, we let $K_t$ be the product of those 2-components $K$ of $C_t$ such that $K/O(K) \in \mathscr{H}$.

THEOREM 4.9. *Assume that $F^*(G)$ is quasisimple and that the elements of $\mathscr{H}$ are either of 2-rank $\geqslant 3$ or simple of 2-rank 2. Then for any involution $t$ of $G$, $K_t$ is either trivial or is a single 2-component.*

(Solomon also allows nonsimple groups of 2-rank 2, but we do not require that result.)

We take $G$ to be a minimal counterexample. It is immediate then that $O(G) = 1$. Also if $t \in \mathscr{I}(G)$ with $t \in Z(F^*(G))$, then as $F^*(G)$ is quasisimple, we have $F^*(G) = L_t$ and $K_t$ has at most one 2-component. Thus the theorem holds for any such involution $t$.

For the argument, Solomon focuses on a maximally "bad" subgroup of $G$. To describe it precisely, he needs the following terminology.

DEFINITION 4.10. For any proper subgroup $K$ of $H$, denote by $K_H$ the product of the 2-components $K$ of $H$ such that for some $x \in \mathscr{I}(G)$ that centralizes $K/O(K)$ and some 2-component $J$ of $K_x$, $J \cap H$ covers $K/O(K)$ (with $K_H = 1$ if no such 2-component $K$ exists).

It is immediate from the definition that

$$K_H \lhd N_G(H). \tag{4.2}$$

Furthermore, using $L$-balance, we easily check that $K_x = K_{C_x}$ for any $x \in \mathscr{I}(G)$.

Since the theorem fails for $G$, there is $t \in \mathscr{I}(G)$ such that $K_t = K_{C_t}$ has at least two 2-components. Then $t \notin Z(F^*(G))$ and so $C_t < G$. We choose a proper subgroup $N$ of $G$ as follows:

(1) The number $r$ of 2-components of $K_N$ is maximal (thus $r \geqslant 2$);

(2) Subject to (1), $m\big(C_N(K_N/O(K_N))\big)$ is maximal;

(3) Subject to (1) and (2), a Sylow 2-subgroup of $C_N\big(K_N/O(K_N)\big)$ has maximal order; and

(4) Subject to (1), (2), and (3), a Sylow 2-subgroup of $N$ has maximal order.

[Note that this ordering is essentially the same as that in the definition of a strongly maximal 2-component (Definition 3.30).]

We denote the 2-components of $K_N$ by $K_1, K_2, ..., K_r$, and we fix $Q \in \mathrm{Syl}_2\big(C_N(K_N/O(K_N))\big)$. Note that as the theorem is violated in some $C_t$, $t \in \mathscr{I}(G)$, maximality of $N$ implies that

$$\text{Either } r \geqslant 3 \text{ or } Q \neq 1. \tag{4.3}$$

The maximality conditions have important consequences for pump-ups.

PROPOSITION 4.11. *If* $y \in \mathscr{I}(N)$ *centralizes* $K_i/O(K_i)$ *for some* $i$, $1 \leqslant i \leqslant r$, *and we set* $J_i = L_{2'}\big(C_{K_i}(y)\big)$ [*whence* $J_i$ *covers* $K_i/O(K_i)$], *then* $J_i$ *is contained in a single 2-component* $L_i$ *of* $K_y$ [*whence* $J_i$ *covers* $L_i/O(L_i)$].

The proof is carried out in several steps.

LEMMA 4.12. $J_i \leqslant K_y$ *and* $\langle J_i^{K_y}\rangle$ *is the product of at most four 2-components of* $K_y$.

Indeed, by definition of $K_N$, there is $x \in \mathscr{I}(G)$ and a 2-component $L$ of $K_x$ such that $L = K_i O(L)$, so also $L = J_i O(L)$. Thus, $\langle x, y\rangle \leqslant C_G\big(L/O(L)\big)$. Replacing $x$ by a conjugate, if necessary, we can assume that $x$ and $y$ lie in the same Sylow 2-subgroup $R$ of $C_G\big(L/O(L)\big)$. Let $z \in \mathscr{I}\big(Z(R)\big)$, so that $z$ centralizes both $x$ and $y$.

Set $J = L_{2'}\big(C_L(z)\big)$, so that $J$ covers $L/O(L)$. By $L$-balance, $I = \langle J^{L_z}\rangle$ is the product of at most two 2-components of $L_z$. Since $L/O(L) \in \mathscr{K} \subseteq \mathscr{L}^*(G)$, it follows that $I \leqslant K_z$. But $y \in C_z$ and $y$ centralizes $J_i$ which covers $J/O(J)$, so $y$ centralizes $J/O(J)$ and thus $y$ leaves $L$ invariant. Set $H = L_2\big(C_L(y)\big)$. If $y$ leaves each 2-component of $L$ invariant, then as $y$ centralizes $J/O(J)$, $y$ centralizes $L/O(L)$, whence $H$ covers $L/O(L)$. In the contrary case, $H$ is a diagonal of $L$ and again as $y$ centralizes $J/O(J)$, we see that $H/O(H) \cong J/O(J)$ is isomorphic to an element of $\mathscr{K}$. Again by $L$-balance, each 2-component of $H$ pumps-up in $L_y$ to at most two 2-components of $K_y$, so $K = \langle H^{K_y}\rangle$ is the product of at most four such 2-components.

But $J_i \cap J$ covers $J_i/O(J_i)$ and $J_i \cap J \leqslant H \leqslant K$. Since $J_i$ is perfect, it follows that $J_i \leqslant K$ and so $\langle J_i^{K_y} \rangle \leqslant K$ is the product of at most four 2-components of $K_y$.

LEMMA 4.13. *Proposition* 4.11 *holds if* $y \in K_j Q$ *for some* $j \neq i$.

Suppose for definiteness that $i = 1$ and $j = r$. Let $J_k = L_2\big(C_{K_k}(y)\big)$, $1 \leqslant k \leqslant r - 1$, and for each such $k$ let $L_k = \langle J_k^{K_y} \rangle$. By the previous lemma, each $L_k$ is the product of at most four 2-components of $K_y$. Furthermore, as $J_k$ centralizes $J_m$ (mod cores), for $m \neq k$, it is immediate that $L_k$ and $L_m$ consist of distinct 2-components of $K_y$. But by our maximal choice of $r$, $K_y$ has at most $r$ 2-components. Hence either the lemma holds or $L_1$ is the product of precisely two 2-components of $K_y$, $L_k$ is a single 2-component for $2 \leqslant k \leqslant r - 1$, and $K_y = L_1 L_2 \ldots L_{r-1}$.

Using our maximal choice of $N$, we show that the latter possibility leads to a contradiction. Since $Q$ centralizes $K_r/O(K_r)$, it centralizes a Sylow 2-subgroup $T$ of $C_{K_r}(y)$. Set $\bar{C}_y = C_y/O(C_y)$. Then for $2 \leqslant k \leqslant r - 1$, $\bar{J}_k = \bar{L}_k$, so $\bar{T}\bar{Q}$ centralizes each such $\bar{L}_k$. Furthermore, $\bar{T}\bar{Q}$ centralizes $\bar{J}_1$, which is a diagonal of $\bar{L}_1$. Hence, as usual, a subgroup $P$ of index at most 2 in $TQ$ centralizes $\bar{L}_1$.

Suppose $P \cap Q \neq 1$, and let $u \in \mathscr{I}(P \cap Q)$. Then $u$ centralizes both $\bar{K}_y$ and $K_r/O(K_r)$; and now a repeat of the first paragraph of the proof yields that $K_u$ has at least $r + 1$ 2-components, contrary to the maximality of $r$. Hence $P \cap Q = 1$ and as $|TQ : P| \leqslant 2$, this forces $|Q| \leqslant 2$. But $P$ centralizes $K_y$ and so by conditions (2) and (3) in the choice of $N$, we must have $|P| \leqslant 2$, whence

$$|TQ| \leqslant 4. \tag{4.4}$$

Finally expand $T$ to a $Q$-invariant Sylow 2-subgroup $S$ of $K_r$. Then $|C_{SQ}(y)| = |TQ| = 4$, so by [I, 1.16, 1.17] $SQ$ is of maximal class and hence dihedral or quasi-dihedral. In particular, $m(S) \leqslant m(SQ) = 2$ and so by definition of $K$, $K_m/O(K_m)$ is simple. Thus $SQ = S \times Q$ and now $m(SQ) = 2$ forces $Q = 1$. However, as $y$ centralizes $K_y$, our maximal choice of $N$ is again contradicted.

LEMMA 4.14. *If* $y$ *centralizes a four-subgroup* $U$ *of* $C_N\big(K_i/O(K_i)\big)$ *such that Proposition* 4.11 *holds for each* $u \in U^\#$, *then the proposition holds for* $y$.

Indeed, set $L = \langle J_i^{K_y} \rangle$, so that by Lemma 4.12, $L$ is the product of 2-components $L_1, L_2, \ldots, L_m$ of $K_y$ for some $m \leqslant 4$. We argue that $m = 1$, so assume false. Now $U \leqslant C_y$ and as $U$ centralizes $K_i/O(K_i)$, $I_i = K_{2'}(C_{J_i}(U))$ covers $J_i/O(J_i)$ [and $K_i/O(K_i)$]. Then $L = \langle I_i^L \rangle$ and so $U$ leaves $L$ invariant. Since $m > 1$, it follows easily from the permutation action of $U$ on the set of $L_i$ that for some $u \in U^\#$, $I = L_{2'}(C_L(u))$ is the product of at least two 2-components and $I = \langle I_i' \rangle$. But by assumption, Proposition 4.11 holds for $u$ and consequently $\langle I_i^{K_u} \rangle = H$ is a single 2-component of $K_u$. However, as $I = \langle I_i' \rangle$, and $I \leqslant L_y$ by $L$-balance, we see that $I \leqslant H$, forcing $I/O(I) \cong I_i/O(I_i)$, which is not the case.

Now the proposition follows at once. Indeed, if we set $X = C_{K_N O_{2'2}(N)}(K_i/O(K_i))$, then clearly every $x \in \mathcal{I}(X)$ centralizes a four-subgroup $U$ of $K_j/O(K_j)$ for any $j \neq i$ and, moreover, $U \leqslant C_N(K_i/O(K_i))$. Thus the proposition holds for $x$ by Lemmas 4.13 and 4.14. Thus it holds for every involution of $X$. But then if $y$ is an arbitrary involution of $N$ centralizing $K_i/O(K_i)$, $y$ leaves $X$ invariant, and one checks in all cases that $y$ centralizes a four-subgroup of $X$. Hence the proposition holds for $y$ by another application of Lemma 4.14.

Using Proposition 4.11, we can establish a key covering property of $K_N$:

LEMMA 4.15. *If $H$ is a proper subgroup of $G$ which covers $K_N Q/O(K_N)$, then*

(i) $L_{2'}(K_N \cap H) \leqslant K_H$; *and*
(ii) $K_N$ *covers* $K_H/O(K_H)$ [*that is,* $K_N/O(K_N) \cong K_H/O(K_H)$].

If (i) holds, then $K_H$ has at least $r$ 2-components, so by the maximality of $N$, the number is exactly $r$, and this implies (ii). Thus it suffices to prove (i).

Set $J_i = L_{2'}(K_i \cap H)$, so that by assumption $J_i$ covers $K_i/O(K_i)$, $1 \leqslant i \leqslant r$. We must show that each $J_i \leqslant K_H$. Since the argument is similar for all $i$, we can assume without loss that $i = 1$.

Since $H$ covers $K_N Q/O(K_N)$, it follows from (4.3) that there is an involution $y$ in $K_r Q \cap H$ that centralizes both $J_1/O(J_1)$ and $J_2/O(J_2)$. Setting $I_j = L_{2'}(C_{J_j}(y))$, $j = 1, 2$, Proposition 4.11 implies that $I_j \leqslant K_y$ with $\langle I_j^{K_y} \rangle$ a single 2-component. It follows that $I_j$ is contained in a 2-component $L_j$ of $L_2(C_H(y))$ with $I_j$ covering $L_j/O(L_j)$, $j = 1, 2$. Hence, by Proposition 3.21, $L_j$ is contained in a $y$-invariant product of one or two 2-components $M_j$ of $H$ with $L_j$ a 2-component of $L_{2'}(C_{M_j}(y))$, and $M_j = \langle L_j^{M_j} \rangle = \langle J_j^{M_j} \rangle$, $j = 1, 2$.

Set $\bar{H} = H/O(H)$, so that $\bar{M}_1, \bar{M}_2$ are components of $\bar{H}$. If $\bar{M}_1 \neq \bar{M}_2$, then there is $y_2 \in \mathcal{I}(J_2)$ which centralizes $\bar{M}_1$ (as $J_2 \leqslant L_2 \leqslant M_2$). Since $J_1 \leqslant M_1$ with $M_1 = \langle J_1^{M_1} \rangle$, it follows easily by another application of

Proposition 4.11 that $L_2(C_{M_1}(y_2))$ is contained in a single 2-component of $K_{y_2}$, whence $\bar{J}_1 = \bar{M}_1$ and it follows that $J_1 \leqslant K_H$, as required.

In the contrary case, $\bar{M}_1 = \bar{M}_2, \bar{M}_1$ is quasisimple, and $\bar{L}_1, \bar{L}_2$ are distinct 2-components of $L_2(C_{\bar{M}_1}(\bar{y}))$. Using Proposition 4.11 and $L$-balance, one easily checks that $\bar{L}_j/O(\bar{L}_j) \in \mathscr{L}^*(\bar{M}_1\langle y\rangle)$. Thus the assumptions of Theorem 4.9 hold in $\bar{M}_1\langle \bar{y}\rangle$, but the conclusion clearly fails. However, as $|\bar{M}_1\langle \bar{y}\rangle| < |G|$, this contradicts our minimal choice of $G$.

These are the basic preliminary results that Solomon needs for his argument, which now breaks up into two major parts, the goal of the first being to establish the following result:

PROPOSITION 4.16. *There exists a maximal subgroup $M$ of $G$ containing $K_N Q$ such that for each $i$, $1 \leqslant i \leqslant r$, and each involution $x$ of $K_i$, we have $C_x \leqslant M$.*

It is the cores of the groups $K_t$, $t \in \mathscr{I}(G)$, which represent the principal obstruction to verifying the proposition, for if each $K_t$ is semisimple (i.e., has quasisimple 2-components), the proof is very easy.

Indeed, under this assumption, it follows first that $K_N$ is semisimple. For if, say, $K_1$ were not quasisimple, we could choose a four-subgroup $U$ of $K_2$, set $J_1 = L_2(C_{K_1}(U))$ [whence $J_1$ covers $K_1/O(K_1)$], and apply [I, 4.13] to conclude that $J_1$ does not centralize $C_{O(K_1)}(u)$ for some $u \in K_1^\#$. But by Proposition 4.11 and the assumed semisimplicity of $K_u$, $J_1$ is a component of $K_u$, which implies that $J_1$ centralizes any subgroup of $C_u$ of odd order which it normalizes, contrary to the action of $J_1$ on $C_{O(K_1)}(u)$.

Next we argue that any maximal subgroup $M$ of $G$ containing $N_G(K_N)$ has the required properties.

For definiteness, take $x \in \mathscr{I}(K_1)$. Then $x$ centralizes $K_i$ for $i \geqslant 2$ and so again by Proposition 4.11 and the assumed semisimplicity of $K_x$, each such $K_i$ is a component of $K_x$. In particular, $K_x \leqslant H = N_G(K_2)$. Also $K_N Q \leqslant H$, whence $K_N$ covers $K_H/O(K_H)$ by Lemma 4.15. But as with $N$ above, also $K_H$ is semisimple, so $K_H = K_N$. On the other hand, it is easy to show by the same argument as in Lemma 4.15 that $K_x$ is a product of 2-components of $K_H$. We conclude that either $K_x = K_2 K_3 \ldots K_r$ or $K_x = K_N$.

In either case if we set $X = N_G(K_x)$, it follows that $K_N Q \leqslant X$. Then as with $H$, we have $K_N = K_x$. But then by (4.2), $K_N \lhd X$ and so $X \leqslant N_G(K_N) \leqslant M$. On the other hand, $K_x \lhd C_x$, again by (4.2), so $C_x \leqslant X$. Thus $C_x \leqslant M$, as asserted, and so the proposition holds under the assumption of semisimplicity.

To deal with the problem of cores, Solomon uses signalizer functor theory. If $B_i, B_j$ are four-subgroups of $K_i, K_j$, respectively, $i \neq j$, such that $A = \langle B_i, B_j \rangle$ is an elementary 2-group of rank $\geq 3$, he verifies the core separation axiom of Goldschmidt [I, 4.66] and then constructs the appropriate $A$-signalizer functor $\theta$ on $G$. Allowing $B_i, B_j, i$, and $j$, to vary, he is able to compare the corresponding completions $\theta(A; G)$. Moreover, under the assumption that some $K_x$ is not semisimple, he forces $\theta(G; A)$ to be nontrivial and then shows that any maximal subgroup $M$ of $G$ containing $N_G(\theta(G; A))$ has the desired properties.

The whole point of signalizer functor theory is to allow one to conclude that $N_G(E) \leq N_G(\theta(G; A))$ for suitable noncyclic elementary 2-subgroups of $G$ with $E$ connected to $A$. In the present situation, because of the allowed variations of $A$, Solomon is able to prove that $K_N O(N) \leq M$ and that $K_x O(C_x) \leq M$ for every $x \in \mathscr{I}(K_i)$, $1 \leq i \leq r$. In effect, this allows him to work with the present $M$ in much the same way as we did with $M$ above in the special case that each such $K_x$ was semisimple.

We note that the argument also implies that $M$ contains a Sylow 2-subgroup of $N$. Also $K_N \leq K_M$ by Lemma 4.15. Hence by our maximal choice of $N$, we have the following additional property of $N$:

LEMMA 4.17. *$N$ contains a Sylow 2-subgroup of $M$.*

Solomon is now in a position to establish the following result.

PROPOSITION 4.18. *Conditions $(a)$, $(b)$, $(c)$ of Theorem 3.39 hold for $M$ with $K_M$ as $L$. In addition, $M = N_G(K_M)$.*

As noted in Section 3.4, Aschbacher shows that these conditions force $M$ to permute the $K_i$ transitively under conjugation, so that Theorem 3.39 is applicable and yields that $K_M$ consists of a single 2-component, contrary to the fact that $r \geq 2$. Thus once Proposition 4.18 is established, it will complete the proof of Theorem 4.9.

To prove the proposition, observe first that as $K_N \leq L = K_M$, $L$ is the product of 2-components $L_i$ with $K_i \leq L_i$ and $L_i = K_i O(L_i)$, $1 \leq i \leq r$. Furthermore, as $M \leq N_G(L)$ by (4.2) and $r \geq 2$, the maximality of $M$ and quasisimplicity of $F^*(G)$ imply that $M = N_G(L)$. Thus we need only show that conditions $(a)$, $(b)$, and $(c)$ hold.

We first prove the following lemma:

LEMMA 4.19. *If $x \in \mathscr{I}(M)$ centralizes $L_i/O(L_i)$ for some $i$, $1 \leq i \leq r$, then $O(C_x) L_x \leq M$.*

Indeed, as $N$ contains a Sylow 2-subgroup of $M$ by Lemma 4.17, we can assume without loss that $x \in N$. Set $I = L_{2'}(C_{L_i}(x))$. Then by Proposition 4.11, $I$ is contained in a 2-component $J$ of $K_x$ with $I$ covering $J/O(J)$. Hence if $u \in \mathcal{I}(I)$, it follows that $C_u$ covers all 2-components of $C_x$ (mod cores) other than $J$, so $IC_u$ covers $L_x/O(L_x)$. Since $C_u \leqslant M$ by Proposition 4.16, we see that $M$ covers $L_x/O(L_x)$. But $I$ contains a four-subgroup $B$ and $C_b \leqslant M$ for each $b \in B^{\#}$. Since $B$ normalizes $O(C_x)$, it follows from [I, 4.13] that $O(C_x) \leqslant M$. In particular, $O(L_x) \leqslant M$, so $O(C_x) L_x \leqslant M$, as asserted.

Now we can make a reduction.

LEMMA 4.20. *If condition* $(c)$ *of Theorem* 3.39 *holds, then so do conditions* $(a)$ *and* $(b)$.

Indeed, suppose $t \in \mathcal{I}(G)$ centralizes $L_i/O(L_i)$ for some $i$, $1 \leqslant i \leqslant r$. Then $t$ centralizes an involution of $L_i$ and so $t \in M$ by Proposition 4.16. Hence $L_t \leqslant M$ by the preceding lemma. But then for any $g \in G$, we have

$$L_{t^g} \leqslant M^g. \tag{4.5}$$

Suppose now that also $t^g$ centralizes $L_j/O(L_j)$ for some $j$, $1 \leqslant j \leqslant r$. Then $I_j = L_{2'}(C_{L_j}(t^g))$ covers $L_j/O(L_j)$ and by $L$-balance, $I_j \leqslant L_{t^g}$, so $I_j \leqslant M^g$ by (4.5). Hence if we put $X = I_j$, we see that $X \leqslant M^g$ and $L_j \leqslant XO(L_j)$. Hence if condition (c) of Theorem 3.39 holds, then $g \in M$ and so condition (a) also holds.

Assume finally that $r \geqslant 3$ and that $t^g$ interchanges suitable $L_j$ and $L_k$, $1 \leqslant j, k \leqslant r$. Setting $I = L_{2'}(C_{L_jL_k}(t^g))$, then again by $L$-balance, $I \leqslant L_{t^g}$ and the normal closure $J$ of $I$ in $L_{t^g}$ is the product of at most two 2-components of $L_{t^g}$. Hence if $x \in \mathcal{I}(I)$, then as $r \geqslant 3$, it follows that $x$ centralizes $L_m/O(L_m)$ and $L_n^g/O(L_n^g)$ [whence $x^{g^{-1}}$ centralizes $L_n/O(L_n)$] for suitable $1 \leqslant m, n \leqslant r$. But then by condition (a), $g^{-1}$ and hence $g$ is in $M$, so (c) also implies (b).

Thus it remains to establish condition (c) of Theorem 3.39. Suppose then that $M^g$ contains a subgroup $X$ such that for some $j$, $1 \leqslant j \leqslant r$,

$$L_j \leqslant XO(L). \tag{4.6}$$

We must show that (4.6) implies $g \in M$. Since $L_j$ is perfect and $X$ covers $L_j/O(L_j)$, we can assume without loss that $X \leqslant L_jO(L_j)$ and that $X$ is perfect.

We consider various possibilities for the embedding of $X$ in $M^g$.

LEMMA 4.21. *If* $X \leqslant L^g$, *then* $g \in M$.

Indeed, set $X_1 = \langle X^{L^g} \rangle$ and let $X_2$ be the product of the 2-components of $L^g$ not in $X_1$ (with $X_2 = 1$ if no such 2-components exist). Then $X_1$ centralizes $X_2/O(X_2)$.

Suppose first that $X_2 \neq 1$ (whence $X_2$ is a nontrivial product of suitable $L_i^g$, $1 \leqslant i \leqslant r$). Let $x \in \mathscr{I}(X) \cap L_j$. Since $x$ centralizes $X_2/O(X_2)$, Lemma 4.19, applied to $M^g$, yields that $O(C_x)L_x \leqslant M^g$. But $x$ centralizes $L_k/O(L_k)$ for all $k \neq j$ and so by $L$-balance and the previous inclusion, $M^g$ covers each such $L_k/O(L_k)$. Since $M^g$ also covers $L_j/O(L_j)$, it follows that $M^g$ covers $L/O(L)$, whence by Lemma 4.15, applied to $M^g$, $L \cap M^g \leqslant L^g$ and covers $L^g/O(L^g)$. Hence $L \cap M^g$ contains a four-subgroup $U$ of $L_1^g$. But then by Proposition 4.16, applied to $M^g$, $C_u \leqslant M^g$ for all $u \in U^\#$, and so $O(L) \leqslant M^g$, again by [I, 4.13]. Hence $L \leqslant M^g$ and we conclude that $L = L^g$, whence $g \in N_G(L) = M$.

Thus we can assume that $X_2 = 1$, whence $X_1 = L^g$. Set $\bar{M}^g = M^g/O(M^g)$. Then for any $k$, $1 \leqslant k \leqslant r$, and any 2-element $\bar{y} \in \bar{L}_k^g$, there exists $x \in \mathscr{I}(X) \cap L_j$ ($x$ depending on $y$) such, that $\bar{x}$ centralizes $\bar{y}$. Since $C_x \leqslant M$ by Proposition 4.16, it follows that $M$ covers each such $\bar{y}$. But as $L_k$ is perfect, the set of such $\bar{y}$ generate $\bar{L}_k^g$. Hence $M$ covers $\bar{L}_k^g$ for all $k$, whence $M$ covers $\bar{L}^g$. But considering a four-subgroup of $X \cap L_j$, we also have that $O(L^g) \leqslant M$. Thus $L^g \leqslant M$ and again we obtain $L^g = L$ and $g \in M$.

LEMMA 4.22. *If $X$ contains a four-subgroup $U$ which leaves some $L_k^g$ invariant, then $g \in M$.*

First of all, it is easy to see that $\langle C_{L_k^g}(u) | u \in U^\# \rangle$ contains a four-subgroup $V$ with each $v \in V^\#$ centralizing an involution $u$ of $U$ ($u$ depending on $v$).

Set $\bar{M} = M/O(M)$. We claim that there is an involution $z$ in $M$ which centralizes both $V$ and $\bar{L}_j$. Indeed, if $O_2(\bar{M}) \neq 1$, we can take $z$ to centralize $V$ with $\bar{z} \in O_2(\bar{M})$. In the contrary case, $Z(\bar{L}_j) = 1$, so $\bar{L}_j$ is simple, and as each $v \in V^\#$ centralizes some $\bar{u} \in \bar{U} \cap \bar{L}_j^\#$, it follows that $V$ leaves $\bar{L}_j$ invariant. But then $V$ leaves $C_M(\bar{L}_j)$ invariant. Since this group has even order (as $r \geqslant 2$), we can take $z \in C_M(\bar{L}_j)$ in this case.

Finally $z \in C_V \leqslant M^g$ by Proposition 4.16 (as $V \leqslant L_k^g$). On the other hand, as $z$ centralizes $\bar{L}_j$, $Y = L_{2'}(C_X(z))$ covers $X/O(X)$. But then by $L$-balance and Proposition 4.11, we conclude, as usual, that $Y \leqslant K_{M^g} = L^g$. Since $X$ is perfect and $X = YO(X)$, it follows that $X \leqslant L^g$, so $g \in M$ by the previous lemma.

It therefore remains to treat the case that $X$ does not contain a four-subgroup leaving some $L_k^g$ invariant. There is then an involution $x \in X$ which conjugates $L_1^g$ into $L_m^g$ for some $m \neq 1$, say $m = 2$. Set $I = L_{2'}(C_W(x))$, where $W = L_1^g L_2^g$, so that $I$ is a diagonal of $W$. Again by Proposition 4.16 and $L$-balance, $I \leqslant L_{2'}(M)$. If $I$ covers $\bar{L}_j$ in $\bar{M} = M/O(M)$, then so does $L^g$ and again $X \leqslant L^g$, whence $g \in M$ by Lemma 4.21. Thus we can assume that $I$ centralizes $\bar{L}_j$.

Let $U$ be a four-subgroup of $X$ with $x \in U$. Then since $X$ covers $\bar{L}_j$, $U$ thus centralizes $I/O(I)$ and so $U$ leaves $W$ invariant. Hence some $u \in U^{\#}$ leaves invariant both $L_1^g$ and $L_2^g$, and consequently $u$ centralizes $L_1^g/O(L_1^g)$. Hence if we set $J = L_{2'}(C_{L_1^g}(u))$, then repeating the argument of the previous paragraph with $u$, $J$ in place of $x$, $I$, we again reduce to the case that $J$ centralizes $\bar{L}_j$.

Finally choosing $z \in \mathscr{I}(J)$, then $C_z \leqslant M^g$ and $z$ centralizes $L_j/O(L_j)$, which, as usual, implies that $L^g$ covers $L_j/O(L_j)$, whence again $X \leqslant L^g$, and again it follows from Lemma 4.21 that $g \in M$.

This establishes Proposition 4.18 and completes the discussion of Solomon's maximal 2-component theorem.

## 4.3. Hypothesis $S$

We shall now describe the precise set of standard form problems whose solutions are needed for the $B$- and $U$-theorems. It is a rather complex list of groups, arising out of specific technical considerations in the proofs of these theorems. However, in the discussion that follows the reader should not lose sight of the central point:

> The proof of the $B$- and $U$-theorems requires the solution of a large number of standard form problems with standard component a group of Lie type of either characteristic 2 or 3, an alternating group, or a sporadic group. $\qquad$ (4.7)

The bulk of these standard form problems occur in connection with an inductive characterization of the groups of Lie type of odd characteristic which Harris needs for solution of the minimal $L_2(q)$ problem. For brevity, we denote the family of quasisimple groups of Lie type of odd characteristic by $\mathscr{L}\mathcal{O}$.

The natural theorem to attempt to prove is the following: if $G$ is a group with $F^*(G)$ simple in which some element of $\mathscr{L}(G)$ is in $\mathscr{L}\mathcal{O}$, then $F^*(G)$ is in $\mathscr{L}\mathcal{O}$. Moreover, the strategy for treating the "general" case—i.e., when $\mathscr{L}(G)$ contains an element $L \in \mathscr{L}\mathcal{O}$ which possesses a classical involution—is clear: simply emulate Thompson's reduction (Theorem 4.1 above) and force $G$ itself to possess a classical involution, so that $F^*(G) \in \mathscr{L}\mathcal{O}$ by Aschbacher's theorem (Theorem 3.64).

This would leave one with certain residual possibilities for $L$. In these cases, beginning with $L$ and using the pumping-up process, the aim would presumably be to show that $\mathscr{L}(G)$ must contain a second element $L^*$ with $L^* \in \mathscr{L}\mathcal{O}$ and $L^*$ having a classical involution, thus reducing to the previous case.

Unfortunately, this second line of argument cannot work in every residual case. For one thing, $L$ may be *maximal*, so that no proper pumping-up takes place. However, there is a more serious problem, due to the exceptional isomorphisms which exist between certain members of $\mathscr{L}\mathcal{O}$ and groups of Lie type of characteristic 2 and alternating groups.

$$
\begin{aligned}
L_2(5) \cong L_2(4) \cong A_5, \qquad & L_2(7) \cong L_3(2) \\
L_2(9) \cong A_6, \qquad & PSp_4(3) \cong U_4(2).
\end{aligned}
\tag{4.8}
$$

As a consequence, if $L$ is one of the four listed groups, $L$ may have a pump-up in $\mathscr{L}(G)$ which is "unambiguously" of Lie type of characteristic 2 or alternating [for example, $L_2(5)$ pumping-up to $L_2(16)$ or to $A_7$]. Likewise there are cases in which $L$ may pump-up to a sporadic group [for example, $L_2(5)$ pumping-up to $M_{12}$]. Moreover, by a *sequence* of pump-ups one can reach still further groups [for example, $L_2(16)$ pumping-up to $L_2(2^8)$ or $L_3(16)$, $A_7$ to $A_{2k+7}$, $k \geqslant 1$, and $M_{12}$ pumping-up to $.3$].

Thus the pumping-up process leads in a natural way to the notion of the *closure* $\mathscr{L}\mathcal{O}^*$ of the family $\mathscr{L}\mathcal{O}$. In effect, Hypothesis $S$ must include a solution of the standard form problem arising from every component in $\mathscr{L}\mathcal{O}^* - \mathscr{L}\mathcal{O}$.

Table 4.1 below gives a complete list of the groups in $\mathscr{L}\mathcal{O}^* - \mathscr{L}\mathcal{O}$ which can be reached by a sequence of pump-ups beginning with the corresponding element of $\mathscr{L}\mathcal{O}$.

Thus $\mathscr{L}\mathcal{O}^* - \mathscr{L}\mathcal{O}$ consists of the set of all groups in the second column, together with their nontrivial central extensions.

The generic argument of the desired inductive characterization of $\mathscr{L}\mathcal{O}$ also breaks down when $L$ is one of a short list of groups defined over $GF(3)$: namely,

$$
\begin{aligned}
& L_3(3), \; U_3(3), \; L_4(3), \; U_4(3), \; PSp_4(3), \; U_4(3)/Z_2, \; P\Omega_7(3), \\
& P\Omega_8^{\pm}(3), \text{ and } G_2(3).
\end{aligned}
\tag{4.9}
$$

**Table 4.1.** Pump-Ups in $\mathscr{L}\mathscr{O}^* - \mathscr{L}\mathscr{O}$ of Elements of $\mathscr{L}\mathscr{O}$

| $\mathscr{L}\mathscr{O}$ | Pump-up in $\mathscr{L}\mathscr{O}^* - \mathscr{L}\mathscr{O}$ |
|---|---|
| $L_2(5)$ | $L_2(4^{2n}), L_3(4^n), n = 2^m, m \geqslant 0, A_n, n$ odd, $n \geqslant 7$, |
| | $M_{12}, J_1, J_2, Ly, He, Suz, ON, .3$ |
| $L_2(7)$ | $L(4^n), n = 2^m, m \geqslant 0, J_2, He, Suz, ON$ |
| $L_2(9)$ | $G_2(2^n), PSp_4(2^n), L_4(2^n), U_4(2^n), n = 2^m, m \geqslant 0$, |
| | $L_5(2^n), U_5(2^n), n = 2^m, m \geqslant 1, A_n, n$ even, |
| | $n \geqslant 8, Mc, HS, .1, F_5$ |
| $L_2(17)$ | $J_3$ |
| $PSp_4(3)$ | $U_4(2^n), L_4(4^n), n = 2^m, m \geqslant 0$ |

Even though $L$ contains a classical involution $t$ in each case, $C_L(t)$ is solvable. As a result, these cases require special treatment. [Note that the groups $SL_4(3) = L_4(3)/Z_2$, $Sp_4(3) = PSp_4(3)/Z_2$, and $SU_4(3) = U_4(3)/Z_4$ are not on our list; even though $C_L(t)$ is again solvable, it turns out that the generic argument works for these three groups].

Finally if $L \cong {}^2G_2(3^n)$, $n$ odd, $n > 1$, $L$ does not contain a classical involution. Moreover, these groups have no proper pump-ups. Hence these cases also require special treatment.

Hence if we let $\mathscr{L}_0(3)$ denote the groups listed in (4.9) plus the family ${}^2G_2(3^n)$, $n$ odd, $n > 1$, we must also include in Hypothesis $S$ solutions of the corresponding standard form problems for elements of $\mathscr{L}_0(3)$.

We make several comments.

1. The family $\mathscr{L}\mathscr{O}^* - \mathscr{L}\mathscr{O}$ is itself closed under the pumping-up process.

2. The groups $L_2(q)$, $q > 3$, also, of course, do not contain a classical involution, so they, too require special treatment. But in essence, this special treatment *is* the content of the inductive characterization of the family $\mathscr{L}\mathscr{O}$.

3. Since the inductive characterization of $\mathscr{L}\mathscr{O}$ is needed primarily for the identification of *proper* sections of a minimal counterexample to the $B$- and $U$-theorems, it suffices to limit the portion of Hypothesis $S$ that deals with $\mathscr{L}\mathscr{O}^* - \mathscr{L}\mathscr{O}$ and $\mathscr{L}_0(3)$ to groups which have the $B$-property.

So far we have been discussing Hypothesis $S$ only as it relates to the needed inductive characterization of the family $\mathscr{L}\mathscr{O}$. However, there are other places in the proofs of the $B$- and $U$-theorems requiring solutions of standard form problems for certain elements of $\mathscr{L}\mathscr{O}^* - \mathscr{L}\mathscr{O}$. Moreover, in those cases, the solution is required *within $G$ itself*, so that only *partial* information concerning the $B$-property is available: namely, in the "neighborhood" of the given standard component $L$. [Cases (B), (C), and (D) of Hypothesis $S$ below give the precise list of such $L$'s together with the appropriate weakened form of the $B$-property.]

The exact formulation of Hypothesis $S$ incorporates all the subtle points raised in the preceding discussion. Here finally is its statement.

HYPOTHESIS S. *Let* $G$ *be a group with* $F^*(G)$ *simple which possesses a standard component* $L$. *Then* $F^*(G)$ *is a known simple group under each of the following conditions*:

(A) $G$ *has the B-property and* $L \in \mathscr{LO}^* - \mathscr{LO}$ *or* $\mathscr{L}_0(3)$.

(B) $L \cong \hat{A}_n, n \geqslant 8$.

(C) $L/Z(L) \cong L_3(4)$.

(D) $L \cong A_n, n \geqslant 9, n$ *odd, Mc, Ly, or He and if* $x$ *is an involution of* $L$ *such that* $I = L(C_L(x))$ *is correspondingly isomorphic to* $A_{n-4}$, $\hat{A}_8, \hat{A}_{11}$, *or* $L_3(4)/Z_2 \times Z_2$, *then* $O(\langle I^{C_x} \rangle) = 1$.

Thus in (B) and (C), the information is restricted to $L$ itself (i.e., $L$ is the only 2-component specified in advance to be quasisimple); while in (D), we are given additional information concerning centralizers (in $G$) of particular involutions $x$ of $L$ [note that as $O(\langle I^{C_x} \rangle) = 1$, the 2-components of $\langle I^{C_x} \rangle$ are necessarily quasisimple].

We conclude with two general comments concerning the implications of Hypothesis $S$ for groups $G$ of component type [with $F^*(G)$ simple] *having the B-property*. First, although Hypothesis $S$ is a statement about standard components, it has a direct consequence for *arbitrary* components $K$ of centralizers of involutions. Indeed, beginning with $K$, the pumping-up process leads ultimately to a component $K^*$ with $K^*/O(K^*) \in \mathscr{L}^*(G)$. It then follows from Aschbacher's component theorem (Theorem 3.31) that $G$ possesses a (strongly) standard component $L$ with $L/O(L) \cong K^*/O(K^*)$ [provided $m(K^*) > 1$]. We see then that if Hypothesis $S$ applies to $L$, then $G$ is determined solely from the initial component $K$. But if $K \in \mathscr{LO}^* - \mathscr{LO}$, then as $\mathscr{LO}^* - \mathscr{LO}$ is a closed family, it follows by repeated use of Hypothesis $S$ that $K^*$ and hence also $L$ is in $\mathscr{LO}^* - \mathscr{LO}$.

Thus Hypothesis $S$ immediately yields the following inductive characterization of the family $\mathscr{LO}^* - \mathscr{LO}$.

THEOREM 4.23. *Assume Hypothesis S. If* $G$ *is a group having the B-property with* $F^*(G)$ *simple and some element of* $\mathscr{L}(G)$ *is in* $\mathscr{LO}^* - \mathscr{LO}$, *then* $F^*(G) \in \mathscr{LO}^* - \mathscr{LO}$.

Finally, an important point concerning the groups of Lie type of odd characteristic: the inductive characterization of the family $\mathscr{LO}$ has as its starting point the existence of a component $K$ in the centralizer of some involution of $G$ with $K/O(K) \in \mathscr{LO}$. Obviously this allows the possibility that the given component $K$ is standard in $G$. Hence, as a corollary of this characterization, we obtain a solution of every standard form problem in which the standard component is in $\mathscr{LO}$. Hence, once the $B$- and $U$-theorems have been reduced to verification of Hypothesis $S$, such an inductive characterization of the family $\mathscr{LO}$ will also provide a reduction of the standard component theorem. Since Aschbacher's component and classical involution theorems (Theorems 3.30 and 3.64) in turn reduce the noncharacteristic 2 theorem to the standard component theorem, it will have direct implications for that theorem as well.

## 4.4. 2-Components of Alternating Type

We shall be considerably sketchier in our discussion of Solomon's work on 2-components of alternating type.

We begin with his intrinsic $\hat{A}_n$ 2-component theorem, which takes the following form to allow for an inductive analysis.

THEOREM 4.24. *Let $G$ be a core-free group with $F^*(G)$ quasisimple and assume that for some $t \in \mathscr{T}(G)$, $C_t$ contains a 2-component $K$ such that either*

(a) *$K$ is intrinsic and $K/O(K) \cong \hat{A}_n$, $n \geqslant 8$; or*
(b) *$K/O(K) \cong Mc$ or $Ly$.*

*Then $K/O(K) \cong \hat{A}_n$ and $F^*(G) \cong \hat{A}_n$, $Mc$ (with $n = 8$), or $Ly$ (with $n = 11$).*

We let $G$ be a minimal counterexample. We first argue as in Theorem 4.9, using $L$-balance and the minimality of $G$:

LEMMA 4.25. *If the hypothesis of Theorem 4.24(a) holds and we choose $t$ and $K$ with $n$ as large as possible, then either*

(i) *$K/O(K) \in \mathscr{L}^*(G)$; or*
(ii) *$n = 8$ or $11$ and there is $x \in \mathscr{T}(G)$ and a 2-component $J$ of $C_x$ with $J/O(J) \cong Mc$ or $Ly$, respectively.*

In view of the lemma, we can divide the analysis into the following two basic cases:

*Case I.* (1) There is $t \in \mathcal{I}(G)$ and an intrinsic 2-component $K$ of $C_t$ with $K/O(K) \cong \hat{A}_n$, $n \geqslant 8$; and

(2) Whenever $t$ and $K$ are chosen in (1) so that $n$ is as large as possible, then $K/O(K) \in \mathcal{L}^*(G)$.

*Case II.* There is $t \in \mathcal{I}(G)$ and a 2-component $K$ of $C_t$ with $K/O(K) \cong Mc$ or $Ly$.

We first treat case I. We fix $n$ in (1) as large as possible. For any $x \in \mathcal{I}(G)$, we denote by $K_x$ the product of all 2-component $J$ of $C_x$ with $J/O(J) \cong \hat{A}_n$ such that if $y \in \mathcal{I}(J)$ with $y$ centralizing $J/O(J)$, then $C_y$ contains an intrinsic 2-component $L$ with $L/O(L) \cong \hat{A}_n$ which covers $J/O(J)$. (We set $K_x = 1$ if $C_x$ contains no such 2-components.) By the assumptions of case I, each such 2-component $J$ of $K_x$ (mod core) is in $\mathcal{L}^*(G)$. Furthermore, by its definition, clearly $K_x \lhd C_x$.

But now if we apply Theorem 4.9 with $\mathcal{K}$ the subset of $\mathcal{L}^*(G)$ consisting of those elements isomorphic to $\hat{A}_n$, we conclude at once:

PROPOSITION 4.26. *For each* $x \in \mathcal{I}(G)$, $K_x$ *is either trivial or is a single 2-component of* $C_x$.

Now fix $t \in \mathcal{I}(G)$ and an intrinsic 2-component $K$ of $C_t$ with $K/O(K) \cong \hat{A}_n$ and with a Sylow 2-subgroup of $C_t$ of largest possible order. By the proposition,

$$K \lhd C_t. \tag{4.10}$$

Furthermore, we have:

LEMMA 4.27. $C_t$ *contains a Sylow 2-subgroup of* $G$.

Indeed, if $T \in \mathrm{Syl}_2(C_t)$ and $T \leqslant S \in \mathrm{Syl}_2(G)$, then $t \in T$ and $Z(S) \leqslant C_t$. By properties of $\hat{A}_n$, any involution $z$ of $Z(S)$ centralizes $K/O(K)$. Hence if we set $J = L_{2'}(C_K(z))$, it follows that $t \in J \leqslant K_z$ and $S \leqslant C_z$. Then $S$ normalizes $K_z$ and as $t$ is the unique involution of $S \cap O_{2'2}(K_z)$, this implies that $S$ centralizes $t$. Hence $S \leqslant C_t$ and so $T = S$, proving the lemma.

By the $Z^*$-theorem, there is $g \in G$ such that

$$y = t^g \in S - \langle t \rangle. \tag{4.11}$$

We fix $g$ and $y$.

The action of $y$ on $K$ is very restricted.

PROPOSITION 4.28.   *Set* $H = K\langle y \rangle$ *and* $\bar{H} = H/O_{2'2}(K)$. *Then* $\bar{H} \cong A_n$
*or* $\Sigma_n$ *and* $\bar{y}$ *fixes at most three points in the natural representation of* $\Sigma_n$.

Indeed, clearly $\bar{H} \cong A_n, \Sigma_n$, or $A_n \times Z_2$ [the last if $y$ centralizes
$K/O(K)$]. If the desired conclusion is false, it follows readily from the
structure of the centralizers of involutions acting on $\hat{A}_n$ that $C_K(y)$ contains
a subnormal subgroup $J$ with $t \in J$ and $J/O(J) \cong \hat{A}_m$, $m \geqslant 4$, where $m$ is the
number of fixed points of $y$ in the natural representation of $\Sigma_n$. [If $m \geqslant 5$, $J$
is, in fact, a 2-component of $C_K(y)$; while if $m = 4$, it is a "solvable" 2-
component.]

Now $J \leqslant C_y$ and so $J$ normalizes $K_y = K^g$. We need only show that $t$
centralizes $K_y/O(K_y)$, for then $I = L_2(C_{K_y}(t))$ covers $K_y/O(K_y)$ and $y \in I$ (as
$y = t^g \in K^g$). Since $K_y$ is intrinsic, it will then follow from the definition of
$K_t$ that $I \leqslant K_t = K$, whence $y = t$, contrary to our choice of $y$.

Assume false and set $N = K_y J$ and $\tilde{N} = N/C_N(K_y/O(K_y))$. Then
$\tilde{K}_y \cong A_n$, $\tilde{N} \cong A_n$ or $\Sigma_n$, $\tilde{J}/O(\tilde{J}) \cong \hat{A}_m$, and $\tilde{J}$ is subnormal in $C_{\tilde{N}}(\tilde{t})$. However,
no involution of $\Sigma_n$ has a centralizer with a subnormal subgroup of this
form.

A closer analysis of the same type, using the structure of the
centralizers of involutions acting on $\hat{A}_n$, yields the following sharper results:

PROPOSITION 4.29.   (i) $\langle t \rangle \in \text{Syl}_2(C_{C_t}(K/O(K)))$; *and*
                      (ii) $n \leqslant 11$.

Thus $C_t/O(C_t)\langle t \rangle \cong A_n$ or $\Sigma_n$, $8 \leqslant n \leqslant 11$. However, using Thompson's
transfer lemma, Solomon shows in the latter case that $G$ has a normal
subgroup $G_0$ of index 2. Then $K \leqslant G_0$ and so $G_0$ satisfies the hypotheses of
the theorem. Hence by the minimality of $G$, $F^*(G_0) = F^*(G)$ satisfies the
conclusion of the theorem, contrary to our choice of $G$. Therefore, we have
the following result:

PROPOSITION 4.30.   *We have* $C_t/O(C_t) \cong \hat{A}_n$, $8 \leqslant n \leqslant 11$.

In particular, the Sylow 2-subgroup $S$ of $G$ is of type $\hat{A}_n$, $8 \leqslant n \leqslant 11$, and now [I, 2.32 and 2.33] forces $n = 8$ or $11$ and $G = F^*(G) \cong Mc$ of $Ly$, respectively, again contradicting our choice of $G$. Thus we have:

PROPOSITION 4.31. *Case II holds.*

Since $Mc$ and $Ly$ are simple, we easily conclude this time from the minimality of $G$:

LEMMA 4.32. $F^*(G)$ *is simple.*

Using the minimality of $G$ and Theorem 4.9 with $\mathscr{K} = \{Mc\}$ or $\{Ly\}$, we easily obtain the following result:

PROPOSITION 4.33. *For any* $x \in \mathscr{I}(G)$, *we have*

(i) *If* $L$ *is a 2-component of* $C_x$ *with* $L/O(L) \cong Mc$ *or* $Ly$, *then* $L/O(L) \in \mathscr{L}^*(G)$; *and*

(ii) $C_x$ *contains at most one 2-component* $L$ *with* $L/O(L) \cong Mc$ *or* $Ly$.

Since we are in case II, there is $t \in \mathscr{I}(G)$ such that $C_t$ has a 2-component $L$ with $L/O(L) \cong Mc$ or $Ly$ (according as $n = 8$ or $11$). Let $K_t$ have the same meaning as before (so that $K_t$ is the product of the 2-components of $C_t$ of type $\hat{A}_n$ which arise from intrinsic 2-components of centralizers of involutions of $G$).

Using Theorem 4.9, a similar argument yields:

LEMMA 4.34. $K_t$ *contains at most one 2-component* $K$ *such that for some* $y \in \mathscr{I}(C_G(K/O(K)))$, *if* $I = L_{2'}(C_K(y))$ *and* $J = \langle I^{L_y} \rangle$, *then* $J/O(J) \cong Mc$ *or* $Ly$.

For brevity, call such a 2-component $K$ of $K_t$ *exceptional*. If $K_t$ has an exceptional 2-component $K$, let $t^* \in \mathscr{I}(K)$ centralize $K/O(K)$ and consider $C_{t^*}$. We check that $L_{2'}(C_K(t))$ and $L_{2'}(C_L(t))$ pump-up in $L_{t^*}$ to 2-components $K^*$, $L^*$ of the same respective types as $K$ and $L$ and, in addition, $t^* \in K^*$. In particular, $K^*$ is exceptional in $K_{t^*}$. Thus, replacing $t$ by $t^*$, if necessary, we can assume henceforth:

$$\text{If } K_t \text{ contains an exceptional 2-component } K, \text{ then } t \in K. \qquad (4.12)$$

Now Solomon brings signalizer functors into play. Let $T \in \mathrm{Syl}_2(L)$, let $A$ be an elementary 2-subgroup of $T\langle t \rangle$ of maximal rank, and set $Z = \Omega_1(Z(T))\langle t \rangle$, so that $Z \leqslant A$. Note that $m(T) = 4$ and hence $m(A) = 5$ in both cases $n = 8$ or $11$. Using Goldschmidt-type signalizer functor theory, Solomon proves the following result:

PROPOSITION 4.35. *If for each $a \in A^{\#}$, we set*

$$\theta(C_a) = \bigl(O(C_a), Z\bigr)\bigl(O(C_Z) \cap C_a\bigr),$$

*then $\theta$ is an $A$-signalizer functor on $G$.*

We make only one observation about the proof. To control where the two factors in the definition of $\theta(C_a)$ "land" in $C_a$, Solomon needs to be able to identify the pump-up $J$ of $I = L_{2'}(C_L(a))$ in $L_a$. If $a = t$, then, of course, $J = I = L$. In the contrary case, $I/O(I) \cong \hat{A}_n$ with $n = 8$ or $11$ according as $L/O(L) \cong Mc$ or $Ly$. If $a \in I$, it follows from the minimality of $G$ and the maximal choice of $n$ that $J$ is the product of one or two 2-components of $C_a$ of type $\hat{A}_n$. Using this result, an easy argument shows that the same conclusion holds when $a \notin I$. This gives Solomon the necessary control to show that $\theta$ is an $A$-signalizer functor on $G$.

As expected, his aim is now to force $\theta$ to be the trivial $A$-signalizer functor. Assume false and set $M = N_G\bigl(\theta(G;A)\bigr)$, where $\theta(G;A)$ denotes the completion of $\theta$. Since $F^*(G)$ is simple and $\theta(G;A)$ has odd order by the signalizer functor theorem, $M$ is, as usual, a proper subgroup of $G$.

Again, the normal closure of $L$ and of $K$ (if it exists) in $L_{2'}(M)$ is, as usual, a single 2-component of the same corresponding type as $K$ and $L$. We denote it by $L_M$ and $K_M$, respectively (with $K_M = 1$ if $K_t$ contains no exceptional 2-component). In particular, $T\langle t \rangle \leqslant M$.

Now, using standard signalizer functor theory arguments together with known properties of the centralizers of involutions $y$ acting on $Mc$ or $Ly$ [if $y$ induces an inner automorphism, the centralizer is $\hat{A}_8$ or $\hat{A}_{11}$, while if $y$ induces an outer automorphism, then necessarily $n = 8$ and $C_{Mc}(y) \cong M_{11}$], Solomon shows that the hypotheses of Aschbacher's strong embedding theorem [I, 4.31] are satisfied:

PROPOSITION 4.36. *The following conditions hold:*

(i) $C_x \leqslant M$ for all $x \in \mathscr{I}\bigl(L_M C_M(L_M K_M)/O(L_M K_M)\bigr)$;
(ii) *If $T\langle t \rangle \leqslant S \in \mathrm{Syl}_2(M)$, then $S \in \mathrm{Syl}_2(G)$;*

*Furthermore, if D denotes the set of involutions of G which are G-conjugate to an involution of $T\langle t \rangle$, then*

(iii) $D \cap S \leqslant T\langle t \rangle$;

(iv) *If $x, y$ are distinct commuting involutions of D, then $xy \in D$; and*

(v) *If $x \in D \cap M$ and $y \in C_D(x)$, then $y \in D \cap M$.*

Thus the general form of Aschbacher's theorem is indeed applicable and yields that either $N_G(D \cap M)$ is strongly embedded in $G$ or else $D \leqslant M$. The first case is excluded by Bender's theorem [I, 4.24]. In the second case, $M$ contains the nontrivial normal subgroup $\langle D \rangle$ of $G$ and so contains $F^*(G)$ as $F^*(G)$ is simple. In particular, $F^*(G) \lhd M$ and so $F^*(G)$ centralizes $O(M)$, forcing $O(M) \leqslant O(F^*(G)) = 1$—contradiction.

Thus $\theta$ is trivial; and this in turn yields:

PROPOSITION 4.37.   $O(C_x) = 1$ *for every* $x \in \mathscr{I}(L\langle t \rangle)$.

Now we can apply Hypothesis $S$ to reach a final contradiction. Indeed, as $O(C_t) = 1$, $L$ is quasisimple and hence $L \cong Mc$ or $Ly$. We claim that $L$ is standard, which will follow at once as in Proposition 3.34 provided we show that $L \lhd C_t$ for every $y \in \mathscr{I}(C_L)$.

Since $L \in \mathscr{L}^*(G)$, Proposition 4.33 implies that $L$ is contained in a 2-component $J$ of $C_y$ with $J/O(J) \cong L$ and $J \lhd C_y$, so we need only show that $J = L$. If false, then $L$ does not centralize $O(J)$. But then for some $a \in (A \cap L)^\#$, $A \cap L$ does not centralize $E = C_{O(J)}(a)$ and hence neither does $I = L(C_L(a))(\cong \hat{A}_n)$. However, $O(C_a) = 1$ by Proposition 4.37. But $I$ is contained in a product of 2-components of $L_a = L(C_a)$ isomorphic to $\hat{A}_n$, so $I$ must centralize $E$—contradiction.

Thus $L$ is indeed standard and now Proposition 4.37 shows that the assumptions of Hypothesis $S$ are satisfied. But by the known structure of centralizers of involutions acting on simple $K$-groups, there is no group $G$ with $F^*(G)$ a simple $K$-group having $Mc$ or $Ly$ as standard component. This completes the discussion of Theorem 4.24.

As a corollary of the theorem, we obtain the following reduction of the *B*- and *U*-theorems:

THEOREM 4.38. *Assume Hypothesis S. If G is a minimal counterexample to the B-theorem and U-theorem and $(x, y, L)$ is an unbalancing triple, then $L/O(L) \not\cong \hat{A}_n$, n odd, $n \geqslant 7$.*

Assume false. We shall argue that for some $t \in \mathscr{I}(G)$, $C_t$ has an intrinsic 2-component $K$ with $K/O(K) \cong \hat{A}_m$, $m \geqslant 7$, $m$ odd. If $m = 7$, then

$m(K) = 1$ and Aschbacher's classical involution theorem (Theorem 3.49) will then yield a contradiction. Thus $m \geqslant 8$, in which case Theorem 4.24 will imply that $F^*(G) \cong Ly$ [as $F^*(G)$ is simple by Theorem 3.3 and $m$ is odd), contrary to Theorem 3.5.

If possible, choose $x$ so that $x \in L_x = L_2(C_x)$. Set $D = O(C_x) \cap C_y$, so that by assumption $D$ does not centralize $L/O(L)$. By $L^*$-balance, [I, 4.81], $D$ leaves $L$ invariant and, as usual, this implies that also $y$ leaves $L$ invariant. Then $\langle x, y \rangle$ centralizes some involution $t \in L$ such that $J = L_2(C_L(t))$ covers $L/O(L)$. Then $t$ normalizes $D$ and so $E = C_D(t)$ does not centralize $J/O(J)$.

By Proposition 3.21 $L = \langle J^{K_t} \rangle$ is either a single $x$-invariant 2-component of $C_t$ or the product of two such 2-components $K_1, K_2$ interchanged by $x$. Then $E$ does not centralize $K/O(K)$ and it follows correspondingly that $(t, y, K)$ or $(t, y, K_1)$ is an unbalancing triple [as $E \leqslant O(C_y) \cap C_t$]. Furthermore, in either case $J$ is a 2-component of $L_2(C_K(t))$.

Using Theorem 3.27(iv) and Theorem 4.1, we conclude in the first case that necessarily $K/O(K) \cong \hat{A}_m$ for some odd $m \geqslant n$. Since $t \in J \leqslant K$, $K$ is intrinsic and we are done.

Assume then that the second case holds. Then $K_1/O(K_1) \cong \hat{A}_n$ as $J$ is a diagonal of $K$. Thus $(t, y, K_1)$ is an unbalancing triple of the same type as $(x, y, L)$. But $t \in J \leqslant K \leqslant L_t$, so by our choice of $x$, we have $x \in L_x$. Hence $x \in L_2(C_{L_x}(t))$ and so by $L$-balance, $x \in L_t$, in which case $x$ leaves each 2-component of $L_t$ invariant, contrary to the fact that $x$ interchanges $K_1$ and $K_2$. Therefore this case does not arise and the proof is complete.

We turn now to Solomon's results on 2-components of simple alternating type. In view of Hypothesis $S$ and Theorem 4.24, the natural result to attempt to prove is the following:

If $G$ is core-free with $F^*(G)$ quasisimple and for some $t \in \mathcal{I}(G)$, $F^*(G)$ contains a 2-component $K$ with $K/O(K) \cong A_n$, $n \geqslant 9$, then $F^*(G) \cong A_m$ for some $m \geqslant n$.

Moreover, the procedure to follow is clear. Pick $t$ and $K$ so that $n$ is maximal, argue that $K/O(K) \in \mathscr{L}^*(G)$, and invoke Theorem 4.9 to conclude that $K$ is the unique 2-component of $C_t$ of type $A_n$. Now emulate the proof of the $Mc$, $Ly$ case II of Theorem 4.24. However, first pick $t$ and $K$ so that a Sylow 2-subgroup of $C_t$ has maximal order (this will give slightly stronger control over pump-ups). Next, for suitable elementary abelian 2-subgroups $A$ of $K\langle t \rangle$ with $t \in A$, construct an appropriate $A$-signalizer functor on $G$. Then by varying $A$, show that either $M = N_G(\theta(G; A))$ satisfies the hypothesis of Aschbacher's strong embedding theorem [I, 4.31] which leads to the same contradiction as before (cf. Proposition 4.36) or else $\theta$ is trivial for each such $A$, in which case $O(C_a) = 1$ for $a \in A^{\#}$. In particular, $K$ will then be standard in $G$.

In the present situation, Solomon shows that the proper $A$'s to take have the form $A = B \times \langle t \rangle$, where $B$ is a four-subgroup of $K$ such that in the natural representation of $\bar{K} = K/O(K)$, $B$ fixes exactly four letters. Since $O(C_a) = 1$ for each $a \in A^{\#}$, Hypothesis $S$ is therefore satisfied and we shall then conclude that $F^*(G) \cong A_{n+2}$ or $A_{n+4}$, as required.

Unfortunately, a problem arises in trying to carry through the preceding argument. Indeed, just as in Proposition 4.33, in order to verify that $\theta$ is, in fact, an $A$-signalizer functor on $G$, it is first necessary to identify the pump-up $J$ of $I = L_2(C_K(a))$ in $L_a$ for $a \in A^{\#}$. In the previous situation, $I/O(I) \cong \hat{A}_n$, $n = 8$ or $11$ (assuming $a \neq t$), and identification was possible within the framework of Theorem 4.24. However, in contrast, we now have $I/O(I) \cong A_{n-4}$. If $n - 4 \geqslant 9$, then $J$ will again be determined within the framework of our desired alternating 2-component theorem. However, if $n - 4 \leqslant 8$, we would require a much broader inductive setup to be able to identify $J$.

More precisely, we would be forced to work with the *closure* $\mathscr{F}$ of the family $A_r, r \geqslant 5$, and to determine $F^*(G)$ when the centralizer of some involution $x$ of $G$ has a 2-component $K$ with $K/O(K)$ isomorphic to an element of $\mathscr{F}$. But as $A_5 \cong L_2(5)$ and $A_6 \cong L_2(9)$, the discussion of the preceding section [cf. (4.8)] shows that $\mathscr{F}$ includes certain families of groups of Lie type of characteristic 2 as well as a number of sporadic groups. Of course, it also includes several families of groups of Lie type of characteristic 3 and 5. Hence just as with an inductive characterization of the groups of Lie type of odd characteristic, a complete solution of the alternating 2-component problem would require a prior inductive characterization of certain families of groups of Lie type of both even and odd characteristic (as well as some sporadic groups). Moreover, it would be necessary to establish this result *without* the assumption that the group $G$ under investigation has the $B$-property.

However, Solomon is able to avoid such a general theorem by restricting himself to the context of the $B$- and $U$-theorems. For, in proving that $\theta$ is an $A$-signalizer functor, one can show that the identification of $J$ is required only in the case that $(a, t, J)$ is an unbalancing triple. But then if $G$ is a minimal counterexample to the $B$- and $U$-theorems. $J/O(J)$ will be determined within the family of known locally unbalanced groups, and one can then continue the argument outlined above. The net effect is to allow Solomon to remain within the family of alternating 2-components.

Here then is his precise result.

THEOREM 4.39. *Assume Hypothesis S. If G is a minimal counter-example to the B-theorem and U-theorem, then there is no involution t of G such that $C_t$ contains a 2-component K with $K/O(K) \cong A_n$, n odd, $n \geqslant 9$.*

We shall show that the situation sets up to accommodate the argument described above.

Suppose false and choose $t \in \mathscr{I}(G)$ with $K/O(K) \cong A_n$, $n$ odd, and $n$ as large as possible. Thus $n \geqslant 9$.

PROPOSITION 4.40. *The following conditions hold:*

(i)  $K/O(K) \in \mathscr{L}^*(G)$; *and*
(ii) $K$ *is the unique 2-component of $C_t$ of type $A_n$.*

Theorem 4.9 and (i) imply (ii), so it suffices to prove (i). If (i) fails, then by the definition of maximal 2-component there is a chain $K = K_1 < K_2 < K_3 < \cdots < K_n = K^*$ of 2-components of centralizers of involutions of $G$ such that $K/O(K) \not\cong K^*/O(K^*)$. Hence, for a suitable value of $i$, $1 \leqslant i \leqslant n-1$, we have $K_i/O(K_i) \cong K/O(K)$ and $K_{i+1}/K_{i+1} \not\cong K/O(K)$. Without loss, we can assume to begin with that $i = 1$, in which case there is $u \in \mathscr{I}(C_t)$ with $u$ centralizing $K/O(K)$ such that if $J$ denotes the normal closure if $I = L_{2'}(C_K(u))$ in $C_u$ and $J_1$ is a 2-component of $J$, then $K/O(K) \not\cong J_1/O(J_1)$.

We know that either $J = J_1$ with $J$ invariant under $t$ or else $J$ is the product of two 2-components $J_1, J_2$ of $C_u$, interchanged by $t$, and in either case if $\bar{C}_u = C_u/O(C_u)$, then $\bar{I}$ is a component of $C_{\bar{J}}(\bar{t})$ (by Proposition 3.21). If $J = J_1 J_2$, then as $\bar{J}_1 \not\cong A_n$, the only possibility is $\bar{J}_1 \cong \hat{A}_n$. But then using Hypothesis $S$ (as in the next paragraph), we conclude easily that for some $v \in \mathscr{I}(G)$ $C_v$ contains an intrinsic 2-component of type $\hat{A}_m$ for some $m \geqslant 9$. Since $F^*(G)$ is simple by Theorem 3.27, Theorem 4.24 implies now that $F^*(G) \cong Ly$ and so is a $K$-group, again contradicting Theorem 3.5.

Hence $\bar{J}$ is quasisimple. Setting $\bar{H} = \bar{J}\langle \bar{t} \rangle$ and $\tilde{H} = \bar{H}/C_{\bar{H}}(\bar{J})$, it follows that $\tilde{J}$ is simple, $\tilde{H} \leqslant \text{Aut}(\tilde{J})$, $\tilde{t} \in \mathscr{I}(\tilde{J})$, and $\tilde{I}$ is a component of $C_{\tilde{H}}(\tilde{t})$ with $\tilde{I} \cong A_n$. By the minimality of $G$, all sections of $\tilde{H}$ have the $B$-property. Hence by Hypothesis $S$ and Theorem 4.23, it follows that $\tilde{J} \cong A_r$, $r = n + 2k$, $k \geqslant 1$. But then $\bar{J} \cong A_r$ or $\hat{A}_r$. However, the latter is excluded as no involution acting on $\hat{A}_r$ has a component isomorphic to $A_n$. Thus $\bar{J} \cong A_r$ with $r > n$, contrary to our maximal choice of $n$.

PROPOSITION 4.41. *Let u be an involution of K which fixes exactly four* *points in the natural representation of $K/O(K)$, set $I = L_{2'}(C_K(u))$ [so that $I/O(I) \cong A_{n-4}$], and denote by $J$ the normal closure of $I$ in $L_u$. If $J$ is a single 2-component of $C_u$ and $(u, t, J)$ is an unbalancing triple, then one of the following holds:*

    (i) $J/O(J) \cong A_m$, $m = n - 4$, $n - 2$, *or* $n$; *or*
    (ii) $n = 9$ *and* $J/O(J)$ *is isomorphic to a covering group of* $L_3(4)$.

Indeed, set $\bar{C}_u = C_u/O(C_u)$, so that $\bar{I}$ is a component of $C_{\bar{J}}(\bar{u})$. Since $\bar{I} \cong A_{n-4}$, $\bar{J} \not\cong \hat{A}_m$. If $\bar{J} \cong A_m$, then it is immediate that $m = (n - 4) + 2k$ for some $k$. But $m \leqslant n$ by the maximality of $m$, so (i) holds. Hence, we can assume that $\bar{J}/Z(\bar{J}) \not\cong A_m$ for any $m$. We can also assume that $\bar{J}$ is not a covering group of $L_3(4)$. Since $(u, t, J)$ is an unbalancing triple, it follows therefore from Theorems 3.27(iv), 4.1, and 4.38 that either

$$\text{(a) } \bar{J} \cong L_2(p^r), p \text{ odd; or}$$
$$\text{(b) } \bar{J} \cong He. \tag{4.13}$$

Since $\bar{I}$ is a component of $C_{\bar{J}}(\bar{u})$ with $\bar{I} \cong A_{n-4}$, we conclude correspondingly from (4.13) that

    (a) $n = 9$, and $\bar{J} \cong L_2(25)$ with $\bar{u}$ inducing a nontrivial field
        automorphism on $\bar{J}$; or                   (4.14)
    (b) $n = 11$ with $\bar{u}$ inducing an outer automorphism on $\bar{J}$.

Solomon must eliminate these two possibilities. Once the situation is set up, the argument is quite pretty. Let $T \in \mathrm{Syl}_2(C_{C_t}(u))$ and set $\tilde{C}_t = C_t/O(C_t)$. By the structure of $A_n$, $C_{\tilde{K}}(\tilde{u})$ contains a normal subgroup of index 2 of the form $Z_2 \times Z_2 \times \tilde{I}$ and $C_{\tilde{K}}(\tilde{u})/\tilde{I} \cong D_8$. Thus $U = C_{T \cap K}(\tilde{I})$ is a four-group and $T$ contains an involution $w$ such that $\langle u \rangle = [U, w]$. In particular,

$$u \in T'. \tag{4.15}$$

Furthermore, $T$ contains a Sylow 2-subgroup $Q$ of $C_{C_t}(K)$ and by the structure of $\Sigma_n$, $T \cap K$ has a complement $R$ in $T$ with $Q \leqslant R$ and $|R : Q| \leqslant 2$ (with equality holding if and only if $\tilde{K}\tilde{T}/\tilde{Q} \cong \Sigma_n$). Moreover, we can choose $R$ so that if we set $H = L_{2'}(C_K(R))$, then

$$\tilde{I} \leqslant \tilde{H} \text{ and } \tilde{H} \cong A_{n-2} \text{ or } A_n \text{ (according as } R > Q \text{ or } R = Q). \tag{4.16}$$

In addition, we have

$$RU = C_T(\tilde{I}). \tag{4.17}$$

Solomon derives a contradiction by examining the structure of $R$ and the embedding of $T$ in $C_u$. He first proves the following result:

LEMMA 4.42. *We have* $R = Q = \langle t \rangle$.

Observe that $T$ normalizes $I$ and hence $J$, so $R$ normalizes $\bar{J}$. Furthermore, as $t$ induces an outer automorphism of $\bar{J}$ and $\mathrm{Out}(\bar{J})$ has elementary Sylow 2-subgroups (of order 4 or 2, respectively), $t \notin \phi(R)$. If $R$ is cyclic or quaternion, then $t$ is the unique involution of $R$ and as $t \notin \phi(R)$, the only possibility is $R = \langle t \rangle$, as asserted.

We can therefore assume that $R$ does not have this form, whence by [I, 1.35], $R$ contains a four-group $V$. Since $V$ normalizes $\bar{J}$ and centralizes $\tilde{I}$ [by (4.17)], we see that some $v \in V^{\#}$ must centralize $\bar{J}$. Set $I_v = L_2(C_I(v))$ and $J_v = L_2(C_J(v))$, so that $I_v$ and $J_v$ cover $I/O(I)$ and $J/O(J)$, respectively. Furthermore, we have that

$$H = \langle I_v^H \rangle \quad \text{and} \quad J_v = \langle I_v^{J_v} \rangle \tag{4.18}$$

Now let $N$ be the normal closure of $J_v$ in $L_v$ and set $C_v^* = C_v/O(C_v)$. Since $N = \langle J_v^N \rangle$, (4.18) implies that $H^* \leqslant N^*$. If $N^*$ is the product of two components, then $N^*/Z(N^*) \cong L_2(25) \times L_2(25)$ or $He \times He$, as $J_v^*$ is then a diagonal of $N^*$. But this is impossible as $N^*$ does not contain a subgroup isomorphic to $H^*(\cong A_{n-2}$ or $A_n$, $n = 9$ or 11, respectively). Thus $N$ is a single $u$-invariant 2-component of $C_v$.

But $(u, t, J)$ is an unbalancing triple, so $D = O(C_t) \cap C_u$ does not centralize $\bar{J}$. Since $v$ normalizes $D$ and centralizes $\bar{J}$, it follows that $E = C_D(v)$ does not centralize $\bar{J}$ and hence does not centralize $N^*$. Thus $(v, t, N)$ is also an unbalancing triple. Since $J_v^* \cong L_2(25)$ or $He$ is a component of $C_{N^*}(u^*)$, we conclude by another application of Theorems 3.27(iv) and 4.1 that $N^* \cong L_2(25)$, $L_2(5^4)$, or $He$. However, as $H^* \leqslant N^*$, this is again a contradiction. This establishes the lemma.

Finally set $Y = C_{C_u}(I)$ and $A = T \cap Y$. By the lemma and (4.17), $C_T(\tilde{I}) = U\langle t \rangle$ and so $A = U\langle t \rangle$. But also as $u \in Z(Y)$ and $Y$ centralizes $\tilde{I}$, we see that $A$ is a Sylow 2-subgroup of $C_Y(t)$. On the other hand, as $t \in Z(T)$, it follows by the structure of $\mathrm{Aut}(L_2(25))$ and $\mathrm{Aut}(He)$ that $T = t\langle T_0 \rangle$, where $T_0 \leqslant JY$. Since $t$ centralizes the projection of $\bar{T}_0$ on $\bar{Y}$ and $\bar{A} \in \mathrm{Syl}_2(C_{\bar{Y}}(\bar{t}))$, this in turn implies that $T_0 = (T \cap J) A$. Thus $T = (T \cap J) U\langle t \rangle$ and so $T/T \cap J$ is abelian. Therefore $T' \leqslant J$ and consequently $u \in J$ by (4.15), contrary to the fact that $\bar{u}$ centralizes $\bar{J}$ and $\bar{J}$ is simple.

Now the stage is set for signalizer functor theory. Taking $A = B \times \langle t \rangle$, where $B$ is a four-subgroup of $K$ which fixes exactly four letters in the natural representation of $K/O(K)$, Solomon defines for $a \in A^{\#}$

$$\theta(C_a) = \big(O(C_a), B\big)\big(O(C_a) \cap O(C_B)\big). \tag{4.19}$$

Using Proposition 4.41, he shows that $\theta$ is an $A$-signalizer functor on $G$ for each such $A$. Now, as we have described above, he is able to emulate case II of the proof of Theorem 4.24 to establish Theorem 4.39.

## 4.5. Groups of Lie Type of Odd Characteristic and the $L_2(q), A_7$ Problem

Our historical summary of the proof of the *B*-theorem in Section 3.3 placed the minimal $L_2(q), A_7$ case as the final stage of the analysis. However, for expository purposes, it is preferable to alter the historical pattern and discuss the entire $L_2(q), A_7$ situation next.

As we have noted in Section 3.3 and more fully in our discussion of Hypothesis $S$, the solution of the $L_2(q)$ problem requires an inductive characterization of the groups of Lie type of odd characteristic, the basic argument being, as in Thompson's theorem 4.1, to force the group $G$ under investigation to possess a classical involution. As we have remarked earlier, Harris and Walter have independently obtained such inductive characterizations. We shall follow Harris's treatment [59, 61] here since it has been designed to mesh with the remaining stages of the proof of the *B*- and *U*-theorems—specifically, with the work of Foote [31], Gilman and Solomon [34], and Griess and Solomon [48].

As we have also pointed out, this inductive characterization is to be applied to proper sections of a minimal counterexample to the *B*- and *U*-theorems, so it suffices to consider the case in which $G$ itself has the *B*-property. But as we have also observed, the *B*-assumption on $G$ is needed primarily in treating elements of $\mathscr{L}(G)$ contained in $\mathscr{LO}^* - \mathscr{LO}$ that arise through the pumping up process. In fact, as Harris's proof shows, once one reaches an *intrinsic* 2-component $L$ with $L/O(L) \in \mathscr{LO}$, the subsequent argument does not require $L$ to be quasisimple, but goes through equally well under the weaker assumption that $L$ is a 2-component. Moreover, the more general result can be used to eliminate a few of the configurations arising in Gilman and Solomon's analysis. However, to simplify the exposition, we shall limit the discussion to groups having the $B$-property. Then at the end of the section, we shall state the more general intrinsic 2-component theorem without comment.

As with the proof of Theorem 4.1, the argument depends upon a number of detailed properties of the family $\mathscr{LO}$. However, we shall make no attempt to list these explicitly here.

The basic idea is very nice and works smoothly if $L$ is of sufficiently high Lie rank $k$ and defined over $GF(q)$ with $q \geqslant 5$.

The first stage consists of a reduction to the *intrinsic* case. Indeed, by the structure of $L$, there is $y \in \mathscr{I}(L)$ such that $L(C_L(y))$ possesses a component $I \in \mathscr{LO}$ of Lie rank $k - 2$ or $k - 1$ with $y \in I$. Hence by $L$-balance and induction, $J = \langle I^{L_y} \rangle$ is either a single $x$-invariant component of $C_y$ of Lie type of odd characteristic defined over $GF(q)$ or the product of two such components $J_1, J_2$ interchanged by $x$. Then $y \in J$ and so in the first case $J$ is an intrinsic component. Hence replacing $L$, $x$ by $J$, $y$, respectively, we can assume that $L$ is intrinsic. Similarly in the second case, if $y \in J_1$, we replace $L$, $x$ by $J_1, y$, respectively, and again $L$ is intrinsic.

There remains the possibility that $y = y_1 y_2$ with $y_i \in \mathscr{I}(Z(J_i))$, $i = 1, 2$. In this case we consider the pump-up $J^*$ of $J_1$ in $C_{y_1}$. Since $y \in J \leqslant L_{y_1}$ by $L$-balance, $J^*$ is necessarily a single $y$-invariant component of $C_{y_1}$, again of Lie type defined over $GF(q)$ of Lie rank at least $k - 2$, and this time we replace $L$, $x$ by $J^*$, $y_1$, respectively. Thus we can assume without loss to begin with that $L$ is intrinsic, whence

$$x \in L. \tag{4.20}$$

The next step is to reduce to the following situation:

> There exists a four-subgroup $Z \leqslant L$ such that
> $L(C_L(Z)) = H_1 H_2$, where $H_1 \cong SL_2(q)$ and $Z = \langle z_1, z_2 \rangle$ (4.21)
> with $z_i \in H_i$, $i = 1, 2$.

Using properties of the family $\mathscr{LO}$ (together with our assumption that the Lie rank of $L$ is sufficiently large), one can show that either $L$ contains such a four-subgroup $Z$, or else for some integer $r$ we have:

> (a) $L/Z(L) \cong L_{2r}(q)$ or $U_{2r}(q)$; and
> (b) correspondingly, $L$ is a homomorphic image of $SL_{2r}(q)$ (4.22)
> or $SU_{2r}(q)$ by a central subgroup of *even* order.

For brevity, let us say that $L$ is of *exceptional* type if (4.22) holds.

To reach (4.21) in the exceptional case, we simply repeat the replacement process of stage one. Indeed, if $L$ is of exceptional type, it is possible to choose the involution $y \in \mathcal{I}(L)$ and the component $I$ of $L(C_L(y))$ such that the components $J, J_1$, or $J^*$, as the case may be, are not of exceptional type, so that when we make the corresponding replacement, the new $L$ is not of exceptional type. Hence without loss we can assume that $L$ possesses a four-subgroup $Z$ satisfying (4.21).

Now we examine $C_{z_1}$. Since $Z$ centralizes $H_1 H_2$, $H_i$ has a pump-up $K_i$ in $L_{z_i}$ for $i = 1$ and 2. Since $x \in Z = \langle z_1, z_2 \rangle \leqslant H_1 H_2$, it again follows that each $K_i$ is a single component. If $K_1 = H_1$, then $K_1 \cong SL_2(q)$ and $z_1 \in K_1$, so $z_1$ is a classical involution and consequently $F^*(G)$ is of Lie type over $GF(q)$ by Aschbacher's theorem. We can therefore assume

$$H_1 < K_1. \tag{4.23}$$

This immediately forces $K_1 = K_2$. Indeed, if $K_1 \neq K_2$, then $z_2 \in K_2$ centralizes $K_1$, so therefore does $Z = \langle z_1, z_2 \rangle$. Since $C_Z \leqslant C_{z_1}$, it follows that $K_1$ is a component of $C_Z$. On the other hand, as $x \in Z$, and $H_1$ is a component of $C_L(Z)$, $H_1$ is, in fact, a component of $C_Z$. But then as $H_1 \leqslant K_1$, we must have $H_1 = K_1$, contrary to assumption. We reach the same contradiction if $z_2 \in Z(K_1)$, so also $z_2 \notin Z(K_1)$.

But now if we set $\bar{K}_1 = K_1/Z(K_1)$, we see that

(a) $\bar{z}_2 \in \mathcal{I}(\bar{K}_1)$;
(b) $\bar{H}_1 \cong L_2(q)$; and $\qquad\qquad$ (4.24)
(c) $\bar{H}_1$ and $\bar{H}_2$ are distinct components of $C_{\bar{K}_1}(\bar{z}_2)$.

Using properties of the groups of Lie type of odd characteristic, these conditions now force

$$\bar{K}_1 \cong P\Omega_m^{\pm}(q) \text{ for some } m. \tag{4.25}$$

Thus, as in Theorem 4.1, we are again reduced to deriving a contradiction from the given orthogonal group configuration; and again this is achieved by using special properties of the orthogonal groups.

In attempting to extend the preceding argument to the general case, several difficulties arise at once, centered primarily around the problem of getting the induction started.

First, to obtain an involution $y$ and intrinsic "subcomponent" $I$ of $L$, we are forced to exclude the groups $L_2(q)$ or $^2G_2(q)$ as possibilities for $L$. For the same reason, certain groups over $GF(3)$ must be omitted: $L_3(3)$, $U_3(3)$, $L_4(3)$, $U_4(3)$, $PSp_4(3)$, $G_2(3)$ [inasmuch as $SL_2(3)$ is solvable].

However, this only postpones, but does not avoid the $L_2(q)$ problem, for if $I \cong SL_2(q)$ [as will necessarily be the case if, say, $L \cong L_3(q)$, $U_3(q)$, $G_2(q)$, $q \geqslant 5$], the problem reappears in attempting to identify the pump-up $J$ (or $J^*$). Indeed, suppose $J$ is a single $x$-invariant component and set $\bar{J}\langle \bar{x} \rangle = J\langle x \rangle / \langle y \rangle = J\langle x \rangle / Z(J)$. Then $\bar{I} \cong L_2(q)$ and $\bar{I}$ is a component of $C_{\bar{J}}(\bar{x})$. Thus we require a solution of the $L_2(q)$ component problem in order to identify $\bar{J}$.

We therefore find ourselves in an unpleasant circular situation: on the one hand, the minimal $L_2(q)$ problem requires an inductive characterization of the groups of Lie type of odd characteristic; on the other, this inductive characterization in turn depends on a solution of the $L_2(q)$ component problem. The easiest way to break the loop is, of course, to prove *both* results simultaneously, focusing on a minimal counterexample to their combined statement. However, considerable care must be taken with the precise formulation of the desired theorem since the $B$- and $U$-theorems require a solution of the $L_2(q)$, $A_7$-problem for groups which do not necessarily have the $B$-property.

An additional problem occurs in attempting to replace $L$ by $J$ (or $J^*$) in stage one [after $J$ (or $J^*$) has been identified]. In the high Lie rank case described above, $J$ (or $J^*$) was also of Lie type (of high Lie rank) and replacement was possible. However, as the discussion of Hypothesis $S$ indicates, in the general situation $J$ (or $J^*$) may well be unambiguously of Lie type of characteristic 2, an alternating group, or a sporadic group; and Hypothesis $S$ has been designed to accommodate this possibility. Indeed, Theorem 4.23 immediately yields then that $F^*(G)$ is a $K$-group.

Finally, certain extra difficulties arise when $L$ is defined over $GF(3)$, due to the solvability of $SL_2(3)$. Recall that the general argument involved a possible *double* replacement of $K$: first, to make $L$ intrinsic and then, if $L$ is linear or unitary of even dimension, to ensure that $O_2(Z(L))$ is as large as possible. The set $\mathscr{L}_0(3)$, defined in (4.9) includes the complete list of groups for which this process breaks down, which explains its inclusion in Hypothesis $S$. Of course, in dealing with elements $L \in \mathscr{L}_0(3)$, we can assume at the outset that $L$ is, in fact, standard, otherwise the pumping-up process would eventually yield an $x^* \in \mathscr{I}(G)$ and a component $L^*$ of $L_{x^*}$ with $L^* \in \mathscr{L}\mathscr{O} - \mathscr{L}_0(3)$, in which case $L$ could have been replaced by $L^*$. Hence, in these cases Hypothesis $S$ yields directly that $F^*(G)$ is a group of Lie type of characteristic 3.

The solvability of $SL_2(3)$ also complicates the analysis because of only the weakened form of $L$-balance available in this case. In particular, identification of the normal closure of $H_1$ in $C_{z_1}$ is more delicate when $H_1 \cong SL_2(3)$ [i.e., when $H_1$ is a solvable component of $C_L(Z)$].

As we have indicated above, once one reaches an intrinsic component $L \in \mathscr{L}\mathscr{O}$, the subsequent reduction to the classical involution theorem applies with no essential change to the case that $L/O(L) \in \mathscr{L}\mathscr{O}$ and $L$ is an intrinsic 2-*component* of $C_x$ and yields Harris's "intrinsic 2-component" theorem (Theorem 4.69 below).

All this having been said, here then is the exact formulation of the result we need.

THEOREM 4.43. *Assume Hypothesis S and let G be a group with* $F^*(G)$ *simple and L a 2-component of the centralizer of some involution of G. Assume further that one of the following two sets of conditions holds*:

(I)  (a) $L/O(L) \cong L_2(q)$, *q odd, or* $A_7$ *and L is a maximal 2-component of G; and*
     (b) *Either G has the B-property or G is a minimal counterexample to the B-theorem and U-theorem; or*
(II) (a) $L/O(L)$ *is of Lie type of odd characteristic; and*
     (b) *G has the B-property.*
*Under these conditions,* $F^*(G)$ *is a K-group.*

For brevity, we refer to the given two sets of conditions as *case I* and *case II*, respectively.

The preceding discussion has been intended as an impressionistic outline of the proof of Theorem 4.43 in case II, *under the assumption that the* $L_2(q)$ *component problem has been solved in every proper section of G.* Since this is indeed the case in a minimal counterexample to the theorem, we can therefore assert:

PROPOSITION 4.44. *It suffices to establish Theorem* 4.43 *under the following conditions*:

(a) *Case I holds; and*
(b) *If X is a proper section of G with* $F^*(X)$ *simple and some element of* $\mathscr{L}(X)$ *is isomorphic to* $L_2(q)$, *q odd, then* $F^*(X)$ *is a K-group.*

Thus, henceforth we can assume that the conditions of Proposition 4.44 hold. As commented earlier, the analysis in this case splits into two parts: first, a reduction (carried out by Foote) to the case that $C_G(L/O(L))$ has cyclic Sylow 2-subgroups and then treatment of this bedrock cyclic case by Harris. We note in passing that Foote's reduction does not require the solution of the $L_2(q)$ component problem in proper sections.

Foote's reduction is itself an elaborate undertaking, similar in spirit to Solomon's alternating 2-component analysis. Using Solomon's maximal 2-component theorem plus signalizer functor theory, his goal is to produce a proper subgroup of $G$ which controls sufficient 2-fusion for either Aschbacher's strong embedding criterion or properties of tightly embedded subgroups to yield a contradiction. Extra difficulties arise in carrying out the analysis when $m(C_G(L/O(L))) \leqslant 2$; as a result, Foote treats the cases $m(C_G(L/O(L))) \geqslant 3$ and $m(C_G(L/O(L))) \leqslant 2$ separately.

Thus Foote first proves the following result:

PROPOSITION 4.45. $C_G(L/O(L)$ has 2-rank $\leqslant 2$.

The proof of the proposition is similar to that of the $Mc$, $Ly$ cases of Theorem 4.24 and of Theorem 4.39. However, because $m(L) = 2$ in the present situation, whereas $m(L) \geqslant 4$ in those previous cases, there are some important distinctions, so we shall give a brief sketch of the argument. It will reinforce a general feature of the classification proof: minor differences in two problems which on the surface appear to be entirely comparable and thus amenable to similar treatment often lead to analyses with widely divergent technical characteristics.

First, again Solomon's maximal 2-component theorem (Theorem 4.9) implies the following:

LEMMA 4.46. For any $t \in \mathcal{I}(G)$, $C_t$ has at most one 2-component $K$ with $K/O(K) \cong L/O(L)$.

If $L$ is quasisimple, it is immediate from the lemma and the maximality of $L$ that $L$ is standard. In particular, $C_G(L)$ is tightly embedded in $G$. When $L$ is not quasisimple, Foote's strategy is to use signalizer functors to construct a perfect subgroup $L^*$ of $G$ with $L \leqslant L^*$ and $L$ covering $L^*/O(L^*)$ such that $C_G(L^*/O(L^*))$ is tightly embedded in $G$. But then he is in a position in both the quasisimple and nonquasisimple cases to invoke Aschbacher's results on tightly embedded subgroups (Theorems 3.47 and 3.48) to derive a final contradiction.

Suppose then that $L$ is not quasisimple, so that by hypothesis $G$ is a minimal counterexample to the $B$- and $U$-theorems. Set $Y = C_G(L/O(L))$ and let $A$ be any $E_8$-subgroup of $Y$. Fix a four-subgroup $B$ of $A$ and, for $a \in A^{\#}$, define the usual Aschbacher–Goldschmidt functor

$$\theta(C_a) = \big(O(C_a), B\big)\big(O(C_a) \cap O(C_B)\big). \qquad (4.26)$$

As $G$ is a minimal counterexample to the $B$- and $U$-theorems, the previous results give Foote sufficient control to show that $\theta$ is, in fact, an $A$-signalizer functor on $G$.

Clearly $O(L) \leqslant \theta(G; A)$, the completion of $\theta$. Since $C_A$ normalizes $\theta(G; A)$ (as is immediate from the definition of $\theta$) and $C_A$ covers $L/O(L)$, it follows that $L$ normalizes, but does not centralize $\theta(G; A)$. Now Foote is able to define the desired subgroup $L^*$: namely,

$$L^* = L_2\big(\theta(G; A)L\big). \qquad (4.27)$$

Hence his goal is to prove that $Y^* = C_G\big(L^*/O(L^*)\big)$ is tightly embedded in $G$.

The problem translates very easily into a 2-local question.

LEMMA 4.47. *Let* $T \in \mathrm{Syl}_2(Y^*)$. *If* $\Gamma_{T,1}(G)$ *normalizes* $L^*$, *then* $Y^*$ *is tightly embedded in* $G$.

Indeed, clearly $A \leqslant Y^*$, so $Y^*$ has even order. Hence to prove tight embedding, we must show that if $|Y^* \cap Y^{*g}|$ is even for $g \in G$, then necessarily $g$ normalizes $Y^*$. Let $Q \in \mathrm{Syl}_2(Y^* \cap Y^{*g})$, so that $Q \neq 1$. Without loss, we can suppose that $Q \leqslant T \cap T^g$. Set $R = N_T(Q)$. Then $R$ is contained in both $\Gamma_{T,1}(G)$ and $\Gamma_{T^g,1}(G)$ and hence by our hypothesis, $R$ normalizes both $L^*$ and $L^{*g}$. Moreover, $R \leqslant Y^*$.

We claim that also $R \leqslant Y^{*g}$. Now $R$ normalizes both $I = L_2\big(C_{L^*}(Q)\big)$ and $J = L_2\big(C_{L^{*g}}(Q)\big)$, which cover $L^*/O(L^*)$ and $L^{*g}/O(L^{*g})$, respectively. But by the maximality of $L$, $L$-balance, and Lemma 4.46, $\bar{I} = \bar{J}$ in $\bar{C}_Q = C_Q/O(C_Q)$. However, $R \leqslant Y^*$ centralizes $L^*/O(L^*)$, so $R$ centralizes $\bar{I} = \bar{J}$ and consequently $R$ centralizes $L^{*g}/O(L^{*g})$. Thus $R \leqslant Y^{*g} = C_G\big(L^{*g}/O(L^*)^g\big)$, as claimed.

Hence $R \leqslant Y^* \cap Y^{*g}$ and as $Q \in \mathrm{Syl}_2(Y^* \cap Y^{*g})$, this forces $R = Q$, whence $Q = T \in \mathrm{Syl}_2(Y^*)$ by [I, 1.11]. Since $Q \leqslant T^g$, it follows that also $Q = T^g$, whence $T = T^g$. But again by hypothesis, $N_G(T)$ normalizes $L^*$, so $g$ does as well. Since $Y^* = C_G\big(L^*/O(L^*)\big)$, clearly $Y^* \lhd N_G(L^*)$ and we conclude that $g$ normalizes $Y^*$, as required.

The proof that $\Gamma_{T,1}(G)$ normalizes $L^*$ is rather delicate. First, using standard signalizer functor methods, Foote proves:

LEMMA 4.48.  *If $U$ is a four-subgroup of $Y^*$ that is connected to $A$ in $Y^*$, then*

$$L^* = \langle L_2\,\big(C_L(u)\,O(C_u)\big)\,|\,u \in U^{\#}\rangle.$$

The lemma shows that $L^*$ is determined "functorially" by any four-subgroup of $Y^*$ connected to $A$ in $Y^*$.

Foote next proves the following result:

LEMMA 4.49,  *$Y^*$ has connected Sylow 2-subgroups.*

The proof involves a careful fusion analysis and depends upon certain generational properties of the groups $L_2(q)$. Furthermore, to eliminate certain minimal configurations involving $L_2(q)$ and $A_7$, in which the required generation fails, Foote is forced to invoke Aschbacher's strong embedding criterion [I, 4.31] to obtain a contradiction. [We note that the argument applies equally well when $L$ is quasisimple, with $Y^* = C_G(L)$.]

As an immediate corollary, together with the functorial property of $L^*$ relative to four-subgroups of $Y^*$, one obtains:

LEMMA 4.50.  *$\Gamma_{T,2}(G)$ normalizes $L^*$.*

Thus Foote is "one step" from his objective; and to prove that $\Gamma_{T,1}(G)$ normalizes $L^*$, it remains to show that $C_t$ normalizes $L^*$ for each $t \in \mathcal{I}(T)$. This last step is not easy.

For $t \in \mathcal{I}(Y^*)$, set

$$L(t) = L_2\,\big(C_{L^*}(t)\big) \quad \text{and} \quad Y(t) = C_G(L(t)/O\big(L(t)\big)). \tag{4.28}$$

Foote next proves:

LEMMA 4.51.  *If $C_t$ does not normalize $L^*$ for $t \in \mathcal{I}(Y^*)$, then $C_{Y(t)}(t)$ has a proper 2-generated core and contains a 2-component $K(t)$ with $K(t)/O\big(K(t)\big) \cong L_2(2^n)$, $U_3(2^n)$, $Sz(2^n)$, or $Sz(8)/Z_2$.*

Indeed, if $C_t$ does not normalize $L^*$, it follows easily by Lemma 4.50 and a Frattini argument that $C_{Y(t)}(t)$ has a proper 2-generated core. But then by (the general form of) Aschbacher's theorem [I, 4.28], either $C_{Y(t)}(t)$ has a 2-component $K(t)$ of the required type or else it has an intrinsic 2-component $I$ with $I/O(I) \cong SL_2(5)$ or $\hat{A}_9$. In the latter cases, one sees that $C_t$ has an intrinsic 2-component of the same type as $I$. But then the classical involution and intrinsic $\hat{A}_n$ theorems imply that $F^*(G)$ is a $K$-group. However, as $G$ is a minimal counterexample to the *B*- and *U*-theorems, this contradicts Theorem 3.27.

Lemma 4.51 gives the obstruction to the desired conclusion. Its elimination proceeds in two stages.

LEMMA 4.52. *If* $C_t$ *does not normalize* $L^*$ *for* $t \in \mathcal{I}(Y^*)$, *then* $C_u$ *normalizes* $L^*$ *for every involution* $u \neq t$ *of* $Y^* \cap K(t)$.

If $E$ is an elementary 2-subgroup of $K(t)$ of maximal rank with $E \leqslant Y^*$ and we expand $E$ to $S \in \text{Syl}_2(N_G(L^*))$, Foote argues that either the lemma holds or else $S \in \text{Syl}_2(G)$ and $E$ is strongly closed in $S$ with respect to $G$. However, in the latter case, [I, 4.128] implies that $F^*(G)$ is a $K$-group—contradiction.

Thus there is essentially a unique class of involutions of $Y^*$ whose centralizers do not normalize $L^*$. In this remaining case Foote argues first that $K(t)/O(K(t)) \not\cong Sz(8)/Z_2$ and then that the hypotheses of Aschbacher's strong embedding criterion [I, 4.31] are satisfied by the conjugacy class of any involution $z$ of $E$, which leads to the same contradiction.

We therefore conclude that $C_t$ normalizes $L^*$ for each $t \in \mathcal{I}(Y^*)$, whence $\Gamma_{T,1}(G)$ normalizes $L^*$. Now Lemma 4.46 yields our objective.

LEMMA 4.53. $Y^*$ *is tightly embedded in* $G$.

Since $N_G(Y^*) = Y^* N_G(T)$ by a Frattini argument, and both factors normalize $L^*$, we also have

$$N_G(L^*) = N_G(Y^*). \tag{4.29}$$

For completeness, if $L$ is quasisimple, we put $L^* = L$ and $Y^* = C_G(L)$. Then Lemma 4.53 and (4.29) hold for $L^*$ and $Y^*$ in this case as well. Set $M = N_G(Y^*)$.

Using Theorem 3.41 concerning tightly embedded subgroups, one obtains the following result in the present situation.

LEMMA 4.54. *For some* $g \in G - M$, *there exist four-groups* $U \leqslant Y^* \cap M^g$ *and* $V \leqslant Y^{*g} \cap M$ *such that* $U$ *centralizes* $V$.

This immediately yields the following result:

LEMMA 4.55. *L\* and L\*ᵍ normalize each other.*

Indeed, by symmetry, it suffices to show that $L^{*g}$ normalizes $L^*$. By the structure of $\mathrm{Aut}(L_2(q))$ and $\Sigma_7$, some $v \in V^{\#}$ induces an inner automorphism on $L^*/O(L^*)$ and hence on $Y^*$. Since $T$ is connected with $m(T) \geqslant 3$, it follows that $v$ centralizes an $E_8$-subgroup $E$ of $Y^*$. But $C_v$ normalizes $L^{*g}$ [as $v \in Y^{*g}$ and $\Gamma_{T^g,1}(G)$ normalizes $L^{*g}$], so $E$ normalizes $L^{*g}$. On the other hand, $\Gamma_{E,1}(G)$ normalizes $L^*$ (as $E \leqslant Y^*$). However, $L^{*g} \leqslant \Gamma_{E,1}(G)$ by generational properties of $L_2(q)$ and $A_7$, so $L^{*g}$ normalizes $L^*$, as required.

Finally as $g \notin M$ and $M = N_G(L^*)$, $L^{*g} \neq L^*$. Hence if we set $H = L^*L^{*g}$, it follows from the preceding lemma that $H/O(H)$ is the direct product of *two* isomorphic components. Furthermore, by Lemma 4.46, $C_G(H/O(H))$ must have odd order. Foote now sets $N = N_G(H)$ and shows that $H$ and $N$ satisfy the conditions of Aschbacher's 2-component fusion theorem, thus obtaining a final contradiction.

This completes the discussion of Proposition 4.45.

Foote now treats the rank $\leqslant 2$ case.

PROPOSITION 4.56. $C_G(L/O(L))$ *has cyclic Sylow 2-subgroups.*

By Proposition 4.45, a Sylow 2-subgroup $R$ of $Y = C_G(L/O(L))$ has 2-rank $\leqslant 2$. We can suppose that $R$ is not cyclic, otherwise we are done. If $R$ is quaternion, it is immediate that $x$ is the unique involution of $Y$ and that $Y$ is tightly embedded in $G$. But in this case, Theorem 4.43 follows from the classical involution theorem. Hence we can also suppose that $R$ is not quaternion, whence $m(R) = 2$ by [I, 1.35].

The argument is technically more difficult than Proposition 4.45 because certain subcases must first be eliminated and then considerable fusion analysis carried out before signalizer functors can be effectively brought into play. We briefly describe the main moves.

Among all choices of $x$ and $L$ with $m(R) = 2$, Foote picks $x$ so that $|R|$ is maximal, and subject to this so that a Sylow 2-subgroup $T$ of $N_G(L)$ has maximal order.

Foote first proves the following result:

LEMMA 4.57. *T is not a Sylow 2-subgroup of G.*

In the contrary case, he applies results from Goldschmidt's strongly closed abelian paper [I: 123] to produce a nontrivial abelian subgroup which is strongly closed in $G$ with respect to $G$. Again [I, 4.128] then yields the possibilities for $F^*(G)$ and these are incompatible with the structure of $C_x$.

Set $Q = L \cap T$. He next proves:

LEMMA 4.58. *If either $Q$ or $R$ is a four-group, then $F^*(G) \cong M_{12}, J_2$, or $A_n$, $n = 9$, 10, or 11.*

Under either of these conditions, he argues that Harada's theorem (Theorem 2.160) is applicable and so $G$ has sectional 2-rank $\leqslant 4$. But then given the structure of $C_x$, the lemma follows from the sectional 2-rank $\leqslant 4$ theorem.

In particular, Theorem 4.43 holds under these conditions and so we are left with the following case:

$$|Q| > 4 \quad \text{and} \quad |R| > 4. \tag{4.30}$$

Foote next proves the following result:

LEMMA 4.59. *$T$ is of index 2 in a Sylow 2-subgroup $S$ of $G$.*

It is this remaining case whose treatment requires signalizer functors. However, as we have observed above, one must first establish detailed information concerning the $G$-fusion of involutions of $Q$ and $R$ in $S$. In particular, if we fix $y \in S - T$, one must first show that

$$QQ^y = Q \times Q^y. \tag{4.31}$$

The goal now is to prove that $Q$ and $Q^y$ are core-separated in the sense of [I, 4.66]. This is equivalent here to the following assertion:

LEMMA 4.60. *If $a \in \mathscr{I}(QQ^y)$ and $K$ is a 2-component of $C_a$, then either $C_Q(a)$ or $C_{Q^y}(a)$ centralizes $K/O(K)$.*

The lemma is a direct consequence of the following statement:

$$\text{If } a \notin Q \text{ or } Q^y, \text{ then } C_a \text{ is solvable.} \tag{4.32}$$

Assuming (4.32) is false, Foote produces a subgroup $X$ of $C_a$ which is tightly embedded in $C_a$ and then uses properties of tightly embedded subgroups to derive a contradiction. To construct $X$, he argues first that there is $f \in \mathscr{T}(T)$ normalizing $Q$ such that either $f$ induces a nontrivial field automorphism on $L/O(L)$ or else $L\langle f \rangle/O(L) \cong \Sigma_7$, and then that $C_a$ is balanced with respect to any $E_8$-subgroup $E$ of $Q\langle f \rangle$. Finally setting $B = E \cap Q \ (\cong Z_2 \times Z_2)$, he shows with the aid of the signalizer functor theorem that

$$X = \langle O(C_{C_a}(b)) \mid b \in B^{\#} \rangle Q \tag{4.33}$$

is the required tightly embedded subgroup of $C_a$.

Now let $A_1, A_2$ be four-subgroups of $Q$ and $Q^y$, respectively, put $A = A_1 \times A_2$, and for $a \in A^{\#}$, set

$$\theta(C_a) = \bigcap_{i=1,2} [O(C_a), A_i](O(C_{A_i}) \cap O(C_a)). \tag{4.34}$$

Since $Q$ and $Q^y$ are core-separated, Foote thus concludes from [I, 4.67] that $\theta$ is an $A$-signalizer functor on $G$.

Now, using the signalizer functor theorem, it follows that $G$ possesses a subgroup $H$ with $\langle L, L^y \rangle \leqslant H$ such that $H/O(H)$ is the product of two components, each isomorphic to $L/O(L)$. Thus we have reached the same configuration as at the end of the proof of Proposition 4.45, and again Aschbacher's 2-component fusion theorem can be shown to yield a contradiction. (Actually it turns out to be easier in this case for Foote to argue that $L^* = \langle L^H \rangle$ is tightly embedded in $G$ and then use properties of tightly embedded subgroups to reach the desired contradiction.)

This completes the discussion of Foote's contributions to the $B$- and $U$-theorems. We now combine Proposition 4.56 with the Fritz–Harris–Solomon theorem (Theorem 2.161). Thus if $L$ has Sylow 2-subgroups of order 4 or 8 [in particular, if $L/O(L) \cong A_7$], we conclude from these results that $F^*(G)$ is a $K$-group, in which case Theorem 4.43 holds.

Therefore the following result together with Propositions 4.44 and 4.56 will complete the proof of Theorem 4.43.

THEOREM 4.61. *Let $G$ be a group with $F^*(G)$ simple and $L$ a 2-component of the centralizer of some involution of $G$, and assume that the following conditions hold:*

(a) *$L/O(L) \cong L_2(q)$ for some $q \equiv \pm 1 \pmod{16}$;*
(b) *$C_G(L/O(L))$ has cyclic Sylow 2-subgroups;*
(c) *Every proper section of $G$ has the B-property; and*

(d) *If X is a proper section of G with $F^*(X)$ simple and some element of $\mathscr{L}(X)$ is isomorphic to $L_2(r)$, r odd, then $F^*(X)$ is a K-group. Under these conditions, $F^*(G)$ is a K-group.*

This is Harris's theorem [59]. As we have earlier indicated, the basic idea is to force the existence of an involution $u$ in $C_x$ which induces a field automorphism on $L/O(L)$ and then to analyze the pump-up $J$ in $L_u$ of the subcomponent $I = L_{2'}(C_L(u))$ $(I/O(I) \cong L_2(q^{1/2}))$. For a suitable choice of the involution $u$, the structure of $C_x$ and $C_u$ are then compared; and Harris ultimately shows that either the 2-fusion determined by these two groups is incompatible or else $G$ possesses a normal subgroup $G_0$ of index 2 containing $x$ (in which case Theorem 4.43 holds in $G_0$ and hence in $G$). (Along the way, some cases are eliminated by appeal to the sectional 2-rank $\leqslant 4$ or classical involution theorems.) In view of the large number of distinct possibilities for the group $J/O(J)$, the complete analysis is very elaborate. We shall therefore limit the discussion to setting up the situation and making a few pointed comments about the proof.

As usual, we proceed by contradiction, taking $G$ to be a minimal counterexample. Let $x \in \mathscr{I}(G)$ with $L$ a 2-component of $C_x$, as in the theorem, with $x$ chosen in such a way that a Sylow 2-subgroup $T$ of $C_x$ has largest possible order. Using the $Z^*$-theorem [I, 4.95] and Harada's self-centralizing $E_8$ theorem (Theorem 2.156), an easy fusion analysis gives the following reduction.

LEMMA 4.62. *The following conditions hold:*

(i) $\langle x \rangle \in \mathrm{Syl}_2(C_G(L/O(L)))$;
(ii) *T splits over $\langle x \rangle$; and*
(iii) *T contains an involution $u$ which induces a nontrivial field automorphism on $L/O(L)$ and such that $C_T(u) \in \mathrm{Syl}_2(C_{C_x}(u))$.*

Note that as $q \equiv \pm 1 \pmod{16}$, (iii) implies that $q = r^2$ for some prime power $r \equiv \pm 1 \pmod 8$.

We set $I = L_{2'}(C_L(u))$. Then

$$I/O(I) \cong L_2(r). \tag{4.35}$$

The following fact will be critical.

LEMMA 4.63. *If $R \in \mathrm{Syl}_2(C_{C_u}(I/O(I)))$ with $\langle u, x \rangle \leqslant R$, then $R$ is dihedral or quasi-dihedral.*

Indeed, by the structure of $\text{Aut}(L_2(q))$ and the fact that $\langle x \rangle \in$ $\text{Syl}_2(C_G(L/O(L)))$, we have $\langle x, u \rangle \in \text{Syl}_2(C_G(I/O(I)))$. Hence $C_R(x) = \langle x, u \rangle$ and the lemma follows from [I, 1.16, 1.17].

Now let $J$ be the normal closure of $I$ in $L_u$, so that as usual by $L$-balance, $J$ is either a single $x$-invariant 2-component of $C_u$ or the product of two such 2-components interchanged by $x$, and in either case $I$ is a 2-component of $C_J(x)$. Setting $\bar{C}_u = C_u/O(C_u)$, assumption (c) of Theorem 4.61 implies that

$$\bar{I} \text{ is a component of } C_{\bar{J}}(\bar{x}). \tag{4.36}$$

The next step is to determine the possibilities for $\bar{J}$. If $\bar{J}$ is the product of two components, then $\bar{J} \cong L_2(r) \times L_2(r)$ or $SL_2(r) * SL_2(r)$. However, in the latter case $\langle \bar{u} \rangle = Z(\bar{J})$ by the previous lemma and $G$ has a classical involution. But then $F^*(G)$ is a $K$-group, contrary to assumption. If $\bar{x}$ centralizes $\bar{J}$, then, of course, $\bar{J} = \bar{I} \cong L_2(r)$, so in determining $\bar{J}$, we can assume that $\bar{J}$ is quasisimple with $\bar{x}$ not centralizing $\bar{J}$.

Set $\bar{X} = \bar{J}\langle \bar{x} \rangle$ and suppose first that $\bar{u} \notin \bar{J}$, in which case $\langle \bar{x} \rangle \in \text{Syl}_2(C_{\bar{X}}(\bar{I}))$, again by the previous lemma. If $Z(\bar{J}) \neq 1$, this forces $\bar{x} \in Z(\bar{J})$ and so $\bar{x}$ centralizes $\bar{J}$, which is not the case. Thus $\bar{J}$ is simple and $\bar{I}$ is standard in $\bar{X}$. Now using assumption (d), we can read off the possibilities for $\bar{J} = F^*(\bar{X})$. Note that as $r \equiv \pm 1 \pmod 8$, $r \neq 5$ or an odd power of 3, thus excluding certain possibilities which exist only for those values of $r$. Likewise $C_L(x)$ contains an involution $t$ such that $I\langle t \rangle/O(I) \cong PGL_2(r)$. Then $\bar{t}$ normalizes $\bar{J}$ and hence $\bar{X}$ and we have $C_{\bar{X}\langle \bar{t} \rangle}\langle \bar{x} \rangle \geqslant \bar{I}\langle \bar{t} \rangle \cong PGL_2(r)$. This condition also rules out several initial possibilities for $\bar{J}$.

Finally if $\bar{u} \in Z(\bar{J})$, set $\tilde{X} = \bar{X}/Z(\bar{J})$ and $\tilde{Q} = \bar{R} \cap \tilde{X}$. Then either $\tilde{Q} = \langle \tilde{x} \rangle$ and $\tilde{I}$ is standard in $\tilde{X}$ or $\tilde{Q}$ is dihedral. Since $\tilde{I} \cong L_2(r)$, we can determine the possibilities for $\tilde{J}$ in either case, again by (d), and from that the possibilities for $\bar{J}$.

The final result is as follows:

PROPOSITION 4.64. *One of the following holds*:

   (i) $\bar{J} \cong L_2(r)$ or $L_2(r) \times L_2(r)$;
   (ii) $\bar{J} \cong L_2(r^2) = L_2(q)$;
   (iii) $\bar{J} \cong L_3(r)$ or $U_3(r)$;
   (iv) $\bar{J}/Z(\bar{J}) \cong L_4(r^{1/2})$ or $U_4(r^{1/2})$; or
   (v) $\bar{J}/Z(\bar{J}) \cong L_3(4)$ or $J_2$ and $r = 7$.

Furthermore, in each case, Harris knows the nature of the action of $\bar{x}$ on $\bar{J}$. In particular, $\bar{x}$ induces an outer automorphism on $\bar{J}$ in all cases.

Harris's next step is to make a more careful choice of the involution $u$. To achieve this, he must first examine $C_x$ more closely. We have $\langle z \rangle = Z(T) \cap L \cong Z_2$ and we put $Z = \langle z, x \rangle$, so that $Z \leqslant Z(T)$. Also set $A = Z\langle u \rangle = \langle z, x, u \rangle (\cong E_8)$.

The following facts are easily verified.

LEMMA 4.65. *The following conditions hold*:

   (i) *If* $a \in A - Z (= uZ)$, *then* $a$ *induces a nontrivial field automorphism on* $L/O(L)$ *and* $C_T(a) \in \mathrm{Syl}_2\big(C_{C_x}(a)\big)$;

  (ii) $C_T(a) = C_T(u) = (T \cap L) \times \langle u, x \rangle$;

 (iii) $|T: C_T(u)| = 1$ *or* 2 *and correspondingly* $Z(T) = A$ *or* $Z$;

 (iv) $|T: C_T(u)| = 2$ *if and only if* $T$ *contains an involution* $t$ *such that* $L\langle t \rangle / O(L) \cong PGL_2(q)$ *and* $[t, u] = z$;

  (v) $C_T(u)$ *char* $T$;

 (vi) $T$ *has sectional 2-rank* $\leqslant 4$;

(vii) $A$ *is the unique normal* $E_8$-*subgroup of* $T$; *and*

(viii) $T$ *contains no normal* $E_{16}$-*subgroups*.

Using the sectional 2-rank $\leqslant 4$ theorem, we also immediately obtain:

LEMMA 4.66. *The following conditions hold*:

  (i) $G$ *has sectional 2-rank* $\geqslant 5$;

 (ii) $T \notin \mathrm{Syl}_2(G)$;

(iii) *If* $T \leqslant S \in \mathrm{Syl}_2(G)$, *then* $S$ *contains a normal* $E_8$-*subgroup* $E$; *and*

(iv) $A = (E \cap T)\langle x \rangle$.

Indeed, (i) holds, otherwise $F^*(G)$ is a $K$-group by the sectional 2-rank $\leqslant 4$ theorem, contrary to assumption. But then (ii) follows from Lemma 4.65(vi). Likewise (iii) follows from (i) and MacWilliams's theorem [I, 1.36]. Finally as $x$ normalizes $E$, $E \cap T = C_E(x)$ has order at least 4 and hence $B = (E \cap T)\langle x \rangle$ is elementary of order 8 or 16. But as $T \leqslant S$ normalizes $E$, $E \cap T$ and hence $B$ is normal in $T$. Now Lemma 4.65(vii) and (viii) force $B = A$, proving (iv).

We need one more preliminary piece of information. We set $Q = C_T(u)$, so that $|T: Q| \leqslant 2$, $Q$ char $T$, and $A = Z(Q)$. Also put $N = N_G(Q)$, so that $T \leqslant N$. Without loss, we can assume that $S \cap N \in \mathrm{Syl}_2(N)$.

LEMMA 4.67. *The following conditions hold*:

(i) $N = (S \cap N) O(N)$;

(ii) $x$ *has 2 or 4 G-conjugates in A and each is a conjugate of x under* $S \cap N$; *and*

(iii) $E \leqslant N$.

Indeed, $N \leqslant N_G(A)$ as $A$ char $Q$, and $C_A(\leqslant C_x)$ has a normal 2-complement by the structure of $C_x$. Hence if we set $\bar{N} = N/C_N(A)$, to prove (i) it will suffice to show that $\bar{N}$ is a 2-group. Assume false. We have $\bar{N} \leqslant \text{Aut}(A) \cong GL_3(2)$. Furthermore, as $Q = (T \cap L) \times \langle u, x \rangle$ with $T \cap L$ dihedral (of order at least 16), $\langle z \rangle = A \cap Q'$ and consequently $z \in Z(N)$. In particular, $|\bar{N}|$ is not divisible by 7, so $\bar{N}$ is a $\{2, 3\}$-group. Hence $N - C_N(A)$ contains a 3-element $d$, and from the structure of $Q$, we see that $[Q, d]$ is a four-subgroup of $A$. Also $S \cap N > T$ as $Q$ char $T < S$, so $x$ is not in the center of a Sylow 2-subgroup of $N$. This implies that $x$ has 6 conjugates in $A$ under $N$, forcing $\bar{N} \cong A_4$ or $\Sigma_4$.

But now if $P$ denotes the pre-image of $O_2(\bar{N})$ in $S \cap N$, by a Frattini argument we can assume $d$ is chosen to normalize $P$. Since $d$ centralizes $z$ and $[\bar{P}, A]$ is $d$-invariant, the only possibility is $[\bar{P}, A] = \langle z \rangle$. In particular, this implies that $C_{\bar{P}}(x) \cong Z_2$. But $C_{\bar{P}}(x) = \bar{T}$ and $\bar{Q} = 1$, so we must have $Q < T$. Thus there is $t \in \mathscr{I}(P)$ with $t \notin Q$ such that $(Q \cap L)\langle t \rangle$ is dihedral (of order at least 32). One checks easily now that $P$ does not admit an automorphism with the properties of $d$ (i.e., with $Z_2 \times Z_2 \cong [Q, d]$ and $\bar{P} = [\bar{P}, d]$). Thus (i) holds.

Since $A = Z(Q)$ char $T \in \text{Syl}_2(C_x)$, it follows directly from (i) that if $x^g = a \in A$ for $g \in G$, then $x^g = x^y$ for some $y \in S \cap N$, which in turn implies that $x$ has 2 or 4 $G$-conjugates in $A$ under $N$. Thus (ii) also holds.

Finally as $E \lhd S$ with $S \cong E_8$ and $|E \cap Q| = |E \cap A| \geqslant 4$ [by Lemma 4.66(iv)], $|QE : Q| \leqslant 2$, so $Q \lhd QE$ and hence $E \leqslant N$, proving (iii).

Now Harris proves the following result:

PROPOSITION 4.68. *There exists an element* $u^* \in uZ$ *with the following properties*:

(i) $u^* \in E$. *In particular, a Sylow 2-subgroup of* $C_{u^*}$ *has index at most 4 in a Sylow 2-subgroup of* $G$;

(ii) $u^*$ *is not G-conjugate to* $x$;

(iii) *If* $P = C_{S \cap N}(u^*)$, *then* $Q \leqslant P$, $P/Q \cong Z_2$ *or* $Z_2 \times Z_2$, *and* $|S \cap N : P| \leqslant 2$; *and*

(iv) *If* $I^* = L_{2'}(C_L(u^*))$, *then some P-conjugate of x does not centralize* $I^*/O(I^*)$.

To prove the existence of the desired element $u^*$, the key facts at Harris's disposal are that $E \lhd S \cap N$ and $|E \cap A| \geqslant 4$, with $x$ having only 2 or 4 $G$-conjugates in $A$, each of which is a conjugate of $x$ under $S \cap N$. However, the argument requires him to separate the analysis into four cases: (a) $T$ not normal in $S \cap N$; (b) $T = Q$; (c) $T \lhd S \cap N$, $T > Q$, and $S \cap N/Q \cong Z_2 \times Z_2$; and (d) $T \lhd S \cap N$, $T > Q$, and $S \cap N/Q \cong Z_4$.

In the first three cases, the given structure of $S \cap N$ yields fairly directly an element $u^* \in (E \cap A) \cap uZ$ with the required properties. On the other hand, in the final case, Harris uses transfer to force $G$ to have a normal subgroup $G_0$ of index 2 containing $x$. Then $x$ and $L$ satisfy the conditions of the theorem in $G_0$, and as $G$ is a minimal counterexample, it follows that $F^*(G_0) = F^*(G)$ is a $K$-group, contrary to the choice of $G$.

Thus without loss Harris can assume to begin with that $u$ is chosen to satisfy the condition of the proposition; and among all such choices, he picks $u$ so that a Sylow 2-subgroup of $C_u$ has maximal order. The given conditions on $u$ give Harris sufficient leverage to eliminate each of the possibilities for $\bar{J}$ in succession. We note that the cases $\bar{J} \cong J_2$, $J_2/Z_2$, $L_3(4)$, and a nontrivial covering group of $L_3(4)$ must be treated individually.

The analysis is very elaborate and highly technical and depends upon detailed properties of the groups $\bar{J}$. The basic thrust of the argument is fusion–theoretic, either producing by transfer a normal subgroup of $G$ of index 2 (which leads, as above, to a contradiction) or showing that the structure of $C_u$ is incompatible with that determined by Proposition 4.68. However, there is one situation, when $\bar{J} \cong L_2(q)$, where Harris is forced to argue that $ux$ satisfies the conditions of Proposition 4.68, but with a Sylow 2-subgroup of $C_{ux}$ having greater order than one of $C_u$, contrary to the maximal choice of $u$.

Finally, there are two places in which assumption (d) of the theorem must be used to identify additional proper sections of $G$. The first occurs when $\bar{J} \cong L_3(r)$, $U_3(r)$, $L_4(r^{1/2})$, or $U_4(r^{1/2})$, whence $H = L_2(C_J(z))$ is isomorphic (mod core) to $SL_2(r)$ or $SL_2(r^{1/2}) * SL_2(r^{1/2})$ (provided $r^{1/2} > 3$). If $K$ denotes the normal closure of $H$ in $L_z$, Harris uses (d) to force $K/O(K)$ to be isomorphic corresponding to a nontrivial cover of $L_3(4)$ (with $r = 7$) or to a central product of two nontrivial covers of $L_3(4)$. The second place arises when $\bar{J}$ is the product of two 2-components $J_1, J_2$ [with $\bar{J}_1 \cong \bar{J}_2 \cong L_2(r)$]. To carry through the analysis, Harris must identify the possibilities for the pump-up of $H_1 = L_2(C_{J_1}(v))$ in $L_v$ for $v \in \mathscr{I}(J_2)$.

There you have a full outline of the $L_2(q), A_7$ story. It is certainly an extremely complex chapter of the theory of groups of component type.

As promised, we conclude the section with the statement of Harris's intrinsic 2-component theorem, which it will be convenient for later application to formulate in the following way.

THEOREM 4.69. *Assume Hypothesis S and let G be a group with $F^*(G)$ simple in which every proper section has the B-property. If the centralizer of some involution of G contains an intrinsic 2-component L with $L/O(L) \in \mathscr{LO}$, then $F^*(G)$ is a K-group.*

## 4.6. Small Unbalancing Triples

Now we return to the main theme. Here is the situation remaining to be analyzed. On the one hand, if $(x, y, L)$ is an unbalancing triple in our minimal counterexample $G$, then by the results of Thompson and Solomon:

$$L/O(L) \cong L_2(q), A_7, He, \text{ or a covering group of } L_3(4). \quad (4.37)$$

On the other hand, by Theorem 4.43(I) [and the fact that $F^*(G)$ is not a K-group], we have

$$\begin{array}{l} \text{If } K \text{ is a maximal 2-component of } G, \text{ then} \\ K/O(K) \not\cong L_2(q), q \text{ odd, or } A_7. \end{array} \quad (4.38)$$

To make effective use of these two conclusions, one must clearly analyze the relationship between "maximal unbalancing triples" and "maximal 2-components." This was done by Gilman and Solomon, and we shall describe their work next.

Their main result asserts that in the presence of (4.37) the two notions are, in fact, identical! Although there are differences in detail, their analysis is very similar to that of Harris in the minimal $L_2(q)$ problem (Theorem 4.61). Again there is a pump-up $J$ of a suitable 2-component $I$, the possibilities for $J/O(J)$ must again be identified, and finally a fusion contradiction must be derived. Unfortunately, the argument is even more elaborate than in Harris's case. Indeed, in order to carry through the required fusion analysis, they are forced to consider a more general situation, abstracted from the given configuration of subgroups, which they then study inductively.

For their argument, Gilman and Solomon need the following general criterion, related to nonembeddedness, for a 2-component of the centralizer of an involution to be a maximal 2-component.

THEOREM 4.70. *Let G be a group, $x \in \mathscr{I}(G)$, and L a 2-component of $C_x$. Assume the following conditions hold:*

(a) $m(L) \geqslant 2$;

(b) *If $t \in C_{C_x}(L/O(L))$ and J is the normal closure of $L_{2'}(C_L(t))$ in $L_t$, then x centralizes $J/O(J)$; and*

(c) *If, in (b), t is in the center of a Sylow 2-subgroup T of $N_{C_x}(L)$, then $T \in \mathrm{Syl}_2(N_{C_t}(J))$.*

*Then $L/O(L) \in \mathscr{L}^*(G)$.*

Note that if $G$ has the *B*-property, (b) asserts that $L$ is nonembedded, whence by Aschbacher's component theorem (Theorem 3.31) and (a), $L$ is standard. Thus, in essence, the theorem asserts that in the presence of (c), standard components are necessarily maximal. Note also that if $x, L$ are part of an unbalancing triple $(x, y, L)$ of $G$, (c) is one of the two conditions in the definition of a maximal unbalancing triple (Definition 3.29).

Although the proof is somewhat technical, the underlying idea is quite simple. Assuming false, there exists a sequence of involutions $x = x_1, ..., x_r$ of $G$ and 2-components $L_i$ of $C_{x_i}$, $1 \leqslant i \leqslant r$, such that for $1 \leqslant i \leqslant r - 1$, we have

(1) $x_i$ centralizes $x_{i+1}$;

(2) $x_{i+1}$ centralizes $L_i/O(L_i)$;

(3) $L_i < L_{i+1}$ [i.e., $L_{i+1}$ is a 2-component of the normal closure of $L_{2'}(C_{L_i}(x_{i+1}))$ in $L_{x_{i+1}}$]; $\qquad$ (4.39)

(4) $L_i/O(L_i) \cong L/O(L)$, $1 \leqslant i \leqslant r - 1$; and

(5) $L_r/O(L_r) \not\cong L/O(L)$.

Among all such sequences of $x_i$ and $L_i$, choose $r$ minimal. In view of assumption (b), we have $r \geqslant 3$. The strategy is to contradict the minimality of $r$ by showing that after replacing the given sequences by suitable $G$-conjugates, one can delete $x_2, L_2$ and still maintain (4.39) for the reduced sequence.

Using assumption (c), the key to the argument is a proof of the existence of an element $g \in G$ such that

(1) $u = x_2^g$ centralizes $L/O(L)$;

(2) $C_T(u) \in \mathrm{Syl}_2(N_{C_u}(K))$, where $K$ denotes the normal closure of $L_{2'}(C_L(u))$ in $L_u$; and $\qquad$ (4.40)

(3) $L_2^g (\leqslant L_u)$ covers $K/O(K)$.

Since $x_3^g$ centralizes $L_2^g/O(L_2^g)$, one argues on the basis of (4.40) that there is $g' \in G$ with $x_3^{g'}$ centralizing $L/O(L)$ such that the sequences

$x = x_1, x_3^{g'}, ..., x_r^{g'}, L = L_1, L_3^{g'}, ..., L_r^{g'}$ satisfy the conditions of (4.39), giving the desired contradiction.

The proof of the existence of the required $g \in G$ satisfying (4.40) is similar in spirit to that of Alperin's fusion theorem [I, 4.86]]. Indeed, one constructs $g$ as a product of suitable elements $g_1, g_2, ..., g_m$ of $G$ such that for $1 \leqslant j \leqslant m$,

$$
\begin{align}
&(1) \ \ x_2^{g_1 g_2 \cdots g_j} \in C_T\big(L/O(L)\big); \text{ and}\\
&(2) \ \ \big(C_T(x_2^{g_1 g_2 \cdots g_{j-1}})\big)^{g_j} \leqslant T.
\end{align}
\tag{4.41}
$$

Here now is Gilman and Solomon's principal result.

THEOREM 4.71. *Assume Hypothesis S and let G be a minimal counterexample to the B-theorem and U-theorem. If $(x, y, L)$ is a maximal unbalancing triple, then L is a maximal 2-component of G.*

The proof is by contradiction. Let $T \in \mathrm{Syl}_2\big(N_{C_x}(L)\big)$ with $\langle x, y \rangle \leqslant T$ and set $R = C_T\big(L/O(L)\big)$, so that $R \lhd T$. Fix this notation. They first verify condition (c) of Theorem 4.70.

LEMMA 4.72. *Let $t \in \mathscr{I}\big(R \cap Z(T)\big)$ and let J be the normal closure of $L_{2'}\big(C_L(t)\big)$ in $L_t$. If J is a single 2-component of $C_t$, then we have*

   (i) *$(z, y, J)$ is an unbalancing triple;*
   (ii) *$J/O(J) \cong L/O(L)$; and*
   (iii) *$T \in Syl_2\big(N_{C_z}(J)\big)$.*

Indeed, as $(x, y, L)$ is an unbalancing triple, $D = O(C_y) \cap C_x$ normalizes, but does not centralize, $L/O(L)$. Since $t \in Z(T)$ centralizes $y$, $t$ normalizes $D$ and it follows, as usual, that $E = C_D(t)$ does not centralize $L/O(L)$. But then $E$ normalizes, but does not centralize, $J/O(J)$, and as $E \leqslant O(C_y) \cap C_t$, we conclude that $(t, y, J)$ is an unbalancing triple in $G$. But $I = L_{2'}\big(C_L(t)\big)$ covers $L/O(L)$ and $I \leqslant J$. Since $(x, y, L)$ is a maximal unbalancing triple, the definition forces $I$ to cover $J/O(J)$, whence $J/O(J) \cong L/O(L)$. Furthermore, as $T$ normalizes $I$, it normalizes $J$ and again the definition of maximal unbalancing triple implies that $T \in \mathrm{Syl}_2\big(N_{C_t}(J)\big)$. Hence all parts of the lemma hold.

Since $L$ is not a maximal 2-component of $G$, Lemma 4.72 and Theorem

4.70 imply that there exists an involution $t \in R$ such that $x$ does not centralize the normal closure of $L_2(C_L(t))$ in $L_t$. Thus we have:

LEMMA 4.73. *There exists a nontrivial 2-subgroup $Q$ of $R$ such that if $I = L_2(C_L(Q))$ and $J$ denotes the normal closure of $I$ in $L_2(C_Q)$, then $x$ does not centralize $J/O(J)$.*

Among all such $Q$, choose $Q$ of maximal order and fix this notation as well. Note that as $Q$ centralizes $J/O(J)$, but $x$ does not, obviously $x \notin Q$. Also set $N = N_G(Q)$ and $\bar{N} = N/O(N)$. Since $C_Q \lhd N$, $J$ is a product of 2-components of $N$ and, as usual, $J$ is either a single such $x$-invariant 2-component or the product of two interchanged by $x$, and in either case $I$ is a 2-component of $C_J(x)$. Since $G$ is a minimal counterexample to the *B*-theorem, we have

$$
\begin{array}{ll}
\text{(a)} & \bar{J} \text{ is semisimple with } O(\bar{J}) = 1; \text{ and} \\
\text{(b)} & \bar{I} \text{ is a component of } C_{\bar{J}}(\bar{x}).
\end{array} \qquad (4.42)
$$

Note also that $\bar{I} \cong L/O(L)$ as I covers $L/O(L)$.

In view of the maximal choice of $Q$, a rather easy argument yields the following key fact:

LEMMA 4.74. *$\bar{I}$ is a standard component of $\bar{J}\langle\bar{x}\rangle$.*

If $\bar{J}$ is a product of two components $\bar{J}_1$, $\bar{J}_2$, then, of course, $\bar{I}$ is a homomorphic image of each $\bar{J}_i$ and so $\bar{J}$ is determined. On the other hand, if $\bar{J}$ is quasisimple, then passing to $\tilde{J}\langle\tilde{x}\rangle = \bar{J}\langle\bar{x}\rangle/Z(\bar{J})$, it follows that $\tilde{I}$ is standard in $\tilde{J}\langle\tilde{x}\rangle$ and $F^*(\tilde{J}\langle\tilde{x}\rangle) = \tilde{J}$ is simple, so $\tilde{J}$ is determined from Hypothesis $S$ if $\tilde{I} \cong He$ or a covering group of $L_3(4)$ and from Theorem 4.43 if $\tilde{I} \cong L_2(q)$ or $A_7$. Hence the possibilities for $\bar{J}$ are determined in this case as well.

Note that this lemma applies whenever $L$ is not a maximal 2-component of $G$ and does not require $L$ to be part of an unbalancing triple.

However, in the present situation Gilman and Solomon are able to substantially reduce the complete list of possibilities by first establishing some additional properties of $N$. Note that as $T$ leaves $L$ invariant, $T$ centralizes an involution $z$ of $L$, and as $Q$ centralizes $z$, we have $z \in I$.

They prove the following result:

LEMMA 4.75. *The following conditions hold*:

(i) $x$ and $xz$ are not conjugate in $N$;

(ii) $Q \cap Q^y = 1$;

(iii) If $\bar{z} \in Z(\bar{I})$, then $\bar{z} \notin Z(\bar{J})$; and

(iv) $Q \in \mathrm{Syl}_2(C_N(\bar{J}))$.

Since $y$ is an involution, $y$ normalizes $Q \cap Q^y$, so if (ii) fails, $y$ centralizes an involution $u$ of $Q$. But then if we set $K = \langle J^{C_u} \rangle$, we have $I \leqslant K$ and it follows by the usual argument that $(u, y, K_1)$ is an unbalancing triple for any 2-component $K_1$ of $K$. However, as $I$ pumps up nontrivially in $J$ and hence in $K$, we see that $(x, y, L)$ is not a maximal unbalancing triple—contradiction. Similarly if (iii) fails, the same argument applies with $z$ in place of $u$, again a contradiction.

The proof of (i) depends on (ii) and involves a careful comparison of $L_x$ and $L_{zx}$. The point is that if $xz = z^g$ for some $g \in N$, (ii) implies that $L^g$ and $L^{gy}$ are distinct 2-components of $C_{xz}$, and it then follows with slight additional argument that $L$ is not normal in $C_x$. Since $y$ leaves $L$ invariant, it leaves invariant the product $L^*$ of the $C_x$-conjugates of $L$ distinct from $L$. Since $zx$ centralizes $L^*/O(L^*)$, there is thus considerable tension between $L_x$ and $L_{xz}$, which is used to derive a contradiction.

In contrast, (iv) follows quickly from the maximal choice of $Q$.

Lemma 4.75(i) eliminates many of the initial possibilities for $\bar{J}$, since one can check in a number of cases that $\bar{x}$ is, in fact, conjugate to $\bar{x}\bar{z}$ in $\bar{J}$. Likewise Lemma 4.75(iii) rules out the possibility that $\bar{I}$ is a nontrivial cover of $L_3(4)$ and $\bar{J} \cong Suz/Z_2$.

One is left with the following precise list of possibilities for $\bar{I}$ and $\bar{J}$.

PROPOSITION 4.76. *One of the following holds*:

(i) $\bar{I} \cong A_n$, $n = 5$, 6, or 7 and $\bar{J} \cong A_{n+2}, A_{n+4}$, or $\hat{A}_n * \hat{A}_n$;

(ii) $\bar{I} \cong L_2(5)$ or $L_2(7)$ and $\bar{J}$ is a nontrivial covering group of $L_3(4)$;

(iii) $\bar{I} \cong L_2(q)$, $q$ odd, $q > 3$, and $\bar{J} \cong L_2(q^2)$, $PSp_4(q^{1/2})$, $SL_2(q) * SL_2(q)$ or a nontrivial covering group of $L_4(q^{1/2})$, $U_4(q^{1/4})$, or $P\Omega_8^-(q^{1/4})$.

Using previous results, Gilman and Solomon eliminate most of the cases in which $|Z(\bar{J})|$ is even. Indeed, if there is $u \in \mathcal{I}(J)$ centralizing $J/O(J)$ and we let $K$ be the pump-up of $L_2(C_J(u))$ in $L_u$, then Theorems 4.24 and 4.43 allow us to identify $K/O(K)$. Moreover, $u \in K$, so if $K$ is a single 2-

component, $K$ is intrinsic. In the contrary case, $K = K_1 K_2$ with $\bar{J}$ a homomorphic image of $K_1/O(K_1)$ and $u = u_1 u_2$ with $u_i \in \mathscr{I}(K_i)$, $i = 1, 2$. But then, as usual, if $K^*$ denotes the normal closure of $L_2(C_{K_1}(u_1))$ in $L_{u_1}$, then $K^*$ is a single intrinsic 2-component of $C_{u_1}$, which can be identified in the same way as $K$.

But now if $K/O(K)$ or $K^*/O(K^*)$, as the case may be, is of Lie type of odd characteristic, Harris's intrinsic 2-component theorem (Theorem 4.69) implies that $F^*(G)$ is a $K$-group, contrary to Theorem 3.27. Likewise, if either of these groups is isomorphic to $\hat{A}_m$, $m \geqslant 7$, $F^*(G)$ is again a $K$-group, either by the classical involution theorem (the case $m = 7$) or by Solomon's intrinsic $\hat{A}_n$ 2-component theorem (the case $m \geqslant 8$), giving the same contradiction.

However, it is easy to check when $|Z(\bar{J})|$ is even that $K$ (or $K^*$) must have one of these forms except possibly when $\bar{J} \cong SL_2(q)$, $q \leqslant 9$ [using the isomorphisms $\hat{A}_5 \cong SL_2(5)$ and $\hat{A}_6 \cong SL_2(9)$].

The case $\bar{J} \cong PSp_4(q^{1/2})$, $q^{1/2} > 9$, is similarly eliminated, by taking $u \in \mathscr{I}(J)$ such that $L(C_{\bar{J}}(\bar{u})) \cong SL_2(q^{1/2}) * SL_2(q^{1/2})$ and repeating the above argument.

Likewise, one can eliminate the case $\bar{J} \cong A_n$, $n$ odd, $n \geqslant 9$; for then if we take $u \in \mathscr{I}(Q)$, define $K$ as above, and let $K_1$ be a 2-component of $K$, Hypothesis $S$ and Theorem 3.21 imply that $K_1/O(K_1) \cong A_m$, $m \geqslant n$, $m$ odd, or $\hat{A}_n$. The latter possibility is again excluded by the intrinsic $\hat{A}_n$ theorem. But, as $G$ is a minimal counterexample to the $B$- and $U$-theorems, Theorem 4.39 shows that the former is also excluded.

Hence together with Proposition 4.76, this argument yields:

PROPOSITION 4.77. *One of the following holds*:

(i) $\bar{I} \cong A_5$ *and* $\bar{J} \cong A_7$;
(ii) $\bar{I} \cong A_6$ *and* $\bar{J} \cong A_8$ *or* $A_{10}$;
(iii) $\bar{I} \cong L_2(5)$ *or* $L_2(7)$ *and* $\bar{J}$ *is a nontrivial cover of* $L_3(4)$;
(iv) $\bar{I} \cong L_2(q)$ *and* $\bar{J} \cong L_2(q^2)$, $q$ *odd*, $q > 3$;
(v) $\bar{I} \cong L_2(q)$ *and* $\bar{J} \cong SL_2(q) * SL_2(q)$, $q = 5, 7$, *or* $9$; *or*
(vi) $\bar{I} \cong L_2(q)$ *and* $\bar{J} \cong PSp_4(q^{1/2})$, $q = 3^2, 5^2, 7^2$, *or* $9^2$.

Using properties of these groups together with Lemma 4.75(ii) and (iv), one immediately obtains the following additional fact.

LEMMA 4.78. *We have* $\langle x \rangle = Z(T) \cap R$.

Elimination of these remaining cases is extremely subtle. Consider an easy case first.

LEMMA 4.79. *Set* $R^* = N_R(Q)$. *If* $y$ *normalizes* $R^*$, *then the following conditions hold*:

  (i) $\langle Q, Q^y \rangle = Q \times Q^y$;
  (ii) *If* $\bar{J}/Z(\bar{J}) \not\cong L_3(4)$ *or* $A_{10}$, *then* $|R^*: Q| = 2$; *and*
  (iii) *If* $|R^*: Q| = 2$, *then*
      (1) $|Q| = 2$ *and* $|R^*| = 4$;
      (2) $R$ *is dihedral or quasi-dihedral*; *and*
      (3) $\bar{J}$ *is standard in* $\bar{N}$.

Indeed, if $y$ normalizes $R^*$, then $Q^y \leqslant R^*$ and so $Q^y$ normalizes $Q$. As $y$ is an involution, likewise $Q$ normalizes $Q^y$. Since $Q \cap Q^y = 1$ by Lemma 4.74(ii), (i) follows. Furthermore, by Lemma 4.74(iv), $R^*/Q$ acts faithfully on $\bar{J}$ and centralizes $\bar{I}$. Considering the various possibilities for $\bar{I}$ and $\bar{J}$ in Proposition 4.77, this immediately yields (ii). Finally if $|R^*: Q| = 2$, then by (i), $|Q^y| = 2$, whence $|Q| = 2$ and $|R^*| = 4$. Since $R^* = N_R(Q)$, $R$ is thus of maximal class and hence dihedral or quasi-dihedral by [I, 1.16, 1.17]. Since $Q \in \mathrm{Syl}_2(C_N(\bar{J}))$ and $|Q| = 2$, clearly $\bar{J}$ is standard in $\bar{N}$. Hence all parts of (iii) hold.

On the other hand, if $y$ does not normalize $R^*$, the trick is to *recover* these conditions in a proper section of $G$. Indeed, as $y$ normalizes $R$, we have $R^* < R$ and we choose $y_0 \in N_R(R^*) - R^*$ with $y_0^2 \in R^*$ ($y_0$ exists by [I, 1.11]). But now if we set $Q_0 = Q \cap Q^{y_0}$, $R_0 = N_R(Q_0)$, $I_0 = L_2(C_L(Q_0))$, $J_0 =$ the pump-up of $I_0$ in $L_{2'}(C_{Q_0})$, $G_0 = N_G(Q_0)$, $\hat{G}_0 = G_0/Q_0$, and $\tilde{G}_0 = \hat{G}_0/O(\hat{G}_0)$, it is not difficult to prove:

LEMMA 4.80. *The following conditions hold*:

  (i) $Q \leqslant G_0$;
  (ii) $F^*(\tilde{G}_0)$ *is simple*;
  (iii) *The conclusions of Lemma 4.75 hold in* $\tilde{G}_0$ *with* $\tilde{Q}$, $\tilde{N}_0 = N_{\tilde{G}_0}(\tilde{Q}_0)$, $\tilde{x}, \tilde{y}_0, \tilde{I}_0, \tilde{J}_0, \tilde{R}_0$ *in place of* $Q, N, x, y, I, J, R$, *respectively*; *and*
  (iv) $\tilde{y}$ *normalizes* $\tilde{R}_0^* = N_{\tilde{R}}(\tilde{Q})$ *and hence the conclusions of Lemma 4.79 also hold for* $\tilde{Q}, \tilde{N}_0, \tilde{x}, \tilde{y}_0, \tilde{I}_0, \tilde{J}_0$, *and* $\tilde{R}_0$.

For uniformity of notation, we set $G = \tilde{G}_0$, $y = y_0$, etc., if $y$ normalizes $R^*$. Thus in all cases, *there exists a section* $\tilde{G}_0$ *satisfying the conditions of Lemmas 4.75 and 4.79*. We also put $\bar{N}_0 = \tilde{N}_0/O(\tilde{N}_0)$. Thus $\bar{J}_0 \cong \bar{J}$.

The existence of $\widetilde{G}_0$ enables Gilman and Solomon to eliminate some of the possibilities for $\bar{J}$.

PROPOSITION 4.81. *We have* $\bar{J} \not\cong A_7, A_{10}$, *or a nontrivial cover of* $L_3(4)$.

To establish the proposition, Gilman and Solomon focus first on $\bar{N}_0$ and $\bar{J}_0 \cong \bar{J}$. First consider the $A_7$ case. Then $\bar{I}_0 \cong A_5$ and by Lemma 4.78, $|\bar{Q}| = 2$, $\bar{J}_0$ is standard in $\bar{N}_0$, and $\bar{Q} = C_{\bar{N}_0}(\bar{J}_0)$. But then $F^*(\bar{N}) \cong A_9$ or $He$ by Theorem 2.160. However, the latter possibility is excluded here as $\bar{I}_0$ is a component of $C_{\bar{J}_0}(\langle \bar{x}, \bar{Q} \rangle)$ and $\text{Aut}(He)$ contains no 2-subgroup whose centralizer has an $A_5$ component. Thus $F^*(\bar{N}) \cong A_9$ and as $I_0$ centralizes $Q_0$, it follows that $C_{Q_0}$ contains a 2-component $K_0$ with $K_0/O(K_0) \cong A_9$.

But now if we consider a chain of subgroups $Q_0 > Q_1 > \cdots > Q_r$ with $|Q_i : Q_{i+1}| = 2$ and $Q_r = \langle u \rangle$ of order 2, then by repeated application of Hypothesis $S$ and the known structure of the centralizers of involutions acting on quasisimple $K$-groups, we can determine the isomorphism type (mod core) of a 2-component $K_i$ of the pump-up of $K_{i-1}$ in $L_{2'}(C_{Q_i})$. Moreover, if any $K_i/O(K_i) \cong \hat{A}_{m_i}$, we easily reduce to Solomon's intrinsic $\hat{A}_n$-theorem, in which case $F^*(G)$ is a $K$-group, contrary to Theorem 3.27. It follows that each $K_i/O(K_i) \cong A_{m_i}$, $m_i$ odd, $m_i \geq 9$. In particular, this is the case for $K_r$, which is a 2-component of $C_u$ and as $G$ is a minimal counterexample to the $B$- and $U$-theorems, this contradicts Theorem 4.39.

The preceding procedure is not available if $\bar{J}_0/Z(\bar{J}_0) \cong L_3(4)$ or $\bar{J}_0 \cong A_{10}$. To eliminate these cases, Gilman and Solomon argue by a delicate fusion analysis that there exists no group $G_0$ satisfying the conditions of Lemmas 4.74 and 4.78 with $\bar{J}_0 \cong A_{10}$ or a nontrivial cover of $L_3(4)$.

Thus by Proposition 4.77 and Lemma 4.79(ii), they are left with the following cases:

(a) $\bar{J} \cong A_8$, $SL_2(q) * SL_2(q)$, $q = 5, 7$, or $9$, $PSp_4(q)$, $q^{1/2} = 3, 5$,
7, or 9, or $L_2(q^2)$, $q$ odd, $q > 3$; and                                           (4.43)
(b) $R^* = N_R(Q) \times \langle x \rangle$.

Gilman and Solomon next manage to eliminate the $L_2(q^2)$ possibility by altering the choice of $Q$.

PROPOSITION 4.82. *We can choose* $Q$ *so that* $\bar{J} \not\cong L_2(q^2)$.

Suppose $\bar{J}$ and hence $\bar{J}_0 \cong L_2(q^2)$. Then again $\tilde{J}_0$ is standard in $\tilde{G}_0$ with

$Z_2 \cong \tilde{Q} \in \mathrm{Syl}_2(C_{\tilde{G}_0}(\tilde{J}_0))$. We conclude at once now from Theorem 4.43 and our conditions that the only possibility is $\tilde{K} = F^*(\tilde{G}_0) \cong PSp_4(q)$. There is then $\tilde{u} \in \mathscr{I}(\tilde{K})$ such that $\tilde{H} = L(C_{\tilde{K}}(\tilde{u})) \cong SL_2(q) * SL_2(q)$ and $\tilde{I}_0 = L(C_{\tilde{H}}(\tilde{x}))$. The basic idea is now to replace $Q$ by $Q^* = Q\langle u \rangle$, where $u$ is an involution of $C_x \cap N_G(Q_0)$ which maps on $\tilde{u}$. If $Q^*$ has the same maximal properties as $Q$, then the corresponding $\bar{J}^* \cong SL_2(q) * SL_2(q)$. On the other hand, if this is not the case, one must suitably expand $Q^*$ to reach a similar conclusion. Hence for a suitable choice of $Q$, the corresponding component $\bar{J} \cong SL_2(q) * SL_2(q)$.

The three remaining cases are by far the most difficult. It is here that the analysis most closely resembles Harris's. First, Gilman and Solomon eliminate the $SL_2(q) * SL_2(q)$ case by an elaborate fusion analysis.

On the other hand, when $\bar{J} \cong A_8$ or $PSp_4(q)$ and $|R| \geqslant 8$, a long argument, again involving fusion analysis, enables them to conclude that

$$F^*(\tilde{G}_0) \cong A_{10}, L_4(q), \text{ or } U_4(q). \tag{4.44}$$

Since $G = \tilde{G}_0$ if $|Q| = 2$, this eliminates the case $|R| \geqslant 8$, $|Q| = 2$.

Now an equally lengthy fusion analysis is required to eliminate the case $|R| \geqslant 8$, $|Q| \geqslant 4$, so they are left with the final possibility

$$|R| = 4, |Q| = 2, \text{ and } G = \tilde{G}_0. \tag{4.45}$$

In this case, setting $R = \langle x, u \rangle$, they argue, on the one hand, that $u \in F^*(G)$ for a suitable choice of $u$, and, on the other, by a third very long fusion analysis and transfer, that $u \notin F^*(G)$.

The details of the total argument are too technical to describe here.

It should be pointed out that *if* there were a classification of groups $G$ with $F^*(G)$ simple in which the centralizer of some involution contained a 2-component $K$ with $K/O(K)$ a pump-up of $A_8$, $SL_2(q)$, or $PSp_4(q)$ (including the trivial pump-up) similar to our previous 2-component characterization theorems, such a result would force $F^*(G)$ to be a $K$-group, giving an immediate contradiction to Theorem 3.27.

We are not quite ready to apply Theorem 4.71—one bolt still remains to be set in place. Recall that, in contrast with the notion of maximal 2-component, the definition of maximal unbalancing triple (Definition 3.29) is independent of the pumping-up process. However, effective application of Theorem 4.71 requires one to first relate the latter two concepts.

Gilman and Solomon prove the following theorem:

THEOREM 4.83. *If G is a minimal counterexample to the B-theorem and U-theorem and $(x, y, L)$ is an unbalancing triple in G, then there exists a maximal unbalancing triple $(x^*, y^*, L^*)$ in G such that $L \ll L^*$.*

We have given the result the status of a theorem, but the proof is straightforward. The idea is to show that if $(x, y, L)$ is not a maximal unbalancing triple, there is $t \in \mathscr{I}(C_x)$ centralizing $L/O(L)$ such that if $J$ is a 2-component of the normal closure of $L_{2'}(C_L(t))$ in $L_t$, then $(t, u, J)$ is an unbalancing triple for some $u \in \mathscr{I}(C_t)$ and one of the following holds:

(1) $|J/O(J)| > |L/O(L)|$;

(2) $|J/O(J)| = |L/O(L)|$ and $m(C_{C_t}(J/O(J))) > m(C_{C_x}(L/O(L)))$;

or

(3) $|J/O(J)| = |L/O(L)|$, $m(C_{C_t}(J/O(J))) = m(C_{C_x}(L/O(L)))$, and a Sylow 2-subgroup of $N_{C_t}(J)$ has greater order than one of $N_{C_x}(L)$.

$(4.46)$

In particular, $L < J$. But then repeating the argument on $J$ if $(t, u, J)$ is not a maximal unbalancing triple, and continuing the process as long as possible, we eventually reach an unbalancing triple $(x^*, y^*, L^*)$ with $L \ll L^*$ such that (4.46) fails for any $t^* \in \mathscr{I}(C_{x^*})$ centralizing $L^*/O(L^*)$, thus forcing $(x^*, y^*, L^*)$ to be a maximal unbalancing triple.

Assuming $(x, y, L)$ is not a maximal unbalancing triple, the only information one needs to establish (4.46) is: (1) either condition (a) or (b) of Definition 3.29 fails for $(x, y, L)$, and (2) if $t \in \mathscr{I}(C_x)$ centralizes $y$ as well as $L/O(L)$, then $(t, y, J)$ *is* an unbalancing triple for any 2-component $J$ of the normal closure of $L_{2'}(C_L(t))$ in $L_t$ (the latter following as in Lemma 4.72).

Now finally Gilman and Solomon have all the pieces in place to derive the following further major reduction in the *B*- and *U*-theorems.

THEOREM 4.84. *Assume Hypothesis S and let G be a minimal counterexample to the B-theorem and U-theorem. Then the following conditions hold:*

(i) *If $(x, y, L)$ is an unbalancing triple in G with $L/O(L) \cong L_2(q)$, q odd, then $q = 5$ or $7$; and*

(ii) *G contains an unbalancing triple $(x, y, L)$ with $L/O(L) \cong He$ or a cover of $L_3(4)$.*

Indeed, suppose $(x, y, L)$ is an unbalancing triple with $L/O(L) \cong L_2(q)$,

$q$ odd, or $A_7$. By Theorem 4.83, there exist unbalancing triples $(x_i, y_i, L_i)$ in $G$, $1 \leqslant i \leqslant n$ with $(x_1, y_1, L_1) = (x, y, L)$ such that $L_i < L_{i+1}$ and $(x_n, y_n, L_n)$ is a maximal unbalancing triple. Choose $n$ minimal. By Theorem 4.71, $L_n$ is a maximal 2-component of $G$. Hence by (4.38), $L_n/O(L_n) \not\cong L_2(q^*)$, $q^*$ odd, or $A_7$, so $L_n/O(L_n) \cong He$ or a cover of $L_3(4)$ by (4.37). In particular, (ii) holds under the given assumption on $L$. Furthermore, $n \geqslant 2$ and it follows again by (4.37) together with the minimality of $n$ that $L_{n-1}/O(L_{n-1}) \cong L_2(5)$, $L_2(7)$, or $A_7$, and we conclude at once that $L_1/O(L_1) = L/O(L) \cong L_2(5)$, $L_2(7)$, or $A_7$. In particular, (i) holds.

On the other hand, if $G$ possesses no unbalancing triples $(x, y, L)$ with $L/O(L) \cong L_2(q)$, $q$ odd, or $A_7$, (ii) follows at once from (4.37).

REMARK. Note that if in the above proof, $L/O(L) = L_1/O(L_1) \cong A_7$, then necessarily $L_n/O(L_n) \cong He$. Hence if $G$ contains an unbalancing triple $(x, y, L)$ with $L/O(L) \cong A_7$, it must possess a maximal unbalancing triple $(x^*, y^*, L^*)$ with $L^*/O(L^*) \cong He$.

As a consequence of Theorem 4.84(i), (4.37) assumes the following sharper form: If $(x, y, L)$ is an unbalancing triple in our minimal counterexample $G$, then

$$L/O(L) \cong L_2(5), \ L_2(7), A_7, He, \text{ or a covering group of } L_3(4). \quad (4.47)$$

Moreover, by Theorem 4.84(ii), for some choice of $(x, y, L)$ we have

$$L/O(L) \cong He \text{ or a covering group of } L_3(4). \quad (4.48)$$

Thus (4.47) and (4.48) give the final list of possibilities that must be eliminated to complete the reduction of the $B$- and $U$-theorems to Hypothesis $S$. However, in their present form, these conditions include a possibility which for technical reasons considerably complicates the subsequent analysis. Indeed, in treating (4.47) and (4.48), it turns out that the most effective signalizer functor with which to work is the 2-*balanced* functor. However, to use this functor, one must know that the group $G$ is 2-balanced, which by [I, 4.64] will be the case if every element of $\mathscr{L}(G)$ is locally 2-balanced. But among the remaining possible elements of $\mathscr{L}(G)$ is the group $A_7$, which unfortunately is not locally 2-balanced [viz., $\Delta_{A_5}(\langle(12)(34), (13)(24)\rangle) = \langle(567)\rangle \cong Z_3$], so *a priori* our group $G$ may not be 2-balanced.

However, using the results they have already established, Gilman and Solomon are, in fact, able to establish this critical property of $G$. This is their final result, which because of its significance for Griess and Solomon's analysis of the *He* and $L_3(4)$ cases, we outline in some detail.

THEOREM 4.85. *Assume Hypothesis S. If G is a minimal counterexample to the B-theorem and U-theorem, then G is 2-balanced.*

Indeed, if false, then by definition there is $x \in \mathscr{I}(G)$ and a four-subgroup $Y$ of $C_x$ such that

$$D = \Delta_G(Y) \cap C_x \nleqslant O(C_x). \tag{4.49}$$

Hence by [I, 4.64], $\overline{C}_x = C_x/O(C_x)$ contains a nonlocally 2-balanced component $\overline{L}$. The argument of [I, 4.58] (of which [I, 4.64] is an extension) implies that $\overline{L}$ can be chosen so that $D$ normalizes $L$ and does not centralize $\overline{L}$. Since $\Delta_G(Y) = \bigcap_{y \in Y^{\#}} O(C_y)$ by definition, it follows that $O(C_y) \cap C_x$ does not centralize $\overline{L}$ for any $y \in Y^{\#}$, whence $(x, y, L)$ is an unbalancing triple in $G$ for each $y \in Y^{\#}$. Hence $\overline{L}$ is isomorphic to one of the groups listed in (4.47). However, one checks directly that $A_7$ is the only one that is not locally 2-balanced. Thus, in fact,

$$\overline{L} \cong A_7. \tag{4.50}$$

More precisely, $\overline{L}$ is not locally 2-balanced "with respect to $\overline{Y}$," which implies that $\overline{Y}$ induces inner automorphisms on $\overline{L}$.

Let $T \in \mathrm{Syl}_2(N_{C_x}(L))$ with $Y \leqslant T$, and among all such choices of $x, Y, L$, assume $T$ is chosen to have maximal order.

The argument emulates portions of the proof of Theorem 4.71. First, as $F^*(G)$ is not a *K*-group, $L$ is not a maximal 2-component of $G$ by Theorem 4.43. But then by Theorem 4.71, $(x, y, L)$ is not a maximal unbalancing triple in $G$. Therefore either (a) or (b) of Definition 3.29 must fail. We claim (a) fails, so assume not, in which case there is $z \in \mathscr{I}(Z(T))$ centralizing $\overline{L}$ such that if $J$ is the normal closure of $L_2(C_L(z))$ in $L_z$, then $J/O(J) \cong \overline{L} \cong A_7$ and $T \notin \mathrm{Syl}_2(N_{C_z}(L))$. But if $E = C_D(z)$, then again as in Lemma 4.72, $E$ normalizes, but does not centralize $J/O(J)$. Hence the triple $x, Y, J$ has the same properties as $x, Y, L$ and we see that our maximal choice of $T$ is contradicted.

Thus (a) fails and we conclude that there is $t \in \mathscr{I}(T)$ centralizing $\overline{L}$ such that if $J$ denotes the pump-up of $L_2(C_L(t))$ in $L_t$, then $x$ does not centralize $J/O(J)$. As noted earlier, Lemma 4.74 can be applied to $(x, y, L)$ under these conditions. Hence with $Q, R, I, J, N$, and $\overline{N}$ as in that lemma [thus $R = C_T(L/O(L))$, $1 \neq Q \leqslant R$, $I = L_2(C_L(Q))$, $N = N_G(Q)$, and $\bar{N} = N/O(N)$], it follows that $\bar{I} \cong A_7$ is standard in $\bar{J}\langle \bar{x} \rangle$, so $\bar{J} \cong A_9, A_{11}$, or $He$ by Theorem 2.160. However, as $G$ is a minimal counterexample to the $B$- and $U$-theorems, the first two possibilities are again ruled out by Theorem 4.39. Thus

$$\bar{J} \cong He. \tag{4.51}$$

But $He$ is a locally 2-balanced group, which puts a critical restriction on the action of $Y$. Indeed, it implies that $Y$ normalizes no nontrivial subgroup of $Q$.

Assume false, in which case $Y$ centralizes an involution $v$ of $Q$. Denote the pump-up of $J$ in $L_v$ by $K$. By $L$-balance and Hypothesis $S$, together with the known structure of centralizers of involutions acting on simple $K$-groups, each component of $K/O(K)$ is necessarily isomorphic to $He$ and so is locally 2-balanced. However, as $v$ centralizes $Y$ and $I \leqslant J \leqslant K$, it follows, as usual, that $E = C_D(v)$ does not centralize $K/O(K)$ and hence $K/O(K)$ must contain a nonlocally 2-balanced component—contradiction. In particular, as $Q \neq 1$, $Y$ does not normalize $Q$.

This in turn forces $R$ to be nonabelian. Indeed, since $Y$ induces inner automorphisms on $\bar{L}$ and $Y \leqslant T$, $Y \leqslant (T \cap L) R = (T \cap L) \times R$. But then if $R$ were abelian, $Y$ would centralize $R$ and hence would normalize $Q$ $(\leqslant R)$, which is not the case.

Gilman and Solomon now complete the proof by showing, to the contrary, that the relationship between $Y$ and $Q$ forces $R$ to be abelian. First of all, it implies that

$$Z(T) \cap R = \langle x \rangle. \tag{4.52}$$

Indeed, $Z = Z(T) \cap R \leqslant N$ and $Z$ centralizes $\bar{I}$. However, by the structure of $\text{Aut}(He)$, $Z = \langle x \rangle \times C_Z(\bar{J})$. But by Lemma 4.74(iv), $Q \in \text{Syl}_2(C_N(\bar{J}))$, so $C_Z(\bar{J}) \leqslant Q$. Since $Z$ centralizes $Y$ and $Y$ normalizes no nontrivial subgroup of $Q$, this forces $C_Z(\bar{J}) = 1$, so $Z = \langle x \rangle$, as asserted.

Now let $z$ be an involution of $Z(T) \cap L = Z(T) \cap I$. Again by the structure of $\text{Aut}(He)$, $x$ and $xz$ are conjugate in $J\langle x \rangle$. Since $\langle x, z \rangle \leqslant Z(T)$ and $T \in \text{Syl}_2(C_x)$, we obtain by a Frattini argument:

$$x^g = xz \qquad \text{for some } g \in N_G(T). \tag{4.53}$$

If $R \cap R^g \neq 1$, then as $R \cap R^g \lhd T$, it would follow that $R \cap R^g$

contains an involution of $Z(T)$. But by (4.52), $x$ is the unique involution of $Z(T) \cap R$ and as $g$ does not centralize $x$, this is a contradiction. We conclude that

$$RR^g = R \times R^g. \tag{4.54}$$

A similar argument shows that also

$$(T \cap L) R^g = (T \cap L) \times R^g. \tag{4.55}$$

Finally as $R = C_T(\bar{L})$, it follows from (4.54) and (4.55) that $R^g$ acts faithfully on $\bar{L}$ and centralizes its Sylow 2-subgroup $T \cap L (\cong D_8)$. Since $\Sigma_7$ has $D_8 \times Z_2$ Sylow 2-subgroups, the only possibility is that $R^g$ is isomorphic to a subgroup of $Z_2 \times Z_2$. We thus conclude that $R^g$, and hence also $R$, is abelian, as required. This completes the discussion of Theorem 4.85.

## 4.7. The Held and $L_3(4)$ Cases

Now Griess and Solomon are poised for the final attack on the *B*- and *U*-theorems, eliminating unbalancing triples of either type *He* or a covering group of $L_3(4)$.

They first prove the following theorem:

THEOREM 4.86. *Assume Hypothesis S and let G be a minimal counterexample to the B-theorem and U-theorem. If $(x, y, L)$ is an unbalancing triple in G, then $L/O(L) \cong L_2(5)$, $L_2(7)$, or a covering group of $L_3(4)$.*

Indeed, assume false. By Theorem 4.84 and the remark following it, $G$ contains an unbalancing triple $(x, y, L)$ with $L/O(L) \cong He$. Furthermore, among all $x \in \mathscr{I}(G)$ such that $C_x$ has a 2-component $L$ with $L/O(L) \cong He$, we choose $x$ and $L$ so that $C_G(L/O(L))$ has maximal 2-rank, and subject to this condition, so that a Sylow 2-subgroup $S$ of $N_G(L)$ has maximal order. They derive a contradiction by the same general pattern of proof as in the *Mc*, *Ly* cases of Theorems 4.24, 4.39, and Proposition 4.45.

First of all, in view of Hypothesis $S$ and the known structure of centralizers of involutions acting on quasisimple *K*-groups, if $K \in \mathscr{L}(G)$ with $K \cong He$, then necessarily $K \in \mathscr{L}^*(G)$. Hence, as usual, Solomon's maximal 2-component theorem (Theorem 4.9) implies that the centralizer of every involution of $G$ has at most one 2-component of type *He*.

Furthermore, as $G$ is 2-balanced by Theorem 4.85, Griess and Solomon can work here with the standard 2-balanced signalizer functor. The analysis is facilitated by the fact that $S$ is 3-*connected* (i.e., any two $E_8$-subgroups of $S$ can be included in a sequence of $E_8$-subgroups of $S$, successive members of which centralize each other). Indeed, as $x \in R = C_S(L/O(L))$, $R \neq 1$ and as $R \lhd S$, it follows that $R \cap Z(S) \neq 1$. Also $S/R$ is isomorphic to a Sylow 2-subgroup of $He$ or $\mathrm{Aut}(He)$ and we check that $S/R$ is connected. Since $R \cap L = 1$, we conclude easily that $S$ is 3-connected.

In particular, any four-subgroup $D$ of $S$ is contained in an $E_{16}$-subgroup $A$ of $S$. The goal is, of course, to show that the 2-balanced $A$-signalizer functor $\theta$ is necessarily trivial. Assuming false and setting $M = N_G(\theta(G; A))$, where $\theta(G; A)$ as usual denotes the completion of $\theta$, one easily proves the following result, using the 2-balance of $G$ and the 3-connectedness of $S$:

LEMMA 4.87. *The following conditions hold*:

  (i) $\Gamma_{S,3}(G) \leqslant M$;
  (ii) $S \in \mathrm{Syl}_2(G)$; *and*
  (iii) $M$ *controls the $G$-fusion of $S$.*

This immediately yields:

LEMMA 4.88. *The following conditions hold*:

  (i) $L \leqslant M$;
  (ii) *If $L^* = \langle L^M \rangle$, then $L^*$ is a 2-component of $M$ and $L$ covers $L^*/O(L^*)$; and*
  (iii) $R \in \mathrm{Syl}_2(C_M(L^*/O(L^*)))$.

Since $S \in \mathrm{Syl}_2(G)$, it is immediate that $L$ is contained in a single 2-component $L^*$ of $M$, and it follows with the aid of Hypothesis $S$ that necessarily $L^*/O(L^*) \cong He$. Hence $L$ covers $L^*/O(L^*)$ and so $R = C_S(L^*/O(L^*))$.

But by Lemma 4.87(iii), $R$ is strongly closed in $S$ with respect to $G$. We can therefore apply Goldschmidt's general strong closure theorem [I, 4.156] to conclude:

LEMMA 4.89. $C_R^{(\infty)}$ *is normal in $G$.*

This yields an immediate contradiction, for as $F^*(G)$ is simple and $R \neq 1$, the lemma forces $C_R^{(\infty)} = 1$. But $C_R$ covers $L^*/O(L^*)$ and so $C_R^{(\infty)} \neq 1$.

We conclude that $\theta$ must be trivial and so we obtain the following result:

PROPOSITION 4.90. $\Delta_G(D) = 1$ *for every four-subgroup $D$ of $S$.*

By the structure of *He*, $S$ contains an $E_8$-subgroup $B$ with $x \in B$ such that if $I = L_2'(C_L(B))$, then $I/O(I) \cong L_3(4)/Z_2 \times Z_2$. Using the proposition, Griess and Solomon prove in the standard way:

PROPOSITION 4.91. *For every* $b \in B^\#, O(\langle I^{L_b} \rangle) = 1$. *In particular, we have $O(L) = 1$.*

It follows now as in the proof of Theorem 4.39 that $L$ is standard in $G$. Hence Hypothesis $S$ is satisfied and yields that $F^*(G)$ is a $K$-group, contrary to Theorem 3.27, thus completing the proof of Theorem 4.86.

Now for the final step! Griess and Solomon eliminate the remaining possibilities for unbalancing triples:

THEOREM 4.92. *Assume Hypothesis $S$ and let $G$ be a minimal coun-terexample to the B-theorem and U-theorem. Then $G$ contains no unbalancing triple $(x, y, L)$ with $L/O(L) \cong L_2(5), L_2(7)$ or covering group of $L_3(4)$.*

Unfortunately, the argument is much more complicated than in the *He* case. Assuming the theorem false, Theorems 4.84 and 4.86 imply that $G$ possesses an unbalancing triple $(x, y, L)$ with $L/O(L)$ a covering group of $L_3(4)$. If one tries to emulate the proof of Theorem 4.86, here are some of the difficulties to be overcome:

1. Covering groups of $L_3(4)$ possess proper pump-ups [$L_3(16)$, *He*, *Suz*, and *ON*]. Even though $L$ is a maximal 2-component of $G$, it may well happen that $L$ pumps-up properly in *some* proper subgroup of $G$ containing $L\langle x \rangle$. Similarly if $K$ is a 2-component of $C_t$ for $t \in \mathcal{I}(G)$ with $K/Z^*(K) \cong L_3(4)$ and $t$, $K$ is not part of an unbalancing triple, then $K$ need not be a maximal 2-component. To control the latter situation, it is necessary to keep the unbalancing involution $y$ in the picture.

2. In contrast to *He*, the centralizer of every 2-subgroup of $L_3(4)$ is 2-constrained.

3. The group $L_3(4)/Z_4$ has 2-rank 3, so it is entirely possible that $G$ itself has 2-rank 3. But in this case the signalizer functor theorem is applicable only if $G$ is *balanced* (rather than 2-balanced).

4. A Sylow 2-subgroup $S$ of $N_G(L)$ need not be 3-connected. This is

obvious if $G$ has 2-rank 3, but it can occur even when $L/O(L) \cong L_3(4)$, in which case $m(G) \geqslant m(L \times \langle x \rangle) = 5$. For example, if $S = (S \cap L)\langle x, y \rangle$, a Sylow 2-subgroup of $C_S(y)$ is a subgroup of $Q_8 \times Z_2 \times Z_2$ [as $y$ induces a unitary automorphism on $L/O(L)$], whence $m(C_S(y)) = 3$. Clearly this implies that $S$ is not 3-connected.

Thus Griess and Solomon have their work cut out for them! For, given all these constraints, effective application of the 2-balanced signalizer functor requires the elimination of a large number of highly complex configurations of subgroups. The argument we present here will be somewhat simpler than the Griess–Solomon published proof. It is based on the observation that if one first reduces to the case in which $G$ is balanced (rather than 2-balanced) with respect to *certain* elementary abelian 2-subgroups, the complexity of the configurations requiring close analysis is correspondingly reduced.

Because of the crucial importance of Theorem 4.92 for completion of the reduction of the $B$- and $U$-theorems to Hypothesis $S$, we discuss its proof in considerable detail.

Choose $(x, y, L)$ so that a Sylow 2-subgroup $S$ of $N_G(L)$ has maximal order. Without loss, $\langle x, y \rangle \leqslant S$. Again put $R = C_S(L/O(L))$; also set $T = (S \cap L)R$ and fix this notation.

Again by Theorem 4.9, the centralizer of every involution of $G$ contains at most one 2-component $K$ with $K/O(K) \in \mathcal{L}^*(G)$ and $K/Z^*(K) \cong L_3(4)$.

Note first that if $m(R) \leqslant 2$, then in view of remark 2 above, $T$ will not contain an $E_8$-subgroup $B$ such that $L_2(C_L(B)) \neq 1$ (as was true in the prior $He$ case). On the other hand, *if* $m(R) \geqslant 3$ and, in addition, $S$ is 3-connected, then the proof of Theorem 4.86 can be repeated provided we first establish an analog of Lemma 4.88. (It is this fact which allows us to reduce to the stated balance conditions.) Indeed, under the assumption of 3-connectedness, it follows as before that $\Delta_G(D) = 1$ for every four-subgroup $D$ of $S$. If also $m(R) \geqslant 3$, there is $E_8 \cong B \leqslant R$ such that $I = L_2(C_L(B))$ covers $L/O(L)$. As before, this forces $O(L) = 1$, which in turn yields that $L$ is standard. But now Hypothesis $S$ implies that $F^*(G)$ is a $K$-group, contradiction.

We shall, in fact, establish a slight extension of Lemma 4.88, which will be needed for the subsequent argument. Lemma 4.88 was predicated on the assumption that $\Gamma_{S,3}(G)$ was contained in a proper subgroup $M$ of $G$ (Lemma 4.87). Our extension deals with $T$ rather than $S$, and we assume that one of the following holds:

(1) $m(T) \geqslant 4$ and $\Gamma_{T,3}(G)$ is contained in a proper subgroup
   $M$ of $G$; or                                                                      (4.56)
(2) $\Gamma_{T,2}(G)$ is contained in a proper subgroup $M$ of $G$.

We establish the extension in three steps. For simplicity, set $k = 3$ or $2$ according as (4.56)(1) or (2) holds. Since $m(T) \geqslant m(T \cap L) \geqslant 3$, we have $m(T) > k$ in all cases.

LEMMA 4.93.    *We have $LS \leqslant M$.*

Indeed, by [I, 2.18], $L_3(4)$ is generated by its minimal parabolics containing a fixed Sylow 2-subgroup [as $L_3(4)$ has Lie rank 2]. It follows therefore that $\Gamma_{T,k}(L)$ covers $L/O(L)$. But also as $m(T) > k$, $O(L) \leqslant \Gamma_{T,k}(L)$ by [I, 4.13]. Since $\Gamma_{T,k}(G) \leqslant M$ by assumption, we conclude that $L \leqslant M$. Likewise $S \leqslant N_G(T) \leqslant M$.

Set $L^* = \langle L^M \rangle$. As before, $L^*$ is either a single $x$-invariant 2-component of $M$ or the product of two such 2-components interchanged by $x$; and in either case $L$ is a 2-component of $C_{L^*}(x)$.

LEMMA 4.94.    *If $L^*$ is a single 2-component, then $C_M(L^*/O(L^*))$ has even order.*

Suppose false, and set $\bar{M} = M/O(M)$, so that $\bar{M} \leqslant \mathrm{Aut}(\bar{L}^*)$, $\bar{L}^*$ is simple, $\bar{M}$ has the *B*-property, and $\bar{L}$ is a component of $C_{\bar{L}^*}(\bar{x})$. Hence by Theorem 4.23 and the known structure of the centralizers of involutions acting on quasisimple *K*-groups, we conclude that

$$\bar{L}^* \cong L_3(16), He, Suz, \text{ or } ON. \tag{4.57}$$

Expand $S$ to $P \in \mathrm{Syl}_2(M)$. In each of the first three cases, we check, using the known structure of $\mathrm{Aut}(\bar{L}^*)$, that $P$ is 3-connected. But now we reach the same conclusions as in Lemma 4.87, with $P$ in place of $S$. In particular, $P \in \mathrm{Syl}_2(G)$. These conditions exclude $\bar{M} \cong \mathrm{Aut}(He)$. Indeed, otherwise $M$ and hence $G$ possesses a normal subgroup $G_0$ of index 2. But $L \leqslant G' \leqslant G_0$ and in the present *He* case $L/O(L) \cong L_3(4)/Z_2 \times Z_2$ with $x \in L$. Since $G$ is a minimal counterexample to the *B*-theorem, it follows that $L$ is standard in $G_0$ and hence by Hypothesis S, $F^*(G_0) = F^*(G) \cong He$, contrary to Theorem 3.27.

Since $\bar{M} \not\leqslant \mathrm{Aut}(He)$, we conclude now in all cases (including $\bar{L}^* \cong ON$) that $O(C_{\bar{M}}(\bar{t})) = 1$ for every $t \in \mathcal{I}(M)$. We claim that this implies that for each $t \in \mathcal{I}(M)$

$$O(C_t) \leqslant O(M). \tag{4.58}$$

Since $O(C_{\bar{M}}(\bar{t})) = 1$, it will suffice to show that $O(C_t) \leqslant M$. But by the known structure of the centralizers of involutions acting on $\bar{L}^*$, it follows that for a suitable $M$-conjugate of $t$, which without loss we can assume to be $t$ itself, $T_1 = C_T(t)$ has rank $>k$. Since $T_1$ acts on $O(C_t)$ and $\Gamma_{T_1,k}(G) \leqslant M$, the desired conclusion follows, as usual, from [I, 4.13].

Observe next that if $\bar{L}^* \cong ON$, then $L/O(L) \cong L_3(4)/Z_4$, $T \leqslant L$, $|S/T| \leqslant 4$, $S \in \mathrm{Syl}_2(C_x)$ and $\langle x \rangle = \Omega_1(Z(S))$. In particular, $\langle x \rangle$ char $Z(S)$, so $S \in \mathrm{Syl}_2(G)$. Thus $P = S \in \mathrm{Syl}_2(G)$ in this case. Likewise $P$ is connected by the structure of $\mathrm{Aut}(ON)$. Hence in all cases $P \in \mathrm{Syl}_2(G)$ and $P$ is connected.

We shall argue finally that $G$ is a balanced group, which as $P$ is connected, will imply that $\Gamma_{P,2}(G) \leqslant M$. But then $O(M) \leqslant O(G) = 1$, by the general form of Aschbacher's theorem [I, 4.28]—contradiction—and the lemma will be proved.

Suppose then that $D = O(C_t) \cap C_u \not\leqslant O(C_u)$ for some pair of commuting involutions $t, u \in G$. Since $M$ contains a Sylow 2-subgroup of $G$, we can assume without loss that $\langle t, u \rangle \leqslant M$. By [I, 4.58], $(u, t, K)$ is an unbalancing triple for some 2-component $K$ of $C_u$. Thus $D$ normalizes $K$, but does not centralize $K/O(K)$. Furthermore, by Theorem 4.86, $K/O(K) \cong L_2(5)$, $L_2(7)$, or a covering group of $L_3(4)$.

It will suffice to show that $M$ contains a Sylow 2-subgroup $Q$ of $C_u$. Indeed, in that case $Q_1 = Q \cap K \in \mathrm{Syl}_2(K)$ and $Q_1 \leqslant M$. But $D \leqslant O(C_t) \leqslant O(M)$ by (4.58) and so $[D, Q_1]$ has odd order. However, given the possibilities for $K/O(K)$, this immediately forces $D$ to centralize $K/O(K)$, contradiction.

Let $Q \in \mathrm{Syl}_2(C_u)$ with $Q_0 = Q \cap M \in \mathrm{Syl}_2(C_M(u))$. Without loss, $Q_0 \leqslant P$. As noted above, $m(Q_0) \geqslant k + 1 \geqslant 3$. But then if $\bar{L}^* \not\cong ON$, we have $N_G(Q_0) \leqslant \Gamma_{P,3}(G) \leqslant M$, whence $Q_0 = N_Q(Q_0)$ and so $Q = Q_0$ by [I, 1.11], as required. We can therefore assume that $\bar{L}^* \cong ON$. If $u$ is a 2-central involution of $M$, then $Q_0 = Q = P$ as $Q_0 \in \mathrm{Syl}_2(C_M(u))$, and again we are done.

Since $ON$ has only one class of involutions, we are left with the single possibility that $\bar{L}^* \cong ON$ with $u$ inducing an outer automorphism on $\bar{L}^*$. But $ON$ also has only one class of outer automorphisms of order 2, with centralizers isomorphic to $J_1$. Thus $I = L_2(C_M(u)) \cong J_1$. On the other hand, $m(P) = 4$ in this case, whence $m(C_u) \leqslant 4$, which implies that $K$ is the unique 2-component of its isomorphism type in $C_u$. Hence $I$ normalizes $K$. But $L_2(5)$, $L_2(7)$, and $L_3(4)$ do not involve $J_1$, so $I$ must centralize $K/O(K)$, whence $m(C_u) > 4$—contradiction. Thus this case does not arise, and so $M$ contains a Sylow 2-subgroup of $C_u$, as required.

Now we can establish the desired extension of Lemma 4.88.

LEMMA 4.95. *The following conditions hold*:

   (i) $L$ covers $L^*/O(L^*)$;
   (ii) $S \in \mathrm{Syl}_2(M)$; *and*
   (iii) $R \in \mathrm{Syl}_2(C_M(L^*/O(L^*)))$.

Again expand $S$ to $P \in \mathrm{Syl}_2(M)$, let $K$ be a 2-component of $L^*$, set $K^* = \langle K^P \rangle$, and let $K = K_1, K_2, \ldots, K_n$ be the 2-components of $K^*$. Suppose first that $n = 1$. Then $L^*$ is not the product of two 2-components interchanged by $x$, so $K = K^* = L^*$. By the previous lemma, $R^* = C_P(L^*/O(L^*)) \neq 1$. Since $R^* \lhd P$, there is an involution $x^* \in Z(P) \cap R^*$ and $x^*$ centralizes $L^*/O(L^*)$. Since $L \leqslant L^*$, $I^* = L_2(C_{L^*}(x^*)) \leqslant L_{x^*}$ by $L$-balance. If $J^*$ is a 2-component of $\langle I^{*L_{x^*}} \rangle$, then as $y \in P \leqslant C_{x^*}$, it follows, as usual, that $(x^*, y, J^*)$ is an unbalancing triple. But $J^*/Z^*(J^*)$ involves $L^*/O(L^*)$ and hence involves $L_3(4)$. We conclude therefore from Theorem 4.86 that $J^*/Z(J^*) \cong L_3(4)$. In particular, $L$ covers $L^*/O(L^*)$. Furthermore, by our maximal choice of $(x, y, L)$, we must have $P = S$, whence $S \in \mathrm{Syl}_2(M)$ and $R = C_S(L^*/O(L^*))$. Hence, all parts of the lemma hold in this case.

Thus, we can assume that $n \geqslant 2$. If $Z^*(K^*)$ has even order, then $Z^*(K^*)$ contains an involution $z$ of $Z(P)$. We now repeat the argument of the previous paragraph with $z$ in place of $x^*$. Using the fact (Theorem 4.71) that a 2-component of type $L_3(4)$ in an unbalancing triple is (mod core) in $\mathscr{L}^*(G)$ together with the transitive action of $P$ on the $K_i$, it will follow that $C_z$ contains a product of $m \geqslant n$ 2-components $J_j$ with $J_j/Z^*(J_j) \cong L_3(4)$ and $J_j/O(J_j) \in \mathscr{L}^*(G)$, contrary to Theorem 4.9.

We therefore conclude that $Z^*(K^*) = O(K^*)$, whence each $K_i/O(K_i) \cong L_3(4)$, $1 \leqslant i \leqslant n$. Since $n \geqslant 2$, it is immediate in this case that $P$ is 3-connected. Thus again as in Lemma 4.87, we have $P \in \mathrm{Syl}_2(G)$ and $\Gamma_{P,3}(G) \leqslant M$. As in Lemma 4.94, we argue now that $G$ is balanced, so that $\Gamma_{P,2}(G) \leqslant M$. Again $O(M) \leqslant O(G) = 1$ by Aschbacher's theorem— contradiction.

By Lemma 4.95 and the discussion preceding Lemma 4.93, we conclude:

PROPOSITION 4.96. *Either $S$ is not 3-connected or $m(R) \leqslant 2$. In particular, $m(R) \leqslant 4$.*

The second statement is a consequence of the first, for if $m(R) \geqslant 5$, then any $t \in \mathscr{I}(S)$ centralizes a four-subgroup of $R$ (as $R\langle t \rangle$ is then connected) and it follows easily that $S$ is 3-connected.

For any $t \in \mathcal{Y}(G)$, let $K_t$ be the product of all unbalancing 2-components $K$ of $C_t$ [i.e., for which there is $u \in \mathcal{Y}(C_t)$ such that $(t, u, K)$ is an unbalancing triple], with $K_t = 1$ if there are no such unbalancing 2-components.

Proposition 4.96 enables us to prove the following result:

LEMMA 4.97.  *For any $t \in \mathcal{Y}(G)$, $K_t$ is a product of at most three 2-components.*

Suppose false and let $K$ be a 2-component of $K_t$. Then $K/O(K)$ centralizes three 2-components of $K_t$, each of which (mod core) is isomorphic to $L_2(5)$, $L_2(7)$, or a covering group of $L_3(4)$, and it follows that $m(C_{C_t}(K/O(K))) \geqslant 7$.

Among all unbalancing triples $(t, u, K)$ of $G$, we choose $t$ and $K$ so that a Sylow 2-subgroup $Q$ of $C_{C_t}(K/O(K))$ has the largest possible rank, and subject to this, so that a Sylow 2-subgroup of $N_{C_t}(K)$ has maximal order. Thus, $m(Q) \geqslant 7$. By the previous proposition $K/O(K)$ is not a covering group of $L_3(4)$, else we could have taken $K$ as $L$, in which case $m(R) \geqslant 7-$ contradiction. Thus $K/O(K) \cong L_2(5)$ or $L_2(7)$.

As usual, Theorems 4.43 and 4.71 imply that $(t, u, K)$ is not a maximal unbalancing triple. But then, as in the proof of Theorem 4.83, there is $a \in \mathcal{Y}(Q)$ such that if $I = L_2(C_K(a))$ and $J = \langle I^{L_a} \rangle$, then $(a, b, J)$ is an unbalancing triple for some $b \in \mathcal{Y}(C_a)$ with $J/O(J)$ a covering group of $L_3(4)$.

We choose $a$ so that $Q_1 = C_Q(a)$ has maximal rank. By the structure of $\text{Aut}(L_3(4))$, $Q_1 = \langle t \rangle C_{Q_1}(J/O(J))$. Using this fact, it is easy to show first that we can take $a$ to centralize a normal four-subgroup $U$ of $Q$ and then to lie in $U$, in which case $m(Q_1) = m(Q) \geqslant 7$. But then $m(C_{Q_1}(J/O(J))) \geqslant 6$. Again, this contradicts the preceding proposition with $J$ as $L$.

We next normalize the choice of $x$. We know that $L_3(4)$ has $Z_4 \times Z_4$ as Sylow 2-subgroup of its Schur multiplier [I, Table 4.1]. Hence if we set $Z = R \cap L$, then as $Z$ maps on $Z(L/O(L))$, it follows that $Z$ is isomorphic to a subgroup of $Z_4 \times Z_4$.

LEMMA 4.98.  *If $Z \neq 1$, we can assume that $x \in Z$, with $z \in \phi(Z)$ if $Z$ is not elementary.*

Indeed, as $Z \lhd S$, if $Z \neq 1$, there is $z \in \mathcal{Y}(Z)$ with $z \in Z(S)$ and $z \in \phi(Z)$ if $Z$ is not elementary. But then if we set $I = L_2(C_L(z))$ and

$J = \langle I^{L_z} \rangle$, it follows that $J/O(J) \cong L/O(L)$, that $(z, y, J)$ is an unbalancing triple, and that $S$ normalizes $J$. Replacing $x$ by $z$, if necessary, the lemma follows.

Henceforth, we assume that if $Z \neq 1$, then $x \in Z$, with $x \in \phi(Z)$ if $Z$ is not elementary. Note that $Z \leqslant Z(T)$.

Our goal is to prove the following modified form of Proposition 4.90.

PROPOSITION 4.99. $\Delta_G(D) = 1$ *for every four-subgroup* $D$ *of* $T$.

We argue by contradiction in a sequence of lemmas. We first prove:

LEMMA 4.100. *If* $m(T) \geqslant 4$ *and* $T$ *is 3-connected, then* $G$ *is balanced with respect to any* $E_8$-*subgroup of* $G$.

Indeed, assume false for some $E_8 \cong B \leqslant T$. Since $m(T) \geqslant 4$ and $T$ is 3-connected, $B \leqslant A \leqslant T$ for some $A \cong E_{16}$. Since $G$ is 2-balanced and Proposition 4.99 is assumed false, the 2-balanced $A$-signalizer functor $\theta$ is nontrivial. It follows that $\Gamma_{T,3}(G) \leqslant M = N_G(\theta(G; A))$; in particular, $M < G$. Thus Lemmas 4.93–4.95 hold in $M$ and as there, we set $L^* = \langle L^M \rangle$.

On the other hand, as $G$ is not balanced with respect to $B$, there are $b, b_0 \in B^{\#}$ and a 2-component $K$ of $K_b$ such that $(b, b_0, K)$ is an unbalancing triple. We have $A \leqslant C_b$ and $K/O(K)$ is of one of the usual three isomorphism types. Moreover, as $b_0 \in A$, $N_A(K)$ does not induce inner automorphisms on $K/O(K)$. One checks easily now that $K \leqslant \Gamma_{A,3}(C_b)$ and hence that $K \leqslant M$. By $L$-balance, $K^* = \langle K^M \rangle$ is a product of 2-components of $M$. However, as $A$ induces inner automorphisms on $L^*/O(L^*) \cong L/O(L)$, the nonbalancing of $K$ with respect to $B$ implies that $L^* \nleqslant K^*$, whence $K^*$ centralizes $L^*/O(L^*)$.

If we take $u \in \mathscr{I}(T \cap L^*)$ with $u$ centralizing $B$, it follows, as usual, that $(u, b_0, J)$ is an unbalancing triple for any 2-component $J$ of the normal closure of $I = L_{2'}(C_{K^*}(u))$ in $L_u$. Since $I$ covers $K^*/O(K^*)$, we conclude at once that each 2-component of $K^*$ is isomorphic (mod core) to $L_2(5), L_2(7)$, or a covering group of $L_3(4)$.

This leads at once to a contradiction of Proposition 4.96. Indeed, $m(\langle x \rangle K^*) \geqslant 3$ and $R \cap K^* \in \mathrm{Syl}_2(K^*)$ [as $R \in \mathrm{Syl}_2(C_M(L^*/O(L^*)))$], so $m(R) \geqslant 3$. Furthermore, it is easy to check that the existence of $K^*$ implies that $S$ is 3-connected, giving the desired contradiction.

We shall derive the same conclusion when $m(T) = 3$ or $T$ is not 3-connected. By the structure of covering groups of $L_3(4)$, if $Z$ is elementary, then $m(L) = m(L_3(4)) + m(Z) = 4 + m(Z)$, and it is easily verified that $T$ is 3-connected. Thus we have the following result:

LEMMA 4.101. *If $m(T) = 3$ or $T$ is not 3-connected, then $Z \cong Z_4$, $Z_4 \times Z_2$, or $Z_4 \times Z_4$.*

We note that the next argument is independent of Proposition 4.99.

LEMMA 4.102. *If $Z$ is not elementary, then $G$ is balanced with respect to any $E_8$-subgroup of $T$.*

Suppose false and let $B$, $b$, $b_0$ and $K$ be as in Lemma 4.99. By our choice of $x$, we have $x \in \phi(Z)$. We first reduce to the case $Z \cong Z_4$.

Assume false and set $N = \langle K^Z \rangle$. Since $Z$ is a 2-group, Lemma 4.97 implies that $N$ is the product of at most two 2-components and, as usual, $N = K$ by Theorem 4.9 if $K/Z^*(K) \cong L_3(4)$. Set $\overline{C}_b = C_b/O(C_b)$, $\overline{Y} = N_{\overline{C}_b}(\overline{N})$, and $\tilde{Y} = \overline{Y}/C_{\overline{Y}}(\overline{N})$, so that $\tilde{Y} \leqslant \mathrm{Aut}(\tilde{N})$ and either $\tilde{N} \cong L_3(4)$, $L_2(q)$, or $L_2(q) \times L_2(q)$, $q = 5$ or $7$. Furthermore, $Z$ normalizes $X = O(C_{b_0}) \cap C_b$ as $Z$ centralizes $\langle b, b_0 \rangle$ and, as usual, $X$ leaves each 2-component of $N$ invariant [whence $\overline{X} \leqslant \overline{Y}$ and $\tilde{X} \leqslant O(C_{\tilde{Y}}(\tilde{b}_0))$]. Since $(b, b_0, K)$ is an unbalancing triple, also $\tilde{X}$ centralizes no component of $\tilde{N}$.

We conclude at once that $\tilde{X} \cong Z_3$ or $Z_3 \times Z_3$, and if we let $\tilde{W} \in \mathrm{Syl}_2(N_{\tilde{Y}}(\tilde{X}))$ with $\tilde{Z} \leqslant \tilde{W}$, then $\tilde{W}$ is isomorphic to a subgroup of $Z_2 \times Z_2$ if $\tilde{N} = \tilde{K} \cong L_2(q)$, a subgroup of $Z_2 \wr Z_2$ if $\tilde{N} \cong L_2(q) \times L_2(q)$, a subgroup of $F \times Z_2$ if $\tilde{N} = \tilde{K} \cong L_3(4)$, where $F$ is quasi-dihedral of order 16. However, as $Z \cong Z_4 \times Z_4$ or $Z_2 \times Z_4$, we conclude in the first two cases that $\tilde{Z} \not\leqslant Z$ and hence some $z \in \mathscr{I}(Z)$ centralizes $\overline{N}$. On the other hand, in the third case we check that some $z \in \mathscr{I}(Z)$ either centralizes $\overline{N} = \overline{K}$ or induces a nontrivial field automorphism on $\tilde{N}$.

Set $K_0 = L_{2'}(C_K(z))$, so that by $L$-balance $K_0 \leqslant L_z$. Furthermore, if we set $L_0 = L_{2'}(C_L(z))$, then $z$ and $x$ are contained in $L_0$. The second inclusion implies that $H = \langle L_0^{L_z} \rangle$ is a single 2-component, while the first, together with Theorem 4.23 and the known structure of centralizers of involutions acting on quasisimple $K$-groups, implies that $H/O(H) \cong L_0/O(L_0) \cong L/O(L)$. Hence, $x \in Z^*(H)$ and consequently $x$ centralizes $K_0/O(K_0)$. Thus, $x$ centralizes $\overline{K}$ if $\overline{K}_0 = \overline{K}$. In the remaining case, $\tilde{N} = \tilde{K} \cong L_3(4)$, $\tilde{K}_0 \cong L_3(2)$, and either $x$ centralizes $\overline{K}$ or else $\tilde{x}$ induces a field automorphism on $\tilde{K}$ of period 2. However, the latter is impossible as $\tilde{x}$ is a square in $\tilde{Z}$ [since $x \in \phi(Z)$]. Thus $x$ centralizes $\overline{K}$ in all cases.

We can therefore take $x$ as $z$ in the above argument, with $K_0$ covering

$\overline{K}$. But now we simply repeat the final portion of the proof of Lemma 4.100 with $C_x$ in place of $M$ to again contradict Proposition 4.96.

This completes the reduction to the case $Z \cong Z_4$. Furthermore, $x$ does not centralize $\overline{K}$ or else we reach the same contradiction. To treat this case, we let $Q$ be a Sylow 2-subgroup of $C_L(b)$, and redefine $N$ to be $\langle K^Q \rangle$. Again by Lemma 4.97 and Theorem 4.9, $N$ has the same general structure as in the previous case. We define $\overline{Y}$ and $\tilde{Y}$ as before; however, this time we let $\tilde{W}$ be a Sylow 2-subgroup $\tilde{N}$ containing $\tilde{Q}$. But it can be shown first that $Q$ has index at most 2 in a Sylow 2-subgroup of $L$ and that $Z = Z(Q)$ [using the fact that $L/O(L) \cong L_3(4)/Z_4$] and then that $\tilde{W}$ does not contain a subgroup of the form or $\tilde{Q}$. Hence, $C_Q(\overline{K}) \neq 1$ and consequently $x$ centralizes $\overline{K}$, giving a final contradiction.

In view of Lemmas 4.100–4.102, $G$ is indeed balanced with respect to every $E_8$-subgroup of $T$. Furthermore, as $T$ contains a normal $E_8$-subgroup, $T$ is connected. Hence if for any four-subgroup $U$ of $T$, we set

$$W_U = \langle O(C_u) | u \in U^{\#} \rangle, \tag{4.59}$$

it follows by the standard argument that $W_U = W_V$ for any two four-subgroups $U$, $V$ of $T$. Furthermore, by the signalizer functor theorem, $W_U$ is of odd order and as Proposition 4.99 is assumed false, it is nontrivial.

Hence if we put $M = N_G(W_U)$, then $M < G$ and

$$\Gamma_{T,2}(G) \leqslant M. \tag{4.60}$$

Again Lemmas 4.93–4.95 are applicable and we let $L^*$ have the usual meaning. In particular, $S \in \mathrm{Syl}_2(M)$.

Observe next that:

If $\langle V, V^g \rangle \leqslant T$ for some $Z_2 \times Z_2 \cong V \leqslant T$ and $g \in G$, then $g \in M$. (4.61)

Indeed, the given conditions imply that $(W_V)^g = W_{V^g} = W_V$, whence $g \in N_G(W_V) = N_G(W_U) = M$.

By the structure of $\mathrm{Aut}(L_3(4))$, every normal four-subgroup of $S$ is contained in $T$. Hence, if we take $Z_2 \times Z_2 \cong V \lhd S$ and let $g \in N_G(S)$, it follows that $\langle V, V^g \rangle \leqslant T$, whence $g \in M$ by (4.61). Thus, $N_G(S) \leqslant M$. Since $S \in \mathrm{Syl}_2(M)$, we conclude that $S \in \mathrm{Syl}_2(G)$. In particular, $x$ is 2-central.

We shall argue finally that the assumptions of Holt's theorem [I, 4.32] are satisfied by $x$ and $M$. We first prove:

LEMMA 4.103. *If* $u \in \mathscr{I}(R)$, *then* $C_u \leqslant M$.

Indeed, if $I = L_2,(C_L(u))$ and $J = \langle I^{L_u} \rangle$, then, as usual, $J/O(J) \cong L/O(L)$ and $J \lhd C_u$. Since $O(C_u) \leqslant W_U$ and $L \leqslant M$, it follows that $J \leqslant M$. But also $T \cap L \in \mathrm{Syl}_2(J)$ (as $u$ centralizes $T \cap L$) and $N_G(T \cap J) \leqslant M$ by (4.60). A Frattini argument now yields the desired conclusion $C_u \leqslant M$.

We argue next that $M$ controls the $G$-fusion of $x$ in $S$.

LEMMA 4.104. *If* $u^g = x$ *for some* $u \in S$ *and* $g \in G$, *then* $g \in M$.

Replacing $g$ by $ga$ for some $a \in C_x$, we can suppose without loss that $C_S(u)^g \leqslant S$. Likewise we can assume that $C_S(u) \in \mathrm{Syl}_2(C_M(u))$. Set $Q = C_S(u)$. If $u \in T$, it follows easily from the structure of $Q \cap T$ that $V^g \leqslant T$ for some four-subgroup $V$ of $Q \cap T$, in which case $g \in M$ by (4.61). Hence we can assume that $u \notin T$.

If $u$ induces a field or transpose inverse map on $L/O(L)$ and we set $I = L_2,(C_L(u))$, then $I/O(I) \cong L_2(5)$, $L_2(7)$, $SL_2(5)$, or $SL_2(7)$. In the first two cases, the preceding argument applies with $V \leqslant I$, so we can assume that $I/O(I) \cong SL_2(5)$ or $SL_2(7)$. In these cases, it can be shown that the involution of $I \cap S$ is the unique involution of $Z \cap Z(S)$. Hence, by our choice of $x$, $x \in I$. But then $v = x^g \in T$. Since $v^{g^{-1}} = x$, we conclude as in the first case, with $v, g^{-1}$ in place of $u, g$, that $g \in M$.

Suppose finally that $u$ induces a unitary automorphism on $L/O(L)$, and consider first the case that $C_R(u)$ contains a four-group $B$. Then $B^g \leqslant S$ and so there is $b \in B^{\#}$ such that either $b^g \in T$ or $b^g$ induces a field or transpose-inverse automorphism on $L/O(L)$. But by the preceding lemma, $C_{b^g} \leqslant M^g$, so $C_M(b^g) \leqslant M^g$. One can easily argue now as in the previous cases that $g \in M$.

We can therefore assume that $m(C_R(u)) = 1$. If $x$ is a square in $Q$, then $x^g \in T$ and, as before, this implies that $g \in M$, so we can also assume that $x$ is not a square in $Q$. In particular, $C_R(u) = \langle x \rangle$ [as $m(C_R(u)) = 1$], so $R\langle u \rangle$ is either dihedral or quasi-dihedral by [I, 1.16 and 1.17]. This gives the structure of $R$. Furthermore, by properties of $\mathrm{Aut}(L_3(4))$, these conditions also force $Z$ to be cyclic and $Q \cap L \cong Q_8$ or $Z_2 \times Z_4$ according as $Z = 1$ or $\neq 1$.

These final cases are treated by Griess and Solomon. They require delicate fusion analyses, which ultimately lead to contradictions unless $g \in M$. We omit the details.

Lemmas 4.103 and 4.104 imply that $x$ is 2-central and, if $N$ is a maximal subgroup of $G$ containing $M$, that $x$ fixes only the coset $N$ itself in the (primitive) permutation representation of $G$ on the right cosets of $N$, so $F^*(G)$ is a $K$-group by [I, 4.32], giving the usual contradiction.

This completes the discussion of Proposition 4.99.

It thus remains to prove the following:

PROPOSITION 4.105. *L is quasisimple and standard in G.*

Indeed, as we have already noted, the result follows if $m(R) \geqslant 3$, so it is the case $m(R) \leqslant 2$ that requires additional work. As usual, it will suffice to prove that $L$ is quasisimple. Lemma 4.97 will enable us to quickly establish the desired conclusion.

Assume false, in which case $L$ does not centralize $O(L)$ and hence $C_S(O(L)) \leqslant R$. Let $U$ be any four-subgroup of $T \cap L$ with $x \notin U$. Since $m(Z) \leqslant 2$, $U \not\leqslant Z$ and so $U \not\leqslant R$. Hence $U$ does not centralize $O(L)$, so by $[I, 4.13]$

$$X = [C_{O(L)}(u), U] \neq 1 \tag{4.62}$$

for some $u \in U^\#$. But now if we set $D = \langle x, u \rangle$, then as $\Delta_G(D) = 1$, we conclude for any nontrivial subgroup $V$ of $X$ that

$$V \not\leqslant O(C_b) \qquad \text{for } b = u \text{ or } ux. \tag{4.63}$$

In particular, $O(C_x) \cap C_b \not\leqslant O(C_b)$, so $G$ is not balanced with respect to any $E_8$-subgroup of $T$ containing $U$. Lemma 4.102 (which as noted is independent of Proposition 4.99) is now applicable and yields a contradiction if $Z$ is not elementary. Thus we have the following result:

LEMMA 4.106. *Z is elementary.*

Set $P = (T \cap L)\langle x \rangle$ and let $B$ be an $E_8$-subgroup of $P$ containing $Z\langle x \rangle$ with the property that $B/Z \leqslant Z(P/Z)$. $B$ exists by the structure of $T \cap L$. In particular, $B \lhd P$. Let $U$ be a complement to $\langle x \rangle$ in $B$ and choose $u \in U^\#$ to satisfy (4.62). Finally set $Q = C_P(u)$. By the choice of $B$ and the structure of $P$, we easily establish the following properties of $Q$:

(a) $Q/\langle x, u \rangle$ has order $2^5$ or $2^6$; and
(b) If $Q_0 \lhd Q$ with $u \in Q_0$ and $U \not\leqslant Q_0$, then $Q/Q_0$ is not a     (4.64)
    subgroup of a quasi-dihedral group.

Since $B \lhd P$ and $\langle x, u \rangle$ centralizes $C_{O(L)}(u)$, $Q$ normalizes $X$. Choose $V$

to be a $Q$-invariant subgroup of $X$ minimal subject to $U$ not centralizing $V$. By [I, 4.2], $Y$ is a $p$-group for some odd prime $p$. Furthermore, by [I, 4.12], $U/\langle u \rangle$ acts nontrivially on the Frattini factor group $V/\phi(V)$. Hence by the minimality of $V$, $U/\langle u \rangle$ necessarily inverts $V/\phi(V)$.

By (4.63), $V \not\leq O(C_b)$ for $b = u$ or $ux$. Hence, there is a 2-component $K$ of $C_b$ such that $V$ normalizes $K$, but does not centralize $K/O(K)$. As in Lemma 4.102, set $N = \langle K^Q \rangle$, $\overline{C}_b = C_b/O(C_b)$, $\overline{Y} = N_{\overline{C}_b}(\overline{N})$, and $\tilde{Y} = \overline{Y}/C_{\overline{Y}}(\overline{N})$. Then $\overline{N}$ and $\tilde{Y}$ have the same general shape as in Lemma 4.102. Again we have $\tilde{V} \cong Z_3$ or $Z_3 \times Z_3$. But now if we set $Q_0 = C_Q(\tilde{V})$, it follows that $Q/Q_0 \cong \tilde{Q}/\tilde{Q}_0$ is isomorphic to a subgroup of $GL_2(3)$ and hence that $Q/Q_0$ is contained in a quasi-dihedral group (of order 16). However, as $u \in Q_0$ and $U \not\leq Q_0$ (as $U/\langle u \rangle$ does not centralize $V$), this contradicts (4.64) and thus completes the proof of Theorem 4.92.

The struggle is over! Theorems 4.86 and 4.92 complete the reduction, thus yielding our main objective:

THEOREM 4.107. *The B-theorem and U-theorem hold under Hypothesis S.*

In view of the discussion at the end of Section 4.3, Theorem 4.107 together with the inductive characterization of the family $\mathscr{LO}$ of groups of Lie type of odd characteristic (Theorem 4.43) give the following reduction for the noncharacteristic 2 theorem (see Table 4.1 for the list of groups in $\mathscr{LO}^* - \mathscr{LO}$).

To state the reduction, we say that $G$ has the *strong B*-property if every section of $G$, including $G$ itself, has the $B$-property.

THEOREM 4.108. *To complete the proof of the noncharacteristic 2 theorem, it suffices to establish the following two results:*

(I) *If $G$ is a group with $F^*(G)$ simple which possesses a standard component $L$, and if $G$, and $L$ satisfy the conditions of Hypothesis S, then $F^*(G)$ is a K-group.*

(II) *If $G$ is a locally balanced simple group with the strong B-property which possesses a standard component $L$ of Lie type of characteristic 2 or sporadic with $L \notin \mathscr{LO}^*$, then $G$ is a K-group.*

This then is the precise formulation of the simplified standard component theorem stated in the introduction of the book.

# 5

# The Classification of Groups of Component Type

In this chapter we complete the discussion of the noncharacteristic 2 theorem. Theorem 4.106 has reduced its validity to the following specific set of standard form problems:

(I) Verification of Hypothesis $S$ [see Section 4.3 for the exact statement of Hypothesis $S$, including the definitions of the families $\mathscr{LO}$, $\mathscr{LO}^*$, and $\mathscr{L}_0(3)$].

(II) Determination of all locally balanced simple groups with the strong $B$-property that contain a standard component of Lie type of characteristic 2 or sporadic, but which is not an element of $\mathscr{LO}^*$.

In both instances, the task is to show that the given group $G$ is a $K$-group. Since the family $\mathscr{LO}^* - \mathscr{LO}$ is closed under pumping-up, the solutions for $F^*(G)$ in (I) consist, as one would anticipate, of the elements of $\mathscr{LO}^* - \mathscr{LO}$ together with the possibilities arising from standard components in $\mathscr{L}_0(3)$. On the other hand, as $G$ is assumed to be simple in (II), the list of solutions in that case is very much shorter (consisting, in fact, entirely of sporadic groups), inasmuch as simple groups of Lie type of characteristic 2 are of *noncomponent* type by the Borel–Tits theorem [I, 1.41]. [Of course, if we assume that $G$ is simple in (I), as will be the case in the noncharacteristic 2 theorem itself, the list of possibilities for $G = F^*(G)$ will be correspondingly reduced.]

Here is the exact statement of (I) and (II).

STANDARD COMPONENT THEOREM (I). *Let $G$ be a group with $F^*(G)$ simple containing a standard component $L$. If $G$ and $L$ satisfy $(A)$, $(B)$, $(C)$, or $(D)$ of Hypothesis $S$, then one of the following holds*:

(i) $F^*(G) \in \mathscr{LO}^* - \mathscr{LO}$; or

(ii) $F^*(G)$ is of Lie type of characteristic 3.

[The precise list of groups arising in (ii) is easily determined; however, it is not required since the statement of the noncharacteristic 2 theorem (second form) allows for all groups of Lie type of odd characteristic (apart from the exceptions listed in its statement).]

As a corollary, we obtain the following precise result in the simple case.

STANDARD COMPONENT THEOREM (I) (The Simple Case). *If $G$ is a simple group with the strong B-property containing a standard component $L \in \mathscr{LO}^* - \mathscr{LO}$ or $\mathscr{L}_0(3)$, then one of the following holds:*

(i) $G$ is of Lie type of characteristic 3;

(ii) $G \cong A_n$, $n \geqslant 9$;

(iii) $G \cong Mc$ and $L \cong \hat{A}_8$;

(iv) $G \cong Ly$ and $L \cong \hat{A}_{11}$;

(v) $G \cong He$ and $L \cong L_3(4)/Z_2 \times Z_2$;

(vi) $G \cong Suz$ and $L \cong L_3(4)$;

(vii) $G \cong ON$ and $L \cong L_3(4)/Z_4$;

(viii) $G \cong .3$ and $L \cong M_{12}$;

(ix) $G \cong .1$ and $L \cong G_2(4)$; or

(x) $G \cong F_5$ and $L \cong HS/Z_2$.

Now for (II):

STANDARD COMPONENT THEOREM (II). *Let $G$ be a locally balanced simple group with the strong B-property containing a standard component $L$. If $L$ is of Lie type of characteristic 2 or sporadic, but $L \notin \mathscr{LO}^*$, then one of the following holds:*

(i) $G \cong Ru$ and $L \cong Sz(8)$;

(ii) $G \cong .3$ and $L \cong Sp_6(2)/Z_2$;

(iii) $G \cong M(22)$ and $L \cong U_6(2)/Z_2$, $Sp_6(2)$, or $P\Omega_8^+(2)$;

(iv) $G \cong M(23)$ and $L \cong M(22)/Z_2$;

(v) $G \cong M(24)'$ and $L \cong M(22)/Z_2$;

(vi) $G \cong F_2$ and $L \cong {}^2E_6(2)/Z_2$ or $F_4(2)$; or

(vii) $G \cong F_1$ and $L \cong F_2/Z_2$.

The preceding two theorems account for 14 of the 18 sporadic groups listed in the statement of the noncharacteristic 2 theorem (second form). The remaining 4 sporadic groups as well as the bulk of the groups of Lie type of odd characteristic (plus the group $A_7$) occurring in its statement arise from the inductive characterization of the family $\mathscr{LO}$, the minimal $L_2(q), A_7$-problem, the 2-rank $\leqslant 2$ theorem, and the sectional 2-rank $\leqslant 4$ theorem.

For completeness, we list the sporadic groups and corresponding standard component $L$ that arise within the latter theorems:

| $G$ | $L$ |
|---|---|
| $M_{12}$ | $L_2(5)$ |
| $J_1$ | $L_2(5)$ |
| $J_2$ | $L_2(7)$ |
| $HS$ | $L_2(9)$ |

[We note that the groups $Mc$, $HS$, $ON$, $Suz$, $F_5$, $M(22)$, and $M(24)'$ also occur as solutions of standard component problems with $G$ having a normal subgroup of index 2 and the given involution inducing an outer automorphism.]

This chapter will be devoted to an outline of the proof of these two major standard form theorems. Although the global strategy exhibits many common features, the sheer number of distinct "types" of components that have to be analyzed (often consisting of a single group or family of groups), each involving its own technical considerations, has made this an enormous undertaking, so that in its totality the complete solution of the involution standard form problem represents one of the longest chapters of finite simple group theory. Indeed, it comprises approximately 70 individual theorems covering somewhere between 1500 and 2000 journal pages!!

In Section 5.1, we shall present an overview of the basic method of attack, while in Section 5.2 we shall divide the total problem into some *twelve* major cases (many of which themselves involve a number of individual subcases). Taken together, these two sections should give the reader a good initial picture of the conceptual basis for the overall solution of the involution standard form problem.

In subsequent sections we shall amplify our preliminary comments on each of these major cases. However, because of the magnitude of the complete analysis, we shall be selective in the cases (or subcases) we choose for systematic discussion. For the same reason, that discussion will be limited primarily to schematic outlines of arguments (including the important auxiliary results used in the analysis). This will be entirely sufficient to convey the nature of the complete solution; to have attempted anything further would have involved an unreasonable level of technical detail.

## 5.1. An Overview of the Involution Standard Form Problem

Although they differ in specifics, the standard component theorems (I) and (II) should be viewed as particular cases of the same general problem. To this end, denote by $\mathscr{K}$ the set of all quasisimple $K$-groups and set

$$\mathscr{K}_0 = (\mathscr{K} - \mathscr{L}\mathscr{O}) \cup \mathscr{L}_0(3).$$

Here then is the general problem:

> Given an "essentially" simple group $G$ with the strong $B$-property (in some cases only with a "partial" $B$-property) containing a standard component $L$ with $L \in \mathscr{K}_0$, determine the possible isomorphism types of $F^*(G)$.

Thus, as in the classical involution theorem, the task is to "identify" $F^*(G)$ from the existence of a single component in the centralizer of an involution, of a given isomorphism type and satisfying suitable embedding conditions. In [I, Chapter 3], we have described a set of internal conditions which suffice to characterize each of the known simple groups, and it is by means of these recognition theorems that the identification of $F^*(G)$ is ultimately to be made. For the groups of Lie type, it is their $(B, N)$-pair description or Curtis–Steinberg relations that we are after; for the alternating groups their presentation by generators and relations, and for the sporadic groups the centralizer of an appropriate involution.

Once it is shown that $F^*(G)$ is a $K$-group, the precise conclusions of the given standard component theorem follow from the known structure of the centralizers of involutions acting on simple $K$-groups.

However, there is an important issue to settle at the very outset: should one attempt to prove the standard component theorems (I) and (II) sequentially or simultaneously? Surely one's natural preference is to give first priority to completing the $B$- and $U$-theorems—they have been dangling long enough!

But, in fact, the distinction between the two theorems is artificial. For example, the cases $L \cong L_2(4^{2n})$, $n = 2^m$, $m \geqslant 0$, fall under (I), while the remaining $L_2(2^r)$ cases, $r \geqslant 3$, come under (II). However, whatever approach one takes for these linear cases will undoubtedly be applicable for all values of $r \geqslant 3$. Similarly, Seitz's approach to the three- and four-dimensional linear groups of characteristic 2 of (I) turns out to work equally well in all dimensions $\geqslant 3$ [105].

There is another, perhaps even more, significant point. The principal difference between the two theorems is the assumed simplicity of $G$ in (II). [The various $B$-assumptions of (I) are minor technical distinctions, which, in fact, have been designed to smoothly accommodate the actual arguments.] However, in treating the Lie type characteristic 2 cases of (II), Seitz proceeds by induction; and to carry through that induction, he must establish (II) without this simplicity asumption [105]. Indeed, he is forced to drop even a simplicity assumption on $F^*(G)$, allowing the direct product of *two* quasisimple groups as an additional possible solution. (This second alternative, referred to as the "wreathed" case, will be described more fully below; see remark 5.1.6.)

The comments clearly suggest that it is best to proceed as though the standard component theorems (I) and (II) have been combined into a single result; and this is the viewpoint we shall adopt. [We note, however, that a few of the Lie type characteristic 2 cases of (II) can be treated independently of the general inductive argument [e.g., $L/O(L) \cong {}^2D_4(2^n)$ or ${}^2F_4(2^n)$]; and in such cases it would, therefore, suffice to show that $G$ has a normal subgroup of index 2, a somewhat easier task than identification of $F^*(G)$. The same remark applies to most of the sporadic cases of (II). [We shall concern ourselves with this point only in those few cases in which solutions have been obtained only under the assumption that $G$ is simple.]

In this section, we make a number of specific comments about the general features of the analysis, but first we introduce some notation, which will be fixed for the entire chapter. By definition of a standard component, $L$ is a component of the centralizer of some involution $x$ of $G$ and $H = C_L = C_G(L)$ is tightly embedded in $G$. In addition, $L$ commutes with none of its $G$-conjugates and $N_G(L) = N_G(H)$. In particular, $L \lhd C_x$. We fix $R \in \mathrm{Syl}_2(H)$ with $x \in R$. Furthermore, we shall assume throughout that $G$ does not possess a classical involution or a tightly embedded subgroup with quaternion Sylow 2-subgroups, since in the contrary case $F^*(G)$ is determined by Aschbacher's theorem. In particular, this implies that $m(L) \geqslant 2$; otherwise $L$ itself would be tightly embedded with quaternion Sylow 2-subgroups. [Hence $L/O(L) \not\cong \hat{A}_7$.]

For later reference, we number our remarks 5.1.1–5.1.13.


5.1.1. We wish first to emphasize a fundamental distinction between the earlier low-rank standard form problems (including some that led to sporadic groups), which often required considerable character theory for their complete solutions, and the present cases $L \in \mathcal{H}_0$, in which the analyses proceeed without recourse to character theory. Instead, all necessary information concerning the subgroup structure of $G$ is developed solely by means of local considerations (involving in some instances determination of centralizers of certain elements of *odd* order).

There are some important distinctions between the cases $m(H) = 1$ and $m(H) > 1$; we therefore separate our comments, beginning with the case $m(H) = 1$. Thus in 5.1.2 through 5.1.7,

$$\text{we assume } m(H) = 1. \tag{5.1}$$

5.1.2. In this case, we have very strong information about the structure of both $H$ and $C_x$. Indeed, as $H$ is tightly embedded in $G$, $R$ is not quaternion, so $R$ is cyclic by [I, 1.35]. But then $H$ and hence also $C_{C_x}(L)$ has a normal 2-complement by Burnside's theorem [I, 1.20]. Thus

$$C_{C_x}(L) = O(C_x)R. \tag{5.2}$$

Furthermore, if we set $\bar{C}_x = C_x/C_{C_x}(L)$, then

$$\bar{L} \text{ is simple and } \bar{C}_x \leqslant \text{Aut}(\bar{L}). \tag{5.3}$$

We see then that $C_x$ is completely determined up to the structure of $O(C_x)$, the order of $R$, and the shape of the outer automorphism group $\bar{C}_x/\bar{L}$.

5.1.3. It is, therefore, reasonable to say at this point that $C_x$ "closely resembles" the centralizer of an involution in $\text{Aut}(G^*)$ for some "known" simple group $G^*$. Thus, in solving the given standard form problems, we shall, in effect, be validating the Brauer principle that the finite simple groups are characterized by the structure of the centralizers of their involutions.

However, as already noted, in most cases that lead to sporadic groups, it is not our intent to "solve" the corresponding standard form problems, but rather to replace the "secondary" centralizers $C_x$ by "primary" centralizer of involution problems, in terms of which we have previously characterized these groups. (This, of course, only serves to reinforce the Brauer principle!)

5.1.4. Just as in prior low-rank theorems, the first step of the analysis always consists of a study of the fusion of the involution $x$ in $C_x$, which requires for its execution a knowledge of the centralizers of involutions acting on $L$. The $Z^*$-theorem, of course, assures that $x$ is $G$-conjugate to an involution of $C_x - \langle x \rangle$.

The case that $L$ is of Lie type of characteristic 2 provides a good illustration of the nature of this initial fusion analysis. It is the work of Seitz [106] (together with Griess and D. Mason [47] when $L$ has Lie rank 1), and is based on previous joint work with Aschbacher [I: 21] on the conjugacy classes of involutions in $\text{Aut}(L)$ and the structure of their centralizers when $L$ is of Lie type of characteristic 2.

Seitz's first result asserts that some conjugate of $x$ induces a nontrivial *inner* automorphism on $L$.

PROPOSITION 5.1. *If $L$ is of Lie type of characteristic 2, then for some $g \in G$,*

$$x^g \in L\langle x \rangle - \langle x \rangle.$$

However, the proposition puts no restriction on the nature of the automorphism which $x^g$ induces on $L/Z(L)$. Seitz's main result gives much more specific information about possible actions of $x^g$, related to the "root" subgroups of $L$.

To describe it, we consider the Lie description of $L$ given in Section 3.5, with $\Sigma$, $\Sigma^+$, $\Sigma^-$, $Q$, $Q_\alpha$, $U_\alpha$ for $\alpha \in \Sigma$, having the same meanings as there, and we let $\alpha_1, \alpha_2, ..., \alpha_r$ be a fundamental root system for $\Sigma$. Choose $i$ so that $\alpha = \alpha_i$ is a long root and set $J = \langle U_\alpha, U_{-\alpha} \rangle$. If $L$ is defined over $GF(q)$, we know by Proposition 3.51 that

$$J \cong SL_2(q). \tag{5.4}$$

Furthermore, if $L$ has Lie rank 2, let $\beta$ be any positive root such that $Q_\beta \neq Q_\alpha$.

In this terminology, Seitz's main result is as follows:

THEOREM 5.2. *If $L$ is of Lie type of characteristic 2, then for some $g \in G$, one of the following holds:*

  (i) *If $r = 1$, then $x^g = ux$ for some $u \in U_\alpha{}^\#$;*
  (ii) *If $r = 2$, then either $x^g = ux$ or $x^g = yx$ for some $u \in U_\alpha^\#$ or $y \in Q_\beta^\#$ [apart from the groups $L/Z(L) \cong L_3(q)$, $G_2(q)$, or $U_5(2)$, in which case somewhat weaker assertions hold]; and*
  (iii) *If $r \geqslant 3$, then for some $y \in C_L(J)^\#$, either $x^g = y$ or $yx$ [apart from the groups $L/Z(L) \cong P\Omega_m^\pm(q)$, in which case a second alternative must be allowed].*

In most cases, one can use the proposition to conclude rather directly that $R = \langle x \rangle$ and that $C_x$ does not contain a Sylow 2-subgroup of $G$.

5.1.5. As we have just observed, the initial fusion analysis very often enables one to conclude that $R = \langle x \rangle$, so that $\langle x \rangle \in \mathrm{Syl}_2(C_{C_x}(L))$ and $\langle x \rangle \in \mathrm{Syl}_2(H)$. Finkelstein [27] has established a rather nice sufficient criterion for this to hold, which he has applied to several sporadic group standard form problems.

PROPOSITION 5.3. *Suppose that for any subgroup* $Y$ *of* $\mathrm{Aut}(L)$ *containing* $\mathrm{Inn}(L)$ *and any* $t \in \mathscr{I}(Y)$, $C_Y(t)/O(C_Y(t))$ *does not contain a normal subgroup isomorphic to* $Z_4$ *or* $Z_2 \times Z_4$. *Under these conditions,* $R = \langle x \rangle$.

The proof proceeds in two stages. Assuming false, Finkelstein first argues that $T = R^g$ normalizes $L$ for some $g \in G - N_G(L)$. The proof of this assertion depends on the $Z^*$-theorem, properties of tightly embedded subgroups, and a rather elementary fusion result of Aschbacher concerning groups $X$ that contain a cyclic subgroup $Z$ of order 4 which is weakly closed in $N_X(Z)$ [I: 41]. Thus $K = H^g \cap N_G(L)$ has Sylow 2-subgroup $T$ isomorphic to $R$.

Now set $Y = LK$, so that $Y$ is a group with $L \lhd Y$. It is easily checked that $K$ is tightly embedded in $Y$. But then, if $t$ is the involution of $T$, it follows from this tight embedding that $C_Y(t)$ necessarily normalizes $K$. On the other hand, as $T$ is cyclic, Burnside's theorem implies, as usual, that $K = O(K)T$. Together these conditions easily force $C_Y(t)/O(C_Y(t))$ to possess a normal subgroup isomorphic to $Z_4$ or $Z_2 \times Z_4$. Passing to $\bar{Y} = Y/C_Y(L)$ $(\leqslant \mathrm{Aut}(\bar{L}))$, the given hypothesis is thus contradicted by $\bar{Y}$ and $\bar{t}$.

We note also that when $|Z(L)|$ is odd (equivalently, when $L\langle x \rangle = L \times \langle x \rangle$), one usually obtains an even stronger conclusion: namely, $\langle x \rangle$ is not a square in $C_x$, which implies that $C_x$ *splits* over $\langle x \rangle$.

5.1.6. Standard form problems of the form $C_x = L \times \langle x \rangle$ always possess a "degenerate" solution referred to as the "wreathed" case: namely,

$$G \cong L \wr Z_2. \tag{5.5}$$

Indeed, $G$ then has the form $G = (L_1 \times L_2)\langle x \rangle$, where $L_1 \cong L_2 \cong L$, and $x$ interchanges $L_1$ and $L_2$ under conjugation. But then $C_{L_1L_2}(x)$ is also isomorphic to $L$ and so if we identify this centralizer with $L$, we see that

$C_G(x) = C_x = L \times \langle x \rangle$, as required. (This example is easily extended to all cases in which $C_x$ splits over $\langle x \rangle$.)

Moreover, as already noted, families of groups which are treated inductively [specifically, the groups of Lie type of characteristic 2 (apart from certain low Lie rank cases) and the alternating groups] require the solution of "lower-dimensional" standard form problems of the same type in proper sections $X$ of the group $G$ under investigation, but without the assumption that $F^*(X)$ is simple. It is therefore necessary to treat this wreathed case as a bona fide possibility, carrying the analysis through until $F^*(G)$ is completely identified.

Three independent steps are required to achieve this objective. First, using the appropriate presentation by generators and relations, one constructs an $x$-invariant subgroup $G^*$ of $G$ of the form $G^* = L_1 \times L_2$, containing $L$ as a "diagonal" of $G^*$ under conjugation by $x$. It is this subgroup $G^*$ which must be shown to equal $F^*(G)$.

Next, one argues that $M = N_G(G^*)$ contains a Sylow 2-subgroup $S$ of $G$ and that $M$ controls suitable fusion of 2-subgroups of $S_i = S \cap L_i$, $i = 1, 2$. To establish these fusion results, the fact that proper sections of $G$ have the $B$-property is critical.

Their exact form depends on which of two possible methods one plans to use for the final step: proving that $M = G$ [which will immediately imply the desired conclusion $G^* = F^*(G)$]. The more efficient method would seem to be that of Seitz [106], in his general Lie type characteristic 2 analysis. Under the assumption that $M < G^*$, he derives a contradiction by appealing to Aschbacher's 2-component fusion theorem (Theorem 3.39) with $G^*$ in the role of $K$. On the other hand, several authors have argued first that $G$ has a normal subgroup $G_0$ of index 2, using Thompson transfer, and then shown that $S_i$ is strongly closed in $S \cap G_0$ with respect to $G_0$, in which case $X_i = C_{G_0}(S_i)^{(\infty)} \lhd G_0$ by Goldschmidt's general strong closure theorem [I, 4.156]. Since $L_j \leqslant X_i$ for $j \neq i$, it follows directly that $G^* \lhd G_0$ and hence that $M = G$. It seems clear that either approach can be used in any given problem.

Of course, when $x \in L$, there is no wreathed solution for $G$, since then $G$ would have a normal subgroup $G_0$ of index 2 not containing $x$, which is impossible as $x \in L \leqslant G' \leqslant G_0$.

5.1.7. When $L$ is a group of Lie type of characteristic 2 (again apart from the same exceptions as in 5.1.6 or an alternating group (of degree $\geqslant 12$) and, in addition, the local subgroup structure is not that of the wreathed case, the basic strategy is once again to construct, using a suitable presentation by generators and relations, a quasisimple subgroup $G^*$ of $G$ containing $L$ (correspondingly of Lie type of characteristic 2 or alternating), which is normalized, but not centralized by $x$.

In practice, two methods have been used in the Lie group characteristic 2 case for the construction of $G^*$ (in the wreathed as well as the quasisimple case)—one involving pushing up maximal parabolic subgroups of $L$ and the other based on a determination of centralizers in $G$ of certain subgroups of $L$ of *odd* order. The latter approach can be viewed as similar to that used in the alternating group case, which depends on a determination of the centralizers in $G$ of *short* involutions (i.e., involutions of $L$ that can be represented as a product of two transpositions). (See Section 2, cases 5.2.4, 5.2.8, 5.2.9, and 5.2.10 for additional details concerning the construction of $G^*$.)

Again two further steps are required to force $G^* = F^*(G)$. One must first prove that $M = N_G(G^*)$ contains a Sylow 2-subgroup $S$ of $G$ and then establish sufficient fusion information to enable one to quote an appropriate fusion classification theorem.

Likewise, as in the wreathed case, appeal has been made to two such fusion theorems, the one with wider applicability being Holt's theorem [I, 4.32]. In this case, one focuses on an involution $z \in Z(S) \cap G^*$ and argues that $C_z \leqslant M$ and $M$ controls the $G$-fusion of $z$ in $S$. If $M < G$, Holt's theorem then yields a contradiction. On the other hand, Miyamoto in the $L \cong U_4(2^n)$, $n \geqslant 2$, $G^* \cong L_4(2^{2n})$ case [84] applies Timmesfeld's theorem on abelian 2-subgroups $A$ of a group $X$ with the property that $A$ is weakly closed in $C_X(a)$ for all $a \in A^\#$ [I, 4.159] (the groups $L_m(2^r)$ possesses such a subgroup $A$). [It should also be pointed out that Miyamoto does not bother to construct $G^*$ (although it is clear he could have done so with slight additional effort), but instead once he has pushed up the maximum parabolics of $L$, he turns immediately to the fusion results needed for Timmesfeld's theorem.]

There are also cases in which an author has avoided most of the fusion analysis, relying instead on a previously established Sylow 2-group characterization theorem. In this approach, one has only to verify that $M$ contains a Sylow 2-subgroup $S$ of $G$ and, in addition, that $S \cap G^*$ is a Sylow 2-subgroup of $G_1 = O^2(G)$. Indeed, in the $L \cong L_m(2^n)$, $G^* \cong L_m(2^{2n})$ case, Seitz [106] quotes the following result of McBride [87] (Collins [15] for the case $m = 3$) to conclude that $G^* = G_1$.

THEOREM 5.4. *If $G$ is a simple group with Sylow 2-subgroups of type $L_m(2^r)$, $m \geqslant 3$, then $G \cong L_m(2^r)$.*

Similarly Yamada [136], [137] obtains the corresponding conclusions in the cases $L \cong G_2(2^n)$ or ${}^3D_4(2^n)$, $G^* \cong G_2(2^{2n})$ or ${}^3D_4(2^{2n})$, respectively, by quoting the following result of Thomas [121], [122].

THEOREM 5.5. *If G is a simple group with Sylow 2-subgroups of type* $G_2(2^r)$, $r \geqslant 2$, ${}^3D_4(2^r)$, $r \geqslant 1$, *then* $G \cong G_2(2^r)$ *or* ${}^3D_4(2^r)$, *respectively.*

But it is clear that this path was taken for convenience rather than necessity, and with some added work these authors could easily have shown that the conditions of Holt's theorem were satisfied.

We turn now to the case $m(H) \geqslant 2$. Thus, in 5.1.8–5.1.11,

$$\text{We assume } m(H) \geqslant 2 \tag{5.6}$$

5.1.8. The structure of $H$ and hence also of $C_{C_x}(L)$ is *a priori* much less restricted in this case. On the other hand, the tight embedding of $H$ takes on powerful force now, dominating the initial fusion analysis, which, as one would expect, focuses on the $G$-fusion of subgroups of $R$ in $N_G(L)$. Aschbacher [I: 11, 12] has studied this situation in considerable generality and his results ultimately place severe restrictions on both $R$ and its fusion in $N_G(L)$.

First of all, as $N_G(L) = N_G(H)$, Theorem 3.47 implies that for some $g \in G - N_G(H)$,

$$T = R^G \cap N_G(L) \neq 1. \tag{5.7}$$

Note that if we set $X = N_{H^g}(L)$, then $T \in \mathrm{Syl}_2(X)$, $T$ acts faithfully on $L$, and $X$ is tightly embedded in $LX$.

However, for the subsequent analysis, one needs to know the extent to which (5.7) implies the stronger conclusion that

$$N_G(L) \text{ contains a Sylow 2-subgroup of } H^g; \tag{5.8}$$

and, in addition, one would like to force $R$ to be elementary abelian.

Aschbacher has shown, in his component paper [I: 11], that failure of (5.8) puts considerable restriction on the structure of both $R$ and $H$ (cf., Theorem 3.48 for an indication of the nature of these restrictions). However, it is unnecessary to pursue these exceptions since Aschbacher has completely settled these two questions under an assumption which is entirely sufficient for the present applications. Indeed, he and Seitz have shown that the given assumption is satisfied by every quasisimple $K$-group and hence by any $L$

considered here. The assumption itself asserts that under suitable side conditions the group $T$ induces *inner* automorphisms on $L$. (We can view this as an analog of Proposition 5.1.)

Aschbacher's precise criterion is incorporated into the following definition (which was mentioned in abbreviated form in [I]; cf., [I, 4.250]).

DEFINITION 5.6. Let $L$ be a quasisimple group and $T$ a noncyclic elementary 2-group acting on $L$. We say that $L$ is of *restricted* type provided the following conditions imply that $T$ induces inner automorphisms on $L$:

(a) There exists a group $A$ of the form $A = TO(A)$ acting on $L$ with $A$ tightly embedded in $LA$; and

(b) $L < \Gamma_{T,1}(L)$ $(= \langle N_L(U) \mid 1 < U \leqslant T \rangle)$.

Aschbacher proves [I: 12] the following theorem:

THEOREM 5.7. *If $L$ is of restricted type, then the following conditions hold*:

(i) *$R$ is elementary abelian*;

(ii) *If $R \cong Z_2 \times Z_2$, then $R^g$ normalizes $L$ for some $g \in G - N_G(H)$; and*

(iii) *If $R \not\cong Z_2 \times Z_2$ and $R^g \cap N_G(L) \neq 1$ for $g \in G - N_G(H)$, then $N_G(L)$ contains a Sylow 2-subgroup of $H^g$.*

Thus when $R \cong Z_2 \times Z_2$, (5.7) only implies (5.8) for *some* $g \in G - N_G(H)$, but not necessarily for all such $g$.

As stated, the theorem combines several of Aschbacher's general results on tightly embedded subgroups. As he himself points out, the proof was strongly influenced by Fischer's work on 3-transposition groups. But the analysis itself also has the flavor of the classical involution proof. In particular, it involves both an analog of the Thompson subgroup (Definition 3.101) and the permutation action of $G$ under conjugation on the set $\Omega$ of $G$-conjugates of $K = O^{2'}(H)$, as well as on the graph $\mathcal{D}$ whose vertices are the elements of $\Omega$ with $I, J \in \Omega$ connected by an edge if and only if $I$ normalizes $J$ and $J$ normalizes $I$.

Finally, as we have remarked above, Aschbacher and Seitz also prove the following result:

THEOREM 5.8. *Every quasisimple K-group is of restricted type.*

The proof requires a detailed knowledge of the structure of the centralizers of involutions acting on quasisimple $K$-groups. The greatest difficulty occurs for the groups of Lie type of characteristic 2, which they study in a separate paper [I: 21]. [Those were the results that Seitz subsequently used in proving his fusion theorem (Theorem 5.2 above).]

5.1.9. On the basis of these tight embedding results, Aschbacher and Seitz were able to establish a major reduction in the $m(H) > 1$ standard problem [I: 22] [the case in which $L/Z(L)$ is a Bender group: $L_2(2^n)$, $U_3(2^n)$, $Sz(2^n)$ is covered by the work of Griess, D. Mason, and Seitz [47], who handled the $m(H) = 1$ case as well].

THEOREM 5.9. *If* $m(H) \geqslant 2$, *then the following conditions hold*:

(i) $R \cong Z_2 \times Z_2$; *and*
(ii) $L \cong A_n$, $n \geqslant 7$, $Sz(8)$, $G_2(4)$, $L_3(4)$, *or* $L_3(4)/Z_2 \times Z_2$.

[The restriction $n \geqslant 7$ in (ii) comes from the assumption that $L \notin \mathscr{LO}$: $A_5 \cong L_2(5)$ and $A_6 \cong L_2(9)$. One can, in fact, restrict to $n \geqslant 8$, since the case $n = 7$ is covered by Foote's maximal $A_7$ 2-component theorem (cf. Proposition 4.56, especially Lemma 4.58).]

The proof involves a case-by-case analysis: the groups of Le type of odd characteristic, those of characteristic 2, the alternating groups, and the sporadic groups being studied separately (Aschbacher and Seitz jointly for all but the alternating groups [I: 22], and Aschbacher alone for the alternating groups themselves [4]). Again detailed properties of the centralizers of involutions are required to eliminate all possibilities except those listed in the theorem.

5.1.10. As a corollary of Theorem 5.9, it again follows that $C_{C_x}(L)$ has a normal 2-complement, so that $C_x$ now has the same general shape as in the $m(H) = 1$ case (but, of course, with fewer possibilities for $L$).

As part of their work, Aschbacher and Seitz go on to solve each of these residual centralizer of involution problems (with the exception of the $Sz(8)$ case, which is due to Dempwolff [19], leading to the Rudvalis group $Ru$). However, for expository purposes, we prefer to separate these two phases of the analysis, thereby emphasizing the fact that the general involution standard form problem reduces in due course to a precise set of centralizer of involution problems.

5.1.11. Finally, we note that the complete analysis in the $m(H) > 1$ case (including both the reduction theorem and the solution of the residual

**Table 5.1.** Simple Groups Containing Sporadic
Standard Components

| $L$ | $G$ |
|---|---|
| $M_{12}$ | .3 |
| $M(22)/Z_2$ | $M(23)$ or $M(24)'$ |
| $HS/Z_2$ | $F_5$ |
| $F_2/Z_2$ | $F_1$ |

centralizer of involution problems) requires the $B$-assumption only on the given standard component $L$; no quasisimplicity assumption is needed for any other 2-component of the centralizer of any involution of $G$.

5.1.12. Two final comments which apply for all values of $m(H)$: First, only *eight* of the sporadic possibilities for $L$ possess solutions with $F^*(G)$ simple. Moreover, only four of these occur with $G$ simple: namely, those listed in Table 5.1.

Hence in the remaining cases, the analysis will lead to a contradiction, obtained ultimately by a suitable fusion argument.

5.1.13. The second remark can be viewed as a partial converse to the discussion of the wreathed case.

LEMMA 5.10. *If $G$ is a minimal counterexample to the standard component theorems $(I)$ and $(II)$, then $G$ does not possess a normal subgroup $G_0$ of index 2 with $R \cap G_0 \neq 1$.*

Indeed, in that case $L \leqslant G_0$ as $L$ is perfect. Since $L$ is nonembedded in $G$, $L$ is a component of $C_{G_0}(x_0)$ for any involution $x_0$ of $R \cap G_0$ [if $|Z(L)|$ is even, $x_0$ can, in fact, be taken in $L$], and it follows at once that $L$ is standard in $G_0$. Hence, by the minimality of $G$, $F^*(G_0) = F^*(G)$ is of one of the desired isomorphism types.

## 5.2. The Major Case Division

As already indicated, despite the common objective to reach appropriate recognition theorems, the specific paths that have been followed to pin down the underlying local structure have varied considerably according to the "type" of the given standard component $L$ and this fact has forced a rather elaborate division of the total set of possibilities for $L$. Moreover, there often turned out to be exceptions within many of these major cases: detailed properties of an individual group or family of groups so dominating the analysis that, in effect, the proof in that single case amounted to its own separate theorem. The situation has been even worse: certain of the major categories consist solely of a collection of individual subcases, so that the overall solution is simply the union of the corresponding subtheorems.

For each of the major categories, we shall include a table consisting of the various subcases, the corresponding solutions with $F^*(G)$ quasisimple, and also the names of the individuals who treated these subcases. A hyphen in the $F^*(G)$ column will indicate that no such solution for $F^*(G)$ exists. Moreover, in view of our assumptions on $G$, we exclude all solutions in which $G$ possesses a classical involution. Likewise, we exclude all wreathed solutions.

The case $m(H) \geqslant 2$ is clearly special, and it is natural to consider it first. As noted in Section 5.1, we treat it in two stages, the first of which is the reduction theorem to a precise set of centralizer of involution problems.

### 5.2.1. The Bender Case

Since the Aschbacher–Seitz reduction theorem (Theorem 5.9) depends, as we have said, on a prior solution of the case in which $L$ is of Lie type of Lie rank 1 over $GF(2^n)$, due to Griess, D. Mason, and Seitz, we begin with a discussion of this case. Moreover, as we have pointed out, they cover both possibilities $m(H) \geqslant 2$ and $m(H) = 1$ within their analysis. In the case $m(H) \geqslant 2$, we state only the reduction portion of their argument, deferring the characterization of $Ru$ to case 5.2.3 below; while in the case $m(H) = 1$, we give the complete solution for $F^*(G)$:

$$\text{If } m(H) \geqslant 2, \text{ then } L \cong Sz(8) \text{ and } R \cong Z_2 \times Z_2. \tag{5.9}$$

On the other hand, if $m(H) = 1$, we have the results shown in Table 5.2.

This is a very difficult case, primarily because a maximal parabolic subgroup $P$ of $L/O(L)$ has such a simple structure $[O_2(P) \in \mathrm{Syl}_2(L/O(L))$ and $P/O_2(P)$ is cyclic of prescribed (odd) order and prescribed action on $O_2(P)]$. As a consequence, the pushing-up process required to pin down a Sylow 2-subgroup of $G$ is very delicate. The case $n = 3$ turns out to be

**Table 5.2.** The Bender Case with $m(H) = 1$

| $L/O(L)$ | $F^*(G)$ |
|---|---|
| $L_2(2^n), n \geqslant 3$ | $L_2(2^{2n})$ or $L_3(2^n)$ |
| $U_3(2^n), n \geqslant 2$ | $L_3(2^{2n})$ |
| $Sz(2^n), n \geqslant 3$ | $PSp_4(2^n)$ |
| $Sz(8)/Z_2$ | — |

exceptional; in particular, it is this case which leads to the single noncyclic solution (5.9).

Furthermore, the structure and embedding of the final "pushed-up" 2-local subgroup $P^*$ of $G$ is "too loose" to allow one to directly construct a $(B, N)$-pair subgroup within $G$ as candidate for $F^*(G)$. Rather it is shown that either $G$ has a normal subgroup of index a power of 2 with Sylow 2-subgroups of class $\leqslant 2$ or containing a strongly closed abelian subgroup (the case $R = \langle x \rangle$), in which case $F^*(G)$ can be determined from the appropriate classification theorem [I, 4.128, 4.129], or else a fusion contradiction is reached (the case $R > \langle x \rangle$ and (5.9) fails).

### 5.2.2. The Aschbacher–Seitz Reduction Theorem

We shall not comment further at this time on Theorem 5.9; we have already sufficiently stressed its reliance on properties of tightly embedded subgroups and the structure of centralizers of involutions acting on quasisimple $K$-groups.

### 5.2.3. The Case $m(H) \geqslant 2$

The solution of the residual centralizer of involution problems from the Aschbacher–Seitz reduction theorem provides a good illustration of some of the observations of Section 5.1, since it splits into a generic case ($A_n$, $n \geqslant 8$) and four exceptional cases. The corresponding table is Table 5.3.

(We note that Aschbacher's argument in the $A_n$ case includes $n = 5$, 6, and 7, but these cases require somewhat special treatment.) For $n \geqslant 8$, his argument is completely uniform. Aschbacher's basic idea is to reconstruct *within* $G$ the so-called "root four-subgroup" graph $\mathscr{D}$ of $A_{n+4}$ and to show that $G$ acts under conjugation as a group of automorphisms of $\mathscr{D}$. Using the

**Table 5.3.** The Case $m(H) \geqslant 2$

| $L/Z(L)$ | $F^*(G)$ | Treated by |
|----------|----------|------------|
| $A_n$ | $A_{n+4}$ | Aschbacher |
| $Sz(8)$ | $Ru$ | Dempwolff |
| $L_3(4)$ | $He$ or $Suz$ | Aschbacher, Seitz, or Reifart |
| $G_2(4)$ | $.1$ | Aschbacher and Seitz |

easily verified fact that $\text{Aut}(\mathcal{D}) \cong \Sigma_{n+4}$, he is then able to identify $F^*(G)$ as $A_{n+4}$. It should also be noted that in the course of the analysis he forces $Z(L) = 1$, so that the $\hat{A}_n$ cases all lead to contradictions.

On the other hand, in the $Sz(8)$, $L_3(4)$, $L_3(4)/Z_2 \times Z_2$, and $G_2(4)$ cases, one pushes up appropriate maximal parabolic subgroups of $L$, the aim being to force the exact structure of the centralizer of a 2-central involution of $G$, in which case one can invoke the corresponding centralizer of involution characterization theorem to identify $F^*(G)$.

Thus from this point on, we can (and do) operate under the following assumption:

$$m(H) = 1, \text{ whence } R \text{ is cyclic}$$
$$\text{(as } G \text{ does not possess a classical involution).} \tag{5.10}$$

### 5.2.4. The Alternating Case

(Again the cases $n = 5$ and $6$ are excluded as $L \notin \mathcal{L}\mathcal{O}$, and $\hat{A}_7$ is excluded as $G$ does not possess a classical involution. Likewise, we can assume that $L/O(L) \ncong A_7$, since in that case $F^*(G)$ is determined by the Fritz–Harris–Solomon theorem (Theorem 2.161) and the sectional 2-rank $\leqslant 4$ theorem. Thus again we can restrict ourselves to the case $n \geqslant 8$ [in particular, $O(L) = 1$].

The complete analysis in this case is due to Solomon [113], [I: 264 and I: 266]. (see Table 5.4).

First of all, in the $\hat{A}_n$ case, easy fusion analysis reduces one to the case $C_x = \hat{A}_n$, in which case $F^*(G)$ is determined from [I, 2.32 and 3.50]. Thus it is the $L \cong A_n$ case which involves the bulk of the effort.

This time the generic argument begins with $n = 12$, with the cases $8 \leqslant n \leqslant 11$ requiring special treatment. In the generic case, one begins with an involution $y \in L$ that can be represented as a product of two

**Table 5.4.** The Alternating Case

| $L$ | $F^*(G)$ |
|---|---|
| $\hat{A}_8$ | $Mc$ |
| $\hat{A}_{11}$ | $Ly$ |
| $A_n, n \neq 8$ or $10$ | $A_{n+2}$ |
| $A_8$ | $A_{10}, L_4(4)$ or $HS$ |
| $A_{10}$ | $A_{12}$ or $F_5$ |

transpositions, so that $I = L\big(C_L(y)\big) \cong A_{n-4}$. The pump-up $K$ of $I$ in $L_y$ is then shown to contain $I$ properly and so by induction and $L$-balance, one obtains either

$$K \cong A_{n-2} \quad \text{or} \quad A_{n-4} \times A_{n-4}. \tag{5.11}$$

[The possibilities $K \cong L_4(4)$ or $HS$ with $n - 4 = 8$ or $F_5$ with $n - 4 = 10$ are easily excluded.] Note also that condition (D) of Hypothesis S ensures that $K$ is semisimple. In the second case, the involution $x$ interchanges the two components of $K$, so that $I$ is a diagonal of $K$. (It is this latter situation which leads, of course, to the wreathed solution.) In the first case, Solomon uses the standard presentation of the symmetric group to prove that

$$G^* = \langle L, K \rangle \cong A_{n+2} \quad \text{and} \quad G^*\langle x \rangle \cong \Sigma_{n+2}. \tag{5.12}$$

Now as observed in Remark 5.1.7, $G$ possesses a normal subgroup $G_0$ of index 2 and Holt's theorem forces $G^* = F^*(G^*) = F^*(G)$. (Actually Solomon verifies Aschbacher's strong embedding criterion [I, 4.31], which Holt's theorem generalizes.)

The exceptional cases require considerably more local analysis, again involving pushing-up of 2-locals. Furthermore, the argument also utilizes the following:

(a) Sylow 2-group characterization theorems to identify the groups $A_m$, $10 \leqslant m \leqslant 13$, $L_4(4)$, and $HS$.

(b) Solutions of the $L_4(4^m)$, $m \geqslant 1$, $PSp_4(4)$, $HS$, $HS/Z_2$, and $F_5$ standard form problems both to identify $L_4(4)$ and $F_5$ [with corresponding standard component $PSp_4(4)$ and $HS/Z_2$] and to eliminate certain configurations. (5.13)

(Thus the alternating case is not completed until *after* these standard problems are themselves solved.)

## 5.2.5. The $\mathscr{L}_0(3)$ Case

Again there is a generic argument (due to Aschbacher [5]), which covers all but four cases, these exceptions requiring individual analyses. Moreover, the argument in the last of these cases $(L \cong {}^2G_2(3^n))$, which is due to Finkelstein [23], applies equally well when $L \cong J_1$, the point being that the Ree groups of characteristic 3 and the group $J_1$ possess isomorphic Sylow 2-normalizers (a split extension of $E_8$ by a Frobenius group of order 21). To avoid duplication, we therefore include the $J_1$ case here. Note also that $PSp_4(3) \cong U_4(2)$, a group of Lie type of characteristic 2. The complete list of solutions is given by Table 5.5.

The various cases treated by Aschbacher had already been studied earlier: Gomi [36] and Foote [32] for $PSp_4(3)$, H. Suzuki [118] for $L_4(3)$, Finkelstein [26] for $U_4(3)$ (covering only a subcase of the general problem) and Walter [129] for $U_4(3)/Z_2$, $P\Omega_7(3)$, and $P\Omega_8^{\pm}(3)$. However, by placing all these problems in an inductive setting, Aschbacher has provided a self-contained argument, considerably shorter than the combined efforts of these individual treatments.

In essence, Aschbacher's argument is modeled after Harris's and Walter's inductive characterization of $\mathscr{L}\mathscr{O}$, the aim once again being to force $G$ to possess a classical involution. Moreover, in contrast to the earlier treatments, he operates under the following assumption:

> If $J$ is a component of $C_y$ for some $y \in \mathscr{I}(G)$ and either $J/O(J) \in \mathscr{L}_0(3)$ or $J/O(J) \cong L_2(9)$ or $L_2(81)$, and $K$ is a component of the pump-up of $J$ in $C_u$ for some $u \in \mathscr{I}(C_y)$ centralizing $J$, then either $K/O(K) \in \mathscr{L}_0(3)$ or $K/O(K) \cong L_2(9)$ or $L_2(81)$. (5.14)

In other words, Aschbacher assumes the Harris–Walter inductive characterization of $\mathscr{L}\mathscr{O}$ [modulo the solution of the $\mathscr{L}_0(3)$ standard form

Table 5.5. The $\mathscr{L}_0(3)$ Case

| $L/O(L)$ | $F^*(G)$ | Treated by |
|---|---|---|
| $L_4(3)$, $U_4(3)$, $U_4(3)/Z_2$, $PSp_4(3)$, $P\Omega_7(3)$, $P\Omega_8^{\pm}(3)$ | $L_4(4)$ | Aschbacher |
| $L_3(3)$ | — | Harris |
| $U_3(3)$ | $G_2(4)$ | Harris |
| $G_2(3)$ | — | Yamada |
| ${}^2G_2(3^n), J_1$ | $ON$ | Finkelstein |

problem] as well as the solution of all standard form problems for elements of $\mathscr{LO}^* - \mathscr{LO}$. This restriction greatly facilitates the analysis. However, to compensate for this simplification, he is forced to manipulate $SL_2(3)$ "subcomponents," which, being solvable, satisfy only a weakened form of $L$-balance, thereby making portions of the analysis very delicate.

We note also that the $PSp_4(3)$ case leads to an exceptional configuration which Aschbacher is eventually able to identify: namely, $F^*(G) \cong L_4(4)$, by appeal to Timmesfeld's weakly closed $T.I.$ theorem [I, 4.161].

In each of the four exceptional cases, one follows the standard 2-local pushing-up procedure. Since $G$ does not contain a classical involution, the $L_3(3)$, $G_2(3)$ cases eventually lead to contradictions (after one forces the structure of a Sylow 2-subgroup of $G$). In the $U_3(3)$ case Harris limits himself to proving that $G$ has a normal subgroup of index 2, which suffices for the noncharacteristic 2 theorem. (See Section 5.7 for a discussion of this point.) In the remaining cases, Finkelstein proves quite easily that $G$ has a normal subgroup of index 2 satisfying the conditions under which O'Nan initially characterized his group (cf. [I, 2.34, 2.35]) and can thus identify $F^*(G)$ from O'Nan's theorem.

Now we come to the case that $L$ is of Lie type of characteristic 2, which by itself breaks up into *six* major subcases, one of which, the Lie rank 1 case, is covered in case 5.2.1 above. Three of the remaining five subcases involve entire families of groups, divided as follows:

1. Nonlinear Lie rank 2.
2. Linear Lie rank $\geqslant 2$.
3. Nonlinear Lie rank $\geqslant 3$.

Unfortunately, there are several cases in which the general analysis

**Table 5.6.** Exceptional Standard Components over $GF(2)$ with Centers of Odd Order

| $L/O(L)$ | $F^*(G)$ | Treated by |
|---|---|---|
| $U_5(2)$ | $L_5(4)$ | Yamada |
| $U_6(2)$ | $L_6(4)$ | Yamada |
| $PSp_6(2)$ | $P\Omega_8^{\pm}(2), U_6(2), U_7(2), L_6(2), PSp_6(4)$ | Gomi |
| $P\Omega_8^+(2)$ | $P\Omega_8^+(4)$ | Egawa |
| $P\Omega_8^-(2)$ | $P\Omega_8^+(4)$ | Alward |
| $^2F_4(2)'$ | $F_4(2)$ | Aschbacher |

**Table 5.7.** Exceptional Standard Components of
Characteristic 2 with Schur Multipliers of Even Order

| $L/O(L)$ | $F^*(G)$ | Treated by |
|---|---|---|
| $L_3(4)/Z_2$ | — | Nah |
| $L_3(4)/Z_4$ | $ON$ | Nah |
| $PSp_6(2)/Z_2$ | — | Seitz and Solomon |
| $U_6(2)/Z_2$ | $M(22)$ | Davis and Solomon |
| $P\Omega_8^+(2)/Z_2$ | — | Egawa and Yoshida |
| $G_2(4)/Z_2$ | — | Seitz and Solomon |
| $F_4(2)/Z_2$ | — | Seitz |
| ${}^2E_6(2)/Z_2$ | $F_2$ | Stroth |

breaks down when $L$ is defined over $GF(2)$. In addition, a few groups over $GF(2)$ [plus $L_3(4)$] possess exceptional Schur multipliers of even order—i.e., not predicted by the general Lie theory. These various cases require separate, individual treatments. They are listed in Tables 5.6 and 5.7.

Since the analyses of the general Lie type characteristic 2 standard form problems proceed by induction, it is clearly necessary to treat these small field exceptional cases first. However, as the arguments in those cases involve primarily variations of the general case analysis, it is preferable for expository purposes to reverse this order.

Thus in discussing cases 1, 2, 3 above, we shall *assume* solutions of the problems listed in Tables 5.6 and 5.7. We shall make no further comments in this section on these exceptional cases.

### 5.2.6. The Nonlinear Lie Rank 2 Case

Since Seitz treats linear groups of characteristic 2 of arbitrary Lie rank $\geqslant 2$ by a uniform argument, the case $L/O(L) \cong L_3(2^n)$, $n \geqslant 2$, is omitted here. Although the proofs are conceptually the same for each of the six families of (nonlinear) groups of Lie rank 2 over $GF(2^n)$, $n \geqslant 2$, nevertheless the actual analyses involve sufficient technical differences that they, too, have received independent treatment (see Table 5.8). [Over $GF(2)$, three of the six cases are included among the exceptions of Table 5.6 $\big(U_4(2), U_5(2), \text{ and } {}^2F_4(2)'\big)$. One is treated within the context of an entire family. Two $\big({}^2D_4(2^n), n \geqslant 1\big)$ are included within the previously considered family $\mathcal{L}_0(3)$ $\big(G_2(2)' \cong U_3(3) \text{ and } U_4(2) \cong PSp_4(3)\big).$]

There is a natural strategy for constructing a $(B, N)$-pair subgroup of $G$ which will turn out ultimately to be $F^*(G)$. Indeed, the groups of Lie type of

**Table 5.8.** Nonlinear Standard Components of
Characteristic 2 and Lie Rank 2

| $L/O(L)$ | $F^*(G)$ | Treated by |
|----------|----------|------------|
| $U_4(2^n)$ | $L_4(2^{2n})$ | Miyamoto |
| $U_5(2^n)$ | $L_5(2^{2n})$ | Miyamoto |
| $PSp_4(2^n)$ | $PSp_4(2^{2n})$ | Gomi |
| $G_2(2^n)$ | $G_2(2^{2n})$ | Yamada |
| $^2D_4(2^n)$ | $^2D_4(2^{2n})$ | Yamada |
| $^2F_4(2^n)$ | $F_4(2^n)$ | Miyamoto and Yamada |

Lie rank $\geqslant 2$ over $GF(2^n)$ are, as we know, generated by their minimal parabolic subgroups containing a given Sylow 2-subgroup [I, 2.18] and hence those of Lie rank 2 are generated by any two such *maximal* parabolic subgroups (since minimal and maximal parabolics are the same in this case). The idea then is to construct with $G$ candidates $P_1^*, P_2^*$ for such a pair of maximal parabolic subgroups by pushing up the two maximal parabolic subgroups $P_1, P_2$ of $L$ containing a given Sylow 2-subgroup of $L$. The goal then is to show that the subgroup $G^* = \langle P_1^*, P_2^* \rangle$ is an (identifiable) split $(B, N)$-pair, its isomorphism type depending on that of $L$. Moreover, either $G^*$ is quasisimple or we have the wreathed case $G^* = L_1 \times L_2$, where $L_1 \cong L_2 \cong L$.

But now as observed in Remarks 5.1.5 and 5.1.6, a further analysis of fusion allows one to invoke an appropriate fusion theorem to force $G^* = F^*(G)$.

### 5.2.7. The Linear Rank $\geqslant 2$ Case

As remarked in Section 5.1, this case was treated by Seitz within a uniform framework. Note the restriction $n \geqslant 2$ when $m = 3$ or $4$ [$L_3(2) \cong L_2(7) \in \mathscr{LO}$, while $L_4(2) \cong A_8$ is already covered within case 5.2.3]. For $m \geqslant 5$, $n$ is an arbitrary positive integer. (See Table 5.9.)

The reason Seitz separates out the linear case is that the methods he uses to treat the general Lie rank $\geqslant 3$ case break down when $L \cong L_m(2^n)$

**Table 5.9.** Linear Standard Components of
Characteristic 2 and Lie Rank $\geqslant 2$

| $L/O(L)$ | $F^*(G)$ |
|----------|----------|
| $L_m(2^n), m \geqslant 3$ | $L_m(2^{2n})$ |

with $m = 4$ or 5. On the other hand, the alternative approach which he develops to handle the latter two families turns out to be equally effective for linear groups of arbitrary Lie rank $\geqslant 2$; he therefore finds it easier to treat all such linear groups together.

In fact, Seitz's argument is essentially the same as in the Lie rank 2 case just described. Although $L$ now possesses $m - 1 \geqslant 2$ maximal parabolic subgroups containing a given Sylow 2-subgroup of $L$, it suffices to focus on just *two* of them, $P_1$ and $P_2$: namely, identifying $L$ with a homomorphic image of $\hat{L} = SL_m(2^n)$ and considering the natural representation of $\hat{L}$ on a vector space $V$ of dimension $2m$ over $GF(2^n)$, he takes $P_1, P_2$ to be the images in $L$ of the subgroup of $\hat{L}$ leaving invariant, respectively, a one-dimensional subspace $V_1$ or a hyperplane $V_{2m-1}$ of $V$, with $V_1 < V_{2m-1}$ (this last condition is required to ensure that $P_1 \cap P_2$ contains a Sylow 2-subgroup of $L$). Again one pushes up $P_1, P_2$ to $P_1^*, P_2^*$ and shows that $G^* = \langle P_1^*, P_2^* \rangle$ is a split $(B, N)$-pair. Here we either have the wreathed case or else $G^* \cong L_m(2^{2n})$. Again the proof is completed by forcing $G^* \lhd G$.

### 5.2.8. The Nonlinear Rank $\geqslant 3$ Case

Again this case is due entirely to Seitz. [Likewise the exceptional cases $U_6(2)$, $PSp_6(2)$, and $P\Omega_8^{\pm}(2)$ are omitted here.] Table 5.10 lists the possible solutions.

In treating this, the general Lie type characteristic 2 case, Seitz drops the parabolic approach and instead works with an equally natural generation of a group of Lie type in terms of the centralizers of suitable *semisimple* elements [over $GF(2^n)$, such elements are necessarily of *odd* order]. This turns out to be the most efficient method of getting at the Curtis–Steinberg

**Table 5.10.** Nonlinear Standard Components of Characteristic 2 and Lie Rank $\geqslant 3$

| $L/O(L)$ | $F^*(G)$ |
|---|---|
| $P\Omega_m^{\pm}(q)$, $m$ even | $P\Omega_m^+(q^2)$ |
| $PSp_m(q)$ | $L_m(q), L_{m+1}(q), P\Omega_{m+2}^{\pm}(q),$ |
|  | $U_m(q), U_{m+1}(q), PSp_m(q^2)$ |
| $U_m(q)$ | $L_m(q^2)$ |
| $E_m(q)$, $m = 6, 7, 8$ | $E_m(q^2)$ |
| $^2E_6(q)$ | $E_6(q^2)$ |
| $F_4(q)$ | $E_6(q), {}^2E_6(q), F_4(q^2)$ |

relations in the present situation, and is essentially the same method that Aschbacher uses in the classical involution theorem [cf. equation (3.42)].

Let $G^* = G(q)$ be a group of Lie type over $GF(q)$ and again consider the Lie description of $G^*$ given in Section 3.5. For clarity, we let $J^*$ be the corresponding $SL_2(q)$-subgroup of Proposition 3.51. Then $J^*$ contains a cyclic subgroup $Y^*$ of order $q + 1$. This is the semisimple subgroup with which Seitz works. (In the orthogonal case, $Y^*$ is defined in a slightly different way.)

Now set

$$X^* = L\big(C_{G^*}(Y^*)\big). \tag{5.15}$$

The same argument that yields (3.42) shows that

$$G^* = \langle X^*, X^{*y} \rangle \tag{5.16}$$

for some $y \in G^*$.

We note parenthetically that $X^*$ can be viewed as a "standard component" for $G^*$ *relative to centralizers of elements of odd order*, and the generational assertion (5.16), in effect, constitutes a solution of the corresponding standard form problem. Such *odd* standard form problems turn out to have a central role in the classification of groups of *characteristic 2 type*, much the same as the present involution standard form problems for the classification of groups of component type.

It is this description of $G^*$ which motivates Seitz's attack. He begins with the given standard component $L$ and first considers the corresponding $SL_2(2^n)$-subgroup $J$ of $L$, semisimple subgroup $Y$ of $J$ of order $2^n + 1$ (again with slight variation in the orthogonal case), and finally the layer $X = L\big(C_L(Y)\big)$.

The critical problem is now the identification of the pump-up $X^*$ of $X$ in $C_Y = C_G(Y)$. (Note that as $Y$ has odd order, we cannot invoke $L$-balance.) Nevertheless, it is not difficult to prove:

$$X \text{ is a component of } C_{C_Y}(x) \text{ and } X \text{ is standard in } C_Y. \tag{5.17}$$

The more serious issue is to show that $X$ pumps-up *properly* in $C_Y$. It is to establish this conclusion that Seitz needs his fusion theorem (Theorem 5.2). Indeed, he knows (again apart from the orthogonal case, in which a second involution fusion pattern must be allowed) that for some element $g \in G$,

$$x^g \in L\langle x \rangle - \langle x \rangle \text{ and } x^g \text{ centralizes } Y. \tag{5.18}$$

Using (5.18) and some additional analysis, Seitz proves

$$X \text{ is not normal in } C_Y. \qquad (5.19)$$

Now he can read off the possibilities for $X^* = F^*(C_Y)$ either by means of induction or in certain low Lie rank cases by appeal to the solutions of previously considered standard form problems (this includes, of course, the proverbial direct product possibility for $X^*$, leading to the wreathed case). With this information, he can now construct the sought-after subgroup $G^*$ of $G$ of Lie type: namely,

$$G^* = \langle X^*, X^{*y} \rangle, \qquad (5.20)$$

with $y$ a specified element of the pre-image of the Weyl group of $L$, relative to the appropriate Borel subgroup. Again the complete analysis takes an entire paper.

In the third and final paper of the series, Seitz carries out the necessary fusion analysis to force $G^* = F^*(G)$; it is a very elaborate argument.

Taken together, Seitz's contribution in these three papers is very impressive and reveals his consummate mastery of the entire Lie machinery.

We come at last to the sporadic groups. Although it has been possible to combine a few cases within similar patterns of argument, in essence every sporadic group and also every one of their nontrivial covering groups has required its own separate analysis, so this, too, constitutes a very elaborate chapter of finite simple theory. The complete solution is the work of many individuals, but the bulk of the effort is due to Finkelstein [23–29] and Solomon [18, 29, 113, 114 and I: 264]. We divide the sporadic standard form problem into two major cases.

### 5.2.9. The Sporadic Groups in Hypothesis $S$

It is natural to single out those sporadic groups that enter into Hypothesis $S$ (omitting $J_1$ which has already been covered under the $\mathcal{L}_0(3)$ case. These are listed in Table 5.11.

### 5.2.10. The Remaining Sporadic Groups

Table 5.12 is the corresponding table for the remaining sporadic group standard form problems.

**Table 5.11.** The Sporadic Groups Occurring in Hypothesis $S$

| $L/O(L)$ | $F^*(G)$ | Treated by |
|---|---|---|
| $M_{12}$ | $.3, Suz$ | Finkelstein, Solomon, Yoshida |
| $M_{12}/Z_2$ | — | Finkelstein and Solomon |
| $J_2$ | $Suz$ | Finkelstein and Solomon |
| $J_2/Z_2$ | — | Finkelstein and Solomon |
| $J_3$ | — | Finkelstein |
| $Mc$ | — | Solomon |
| $Ly$ | — | Solomon |
| $HS$ | — | Solomon |
| $HS/Z_2$ | $F_5$ | Finkelstein and Harada |
| $He$ | — | Griess and Solomon |
| $Suz$ | — | Finkelstein and Solomon |
| $Suz/Z_2$ | — | Finkelstein, Solomon, and Wright |
| $ON$ | — | Solomon |
| $.3$ | — | Finkelstein and Solomon |
| $.1$ | — | Solomon |
| $.0 (= .1/Z_2)$ | — | Finkelstein |
| $F_5$ | — | Solomon |

## 5.3. The Bender Case

We briefly outline the Griess–Mason–Seitz solution of the rank 1 characteristic 2 standard form problem. However, we first fix some additional notation, which we preserve for the remainder of the chapter. We let $Q \in \mathrm{Syl}_2(L)$ (so that $Q$ centralizes $R$) and we let $T \in \mathrm{Syl}_2(N_G(L))$ with $QR \leqslant T$.

**Table 5.12.** The Sporadic Groups not in Hypothesis $S$

| $L/O(L)$ | $F^*(G)$ | Treated by | $L/O(L)$ | $F^*(G)$ | Treated by |
|---|---|---|---|---|---|
| $M_{11}$ | $Mc$ | Gorenstein and Harada | $M(22)/Z_2$ | $M(23), M(24)'$ | Hunt, Davis and Solomon |
| $M_{22}$ | — | Finkelstein | $M(23)$ | $M(24)'$ | Solomon |
| $M_{22}/Z_2$ | — | Harada | $M(24)'$ | — | Solomon |
| $M_{22}/Z_4$ | — | Griess | $Ru$ | — | Solomon |
| $M_{23}$ | — | Finkelstein | $Ru/Z_2$ | — | Finkelstein |
| $M_{24}$ | — | Egawa | $F_3$ | — | Syskin |
| $J_4$ | — | Finkelstein | $F_2$ | — | Solomon |
| $.2$ | — | Manferdelli | $F_2/Z_2$ | $F_1$ | Davis and Solomon |
| $M(22)$ | — | Solomon | $F_1$ | — | Solomon |

Even in this low-rank problem, the 2-local pushing-up process requires an inductive approach, so that they are forced to establish a more general theorem (which we shall not attempt to state here), in which the assumption that $F^*(G)$ is simple is dropped (allowing a wreathed solution) and, in addition, $L$ need not be quasisimple, but simply a 2-component [however, the critical condition that $C_G(L/O(L))$ is tightly embedded is retained].

As noted in Section 5.2.1, their aim is to show that either $R \cong Z_2 \times Z_2$ and $L \cong Sz(8)$ or else $R = \langle x \rangle$ and $O^2(G)$ has Sylow 2-subgroups that are either of class $\leqslant 2$ or contain a nontrivial strongly closed abelian subgroup.

The proof is by contradiction, with $G$ taken as a minimal counterexample. We begin by describing the maximal parabolic subgroup $P$ of $L$ containing $Q$. However, for simplicity of notation, we omit the $Sz(8)/Z_2$ and $Sz(8)/Z_2 \times Z_2$ cases (they are, in fact, eliminated as part of the initial fusion analysis), and also assume that $O(L) = 1$ (whence $L$ is simple). In the present case, $P$ is a Borel subgroup of $L$ and we have

$$Q = O_2(P) \text{ and } P = AQ, \text{ where } A \text{ is a cyclic complement to } Q. \quad (5.21)$$

We put $U = \Omega_1(Q)$ and set $q = 2^n$.

The following lemma gives a more precise picture of the structure of $P$.

LEMMA 5.11. *The following conditions hold*:

(i) $U = Z(Q)$ *is elementary of order* $q$;

(ii) $A/C_A(U)$ *has order* $q - 1$ *and permutes the involutions of* $U$ *trasitively*;

(iii) *One of the following holds*:

(1) $L \cong L_2(q)$, $Q = U$ *is elementary, $A$ has order* $q - 1$, *and $P$ is a Frobenius group*;

(2) $L \cong Sz(q)$, $Q$ *has order $q^2$ and class 2, $A$ has order* $q - 1$, *and $P$ is a Frobenius group; or*

(3) $L \cong U_3(q)$, $Q$ *has order $q^3$ and class 2, $A$ has order $(q^2 - 1)/d$, where $d =$ g.c.d.$(3, q + 1)$, and $P/U$ is a Frobenius group*.

Because they are treating both the $m(H) = 1$ and $m(H) > 1$ cases, the initial fusion analysis requires both Theorem 5.2 as well as Aschbacher's general results on tightly embedded subgroups (Theorem 5.7). To obtain all the desired conclusions, they also invoke Goldschmidt's strongly closed abelian 2-group theorem [I, 4.128]. Here is the precise result.

PROPOSITION 5.12. *The following conditions hold*:

(i) $T \notin \mathrm{Syl}_2(G)$;

(ii) $N_G(T)$ *normalizes QR and hence UR*;

(iii) $R$ *is elementary*;

(iv) $R^g \cap T \neq 1$ *for some* $g \in G - N_G(L)$ *(with* $R^g \leqslant T$ *unless* $R \cong Z_2 \times Z_2$*); and*

(v) *No involution of $R$ is G-conjugate to an involution of $U$.*

Now Griess, Mason, and Seitz are ready to begin the pushing up process. We have $H = C_L$ and $R \in \mathrm{Syl}_2(H)$. Set $V = UR$ and $N = N_G(V)$. Their basic strategy is to analyze the action of $N$ on the set $\Omega$ of $G$-conjugates of $R$ which lie in $V$. Clearly $N$ acts by conjugation on $\Omega$. Hence if we let $N^0$ be the subgroup of $N$ acting trivially on $\Omega$, then the group $\bar{N} = N/N^0$ is a *permutation group on the set $\Omega$*. We also let $N^R = N_N(R)$, so that $\bar{N}^R$ is the stabilizer of $\{R\}$ in $\bar{N}$. Clearly, $N^R$ normalizes $L$, so $P \lhd N^R$. Moreover, it follows easily that

$$\bar{N}^R / \bar{P} \cong N_G(L)/LH, \qquad (5.22)$$

and hence is isomorphic to a (cyclic) group of outer automorphisms of $L$.

As a direct consequence of the proposition, they obtain the following basic properties of $N$.

PROPOSITION 5.13. *The following conditions hold*:

(i) $|\Omega| = q$;

(ii) *The elements of $\Omega$ are disjoint subgroups of $V$ and their union is precisely* $V - U^{\#}$;

(iii) *$U$ is strongly closed in $V$ with respect to $G$ (in particular, $U \lhd N$)*;

(iv) *$\bar{N}$ acts doubly transitively on $\Omega$; and*

(v) *$\bar{A} \lhd \bar{N}^R$, $\bar{A}$ has order $q - 1$ and acts transitively on $\Omega - \{R\}$, and $\bar{N}^R/\bar{A}$ is cyclic.*

In particular, every $G$-conjugate of $R$ contained in $V$ is an $N$-conjugate of $R$, so $N$ controls the fusion of $R$ in $V$.

Since $\bar{A}$ is cyclic, it is clearly a *nilpotent* normal subgroup of $\bar{N}^R$. But then as $\bar{A}$ acts transitively on $\Omega - \{R\}$, $\bar{N}$ is, in fact, a *split $(B, N)$-pair of rank* 1. Hence the structure of $\bar{N}$ is completely determined by (the general form of) the classification theorem concerning such groups [I, 3.39], which asserts that if $X$ is an arbitrary such split $(B, N)$-pair of rank 1 acting on a set $\Omega$ of cardinality $q$ and $X^1$ is a one-point stabilizer, then either

(1) $X = X^1 Y$, where $Y$ is a normal subgroup of $X$ of order $q$ which acts transitively (and hence regularly) on $\Omega$; or

(2) $X \cong L_2(r)$, $U_3(r)$, $Sz(r)$, or $^2G_2(r)$ for some prime power $r$.

(5.23)

However, in our case $(X = \bar{N})$, the nilpotent normal subgroup of $X^1$ is, in fact, cyclic, which excludes all but the single possibility $q = 8$ and $X \cong L_2(7)$ $(\cong L_3(2))$ in (2). Since also $q = 2^n$ in our case, we thus conclude:

PROPOSITION 5.14. *One of the following holds*:

(i) $\bar{N} = \bar{N}^R O_2(\bar{N})$, *where* $O_2(\bar{N})$ *is elementary of order* $q$ *and acts regularly on* $\Omega$; *or*

(ii) $q = 8$ *and* $\bar{N} \cong L_3(2)$.

[In fact, $\bar{A}O_2(\bar{N})$ is easily also seen to be a Frobenius group.]

It is the exceptional case (ii) of the proposition which will eventually lead to Rudvalis's group. However, Griess, Mason, and Seitz argue that the exceptional case cannot occur at this, the first stage of the pushing up process. Indeed, if (ii) holds, then they prove: $R \cong Z_2$, $L \cong L_2(8)$, and $G$ has a normal subgroup $G_0$ of index 2 not containing $x$. But then $U = Q \in \mathrm{Syl}_2(C_{G_0}(U))$ and as $U \cong E_8$, Harada's self-centralizing $E_8$ theorem (Theorem 2.156) is therefore applicable and yields that $G_0$ has sectional 2-rank $\leqslant 4$. Thus the possibilities for $G_0$ and hence also for $G$ are determined from that classification theorem. But it is easily checked that in no case does $G$ contain both an involution $x$ whose centralizer has an $L_2(8)$ component and a 2-local subgroup of the structure of $N$. Thus we have:

PROPOSITION 5.15. *Proposition* 5.14(i) *holds*.

Now Griess, Mason, and Seitz are ready for the second stage of the pushing-up process. For simplicity, we limit the discussion to the linear case, adding some brief comments about the Suzuki and unitary cases. Thus through Proposition 5.22 we assume

$$L \cong L_2(q). \tag{5.24}$$

In this case, we have $P = AQ = AU$ is a Frobenius group of order $q(q-1)$ and $N^0 = V = UR$. Let $V_1$ be the pre-image of $O_2(\bar{N})$ in $N$, so that $V_1$ is an $A$-invariant 2-group with $V_1/V = \bar{V}_1$ elementary of order $q$. In view of the transitive action of $A$ on $\bar{V}_1^\#$, it is not difficult to prove that $V_1/U$ splits over $V/U$. Hence we obtain (in the linear case) the following result:

LEMMA 5.16. *We have* $V_1 = U_1 R$, *where*

(i)   $U_1 \lhd V_1 A$;
(ii)  $U_1 \cap R = 1$;
(iii) $|U_1| = q^2$; *and*
(iv)  $U_1 A$ *is a Frobenius group.*

This time they consider $N_1 = N_G(V_1)$ and $\tilde{N}_1 = N_1/U$ and they let $\Omega_1$ denote the set of $\tilde{N}_1$-conjugates of $\tilde{R}$ in $\tilde{V}_1$. Similarly, let $\tilde{N}_1^0$ be the subgroup of $\tilde{N}_1$ acting trivially on $\Omega_1$, $\tilde{N}_1^R$ the subgroup of $\tilde{N}_1$ leaving $\tilde{R}$ invariant, and put $\bar{N}_1 = \tilde{N}_1/\tilde{N}_1^0$, so that again $\bar{N}_1$ is a permutation group on $\Omega_1$ and $\bar{N}_1^R$ is the stabilizer of $\tilde{R}$ in $\bar{N}_1$.

Analyzing this configuration, they next prove:

PROPOSITION 5.17. *One of the following holds*:

(i)  *A Sylow 2-subgroup of* $V_1 N^R$ *is a Sylow 2-subgroup of* $G$; *or*
(ii) *Proposition 5.13 holds with* $\Omega_1, \bar{N}_1, \bar{N}_1^R, \tilde{V}_1, \tilde{U}_1$, *and* $\tilde{R}$ *in the roles of* $\Omega, \bar{N}, \bar{N}^R, V, U$, *and* $R$, *respectively.*

In this way, Griess, Mason, and Seitz develop an inductive procedure, constructing a *sequence* of doubly transitive permutation groups $\bar{N} = \bar{N}_0$, $\bar{N}_1, \bar{N}_2,...$ acting on corresponding sets $\Omega, \Omega_1, \Omega_2$, of conjugates of $\tilde{R}$ in suitable 2-groups $\tilde{V} = V$, $\tilde{V}_1 = V_1/U_1$, $\tilde{V}_2 = V_2/U_2,...$.

There are then two ways in which the process will terminate at a given stage $m \geqslant 2$: namely, either

$$(1)\ \ \bar{N}_m \cong L_3(2);\ \text{or}$$
$$(2)\ \ \text{A Sylow 2-subgroup of } V_m N^R \text{ is a Sylow 2-subgroup of } G. \tag{5.25}$$

To complete the analysis, they must clearly first pin down the structure of the 2-groups $V_m$ and $U_m$. Inductively, the same argument as in the $V_1$ case yields

LEMMA 5.18. *Lemma* 5.16 *holds with* $V_m, U_m$ *in place of* $V_1, U_1$, *respectively, but with* $|U_m| = q^{m+1}$. *In particular,* $V_m = U_m R$ *with* $U_m \triangleleft V_m$ *and* $U_m \cap R = 1$.

This is achieved by appeal to the following general result, proved independently by Finkelstein [23] and Landrock and Solomon [75]:

LEMMA 5.19. *Let* $U^*$ *be a 2-group acted on by a group* $A^* \times \langle x^* \rangle$, *where* $A^*$ *is cyclic of order* $q - 1 = 2^n - 1$ *for some n and* $x^*$ *is an involution. If* $C_{U^*}(x^*)$ *is elementary of order q with* $A^*$ *transitively permuting the involutions of* $C_{U^*}(x^*)$, *then one of the following holds:*

   (i) $U^*$ *is isomorphic to a Sylow 2-subgroup of* $L_3(q)$ *or* $U_3(q)$;
  (ii) $U^*$ *is homocyclic abelian of rank n and either* $x^*$ *inverts* $U^*$ *or every involution of* $U^* x^*$ *is* $U^*$-*conjugate to* $x^*$; *or*
 (iii) $U^*$ *is elementary abelian of rank 2n and every involution of* $U^* x^*$ *is* $U^*$-*conjugate to* $x^*$.

*In particular,* $U^*$ *has class* $\leqslant 2$.

The lemma applies with $U_m, A$, and any involution of $R$ in the roles of $U^*, A^*$, and $x^*$. In particular, it gives the possible structures of $U_m$ and also implies the following:

LEMMA 5.20. *We have* $R = \langle x \rangle$. *(Thus* $V_m = U_m \langle x \rangle$.)

Griess, Mason, and Seitz now consider the two cases of (5.25) separately, first proving the following result:

PROPOSITION 5.21. *We have* $\bar{N}_m \not\cong L_3(2)$ *for any* $m \geqslant 2$.

Indeed, if not, they argue that $n = 3$, that $U_m$ is necessarily homocyclic abelian of rank 3, and that $N_m$ contains a Sylow 2-subgroup of $G$. Then by Thompson transfer, $G$ possesses a normal subgroup $G_0$ of index 2 and it follows that $N_m \cap G_0$ is a 2-local subgroup of the form studied by O'Nan in [I, 2.34, 2.35]. Since $m \geqslant 2$, they conclude from his theorem that $G_0 \cong HS$ or $ON$. However, neither of these groups admits an automorphism of order 2 whose centralizer has an $L_2(8)$ component.

Finally they prove:

PROPOSITION 5.22. *We have* $F^*(G) \cong L_2(q) \times L_2(q)$, $L_2(q^2)$, $L_3(q)$, *or* $U_3(q)$.

By the previous proposition, $(5.25)(2)$ holds and hence $V_m N^R = U_m \langle x \rangle N^R$ contains a Sylow 2-subgroup $S$ of $G$ ($S/V_m$ is cyclic, inducing a group of field automorphisms on $L$). In this case, a delicate fusion analysis shows that one of the following three possibilities must occur:

(1) $S$ contains a nontrivial abelian subgroup that is strongly closed in $S$ with respect to $G$;

(2) $G$ contains a normal subgroup $G_0$ with Sylow 2-subgroup $U_m$; or

(3) $m = 3$, $U_m$ is isomorphic to a Sylow 2-subgroup of $L_3(q)$ or $U_3(q)$, and $C_x$ contains an involution $t$ which induces a nontrivial field automorphism on $L$ (whence $n$ is even).

If (1) holds, then $F^*(G)$ is determined from Goldschmidt's theorem [I, 4.128], while if (2) holds, then $G$ has Sylow 2-subgroups of class $\leqslant 2$ and $F^*(G)$ is determined from the corresponding classification theorem [I, 4.129]. In either case, $F^*(G)$ has one of the listed forms.

On the other hand, if (3) holds, they set $L_0 = L(C_L(t))$ [whence $L_0 \cong L_2(q^{1/2})$] and $\bar{C}_t = C_t/\langle t \rangle$. Then $\bar{L}_0$ is a component of $C_{\bar{C}_t}(\bar{x})$ and they argue that $\bar{L}_0$ is standard in $\bar{C}_t$. Assuming as they may that $G$ is a minimal counterexample to the proposed theorem, they now identify $F^*(\bar{C}_t)$ and then derive a fusion contradiction.

The Suzuki and unitary cases are technically considerably more complicated, since now we have $Q > U$. In these cases, the analog of Lemma 5.16 reads [here $V_1$ again denotes the pre-image of $O_2(\bar{N})$ in $N$, which we assume to be $A$-invariant]:

LEMMA 5.23. *We have* $V_1 = W_1 R$, *where*

(i) $W_1 \lhd V_1 A$;
(ii) $W_1 \cap R = 1$;
(iii) $W_1 = Q U_1$, *where* $U_1 \lhd V_1 A$;
(iv) $Q \cap U_1 = U$ *and* $|U_1| = q^2$; *and*
(v) $U_1 A / C_A(U_1)$ *is a Frobenius group with cyclic complement $A/C_A(U_1)$ of order $q - 1$.*

[In the Suzuki case, $C_A(U_1) = 1$, while in the unitary case $|C_A(U_1)| = (q + 1)/d$, $d = 1$ or 3.]

We see then that in the Suzuki and unitary cases, the pushing-up process will yield a sequence of *partial* complements $U_1, U_2, \ldots$ to $Q$ (with $U_i \cap Q = U$ for each $i$), and again the structure of each $U_i$ will be determined from Finkelstein's Lemma 5.19. Of course, to reach even the group $U_2$, one must first prove double transitivity of the corresponding permutation group $\bar{N}_1$ and then eliminate the possibility $\bar{N}_1 \cong L_3(2)$. Both of these assertions are considerably more difficult to establish than in the linear case.

Furthermore, in contrast to the linear case, the pushing-up process proceeds only as far as stage *three*. Indeed, Griess, Mason, and Seitz focus their analysis on the group $V_2 = W_2 R = Q U_2 R$. Determination of its structure depends on both fusion analysis and a study of the normalizers in $G$ of certain nontrivial subgroups of $W_2$.

Note that in the Suzuki case, $\text{Out}(L)$ has odd order, so $QR \in \text{Syl}_2(N_G(L))$. In this case, they prove:

PROPOSITION 5.24. *If $L \cong Sz(q)$, then one of the following holds*:

(i) $q = 8$ and $R \cong Z_2 \times Z_2$; or
(ii) $V_2 = W_2 R \in \text{Syl}_2(G)$ and $W_2 = Q U_2$ has class 2.

The case $q = 8$, $R \cong Z_2 \times Z_2$, is, of course, one of the allowed conclusions of the reduction Theorem 5.2 (leading to the group $Ru$). On the other hand, if (ii) holds, a contradiction is reached (a) when $R = \langle x \rangle$, by showing that $G$ has a normal subgroup of index 2 with Sylow 2-subgroup $W_2$ and then invoking the class 2 Sylow 2-group classification theorem, and (b) when $R > \langle x \rangle$, by analyzing the fusion of $R$ in $V_2$. [In particular, note that the $m(H) \geqslant 2$ case is not eliminated until the very end of the proof.]

The unitary case is even more difficult than the Suzuki case, partly because of the more complicated structure of $P [ |Q| = q^3$ and $|A| = (q^2 - 1)/d]$ and partly because $N_G(L)$ can now contain elements inducing field automorphisms on $L$ of *even* order. The general flavor of the proof is similar to that of the Suzuki case. We make only a single observation: namely, the reduction to the case $R = \langle x \rangle$ occurs at the outset of the analysis.

Indeed, $A$ contains a subgroup $A_0$ of order $(q + 1)/d$ and we have

$$L_0 = L\big(C_L(A_0)\big) \cong L_2(q). \tag{5.26}$$

Setting $\bar{C} = C_{A_0}/O(C_{A_0})$, they argue first that $\bar{L}_0$ is standard in $\bar{C}$ with $\bar{H} \leqslant C_{\bar{C}}(\bar{L}_0)$ and $\bar{L}_0$ not normal in $\bar{C}$. But then if $R > \langle x \rangle$, they can conclude by induction together with the prior characterization of $\mathscr{L}\mathscr{O}$ that $q = 4$ [whence $L \cong L_2(4) \cong L_2(5)$] and

$$F^*(\bar{C}) \cong A_9, J_2, \text{ or } M_{12}. \tag{5.27}$$

The first possibility is shown to be incompatible with previously established results concerning the fusion of $R$ in $N_G(L)$, and the latter two possibilities with the structure of $C_x$.

## 5.4. The Aschbacher–Seitz Reduction Theorem

We turn now to the Aschbacher–Seitz reduction theorem (Theorem 5.9). Thus we assume

$$m(R) \geqslant 2. \tag{5.28}$$

[As above, $R \in \mathrm{Syl}_2(H)$ and $H = C_L$.]

In this situation, we can *prove* that $F^*(G)$ is necessarily simple, provided we assume:

$$\begin{array}{l}(1) \ L \text{ is not normal in } G; \text{ and} \\ (2) \ O(G) = 1.\end{array} \tag{5.29}$$

The result depends on the following direct consequence of the definition of standard component:

LEMMA 5.25. *L is normal in $C_y$ for every involution $y \in R^\#$.*

Indeed, if false, then for some such $y$, there is $g \in C_y$ with $L^g \neq L$. Thus $g \notin N_G(L) = N_G(H)$. Since $H$ is tightly embedded in $G$, it follows that $|H \cap H^g|$ is odd. However, as $g$ centralizes $y \in R \leqslant H$, $y \in H \cap H^g$, so $|H \cap H^g|$ is even—contradiction.

PROPOSITION 5.26. *$F^*(G)$ is simple.*

Indeed, as $O(G) = 1$, it follows by $L$-balance that either $L$ is contained in a component $K$ of $G$ or in the product of two components $K_1, K_2$ of $G$ that are interchanged by some involution $x$ of $R$, with $L$ the corresponding "diagonal" of $K_1 K_2$. Consider the latter possibility. Since $R$ centralizes $L$ and induces a permutation on the components of $G$, $R$ leaves $K_1 K_2$ invariant. Since $m(R) \geqslant 2$, some involution $y$ of $C_R(x)$ must therefore leave both $K_1$ and $K_2$ invariant. However, the projection of $L$ on $K_i/Z(K_i)$ covers $K_i/Z(K_i)$ for both $i = 1, 2$, as $L$ is a diagonal of $K_1 K_2$, so $y$ centralizes $K_i/Z(K_i)$ and hence $K_i$, $i = 1, 2$. Thus $K_1 K_2 \leqslant C_y$ and as $L$ is not normal in

$K_1 K_2$, it follows that $L$ is not normal in $C_y$, contrary to the previous lemma. We conclude that $K$ is single component.

If $L = K$, then as $L$ commutes with none of its $G$-conjugates, $L$ must be normal in $G$, contrary to assumption. Thus $L < K$. If $|C_K|$ is even, then as $R \in \mathrm{Syl}_2(H)$ with $H = C_L$, it follows that $R \cap C_K \in \mathrm{Syl}_2(C_K)$, whence some involution $t$ of $R$ centralizes $K$. But then $K \leqslant C_t$ and as $L$ is not normal in $K$, the preceding lemma is again contradicted. Hence $|C_K|$ must be odd. In particular, $K \lhd G$. Since $O(G) = 1$, it follows that $G \leqslant \mathrm{Aut}(K)$ and $K = F^*(G)$. Thus $F^*(G)$ is simple, as asserted.

This is the situation that Aschbacher and Seitz must analyze. As indicated in Section 5.1, their analysis depends heavily on properties of tightly embedded subgroups. We therefore begin by restating Theorem 5.7 in the slightly sharpened form that Aschbacher has established.

We split the full result into two parts.

PROPOSITION 5.27. *The following conditions hold*:

  (i) *$R$ is elementary abelian*;
  (ii) *$RO(H)$ is tightly embedded in $G$; and*
  (iii) *$C_H(y)$ is solvable for each $y \in R^{\#}$.*

Observe next that as $N_G(L) = N_G(H)$, any subgroup of $N_G(L)$ normalizes $H$ and so possesses a Sylow 2-subgroup leaving $R$ invariant.

PROPOSITION 5.28. *Let $g \in G - N_G(L)$ and let $V$ be a Sylow 2-subgroup of $N_{H^g}(L)$ that leaves $R$ invariant. Then the following conditions hold*:

  (i) *We can choose g so that $V \in \mathrm{Syl}_2(H^g)$;*
  (ii) *If $|R| \geqslant 8$ and $V \neq 1$, then $V \in \mathrm{Syl}_2(H^g)$; and*
  (iii) *If $V \neq 1$, then $C_R(V) \cong V$. In particular, if $V \in \mathrm{Syl}_2(H^g)$, then $V$ centralizes $R$.*

Note also that $V$ acts faithfully on $L$. Indeed, $V_0 = C_V(L) \leqslant H = C_L$ and $V \leqslant H^g$, so $V_0 \leqslant H \cap H^g$. Since $H$ is tightly embedded in $G$ and $g \in G - N_G(H)$, this forces $V_0 = C_V(L) = 1$, so $V$ acts faithfully on $L$.

Notice that Proposition 5.27(ii) is closely related to condition (a) of Definition 5.6. Indeed, if $g$ and $V$ are as in Proposition 5.28(i) and we set $X = V(N_{O(H^g)}(L))$, then by Proposition 5.27(ii), $X$ is tightly embedded in $LX$. Thus we have the following result:

LEMMA 5.29. *If* $g \in G - N_G(L)$ *is such that* $N_{H^g}(L)$ *contains a Sylow* 2-*subgroup* $V$ *of* $H^g$, *then* $V$ *is a noncyclic elementary abelian* 2-*group and there exists a subgroup* $A$ *of* $G$ *acting on* $L$ *of the form* $A = VO(A)$ *with* $A$ *tightly embedded in* $LA$.

Hence $V$ and $A$ indeed satisfy the conditions of Definition 5.6(a).

Moreover, one can also show that $V$ satisfies condition (b) of Definition 5.6. We first prove:

LEMMA 5.30. *If* $g$, $V$, *and* $A$ *are as in the preceding lemma, then*

$$\Gamma_{V,1}(L) \leqslant N_L(A).$$

Indeed, it suffices to show that $N_L(W) \leqslant N_L(A)$ for all $1 \neq W \leqslant V$. But if $w \in N_L(W)$, then $W \leqslant A \cap A^w$, so as $A$ is tightly embedded in $LA$ and $|W|$ is even, $w \in N_{LA}(A) \cap L = N_L(A)$, and the assertion follows.

LEMMA 5.31. *If* $g$, $V$, *and* $A$ *are as in Lemma* 5.29, *then*

$$L < \Gamma_{V,1}(L).$$

Indeed, if false, then $L$ normalizes $A$ by the previous lemma, whence $[L, V]$ is a subgroup of $A$ and hence solvable. But $V$ normalizes $L$, so by [I: 130, Theorem 2.2.1(iii)], $[L, V] \lhd L$. Since $L$ is quasisimple and $[L, V]$ is solvable, this forces $[L, V] \leqslant Z(L)$. Thus $V$ centralizes $L/Z(L)$ and consequently $V$ centralizes $L$ (cf. [I, p. 186]), contrary to the fact that $V$ acts faithfully on $L$.

Lemmas 5.29 and 5.31 explain the motivation for the two assumptions in the definition of a quasisimple group of restricted type. Since our group $L$ is a $K$-group, we conclude therefore from Theorem 5.8:

PROPOSITION 5.32. *If* $g$ *and* $V$ *are as in Lemma* 5.29, *then* $V \leqslant LH$ (*i.e.*, $V$ *induces inner automorphisms on* $L$).

Aschbacher and Seitz now divide the proof of their reduction theorem into four major cases according to the isomorphism type of $L$.

Throughout we assume $g$ and $V$ are chosen as in Proposition 5.28(i) and we fix $A$ as in Lemma 5.29. We can also assume, without loss, that $V = R^g$.

PROPOSITION 5.33. *We have* $L \notin \mathscr{L}_0(3)$.

Their argument is quite general and applies, in fact, to any $L$ of Lie type of odd characteristic, excluding $L_2(q)$ or ${}^2G_2(q)$. Indeed, using the Lie description given in Section 3.5 with $L$ as $G(q)$, $q = p^n$, $p$ an odd prime, and taking $J \cong SL_2(q)$ as in Proposition 3.52 and $U \in \mathrm{Syl}_p(J)$, it follows directly from that proposition that

$$L = \langle J, O_p(N_L(U)) \rangle. \tag{5.30}$$

They next argue that the subgroup $Y$ of $L$ generated by the $V$-conjugates of $J$ is a central product of copies of $J$ (if $J$ is $V$-invariant, then, of course, $Y = J$). By its definition, $Y$ is $V$-invariant. Now, using the tight embedding of $A$ in $LA$, they easily force $V$ to centralize $Y$, whence $V$ centralizes $J$ and hence $U$. But then $V$ acts on $W = O_p(N_L(U))$. Since $p$ is odd and $V$ is noncyclic abelian, $[I, 4.13]$ now yields that

$$W = \Gamma_{V,1}(W). \tag{5.31}$$

Since $V$ centralizes $J$, (5.30) and (5.31) together imply that

$$L = \langle J, W \rangle = \Gamma_{V,1}(L), \tag{5.32}$$

contrary to Lemma 5.31.

To complete the proof of the proposition, we are therefore left to eliminate the possibility $L \cong {}^2G_2(q)$, $q = 3^n$, $n$ odd, $n > 1$. However, in this case one easily checks that $L = \Gamma_{V,1}(L)$ for *any* four group $V$ acting on $L$, giving the same contradiction as above. [We note parenthetically that the groups $L_2(q)$, $q$ odd, $q > 9$, are similarly generated, for any four-group acting on them, but this is not true of $L_2(5)$, $L_2(7)$, or $L_2(9)$.]

Next consider the case $L/Z(L) \cong A_n$, which as noted in Remark 5.1.8 was treated by Aschbacher alone. If $L/O(L) \cong \hat{A}_n$ and $x$ is the central involution of $L$, then $L$ is a component of $C_x$ by Lemma 5.25. Thus $L$ is an intrinsic 2-component of type $\hat{A}_n$, so by Solomon's theorem (Theorem 4.24), $F^*(G) \cong Mc$ or $Ly$, in which cases $H = \langle x \rangle$, contrary to our present hypothesis $m(H) \geq 2$. Thus, in fact, we have $L/O(L) \cong A_n$. [Using properties of $LA$ together with the structure of the group $\hat{A}_n$, Aschbacher quickly eliminates the $\hat{A}_n$ case without invoking Solomon's result.]

Furthermore, as also noted in Remark 5.1.8, it suffices to treat the case $n \geq 8$ (whence $L \cong A_n$ by $[I, \text{Table 4.1}]$). Hence the alternating case is completely covered by the following result, which for later purposes we state in slightly stronger form than required here.

PROPOSITION 5.34. *If* $L \cong A_n$, $n \geqslant 8$, *then we have*

(i) $R \cong Z_2 \times Z_2$; *and*
(ii) $V \leqslant L$ *and* $V$ *moves exactly four letters in the natural represen-tation of* $L$.

Note first that as $V \leqslant LH$ and $|\mathrm{Out}(L)| = 2$, $A = VO(A) \leqslant LH$. Let $B$, $U$ be the projections of $A$, $V$, respectively, on $L$, so that $B = UO(B)$. Furthermore, as $V$ is faithful on $L$, $U \cong V$.

Next set $J = \Gamma_{V,1}(L)$. Then $J$ normalizes $A$ by Lemma 5.30 and conse-quently $J$ normalizes $B$. But also clearly $J = \Gamma_{U,1}(L)$, so we can analyze the situation entirely in $L$. Furthermore, we know from Lemma 5.31 that $J < L$. One can now determine by direct computation the possibilities for $J$ and $U$ satisfying these conditions; one finds that either

$$
\begin{aligned}
&(1)\ U \cong Z_2 \times Z_2,\ J/C_L(U) \cong \Sigma_3,\ \text{and}\ U\ \text{moves exactly four}\\
&\quad\ \text{letters in the natural representation of}\ L;\ \text{or}\\
&(2)\ n = 8\ \text{or}\ 9,\ U \cong E_8,\ J/C_L(U) \cong L_3(2),\ \text{and}\ U\ \text{moves}\\
&\quad\ \text{exactly eight letters in the natural representation of}\ L.
\end{aligned}
\tag{5.33}
$$

We claim now that $U = V$. Indeed, let $U_0$ be the projection of $V$ on $H$, so that $V \leqslant UU_0$ and $|U_0| \leqslant |V|$. But corresponding to the two possibilities of (5.33), $J$ contains a subgroup $D$ of order 3 or 7 acting transitively on $U^{\#}$. Since $D \leqslant L$ centralizes $U_0$, it follows that $U$ and possibly $U_0$ are the only $D$-invariant subgroups of $UU_0$ of order $|V|$. But $D \leqslant J$ normalizes $A$ and hence $D$ normalizes $V = A \cap UU_0$ [as $UU_0$ is a 2-group containing $V$ and $A = VO(A)$], thus forcing $V = U$ or $U_0$. However, as $V$ acts faithfully on $L$, the latter is excluded and our claim is proved.

Hence to complete the proof, it remains to eliminate the second alter-native of (5.33). But in that case by the structure of $L$ ($\cong A_8$ or $A_9$), there exists an element $x \in L$ such that if we set $W = N_V(V^x)$, then $W \neq 1$ and $C_V(W) \not\leqslant W$ (i.e., $W$ is either of order 2 or is a four-group centralizing a subgroup of $V$ of order 2). Hence if we put $g_1 = gxg^{-1}$ and $V_1 = V^{xg^{-1}} = R^{g_1}$, it follows that $V_1 \in \mathrm{Syl}_2(N_{H^{g_1}}(L))$, $V_1 \neq 1$, $V_1$ leaves $R$ invariant, and $C_R(V_1) \not\leqslant V_1$. However, this contradicts Proposition 5.28(iii).

We turn now to Aschbacher and Seitz's analysis of the case in which $L$ is of Lie type of characteristic 2. In view of Propositions 5.33 and 5.34

together with the Griess–Mason–Seitz solution of the Bender case and the inductive characterization of the groups of Lie type of odd characteristic, we can assume

$$L/Z(L) \not\cong L_2(2^n), U_3(2^n), Sz(2^n), L_3(2) \left(\cong L_2(7)\right),$$
$$G_2(2)' \left(\cong U_3(3)\right), U_4(2) \left(\cong PSp_4(3)\right), \text{ or } L_4(2) \left(\cong A_8\right). \tag{5.34}$$

The argument is inductive and conceptually very similar to the first portion of Seitz's analysis of the general nonlinear Lie rank $\geqslant 3$ characteristic 2 case when $m(H) = 1$, which we have briefly described in Section 5.2.8. (For the record, it should be noted that the present Aschbacher–Seitz work preceded that of Seitz's characteristic 2 case analysis.) Throughout we assume that

> $G$ is a minimal counterexample to the solution of the standard form problem with $L$ of Lie type of characteristic 2 and $m(H) > 1$.

Again the first step in the analysis is to establish an analog of Seitz's fusion theorem (Theorem 5.2), which amounts, in effect, to a determination of the possible embeddings of the projection of $V$ on $L$. Its proof depends very strongly on the detailed properties of both the centralizers and conjugacy classes of involutions in groups of Lie type of characteristic 2 that they established in [I: 21].

Since the precise result is rather complicated, we shall state it in somewhat simplified form (and then only for the classical groups). This will be facilitated by first disposing of the cases in which $|Z(L)|$ is even. Likewise it will be easier if we exclude the case $L/Z(L) \cong L_3(4)$ from the present discussion. In view of (5.34), [I, Table 4.1] thus implies that

(1) $L/Z(L) \cong U_6(2), PSp_6(2), P\Omega_8^+(2), G_2(4), F_4(2), \text{ or } {}^2E_6(2)$; and
(2) Either $Z(L) \cong Z_2$ or $L/Z(L) \cong {}^2E_6(2)$ and $Z(L) \leqslant Z_3 \times Z_2 \times Z_2$.
$$\tag{5.35}$$

Elimination of these cases again depends on detailed properties of the involutions of $L$. For example, in the first three cases, if $W$ denotes the pre-image in $L$ of the projection of $V$ on $L/Z(L)$, then $W$ is elementary abelian and so if $U$ is a complement to $Z(L)$ in $W$, then $U \cong V$ as $V$ acts faithfully on $L$. Using the tight embedding of $A$ in $LA$, Aschbacher and Seitz prove that for any four-subgroup $U_1$ of $U$ and any $u \in U_1^{\#}$:

$$
\begin{aligned}
&(1) \ U_1 \leqslant O_2(C_L(u)); \text{ and}\\
&(2) \ \text{Either } U_1 \lhd O_2(C_L(u)) \text{ or else } U_1 \text{ centralizes} \qquad (5.36)\\
&\quad\ \ \text{every } L\text{-conjugate of } u \text{ contained in } C_L(u).
\end{aligned}
$$

They then check by direct computation that none of the groups $U_6(2)/Z_2$, $PSp_6(2)/Z_2$, or $P\Omega_8^+(2)/Z_2$ possesses a four-subgroup satisfying the two conditions of (5.36).

Thus they conclude:

$$Z(L) \text{ has odd order.} \qquad (5.37)$$

We can now state the Aschbacher–Seitz fusion result. By (5.37) we can consider the projection $U$ of $V$ on $L$. We assume $L$ is defined over $GF(q)$ and we fix $u \in U^{\#}$.

PROPOSITION 5.35. *If* $L$ *is either a linear, unitary, symplectic, or orthogonal group, then one of the following two sets of conditions holds*:

(i) (1) $|V| \leqslant q$;
  (2) *If* $D$ *denotes the subgroup generated by the set of* $L$-*conjugates of* $u$ *contained in* $C_L(u)$ *and we put* $W = O_2(C_L(u)) \cap C_L(D)$, *then* $U \leqslant W$;
  (3) $W$ *is elementary abelian of order* $q$ *or* $q^2$ *and correspondingly* $N_L(W)/C_L(W) \cong Z_{q-1}$ *or* $Z_{q-1} \times Z_{q-1}$; *and*
  (4) *All involutions of* $U$ *are conjugate in* $N_L(W)$; *or*
(ii) (1) $|V| = 4$;
  (2) $U = V$; *and*
  (3) $L \cong L_n(2)$ *or* $PSp_n(2)$ *and* $u$ *is contained in a uniquely determined conjugacy class of involutions of* $L$.

Aschbacher and Seitz prove similar results about the embedding of $V$ when $L$ is an exceptional group of Lie type, which they establish on a case-by-case basis. We omit the statements, as they are somewhat technical.

In the case $R\langle x \rangle$ and $L$ has Lie rank $\geqslant 3$, Seitz's fusion theorem implies that there is a conjugate $y$ of $x$ that acts on $L$ and centralizes the $SL_2(q)$-subgroup $J$ of (5.3) (with a slight variation of this conclusion when $L$ is

**Table 5.13.** The Structure of $Y$ and $X$

| Classical | | | Nonclassical | | |
|---|---|---|---|---|---|
| $L/Z(L)$ | $Y$ | $X/Z(X)$ | $L/Z(L)$ | $Y$ | $X/Z(X)$ |
| $L_3(q)$ | $Z_{q-1}$ or $Z_{(q-1/3)}$ | $L_2(q)$ | $G_2(q)$ | $Z_{q-1}$ | $L_3(q)$ |
| $L_4(q)$ | $Z_{q-1} \times L_2(q)$ | $L_2(q)$ | $^3D_4(q), q=2^{3n}$ | $Z_{q-1}$ or $Z_{q^2+q+1}$ | $L_3(q)$ or $L_2(q^2)$ |
| $U_4(q)$ | $Z_{q+1} \times L_2(q)$ | $L_2(q)$ | $^2F_4(q), q=2^{2n+1}$ | $Z_{q-1}$ | $L_2(q)$ |
| $L_n(q), n \geqslant 5$ | $L_2(q)$ | $L_{n-2}(q)$ | $F_4(q)$ | $Z_{q-1}$ | $PSp_6(q)$ |
| $U_n(q), n \geqslant 5$ | $L_2(q)$ | $U_{n-2}(q)$ | $^2E_6(q)$ | $Z_{q-1}$ | $U_6(q)$ or $P\Omega_8^-(q)$ |
| $PSp_n(q)$ | $L_2(q)$ | $PSp_{n-2}(q)$ | $E_6(q)$ | $Z_{q-1}$ | $L_6(q)$ |
| $P\Omega_8^\varepsilon(q), \varepsilon = \pm 1$ | $L_2(q) \times \Omega_4^\varepsilon(q)$ | $L_2(q)$ | $E_7(q)$ | $Z_{q-1}$ | $P\Omega_{12}^+(q)$ |
| $P\Omega_n^\varepsilon(q), n \geqslant 9$ | $L_2(q) \times L_2(q)$ | $P\Omega_{n-4}^\varepsilon(q)$ | $E_8(q)$ | $Z_{q-1}$ | $E_7(q)$ |

orthogonal). Considering a cyclic subgroup $Y$ of $J$ of order $q + 1$ (again with a slight variation in the orthogonal case) and the layer $X = L(C_L(Y))$, Seitz then uses the existence of $y$ to prove that $X$ pumps-up properly in $C_Y$.

Aschbacher and Seitz use Proposition 5.35 (and the analogous results when $L$ is an exceptional group) in a similar way. However, because $L$ may now have Lie rank 2 and also because of the greater complexity of the present fusion results, there are more possibilities for the structure of the group $X$. Ultimately, by a delicate case-by-case analysis, on the basis of these fusion results [directly in the case $L/Z(L) \cong L_3(4)$], they produce for a suitable choice of $g$ and $V$ a nontrivial subgroup $Y$ of $C_L(V)$ such that $X = L(C_L(Y))$ is quasisimple, with $Y$ and $X/Z(X)$ given by Table 5.13.

The proof that $X$ pumps-up properly in $C_Y$ depends on the following key properties of the subgroups $V, R, Y,$ and $X$, which again require detailed information about both the centralizers and conjugacy classes of involutions in $\mathrm{Aut}(L)$ for their verification: [recall that $V$ centralizes both $R$ and $Y$, so $RY$ acts on $L^g$, which is normal in $N_G(V)$, and $V$ acts on $X$, which is normal in $C_L(Y)$]:

LEMMA 5.36. *The following conditions hold:*

(i) $V \cap X = 1$ *and* $V$ *induces inner automorphisms on* $X$ [*except possibly in the case* $L/Z(L) \cong G_2(q)$, *in which case these conditions hold for some choice of* $g$ *and* $V$];

(ii) $R$ *is a Sylow 2-subgroup of* $N_G(R) \cap C_{YX}$; *and*

(iii) $C_{L^g}(Y)$ *does not normalize* $R$.

We limit ourselves to a few observations about the proof of (ii) and (iii). First, from the embedding of $YX$ in $L$, we obtain:

$$\text{If } t \text{ is a 2-element of } N_G(L) \text{ that centralizes } YX, \tag{5.38}$$
$$\text{then } t \text{ centralizes } L.$$

Since $L \lhd N_G(R)$ and $R$ is a Sylow 2-subgroup of $C_L = H$, (5.38) immediately yields (ii).

In most cases, (iii) is established by showing that $C_{L^g}(Y) \cong C_L(Y)$. This in turn implies that $Y$ induces inner automorphisms on $L^g$. But $R$ centralizes $Y$, so if (iii) fails, we can then apply (5.38) to $N_G(L^g)$ and its subgroup $YC_{L^g}(Y)$ to conclude that $R$ centralizes $L^g$, whence $R \leqslant H^g$. But $R$ centralizes $V$ and $V \in \mathrm{Syl}_2(H^g)$, forcing $R = V$, contrary to the fact that $V$ is faithful on $L$, while $R$ centralizes $L$.

Aschbacher and Seitz are now able to establish the following completely general result about subgroups $X$ and $Y$ satisfying the above conditions. Moreover, their argument covers the Seitz situation $R \cong Z_2$ as well.

PROPOSITION 5.37. *Let $Y$ be a nontrivial subgroup of $L$ such that $X = L(C_L(Y))$ is quasisimple. If for some $g \in G$, $V = R^g$ normalizes $L$ with $V$ centralizing $Y$ and if the conditions of Lemma 5.36 are satisfied, then*

  (i) *$X$ is standard in $C_Y$; and*
  (ii) *$X$ is not normal in $C_Y$.*

Because of its crucial importance, we prove the proposition in detail. Set $C = C_Y$, $B = RO(H)$, and $K = C_C(X)$. Since $XY \leqslant L$ and $B \leqslant H = C_L$, $B \leqslant K$. By Lemma 5.36(ii), $R$ is a Sylow 2-subgroup of $N_C(R) \cap N_C(X) = N_K(R)$ and so by [I, 1.11] $R \in \mathrm{Syl}_2(K)$. Furthermore, by Proposition 5.27(ii) $B$ is tightly embedded in $G$ and hence in $C$.

We claim that these last two condition imply that $K$ is tightly embedded in $C$. First, $L \lhd C_B(\leqslant C_R)$ by Lemma 5.25, and as $L$ commutes with none of its $G$-conjugates, it follows that $L \lhd N_G(B)$. Since $X = L(C_L(Y)) = L(L \cap C)$, this implies that

$$X \lhd N_C(B). \tag{5.39}$$

Suppose now that $|K \cap K^c|$ is even for some $c \in C$. By Sylow's theorem, $R^k \cap R^{kc} \neq 1$ for some $k \in K$, and replacing $c$ by $ck$, we can assume without loss that $R \cap R^c \neq 1$. But then $|B \cap B^c|$ is even, so as $B$ is tightly embedded in $C$, $c \in N_C(B)$. It follows therefore from (5.39) that $c$ normalizes $X$, whence $c$ normalizes $C_C(X) = K$, and we conclude from the definition that $K$ is tightly embedded in $C$, as claimed.

Observe next that as $R \in \mathrm{Syl}_2(K)$ and $R \in \mathrm{Syl}_2(B)$, $N_C(K) = K N_C(R)$ and $N_C(B) = B N_C(R)$ by the Frattini argument. Since $B \leqslant K$, these two equalities imply that $N_C(K) = K N_C(B)$. Since $N_C(B)$ normalizes $X$ by (5.39) and $K = C_C(X)$, it follows that $N_C(K) \leqslant N_C(X)$. The reverse inclusion is clear as $K = C_C(X)$, so

$$N_C(X) = N_C(K). \tag{5.40}$$

Hence by definition, either $X$ is standard in $C$ or else $X$ centralizes $X^c$ for some $c \in C$. Consider the latter possibility, whence $X^c \leqslant K = C_C(X)$. We claim that

$$X^c \vartriangleleft K. \tag{5.41}$$

Indeed, let $W \in \mathrm{Syl}_2(X)$, so that $W$ centralizes $X^c$ and hence $W \leqslant K^c$. But as $K$ is tightly embedded in $C$, so is $K^c$ and hence $C_C(W) \leqslant N_C(K^c)$. Since $K = C_C(X) \leqslant C_C(W)$ and $N_C(K^c) = N_C(X^c)$, by (5.40), this in turn implies that $K \leqslant N_C(X^c)$. Thus $K$ normalizes $X^c$ and as $X^c \leqslant K$, our claim follows.

In particular, as $R \in \mathrm{Syl}_2(K)$, $R \cap X^c \in \mathrm{Syl}_2(X^c)$. Since $R$ is elementary abelian, Bender's strong embedding theorem [I, 4.24] now yields that either

$$\begin{array}{l} (1) \;\; X^c \leqslant \Gamma_{R,1}(C); \text{ or} \\ (2) \;\; X^c \cong L_2(2^n) \text{ and } R \in \mathrm{Syl}_2(X^c). \end{array} \tag{5.42}$$

In the first case, as $R \in \mathrm{Syl}_2(B)$ and $B$ is tightly embedded in $C$, it follows as in Lemma 5.30 that $\Gamma_{R,1}(C)$ and hence $X^c$ normalizes $B$, whence $[X^c, R] \leqslant B$. But as $R \cap X^c \in \mathrm{Syl}_2(X^c)$, we have $X^c = [X^c, R \cap X^c]$, so $X^c \leqslant B$, contrary to the fact that $B = RO(B)$ is solvable.

On the other hand, in the second case, $X \cong X^c \cong L_2(2^n)$, so $L/Z(L) \not\cong G_2(q)$ by Table 5.13. Furthermore, $R \in \mathrm{Syl}_2(X^c)$, so $R^{c^{-1}} \leqslant X$ and, in particular, $R^{c^{-1}}$ centralizes $Y$. But now if we take $g = c^{-1}$ and $V = R^g = R^{c^{-1}}$, Lemma 5.36(i) implies that $V \cap X = 1$, contrary to the fact that $V \leqslant X$.

Hence no such $X^c$ exists and we conclude that $X$ is standard in $C$, proving (i).

Suppose now that (ii) fails. It will suffice to prove

$$R \lhd C_C(V). \tag{5.43}$$

Indeed, as $V$ centralizes $L^g$, this will imply that $C \cap L^g = C_{L^g}(Y)$ normalizes $R$, contrary to Lemma 5.36(iii).

By Lemma 5.36(i), $V$ induces inner automorphisms on $X$, so $V \leqslant XK$. Moreover, $V$ centralizes $R$ [by Proposition 5.28(iii) if $|R| > 2$ and directly if $|R| = 2$]. Since $X$ centralizes $R$, it follows that $V \leqslant C_{XK}(R) = XC_K(R)$, so $V \leqslant O^{2'}(XC_K(R)) = XO^{2'}(C_K(R))$ [as $X = O^{2'}(X)$]. But as $R \in \mathrm{Syl}_2(K)$ with $R$ abelian, $C_K(R) = O(C_K(R)) \times R$ by Burnside's theorem [I, 1.20(ii)], whence $O^{2'}(C_K(R)) = R$. Thus $V \leqslant XR$ and so $XV \leqslant XR$. But $V = R^g \cong R$ and $V \cap X = 1$ by Lemma 5.36(i), which together force

$$XV = XR. \tag{5.44}$$

Finally, $R \in \mathrm{Syl}_2(K)$ and $XR \cap K = RZ(X) = R \times O(Z(X))$, and consequently $R = O_2(XR \cap K)$. Thus $R = O_2(XV \cap K)$ by (5.44). On the other hand, as (ii) fails, $X \lhd C$ and hence also $K = C_C(X) \lhd C$. Therefore $C_C(V)$ normalizes $XV \cap K$ and so normalizes $R = O_2(XV \cap K)$. Since $R \leqslant C_C(V)$, we conclude that $R \lhd C_C(V)$, proving (5.43), as required. This establishes (ii) and completes the proof of the proposition.

Now set $\bar{C} = C/O(C)$, where $C = C_y$, as above. Then $\bar{X}$ is standard in $\bar{C}$, $\bar{X}$ is not normal in $\bar{C}$, $\bar{K} = C_{\bar{C}}(\bar{X})$, and $m(\bar{K}) = m(R) > 1$. Considering the various possibilities for $X/Z(X)$ in Table 5.13, Aschbacher and Seitz therefore conclude from the minimality of $G$:

PROPOSITION 5.38. *We have* $R \cong Z_2 \times Z_2$ *and one of the following holds*:

(i) $L/Z(L) \cong L_3(4)$ *and* $\bar{X} \cong L_2(4)$;

(ii) $L \cong G_2(4)$ *and* $\bar{X} \cong L_3(4)$;

(iii) (1) $L \cong L_4(4)$, $U_4(4)$, $PSp_4(4)$, *or* $P\Omega_8^\varepsilon(4)$; *and*

    (2) $\bar{X} \cong L_2(4)$ *and* $F^*(\bar{C}) \cong A_9$, $M_{12}$, *or* $J_2$;

(iv) (1) $L \cong L_5(4)$ *or* $^3D_4(4)$; *and*

    (2) $\bar{X} \cong L_3(4)$ *and* $F^*(\bar{C}) \cong Suz$; *or*

(v) (1) $L \cong L_6(2)$; *and*

    (2) $\bar{X} \cong L_4(2)$ ($\cong A_8$) *and* $F^*(\bar{C}) \cong A_{12}$.

It remains for Aschbacher and Seitz to eliminate the various possibilities of (iii), (iv), and (v), for then $L/Z(L) \cong L_3(4)$ or $L \cong G_2(4)$ and the desired reduction will be complete.

Observe first that from the action of $R$ on $O(C)$, it follows easily that $X$ centralizes $O(C)$, which immediately implies that $L(C)$ maps on $F^*(\bar{C})$. Thus if we set $J = L(C)$, we have $J/Z(J) \cong F^*(\bar{C})$. Also $XR \leqslant J$.

If $J \cong A_9$ or $A_{12}$, then by the structure of $J$, $R^x \leqslant X$ for some $x \in J$. But then taking $x = g$ and $V = R^g$, we see that Lemma 5.36(i) is contradicted. In particular, (v) does not hold.

Suppose next that $X \cong L_2(4)$, in which case $L \cong L_4(4)$, $U_4(4)$, $PSp_4(4)$, or $P\Omega_8^\varepsilon(4)$ and $J \cong M_{12}$ or $J_2$ (as $J \not\leqslant A_9$). If $J \cong M_{12}$, then $\bar{C} \cong \mathrm{Aut}(M_{12})$ and $C$ contains an involution $x$ which normalizes, but does not centralize, $R$ and such that $X\langle x \rangle \cong \Sigma_5$. Then $x$ normalizes $L$ and centralizes $Y$ (as $x \in C$). However, one easily checks that none of the four possibilities for $L$ possesses an automorphism with the properties of $x$. Thus $J \cong J_2$.

If $L \cong L_4(4)$ or $U_4(4)$, set $Y_1 = O(Y)$ ($\cong Z_3$ or $Z_5$, respectively), while if $L/Z(L) \cong P\Omega_8^\varepsilon(4)$, let $Y_1$ be the normal subgroup of $Y$ isomorphic to $L_2(4)$. Then by Table 5.13, $Y = Y_1 \times Y_0$, where $Y_0 \cong L_2(4)$, $L_2(4)$, or $P\Omega_4^\varepsilon(4)$, respectively. Set $C_1 = C_{Y_1}$. Since $J \leqslant C = C_Y$, we have $Y_0 \times J \leqslant C_1$. Also as $R \leqslant J$, we see that $C_{C_1}(J) = Y \cap C_1$ ($=Y$, $Y$, or $Y_0$, respectively). Hence if we put $\tilde{C}_1 = C_1/O(C_1)$, it follows that $\tilde{Y}_0$ contains a Sylow 2-subgroup of $C_{\tilde{C}_1}(\tilde{J})$ and consequently $m(C_{\tilde{C}_1}(\tilde{J})) \geqslant 2$. Furthermore, $\tilde{J} \cong J_2$.

To eliminate these cases, Aschbacher and Seitz now argue that $\tilde{J}$ is standard and non-normal in $\tilde{C}_1$, so that $F^*(\tilde{C}_1)$ is determined by the minimality of $G$. However, as there are no solutions for $F^*(\tilde{C}_1)$ with standard component isomorphic to $J_2$, this is a contradiction.

On the other hand if $L \cong PSp_4(4)$, then $X$ and $Y$ are each isomorphic to $L_2(4)$ and by the symmetry between them, $C_X$ likewise contains a subgroup $K \cong J_2$. Considering a subgroup $X_1$ of $X$ of order 5 and this time putting $C_1 = C_{X_1}$ and $\tilde{C}_1 = C_1/O(C_1)$, Aschbacher and Seitz argue, on the one hand, that $\tilde{K}$ is not normal in $\tilde{C}_1$ and, on the other hand, that $\tilde{Y}$ is standard and non-normal in $\tilde{C}_1$ [with $\tilde{R} \in \mathrm{Syl}_2(C_{\tilde{C}_1}(\tilde{Y}))$]. But then by the minimality of $G$, $F^*(\tilde{C}_1) \cong A_9$, $M_{12}$, or $J_2$. Since $J_2 \cong \tilde{K} \leqslant \tilde{C}_1$, the only possibility is that $F^*(\tilde{C}_1) \cong J_2$ and $F^*(\tilde{C}_1) = \tilde{K}$, so $\tilde{K} \lhd \tilde{C}_1$—contradiction.

Hence $X \not\cong L_2(4)$ and so all possibilities in (iii) are excluded. Suppose finally that (iv) holds, whence $L \cong L_5(4)$ or $^3D_4(4)$, $\bar{X} \cong L_3(4)$, and $\bar{J} \cong Suz$. In the first case, we check that $C_L(Y) \cong GL(3,4)$ and in the second that $C_L(Y)/Y$ contains a subgroup isomorphic to $GL(3,4)$. Since $\bar{J} = F^*(\bar{C}) \cong Suz$ and $\mathrm{Out}(Suz) \cong Z_2$, it follows that $\bar{J}$ contains a subgroup $\bar{D}$ of order 3 centralizing $\bar{R}$ and such that $\bar{X}\bar{D} \cong GL(3,4)$. However, $Suz$ does not possess a subgroup with these properties.

Thus Aschbacher and Seitz have obtained their objective:

PROPOSITION 5.39. *If $L$ is of Lie type of characteristic 2 and $G$ is a minimal counterexample to the standard component problem with $m(H) > 1$, then $R \cong Z_2 \times Z_2$ and either $L/Z(L) \cong L_3(4)$ or $L \cong G_2(4)$.*

Aschbacher and Seitz are therefore left to treat the cases in which $L/Z(L)$ is a sporadic group. Again the analysis involves detailed information concerning both the centralizers and conjugacy classes of involutions of $L$ (as well as other structural properties), most of which had been previously developed in the course of studying these groups.

Underlying the analysis are the following addition properties of $V$, which follow easily from Propositions 5.27 and 5.28 and which hold for any standard component $L$ such that the commutator subgroup of $\mathrm{Out}(L)$ has odd order. Note that any group $X$ possesses a unique smallest normal subgroup $Y$ such that $X/Y$ is solvable with Abelian Sylow 2-subgroups, as is easiy checked.

LEMMA 5.40. *If $[\mathrm{Out}(L), \mathrm{Out}(L)]$ has odd order, then*

(i) *For every $v \in V^{\#}$, $V$ centralizes the unique smallest normal subgroup of $O^2(C_L(v))$ such that the factor group is solvable with abelian Sylow 2-subgroups; and*

(ii) *Either $V \cong Z_2 \times Z_2$ or $N_V(V^x) = C_V(V^x)$ for every $x \in L$.*

[Aschbacher and Seitz refer to the conclusion of the lemma by saying that $L$ is *V-admissible*. The notion plays a key role throughout their analysis; in particular, the second condition is critical for the proof of Proposition 5.35.]

Since $|\mathrm{Out}(L)| \leqslant 2$ for every sporadic group [I, 4.239], the assumption of the lemma holds in these cases. Aschbacher and Seitz now show by direct calculation that most sporadic groups do not admit the action of a subgroup $V$ satisfying the conclusions of the lemma. (At the time their paper was written, Janko's group $J_4$ had not yet been discovered, but using its properties, one can show that it, too, does not admit such a subgroup $V$.)

PROPOSITION 5.41. *If $L/Z(L) \cong M_{11}$, $M_{12}$, $M_{22}$, $M_{23}$, $J_1$, $J_3$, $J_4$, $HS$, $Mc$, $Ly$, $ON$, $.2$, $.3$, $M(22)$, $M(23)$, $M(24)'$, $F_1$, $F_2$, $F_3$, or $F_5$, then $L$ does not admit a noncyclic elementary abelian 2-subgroup $V$ inducing faithful inner automorphisms on $L$ and satisfying the conditions of Lemma 5.40.*

Thus they are reduced to eliminating the following possibilities:

$$L/Z(L) \cong M_{24}, J_2, He, Suz, Ru, \text{ or } .1. \tag{5.45}$$

In these cases, the corresponding calculations yield:

PROPOSITION 5.42. *The following conditions hold*:

  (i) $Z(L)$ *is of odd order*;
 (ii) $V \cong Z_2 \times Z_2$ *and the projection of $V$ on $L$ is uniquely determined up to conjugacy*;
(iii) *If* $L/Z(L) \cong M_{24}$, *He, Suz, or* .1, *then* $V \leqslant L$;
 (iv) *If* $L/Z(L) \cong J_2$ *or Ru, then either* $V \leqslant L$ *or* $V \cap L = 1$.

If $L/Z(L) \cong M_{24}$ or $He$, then $V \leqslant L$ and Aschbacher and Seitz know from the embedding of $V$ that $V \leqslant D = C_L(V)^{(\infty)}$. But by the structure of $N = N_G(R)$, $L \lhd N$ and $N/L$ is solvable, so likewise $N^g/L^g$ is solvable. Since $V = R^g$, we have $D \leqslant N^g$ and as $D = D^{(\infty)}$, it follows that $D \leqslant L^g$, whence also $V \leqslant L^g$. However, $V \leqslant H^g$ centralizes $L^g$, so $V \leqslant Z(L^g)$, contrary to the fact that $Z(L)$ and hence $Z(L^g)$ has odd order by Proposition 5.42(i). Hence these two cases are also excluded, and consequently

$$L/Z(L) \cong J_2, Suz, Ru, \text{ or } .1. \tag{5.46}$$

Elimination of these last cases depends on an analysis of the set $\mathscr{B}$ of all $G$-conjugates of $R$ contained in $C_{RV}$. As with $V$, every element of $\mathscr{B}$ other than $R$ induces faithful inner automorphisms on $L$. Aschbacher and Seitz establish a general lemma concerning the subset $\mathscr{B}_0$ of those $B \in \mathscr{B}$ that centralize every element of $\mathscr{B}$ [in particular, $\{R, V\} \subseteq \mathscr{B}_0$] and the subgroup $D = \langle B \mid B \in \mathscr{B}_0 \rangle$ [clearly $D$ is an elementary abelian 2-group and our conditions also imply that $D = R(D \cap L) = V(D \cap L^g)$], and which they use at various places throughout the analysis.

LEMMA 5.43. *Set* $N = \langle O^2(N_L(D)), O^2(N_{L^g}(D)) \rangle$. *If* $\mathscr{B}_0 \supset \{R, V\}$ *and* $N_L(D)/C_L(D)$ *contains a characteristic cyclic subgroup acting transitively on* $(L \cap D)^\#$, *then we have*

  (i) $N$ *acts doubly transitively on* $\mathscr{B}_0$; *and*
 (ii) *If* $D_0 = N_L(D)/C_L(D)$, *then* $D - D_0 = \bigcup_{B \in \mathscr{B}_0} B^\#$.

Aschbacher and Seitz show that the lemma is applicable in the present situation. If $V \leqslant L$, they use the lemma first to produce a $G$-conjugate $W$ of $R$ centralizing $R$ with $W \neq R$ and $W \cap L = 1$. But by Proposition 5.42(ii), the projection of $W$ on $L$ centralizes an $L$-conjugate of $V$ (as $V \leqslant L$), so without loss we can assume that $W \leqslant VR \leqslant D$, whence $W \in \mathcal{B}_0$. However, as $V \leqslant L$ and $W \cap L = 1$, this is incompatible with the conclusions of the lemma.

Hence by (5.46) and Proposition 5.42, $V \cap L = 1$ and $L/Z(L) \cong J_2$ or $Ru$. In these cases, they compute that $|D| = 64$ or 128, respectively, and as $V \cap L = 1$, that correspondingly $|\mathcal{B}_0| = 13$ or 25. Furthermore, setting $N^R = N_N(R)$ and $\overline{N} = N/C_N(D)$, then $\overline{N}$ is a doubly transitive permutation group on $\mathcal{B}_0$ and $\overline{N}^R$ is the stabilizer of $R$ in $\overline{N}$. They now check that $O_2(\overline{N}^R) = Z_2 \times Z_2$ or $E_8$, respectively, and acts semiregularly on $\mathcal{B}_0 - \{R\}$. However, under these conditions, Shult's theorem [I: 251] is applicable and yields that $|\mathcal{B}_0| = 1 + 2^n$ for some $n$, contrary to the fact that $|\mathcal{B}_0| = 13$ or 25. (Shult's theorem is used at other places in the analysis as well.)

This completes our outline of the Aschbacher–Seitz reduction theorem.

## 5.5. The Case $m(H) \geqslant 2$

We now consider the residual cases of the Aschbacher–Seitz reduction theorem. In particular, we have

$$R \cong Z_2 \times Z_2. \tag{5.47}$$

Throughout we assume that $G$ is a minimal counterexample to the proposed theorem.

We begin with the *generic* case:

$$L \cong A_n. \tag{5.48}$$

As noted in Remark 5.1.10, we can assume that $n \geqslant 8$.

Aschbacher's solution in this case is very beautiful, his aim being to prove that

$$F^*(G) \cong A_{n+4}. \tag{5.49}$$

He begins with a description of the root four-subgroup graph of $A_{n+4}$, which as noted in Remark 5.1.3 is the basis for the final identification of $F^*(G)$.

Let then $X = A_m$ acting on the set $\Lambda = \{1, 2,..., m\}$. We are interested here in the case $m \geq 12$, although the key result is valid for $m = 10$ and $11$ as well.

DEFINITION 5.44. For any subgroup $Y$ of $X$, the *support* $\Lambda(Y)$ of $Y$ is the subset of $\Lambda$ moved by $Y$—i.e., the set of $i \in \Lambda$ such that $(i)y \neq i$ for some $y \in Y$. A four-subgroup $U$ of $X$ is called a *root* four-subgroup if $|\Lambda(U)| = 4$ [whence $U = \langle (ij)(hk), (ih)(jk) \rangle$ for suitable distinct points $h, i, j, k \in \Lambda$].

Let $\Omega$ be the collection of root four-subgroups of $X$ and define a graph $\mathscr{D}$ having vertex set $\Omega$ with two distinct vertices $U, V \in \Omega$ connected by an edge if and only if $[U, V] = 1$.

Thus $\mathscr{D}$ is a subgraph of the full commuting graph of four-subgroups of $X$, defined in [I, p. 47] for any group $X$.

Observe that each root subgroup of $X$ is uniquely determined by its support and two distinct such subgroups commute elementwise if and only if their supports are disjoint. Thus we have the following alternate description of $\mathscr{D}$.

PROPOSITION 5.45. $\mathscr{D}$ *is equivalent to the graph* $\mathscr{D}^*$ *on* $\Lambda$ *whose vertices are the subsets of* $\Lambda$ *of cardinality* 4 *with two such subsets connected by an edge if and only if their intersection is empty.*

Clearly the symmetric group $\Sigma(\Lambda)$ on $\Lambda$ permutes the elements of $\Omega$ transitively and preserves incidence in $\mathscr{D}$, so $\Sigma(\Lambda)$ is contained in the automorphism group Aut($\mathscr{D}$) of $\mathscr{D}$. Moreover, we can identify $X$ with a subgroup of $\Sigma(\Lambda)$.

Aschbacher's basic result here is the following:

PROPOSITION 5.46. *If* $m \geq 10$, *then* Aut($\mathscr{D}$) $= \Sigma(\Lambda) \cong \Sigma_m$.

Indeed, for $U \in \Omega$, let $\Omega_0(U)$ be the set of $V \in \Omega$ having support disjoint from that of $U$. Then each $V \in \Omega_0(U)$ centralizes $U$ and we see that $\langle \Omega_0(U) \rangle \cong A_{m-4}$ (on the $m - 4$ letters fixed by $U$). Hence if we set $\mathscr{D}(U) = \{U\} \cup \mathscr{D}_0(U)$, it follows that

$$\langle \mathscr{D}(U) \rangle \cong Z_2 \times Z_2 \times A_{m-4}. \tag{5.50}$$

It is immediate now that if $\Sigma(\Lambda)_U$ denotes the subgroup of $\Sigma(\Lambda)$ leaving $\mathscr{D}(U)$ pointwise fixed, then

$$\Sigma(\Lambda)_U \cong \Sigma_4 \quad \text{and} \quad U = O_2\big(\Sigma(\Lambda)_U\big). \tag{5.51}$$

Now set $A = \text{Aut}(\mathscr{D})$, so that $\Sigma(\Lambda) \leqslant A$, and for $U \in \Omega$, let $A_U$ be the subgroup of $A$ leaving $\mathscr{D}(U)$ pointwise fixed. The crucial fact that Aschbacher needs is the following:

$$A_U = \Sigma(\Lambda)_U, \tag{5.52}$$

which he derives from elementary properties of $\mathscr{D}$ (under the assumption $m \geqslant 10$).

Finally set $B = \langle A_U \mid U \in \Omega \rangle$. Since $A$ transitively permutes the elements of $\Omega$, $B$ is a *normal* subgroup of $A$. But by (5.52) each $A_U$ and hence $B$ is contained in $\Sigma(\Lambda)$. Furthermore, by (5.51), each $U \leqslant \Sigma(\Lambda)_U$, so $X = \langle U \mid U \in \Omega \rangle \leqslant B$. Hence, either $B = X$ or $B = \Sigma(\Lambda)$. In addition, $C_A(B)$ fixes each $U \in \Omega$ and hence by definition of $A$, we have $C_A(B) = 1$. Since $B \lhd A$, it therefore follows that

$$A \leqslant \text{Aut}(B). \tag{5.53}$$

But it is well known that $\text{Aut}(A_m) = \text{Aut}(\Sigma_m) = \Sigma_m$ for all $m > 6$. Since $\Sigma(\Lambda) \leqslant A$ and $B \cong A_m$ or $\Sigma_m$, we thus conclude from (5.53) that $A = \Sigma(\Lambda)$, as asserted.

There is a further property of $\mathscr{D}$ that plays a central role in Aschbacher's analysis, and is immediate from the structure of $A_m$.

PROPOSITION 5.47. *If $U, V \in \Omega$, then*
(i) $\langle U, V \rangle \cong E_{16}, A_7, \Sigma_4, A_5,$ *or* $Z_2 \times Z_2$; *and*
(ii) *Correspondingly* $|\Lambda(U) \cap \Lambda(V)| = 0, 1, 2, 3$ *or* $4$.

[Clearly if $|\Lambda(U) \cap \Lambda(V)| = 0$ or $4$, then $V$ centralizes $U$, with $V = U$ in the latter case.]

The thrust of Aschbacher's analysis is to show that the commuting graph $\mathscr{D}$ defined on the set $\Omega$ of $G$-conjugates of $R$ is isomorphic to the root four-subgroup graph of $A_{n+4}$. But $G$ acts by conjugation on $\Omega$ and preserves incidence in $\mathscr{D}$. Moreover, as $L$ is tightly embedded in $G$ and $O(G) = 1$, this action is faithful, so $G \leqslant \text{Aut}(\mathscr{D})$. Hence once this graph isomorphism is established, Proposition 5.46 will imply that $G$ is isomorphic to a subgroup of $A_{n+4}$. The embedding of $L \times R$ in $G$ will then yield the desired conclusion $F^*(G) \cong A_{n+4}$.

We describe Aschbacher's principal moves. Set $N = N_G(L)$. First, by Proposition 5.34 there is $g \in G$ such that $R^g$ is a root four-subgroup of $L$, and he derives the following analog of Proposition 3.88 in the proof of the classical involution theorem.

PROPOSITION 5.48. *We have* $F^*(G) = \langle LR, (LR)^g \rangle$.

Indeed, assume false. Then setting $X = \langle LR, (LR)^g \rangle$, we have $X < G$, and so by the minimality of $G$, $X/O(X) \cong A_{n+4}$. Since $C_t \leqslant N$ and $N/C_L \leqslant \Sigma_n$ for $t \in R^\#$, we see that $L$ centralizes each $C_{O(X)}(t)$, and as $O(X) = \langle C_{O(X)}(t) \mid t \in R^\# \rangle$, it follows that $L$ and hence $X$ centralizes $O(X)$. Thus $X$ is, in fact, quasisimple, whence $X \cong A_{n+4}$. Again by the tight embedding of $L$, he easily concludes that

$$\begin{align} &(1) \ N_G(X) = XC_X; \text{ and} \\ &(2) \ X \text{ contains a Sylow 2-subgroup } S \text{ of } G. \end{align} \tag{5.54}$$

In particular, setting $M = N_G(X)$, it follows likewise for any $t \in R^\#$ that $C_t \leqslant M$ and that $M$ controls the $G$-fusion of $t$ in $S$. Aschbacher now derives a contradiction from his strong embedding criterion [I, 4.31]. To apply it, he must first show that $C_{tt^y} \leqslant M$ whenever $t^y$ centralizes $t$ with $t^y \neq t$, $y \in G$. His proof of this assertion depends on Goldschmidt's strongly closed abelian 2-group theorem [I, 4.28]. Once this is established, he concludes that $M$ is strongly embedded in $G$, which in turn implies that $M$ has only one conjugacy class of involutions, contrary to the fact that $M = X \times O(M)$ with $X \cong A_{n+4}$ has more than one class of involutions.

The proposition has several important direct consequences.

PROPOSITION 5.49. *The following conditions hold*:

    (i) $\mathscr{D}$ *is connected*;
    (ii) $O(N) = 1$; *and*
    (iii) $N$ *is isomorphic to the normalizer of a root four-subgroup of* $A_{n+4}$.

Note that as $U = R^g \leqslant L$, $N_L(U)/C_L(U) \cong \Sigma_3$, so once (ii) is proved, it follows at once that $N/L \cong \Sigma_4$ and hence that $N$ has the structure specified in (iii).

Aschbacher needs one other important property of $\mathscr{D}$.

PROPOSITION 5.50. $\mathscr{D}$ *has diameter* 2—*i.e., for any two elements* $U, V \in \Omega$, *there exists an element* $W \in \Omega$ *such that* $U, W$ *and* $W, V$ *are each connected by an edge.*

Indeed, if false, there exists a connected chain in $\mathscr{D}$ of four vertices $U, X, W, V \in \Omega$ such that $U$ and $V$ are not "adjacent" in $\mathscr{D}$ to a common vertex of $\Omega$. In particular, $U, X, W, V$, are all distinct and $X$ does not centralize $V$. Since $X = R^x$ for some $x \in G$, we can assume by conjugating by $x^{-1}$, if necessary, that $X = R$. Thus $U \leqslant L$ and

$$\langle U, R \rangle \cong E_{16}. \tag{5.55}$$

On the other hand, $W = R^w$ for some $w \in G$ and as $R \neq W \neq V$, it follows that $\langle R, V \rangle \leqslant L^w$. But now applying Proposition 5.47 in $L^w$ and using the fact that $[R, V] \neq 1$, it follows that

$$\langle R, V \rangle \cong A_7, \Sigma_4, \text{ or } A_5. \tag{5.56}$$

In each case, using (5.55) and the tight embedding of $R$, Aschbacher now produces a vertex of $\Omega$ which centralizes *both* $U$ and $V$, contrary to assumption.

Now Aschbacher is ready to tackle the global structure of $\mathscr{D}$. First, identify $L$ with the alternating group on the set $\Lambda_n = \{1, 2, ..., n\}$ with $R^g = \langle (12)(34), (13)(24) \rangle$, so that $R^g$ is a root four-subgroup of $L$ in this representation. Observe next that $C_L \cong A_4$ possesses precisely *four* Sylow 3-subgroups. Denote the set of these by $\Lambda_4$ and designate the elements of $\Lambda_4$ by $\{n + 1, n + 2, n + 3, n + 4\}$. Finally put

$$\Lambda = \Lambda_n \cup \Lambda_4, \tag{5.57}$$

so that $|\Lambda| = n + 4$.

Let $\Omega^*$ be the set of subsets of $\Lambda$ of cardinality 4, and let $\mathscr{D}^*$ be the graph having $\Omega^*$ as vertex set with two elements of $\Omega^*$ connected by an edge if and only if their intersection is empty. In view of Proposition 5.45, Aschbacher's task is thus to show that $\mathscr{D}$ and $\mathscr{D}^*$ are equivalent graphs.

Clearly to accomplish this, he must first define a correspondence between the elements of $\Omega$ and $\Omega^*$, and then show that it preserves incidence in $\mathscr{D}$. As a preliminary step, he first establishes a more intrinsic group-theoretic description of $\mathscr{D}$, which is motivated by Proposition 5.47 and verified by means of Proposition 5.50.

Thus for $0 \leqslant k \leqslant 4$, define $\Omega_k(R) = $ set of $U \in \Omega$ such that

$$\langle U, R \rangle \cong \begin{cases} E_{16} & \text{if} \quad k = 0 \\ A_7 & \text{if} \quad k = 1 \\ \Sigma_4 & \text{if} \quad k = 2 \\ A_5 & \text{if} \quad k = 3 \\ Z_2 \times Z_2 & \text{if} \quad k = 4 \end{cases} \tag{5.58}$$

In particular, $\Omega_4(R) = \{R\}$. Moreover, his prior analysis of the tight embedding of $R$ implies that $\Omega_0(R)$ is precisely the set of root four-subgroups of $L$; and by definition of $\mathscr{D}$, $\Omega_0(R)$ is the set of vertices in $\Omega$ connected to $R$ by an edge.

With the aid of Proposition 5.50, Aschbacher now easily proves the following result:

PROPOSITION 5.51. *We have* $\Omega = \displaystyle\bigcup_{k=0}^{4} \Omega_k(R)$.

We shall refer to the unique integer $k$ for which $U \in \Omega_k(R)$, $0 \leqslant k \leqslant 4$, as the *valence* of the element $U \in \Omega$.

Now Aschbacher can define the required correspondence between $\Omega$ and $\Omega^*$. For each $U \in \Omega$, he associates a subset $\Lambda(U)$ of $\Lambda$, which is composed of two parts: $\Lambda_n(U)$, a subset of $\Lambda_n$, and $\Lambda_4(U)$, a subset of $\Lambda_4$. Moreover, the definition of $\Lambda_4(U)$ depends on the valence $k$ of $U$. Specifically Aschbacher sets

$$\Lambda_n(U) = \{\text{set of fixed points of } \langle \Omega_0(U) \cap \Omega_0(R) \rangle \text{ on } \Lambda_n\}. \tag{5.59}$$

[Note that as $\Omega_0(R)$ is the set of root subgroups of $L$, $\langle \Omega_0(U) \cap \Omega_0(R) \rangle \leqslant L$ and so $\Lambda_n(U)$ is well-defined.]

$$\Lambda_4(U) = \begin{cases} \varnothing & \text{if} \quad k = 0 \\ \{C_H(U)\} & \text{if} \quad k = 1 \\ \text{the support of } N_U(R) \text{ on } \Lambda_4 & \text{if} \quad k = 2 \\ \Lambda_4 - \{N_H(U)\} & \text{if} \quad k = 3 \\ \Lambda_4 & \text{if} \quad k = 4 \end{cases} \tag{5.60}$$

For $\Lambda_4(U)$ to be well-defined, he must show when $k = 1$ or $3$ that $C_H(U)$, $N_H(U)$, respectively, has order 3.

Finally, as we have said, he defines

$$\Lambda(U) = \Lambda_n(U) \cup \Lambda_4(U). \tag{5.61}$$

To show that the map $U \mapsto \Lambda(U)$ for $U \in \Omega$ transforms $\Omega$ to $\Omega^*$, Aschbacher must obviously show that $|\Lambda(U)| = 4$ for every $U \in \Omega$. He, in fact, proves the following stronger result:

PROPOSITION 5.52. (i) *For any $U \in \Omega$, $|\Lambda(U)| = 4$ and $\Lambda_4(U) =$ the valence of $U$. [In particular, $\Lambda(U) \in \Omega^*$.]*
(ii) *If $\Omega_0^* = \{\Lambda(U) \mid U \in \{R\} \cup \Omega_0(R)\}$, then the subgraph $\mathscr{D}_0^*$ of $\mathscr{D}^*$ with vertex set $\Omega_0^*$ is isomorphic to the graph of the normalizer of a root four-subgroup of $A_{n+4}$ in its action on root subgroups of $A_{n+4}$.*

Again the result depends critically on properties of the subgroups $\langle U, V \rangle$ for $U, V \in \Omega$. So likewise does Aschbacher's final result, which completes the identification of $\mathscr{D}$.

PROPOSITION 5.53. *Two distinct vertices $U, V \in \Omega$ are connected by an edge if and only if $\Lambda(U) \cap \Lambda(V) = \varnothing$.*

By the proposition, the map $U \mapsto \Lambda(U)$ for $U \in \Omega$ is incidence-preserving and so induces an isomorphism of $\mathscr{D}$ on $\mathscr{D}^*$, as required, and the desired conclusion $F^*(G) \cong A_{n+4}$ follows.

We turn now to the four *exceptional* cases of the Aschbacher–Seitz reduction theorem:

$$\begin{aligned} &(1)\ L \cong Sz(8);\ \text{or} \\ &(2)\ L/O(L) \cong L_3(4),\ L_3(4)/Z_2 \times Z_2,\ \text{or}\ G_2(4), \end{aligned} \tag{5.62}$$

beginning with Dempwolff's solution of the $Sz(8)$ case, one of the most technically difficult of all standard component problems [19]. We can do little more than sketch his principal moves.

We note that in his paper, Dempwolff assumes, in addition, that $C_y = L \times R$ for every $y \in R^{\#}$ [equivalently, that each $O(C_y) = 1$]. However, his analysis applies without this restriction, and the condition itself can be derived as a consequence of the balanced, connected theorem [I, 4.54].

As described in Section 5.2.3, Dempwolff's goal is to pin down the centralizer of a 2-central involution of $G$, in which case he can quote Parrott's classification theorem [94] to conclude that

$$G = F^*(G) \cong Ru. \tag{5.63}$$

Unfortunately the centralizer $C$ of a 2-central involution of $Ru$ is quite complicated; it is a 2-constrained group with the following properties:

(1) $O_2(C)$ has order $2^{11}$ and class 3;
(2) $C/O_2(C) \cong \Sigma_5$; and                                          (5.64)
(3) $C_{O_2(C)}(D) \cong Q_8$ for any Sylow 3-subgroup $D$ of $C$.

[These conditions form the initial hypotheses of Parrott's theorem.]

Starting from the standard component $L \cong Sz(8)$, it is a long and painful path to show that the centralizer of a 2-central involution of the given group $G$ has this general structure. Indeed, in carrying out the analysis, at various places in the argument Dempwolff uses block theory for the prime 2, delicate arithmetic estimates on the number of involutions in various 2-groups, Lie ring methods for determination of the structure of suitable 2-groups, including Higman's transitive involution theorem [I: 170] as well as explicit sets of generators and relations; and, of course, underlying it all is an elaborate analysis of the fusion of involutions in $G$. In particular, to eliminate one especially troublesome configuration, Dempwolff is forced to compute the order of $G$ by means of the Thompson order formula and then apply Sylow's theorem for an appropriate odd prime to reach a contradiction. Again we assume that $G$ is a minimal counterexample to the proposed theorem.

We follow the same notation as Griess, Mason, and Seitz in Section 5.3. Dempwolff's first step is the same as theirs: namely, to determine the structure of $N = N_G(V)$, where $V = UR$, $U = \Omega_1(Q) \cong E_8$. The desired result is stated in Lemma 5.23, which gives the general structure of $V_1 = O_2(N)$ and the action of $A$ on $V_1$ [here $P = AQ$, where $A \cong Z_7$ and $Q \in \mathrm{Syl}_2(L)$ has class 2 and order $2^6$]. In particular, $V_1$ possesses an $A$-invariant complement $W_1$ to $R$ or order $2^9$ and $W_1 = QU_1$, where $U_1 \lhd N$, $Q \cap U_1 = U$, and $|U_1| = 2^6$.

Dempwolff analyzes the structure of $W_1$ and the action of $A$ on $W_1$ more closely, arguing that $U_1$ is either elementary or homocyclic abelian of exponent 4 and that either $Q$ centralizes $U_1$ or $[Q, U_1] = U$. Thus he concludes:

PROPOSITION 5.54. *The group $W_1 A$ has one of four general structures. Moreover, in all cases, $U \leqslant Z(V_1)$ and $V_1/U$ is elementary of order $2^8$.*

In *Ru* itself, one has

$$(1) \ \ U_1 \ \text{is homocyclic of exponent 4; and}$$
$$(2) \ \ [Q, U_1] = U. \tag{5.65}$$

Dempwolff next proves the following result:

PROPOSITION 5.55. *We have $V_1 \notin Syl_2(G)$.*

Now set $N_1 = N_G(V_1)$, $\overline{N}_1 = N_1/V_1 O(N_1)$, and let $S_1 \in \mathrm{Syl}_2(N_1)$. By a long and difficult argument, he forces the structures of $\overline{N}_1$ and its action on $V_1$.

PROPOSITION 5.56. *The following conditions hold:*

(i) $\overline{N}_1 \cong GL_3(2)$;
(ii) $\overline{N}_1$ *acts faithfully on both $U$ ($\cong E_8$) and $V_1/U$ ($\cong E_{2^8}$); and*
(iii) $V_1/U$ *affords the regular representation for $\overline{S}_1$ ($\cong D_8$).*

In particular, as $V_1 = W_1 R$ has order $2^{11}$,

$$|S_1| = 2^{14}. \tag{5.66}$$

Proposition 5.55 has the following important consequences:

PROPOSITION 5.57. *The following conditions hold:*

(i) $S_1 \in \mathrm{Syl}_2(G)$;
(ii) $U_1$ *is homocyclic abelian of exponent 4 and $[Q, U_1] = U$;*
(iii) *If $D_1 \in \mathrm{Syl}_3(N_1)$, then $C_{V_1}(D_1) = C_{W_1}(D_1) \cong Q_8$; and*
(iv) *The structure of $W_1$ is uniquely determined.*

In view of (ii) and (iv), and (5.65), $W_1 A$ is thus isomorphic to the corresponding subgroup of *Ru*. Furthermore, as $S_1 \in \mathrm{Syl}_2(G)$ and $U \lhd S_1$ with $\overline{N}_1$ acting transitively on $U^\#$, every involution of $U$ is 2-central. We fix $z \in Z(S_1) \cap U^\#$.

At this point, one can also easily force $O(C_y) = 1$ for every $y \in R^\#$, and reduce to Dempwolff's initial hypothesis. indeed, $O(N_1) = \langle O(C_y) \mid y \in R^\# \rangle$, as is easily verified, so if the assertion is false, then $O(N_1) \neq 1$ and it follows that $M = N_G(O(N_1))$ is a proper subgroup of $G$ containing both $N_1$ and $L$. But then by the minimality of $G$, $M/O(M) \cong Ru$. This immediately implies that $G$ has precisely two conjugacy classes of involutions, represented by $x$ and $z$ and $C_z$ is 2-constrained. Since $Sz(8)$ is locally balanced, [I, 4.58] yields now that $G$ is balanced. Since $S_1$ is connected, we conclude therefore from [I, 1.54] that $O(C_y) = 1$ for every $y \in R^{\#}$ and hence that $O(N_1) = 1$—contradiction. Thus for each $y \in R^{\#}$, we have

$$C_y = L \times R. \tag{5.67}$$

Dempwolff now undertakes an elaborates analysis of the involution fusion pattern of $G$. In particular, at one point he requires Timmesfeld's weakly closed theorem [I, 4.161]. The final result is as follows:

PROPOSITION 5.58. *The following conditions hold*:

(i) *$G$ has exactly two conjugacy classes of involutions, represented by $x$ and $z$; and*
(ii) *One of the following holds*:
   (1) $C_z \leqslant N_1$; *or*
   (2) $C_z/O_2(C_z) \cong \Sigma_5$, *(whence certainly $C_z \not\leqslant N_1$), and $C_z$ is isomorphic to the centralizer of a 2-central involution of $Ru$.*

It therefore remains only for Dempwolff to rule out the possibility $C_z \leqslant N_1$ to reach Parrott's hypothesis. But in this case, the structure of $C_z$ is completely determined by $N_1$ and, in addition, he knows the involution fusion pattern in $G$ (there are two possible patterns, according as $S_1 - V_1$ contains or does not contain a $G$-conjugate of $z$), so he can compute the order of $G$ by the Thompson order formula [I, 2.43]. Corresponding to the respective fusion patterns, he obtains the following proposition:

PROPOSITION 5.59. *If $C_z \leqslant N_1$, then either*

$$|G| = 2^{14} \cdot 3 \cdot 5^2 \cdot 7 \cdot 13 \quad or \quad 2^{14} \cdot 3 \cdot 5 \cdot 7^2 \cdot 13 \cdot 31.$$

Finally Dempwolff determines the order $h$ of a Sylow $p$-normalizer in $G$, where $p = 7$ or 13, respectively. In the first case, $h = 2^3 \cdot 3 \cdot 7$ and in the second case, either $h = 2^4 \cdot 13$ or $2^4 \cdot 3 \cdot 13$. Given the values of $|G|$ in Proposition 5.59, we check in each case that $n = |G|/h \not\equiv 1 \pmod{p}$. But $n$ is the number of Sylow $p$-subgroups of $G$ and so by Sylow's theorem, we must have $n \equiv 1 \pmod{p}$—contradiction.

There is a global picture of Dempwolff's determination of the structure of a 2-central involution of $G$.

We conclude with a few words about Parrott's proof, which follows the general lines of argument described in [I, Chapter 2], the aim being to prove:

> (1) $G$ is a rank 3 primitive permutation group of degree 4,060 in which a one-point stabilizer $G_1 \cong {}^2F_4(2)$; and
> (2) In its permutation action, the nontrivial orbits of $G_1$ have lengths 1,755 and 2,304.

$$(5.68)$$

The desired conclusion $G \cong Ru$ will then follow from the Conway–Wales characterization theorem [I: 71].

In succession, Parrott determines the exact involution fusion pattern of $G$ (two conjugacy classes), generators and relations for a Sylow 2-subgroup of $G$, the order of $G$ by the Thompson order formula, the distinct conjugacy classes of $G$ (a total of 36 classes), and much of its local structure. Then using block theory for the various prime divisors of $|G|$, he determines the degrees of each of the 36 irreducible characters of $G$, and finally calculates the values of one of these irreducible characters $\chi$ (of degree 406) on each conjugacy class ($\chi$ is integral-valued).

Now Parrott is in a position to apply the "Brauer trick" [I, p. 98] to two suitably chosen subgroups $A, B$ of $G$, where $A = O_2(C_z) N_{C_z}(F)$ for an appropriate Sylow 5-subgroup $F$ of $C_z$ and $B \leqslant N_G(L)$, satisfying

> (1) $A \cap B \in \mathrm{Syl}_2(A)$; and
> (2) $|B : A \cap B| = 3$,

$$(5.69)$$

to conclude that $\langle A, B \rangle$ is a proper subgroup of $G$.

Furthermore, from the structure of $C_z$, he knows there is an element $t \in C_z$ of order 4 that normalizes both $A$ and $B$ with $t^2 \in \langle A, B \rangle$. Hence $G_1 = \langle t, A, B \rangle$ is also a proper subgroup of $G$. It is this subgroup which Parrott identifies as ${}^2F_4(2)$. Indeed, $A\langle t \rangle$ is isomorphic to the centralizer of a 2-central involution of ${}^2F_4(2)$ and hence to the maximal parabolic subgroup of ${}^2F_4(2)$, whose Levi factor is $Sz(2)$, a Frobenius group of order 20.

Parrott argues that $A\langle t \rangle = C_{G_1}(t)$ and then by an analysis of the fusion of involutions in $G_1$ shows that $G_1$ has a normal subgroup $G_0$ of index 2 not containing $t$. It follows that $C_{G_0}(z)$ satisfies the conditions of his prior centralizer of involution characterization of the Tits group $^2F_4(2)'$ [I: 233], thus yielding the desired conclusion $G_0 \cong {}^2F_4(2)'$ and

$$G_1 \cong {}^2F_4(2). \qquad (5.70)$$

Finally with the aid of his previously derived local and character-theoretic information, Parrott derives the required properties of the permutation representation of $G$ on the right cosets of $G_1$ (rank 3 with subdegrees 1,755 and 2,304).

We turn finally to the Aschbacher–Seitz solutions of the three cases of (5.62)(2). Their objective is to prove correspondingly that

$$F^*(G) \cong Suz, He, \text{ or } .1. \qquad (5.71)$$

(We note that $Suz$ had previously been characterized by Reifart [122] in terms of the exact structure of the centralizer of a non-2-central involution.)

As in the $Sz(8)$ case, just considered, it suffices for them to pin down the centralizer $C_0$ of a 2-central involution of $G_0 = O^2(G)$: namely,

(1) $O_2(C_0)$ is extra-special of order $2^7, 2^7$, or $2^9$,
respectively; and
(2) $C_0/O_2(C_0) \cong PSp_4(3), L_3(2), \text{ or } P\Omega_8^+(2)$,
respectively, $\qquad (5.72)$

for then they can invoke the appropriate centralizer of involution theorem to determine the isomorphism type of $G_0$—F. Smith [110] and Patterson and S. K. Wong [95] for $Suz$, Held [I: 165] and Dempwolff and S. K. Wong [20] for $He$, and Patterson [I: 235] for $.1$. Just as Parrott's argument is, in effect, a reduction to the Conway–Wales rank 3 permutation group characterization of $Ru$, so likewise each of these analyses consists of a reduction to an appropriate permutation group recognition theorem. See [I, Chapter 2 and Chapter 3] for additional details concerning these various latter results. We shall not discuss them further here.

The group $G_2(4)$ contains a subgroup isomorphic to $SL_3(4)$ [it is the group generated by all long root subgroups in a fixed system of root subgroups of $G_2(4)$]. Because of this, Aschbacher and Seitz are able to carry out the initial portion of their analysis for all three cases simultaneously. In the $G_2(4)$ case, let $L_1$ be such an $SL_3(4)$-subgroup of $L$, chosen so that $Q_1 = Q \cap L_1 \in \mathrm{Syl}_2(L_1)$ and in the remaining two cases, set $L_1 = L$ and $Q_1 = Q$. Hence in all cases,

$$L_1/Z(L_1) \cong L_3(4). \qquad (5.73)$$

~    A Sylow 2-subgroup of $L_3(4)$ is special of order 64 and is the product of two normal $E_{16}$-subgroups that intersect in its center. Moreover, whether or not $Q_1 R$ splits over $R$, it can be shown that the pre-images of these $E_{16}$-subgroups of $Q_1 R / R$ are elementary. Hence $Q_1 R$ is the product of two normal $E_{64}$-subgroups $V_1, V_2$ with $R \leqslant V_1 \cap V_2 \cong E_{16}$. In addition, the fusion analysis implies that

$$V_1 \cap V_2 = R \times R^g \tag{5.74}$$

for a suitable $g \in G - N_G(L)$.

Because $L$ now has two maximal parabolics containing $Q$, there is a richer structure here than in the Dempwolff situation. Indeed, motivation for many of their arguments comes from a certain "three-dimensional" geometry related to $V_1$ and $V_2$. The set $\Omega$ of all $G$-conjugates of $V_1$ and $V_2$ are its *planes*, of all $G$-conjugates of $R$ are its *points*, and of all $G$-conjugates of $V_1 \cap V_2$ are its *lines*. [Our later discussions of other Lie type characteristic 2 standard form problems with $m(R) = 1$ can be viewed from analogous geometric perspectives.]

Here are the common results that Aschbacher and Seitz establish. Set $N_i = N_G(V_i)$ and $\bar{N}_i = N_i / C_{V_i}$, $i = 1, 2$. Also put $N_i^R = N_{N_i}(R)$ and note that in all cases $N_i^R \cap L = N_i \cap L$ with $\bar{N}_i \cap L \cong A_5$.

PROPOSITION 5.60. *For $i = 1$ and 2, we have*

  (i) *If $|Z(L)|$ is odd, then we have $\bar{N}_i = O_2(\bar{N}_i)\bar{N}_i^R$, $O_2(\bar{N}_i) \cong E_{16}$, and $V_i \cap Q_1 \lhd N_i$.*

  (ii) *If $|Z(L)|$ is even [whence $L/O(L) \cong L_3(4)/Z_2 \times Z_2$], then we have $F^*(\bar{N}_i) \cong A_6/Z_3$.*

The distinction in the cases arises from the fact that correspondingly $V_i$ is a *decomposable* or *indecomposable* $GF(2)$-module for $\bar{N}_i \cap L$. As a result, there are distinct numbers of $G$-conjugates of $R$ contained in $V_i$ in the two cases (16 versus 6). Moreover, in the first case, the argument is similar to that of Seitz's Theorem 5.126 below.

Using the minimality of $G$, they also prove:

PROPOSITION 5.61. *If $O(C_L) \neq 1$, then*

$$N_G(O(C_L))/O(C_L) \cong Suz, He, or \cdot 1.$$

Their final common result is a sufficient condition for (5.72) to hold in

$G$. For simplicity, we combine it with some of their subsequent analysis. Let $S \in \mathrm{Syl}_2(G)$, fix $z \in Z(S)$, and set $\bar{C}_z = C_z/O(C_z)$.

PROPOSITION 5.62. *Suppose that $S$ contains an extraspecial subgroup $X$ of order $2^7$ or $2^9$ with the following properties:*

(a) $\langle z \rangle \in \mathrm{Syl}_2(C_X)$;

(b) *$X$ is weakly closed in $S$ with respect to $C_z$;*

(c) *If $m(X \cap X^y) > 1$ for $y \in C_z$, then $y$ normalizes $X$;*

(d) *$F^*\big(N_{\bar{C}_z}(\bar{X})/\bar{X}\big) \cong PSp_4(3)$ or $L_3(2)$ (with $|X| = 2^7$) or $P\Omega_8^+(2)$ (with $|X| = 2^9$).*

*Under these conditions, $X = O_2(C_z)$ and if $G_0 = O^2(G)$, then $C_{G_0}(z)$ is isomorphic to the centralizer of a 2-central involution of Suz, He, or .1, respectively.*

The proof that $\bar{X} = O_2(\bar{C}_z)$ depends on Timmesfeld's results on weakly closed 2-subgroups [I, 4.161] and Goldschmidt's results on strongly closed abelian 2-subgroups [I, 4.128], and then uses Holt's theorem [I, 4.32] to conclude that $O(C_z) = 1$.

Thus Aschbacher and Seitz are reduced to producing the extraspecial 2-subgroup $X$ satisfying the conditions of the proposition. They accomplish this by a delicate case-by-case 2-local pushing-up process. We shall indicate the principal moves leading to the definition of $X$. However, we note first that if $O(C_L) \neq 1$, then by Proposition 5.61, $X$ is already visible in $N_G\big(O(C_L)\big)$. Hence without loss we can assume throughout that

$$O(C_L) = 1. \tag{5.75}$$

In particular, $O(L) = 1$.

We first consider the two cases in which $Z(L) = 1$ [whence $L \cong L_3(4)$ or $G_2(4)$]. Here we have $Q_1 R = Q_1 \times R$ and $V_i = U_i \times R$ with $E_{16} \cong U_i \leqslant Q_1$, $i = 1, 2$. Likewise $Z = U_1 \cap U_2 = Z(Q_1) \cong Z_2 \times Z_2$. Moreover, Proposition 5.59(i) holds for $N_i$, $i = 1, 2$. Note also that as $V_i \lhd Q_1 R$, $Q_1 \leqslant N_i$ and we let $V_i^*$ be a $Q_1$-invariant Sylow 2-subgroup of the pre-image of $O_2(\bar{N}_i)$ in $N_i$, so that

$$V_i \lhd V_i^* \quad \text{and} \quad V_i^*/V_i \cong E_{16}. \tag{5.76}$$

It follows easily from Proposition 5.60 that $Z \leqslant Z(V_i^*)$, $i = 1, 2$. Finally let $Y_i$ be the pre-image of $C_{V_i^*/Z}(R)$ in $V_i^*$ and let $W_j$ be the subgroup $V_j Y_i$, $j \neq i$, $1 \leqslant i, j \leqslant 2$. We fix this notation.

First consider the case

$$L \cong L_3(4) \tag{5.77}$$

(whence $Q = Q_1$). Here Aschbacher and Seitz focus first on the group $K = \langle V_1^*, V_2^* \rangle$ and its relation to $W_1$, establishing all the following facts.

PROPOSITION 5.63. *The following conditions hold*:

(i) $K$ *normalizes* $W_1$;
(ii) $K/W_1 \cong A_5$;
(iii) $W_1$ *is the central product of two copies of the Sylow 2-subgroup of* $L_3(4)$. *In particular*, $W_1$ *is special with center* $Z$; *and*
(iv) $W_1/Z$ *is the sum of two natural modules for* $K/W_1$ *as a* GF(2)-*module*.

On the basis of this information, they next argue that $V_1$ and $V_2$ are $G$-conjugate (whence $\Omega = \{V_1^g \mid g \in G\}$) and then go on to pin down the structure of $M_1 = N_G(V_1^*)$. Indeed, setting $M_0 = O^2(M_1)$ and $\bar{M}_1 = M_1/V_1^* O(M_1)$, they prove:

PROPOSITION 5.64. *The following conditions hold*:

(i) $\bar{M}_0 \cong A_6/Z_3$;
(ii) $U_1 \lhd M_0$ *is the natural module for* $M_0/C_{M_0}(U_1) \cong A_6$; *and*
(iii) $\bar{M}_0$ *acts irreducibly on* $V_1^*/U_1 \cong E_{64}$.

This in turn yields:

PROPOSITION 5.65. *The following conditions hold*:

(i) $M_1$ *contains a Sylow 2-subgroup* $S_1$ *of* $G$;
(ii) $S_0 = S_1 \cap M_0$ *has index* $\leqslant 2$ *in* $S_1$;
(iii) $S_1$ *can be chosen to contain* $W_1$ *as a normal subgroup; and*
(iv) $N_{M_1}(W_1)$ *contains a subgroup* $D$ *of order* 3 *permutable with both* $S_0$ *and* $K$ *such that* $KDS_0/W_1 \cong \Sigma_3 \times A_5$.

In particular, it follows from (i) and (ii) and Proposition 5.64 that a Sylow 2-subgroup of $G$ has order $2^{13}$ or $2^{14}$.

Finally the extra-special subgroup $X$ emerges from the analysis.

PROPOSITION 5.66. $S_0$ *possesses a normal extra-special subgroup* $X$ *of order* $2^7$ *with the following properties*:

(i) $X$ *is* $K$-*invariant*; *and*
(ii) $|U_1 : U_1 \cap X| = 2$.

In fact, $X$ is shown to be uniquely determined, subject to the additional requirement that it be invariant under a specified subgroup of $M_0$ of order 3 contained in the pre-image of $Z(\bar{M}_0)$.

Letting $N_0$ be the normal closure of $S_0$ in $N_G(X)$, they prove the following additional result:

PROPOSITION 5.67. *We have* $N_0/XO(N_0) \cong PSp_4(3)$.

Now Aschbacher and Seitz argue easily that the various conditions of Propositions 5.61 hold with $S = S_1$ and $\langle z \rangle = Z(X)$ and with $F^*\left(N_{\bar{C}_z}(\bar{X})/\bar{X}\right) \cong PSp_4(3)$. In view of our earlier discussion, they thus conclude that $F^*(G) \cong Suz$, as required.

We next consider the case

$$L \cong G_2(4), \tag{5.78}$$

whence $L_1 \cong SL_3(4)$. Setting $D = Z(L_1) \cong Z_3$, Aschbacher and Seitz first prove:

PROPOSITION 5.68. *We have* $C_D \cong Suz/Z_3$.

Indeed, $L_1$ is standard in $C_D$ and $R \in \mathrm{Syl}_2\left(C_{C_D}(L_1)\right)$. Moreover, their information on the $G$-conjugates of $R$ implies that $L_1$ is not normal in $C_D$ (cf. Proposition 5.37). Hence all the results obtained in the previous case apply with $\bar{C}_D = C_D/O(C_D)$, $\bar{L}_1$, $\bar{R}$ in the roles of $G, L, R$, respectively, and we conclude that $F^*(\bar{C}_D) \cong Suz$. But it is also easy to check that $O(C_D) = DO(C_L)$, whence $O(C_D) = D$ by (5.75). Since $L_1$ does not split over $D$, we must have $D \leqslant F^*(C_D)$, whence $F^*(C_D) \cong Suz/Z_3$. Finally outer automorphisms of $Suz$ of even order $\left(\mathrm{Out}(Suz) \cong Z_2\right)$ are known to invert the three-part of the Schur multiplier of $Suz$, forcing $C_D = F^*(C_D)$, so $C_D \cong Suz/Z_3$, as asserted.

The proposition goes "three-quarters" of the way towards identification of the required extra-special group of order $2^9$. Indeed, if $X$ and $z$ are defined in $C_D$ exactly as in the preceding case [whence $X \cong D_8 * D_8 * D_8$ is extra-special of order $2^7$ and normal in $C_0 = C_{C_D}(z)$ with $C_0/XO(C_0) \cong PSp_4(3)$],

Aschbacher and Seitz argue in a sequence of steps that $O_2(N_L(U_1))$ contains a $C_0$-invariant quaternion subgroup $X_1$ with $z \in X_1$. It follows that $X$ ($\leqslant C_0$) centralizes $X_1$. The group

$$X^* = X * X_1 \tag{5.79}$$

is the required extra-special group of order $2^9$.

We note two results that Aschbacher and Seitz need to reach $X_1$, the first of which follows directly from the structure of $G_2(4)$. Set $Y_1 = O_2(P_1)$ and $\bar{P}_1 = P_1/Y_1$.

> **PROPOSITION 5.69.** *The following conditions hold*:
>
> (i) $U_1 = Z(Y_1)$ *with* $Y_1/U_1 \cong E_{64}$;
> (ii) $\bar{P}_1 \cong GL_2(4)$; *and*
> (iii) $\bar{P}_1$ *acts indecomposably on* $Y_1/U_1$ *with precisely described action*.

> **PROPOSITION 5.70.** *The following conditions hold*:
>
> (i) $Y_1 = O_2(C_{V_1})$; *and*
> (ii) $C_{V_1}/Y_1$ *has odd order* [*whence* $Y_1 \in \mathrm{Syl}_2(C_{V_1})$].

Observe next as $C_0$ normalizes both $X$ and $X_1$, $C_0 \leqslant N^* = N_G(X^*)$. Setting $\bar{N}^* = N^*/X^*O(N^*)$, it then follows that $C_0 = Z_3 \times \Omega_6^-(2)$ [$P\Omega_6^-(2) \cong PSp_4(3)$ is of index 2 in $\Omega_6^-(2)$]. On the other hand, as $X^* \cong (D_8)^3 * Q_8$, it is known that $\mathrm{Out}(X^*) \cong \mathrm{Aut}(P\Omega_8^+(2))$ [which contains $P\Omega_8^+(2)$ as a subgroup of index 2]. It is also easy to see that $Z_3 \times P\Omega_6^-(2)$ is maximal in $P\Omega_8^+(2)$. Aschbacher and Seitz now argue that $\bar{C}_0 < \bar{N}^* \leqslant P\Omega_8^+(2)$, whence they obtain the following result:

> **PROPOSITION 5.71.** *We have* $\bar{N}^* \cong P\Omega_8^+(2)$.

However, they still require an additional piece of information before they can show that $C_z$ satisfies the conditions of Proposition 5.62—namely, the structure of a Sylow 2-subgroup $S$ of $G$.

To pin down $S$, they first use the structure of $N^*$ to show that $N_0^* = C_{N^*}(Z)$ contains a normal subgroup $S^*$ with the following properties:

> (1) $S^*$ is special with center $Z$ and is the central product of
>      three copies of a Sylow 2-subgroup of $L_3(4)$; and
> (2) $\bar{N}_0^* \cong P\Omega_6^+(2)$.
>
> (5.80)

Setting $M^* = N_G(S^*)$ and $\bar{M}^* = M^*/S^*O(M^*)$, they finally prove the following result:

PROPOSITION 5.72. *The following conditions hold*:

(i) $\bar{M}^* \cong \Sigma_3 \times P\Omega_6^+(2)$;
(ii) $S^* = O_2(C_Z)$; *and*
(iii) $M^*$ *contains a Sylow* 2-*subgroup of G*.

Now they can verify the conditions of Proposition 5.62 and thus obtain the desired conclusion $F^*(G) \cong .1$.

The final case

$$L \cong L_3(4)/Z_2 \times Z_2 \qquad (5.81)$$

is similar. However, as $N_1$ and $N_2$ now have the "richer" structure of Proposition 5.60(ii), the analysis is shorter. Indeed, most of the needed information is contained in the group $M = N_G(V_1 V_2)$. Setting $\bar{M} = M/V_1 V_2 O(M)$, Aschbacher and Seitz first prove:

PROPOSITION 5.73. *The following conditions hold*:

(i) $M$ *contains a Sylow* 2-*subgroup* $S$ *of* $G$;
(ii) $\bar{M} \cong \Sigma_3 \times \Sigma_3$ *or* $\Sigma_3 \wr Z_2$; *and*
(iii) *Correspondingly*, $S$ *has order* $2^{10}$ *or* $2^{11}$ *and is of type He or* Aut($He$).

In particular, it follows from (iii) that $S$ contains a unique normal extra-special subgroup $X \cong D_8 * D_8 * D_8$ with $S/X \cong D_8$ or $D_{16}$, respectively.

Finally set $N = N_G(X)$ and $\bar{N} = N/XO(N)$. Using their 2-local information and the fact that $\bar{S} \in \mathrm{Syl}_2(\bar{N})$ is dihedral, Aschbacher and Seitz next prove:

PROPOSITION 5.74. *According as* $|S| = 2^{10}$ *or* $2^{11}$,

$$\bar{N} \cong L_3(2) \quad \text{or} \quad PGL_2(7).$$

[Note that $L_3(2) \cong L_2(7)$.]

Once again they are now in a position to verify that the conditions of Proposition 5.62 hold in $C_z$, where $\langle z \rangle = Z(X)$, thus obtaining the desired conclusion $F^*(G) \cong He$.

This completes our outline of the solutions of standard form problems with $m(H) \geqslant 2$.

## 5.6. The Alternating Case

In view of the results of Sections 5.3–5.5, for the remainder of the chapter:

$$\text{We assume } m(H) = 1 \text{ and } R \text{ is cyclic.} \tag{5.82}$$

In this section, we discuss Solomon's solution of the standard component problem

$$L \cong A_n, \qquad n \geqslant 8. \tag{5.83}$$

In view of (5.13)(b), we shall operate under the assumption that the following standard component problems have been solved:

$$L_4(4^m), PSp_4(4), HS, HS/Z_2, \text{ and } F_5. \tag{5.84}$$

Under these assumptions and Hypothesis $S$, and allowing the wreathed solution, Solomon's goal therefore is to prove that

$$\begin{aligned} &F^*(G) \cong A_{n+2}, L_4(4) \text{ or } HS \text{ (with } n = 8), \\ &F_5 \text{ (with } n = 10), \text{ or } A_n \times A_n. \end{aligned} \tag{5.85}$$

Note that as $A_8 \cong L_4(2)$, this will, in particular, solve the $L_4(2)$ standard component problem.

We assume throughout that $G$ is a minimal counterexample to the proposed theorem. We regard $N_G(L)/H$ ($\cong A_n$ or $\Sigma_n$) as a permutation group on the set $\Lambda = \{1, 2, ..., n\}$.

Solomon first proves the following result:

PROPOSITION 5.75. *The following conditions hold*:

(i) $R = \langle x \rangle$; *and*

(ii) $C_x$ *splits over* $\langle x \rangle$.

Finkelstein's argument in Proposition 5.3 gives a quick proof of (i).

Indeed, if false, it yields that there is $g \in G - N_G(L)$ such that $R^g$ acts faithfully on $L$ and that if we set $Y = LN_{H^g}(L)$ and $t = x^g$, then $C_Y(t)/O\big(C_Y(t)\big)$ contains a normal subgroup isomorphic to $Z_4$ or $Z_4 \times Z_2$. But then by the structure of $\Sigma_n$, $t$ must act on $\Lambda$ as a product of two transpositions. This in turn forces

$$\begin{array}{ll}(1) \ R \cong Z_4; \text{ and} \\ (2) \ I = L\big(C_L(t)\big) \cong A_{n-4}.\end{array} \qquad (5.86)$$

In particular, $t$ induces an inner automorphism on $L$ and so centralizes some Sylow 2-subgroup of $H$, which without loss we can assume to be $R$. By (5.86) and the structure of $C_{\Sigma_n}(t)$, it follows that

$$C_G(R\langle t\rangle) \text{ contains a } Z_2 \times Z_2 \times Z_4 \times A_{n-4} \text{ subgroup.} \qquad (5.87)$$

On the other hand, by $L$-balance, $I \leqslant L^g = L(C_t)$ and $x$ does not centralize $L^g$, so $R$ ($\leqslant C_t$) acts faithfully on $L^g$. Since $R$ centralizes $I$ and $R^g \in \mathrm{Syl}_2(H^g)$ with $R^g \cong Z_4$, we conclude from the structure of $C_{\Sigma_n}(R)$ that $C_G(R\langle t\rangle)$ is contained (mod core) in a $Z_4 \times Z_4 \times A_{n-4}$ subgroup of $G$, contrary to (5.87).

The proof of (ii) is similar in spirit, but more delicate. By the $Z^*$-theorem, $t = x^g \in T - \langle x \rangle$ for some $g \in G$. But now if (ii) is false, then $\Omega_1(T) = (T \cap L)\langle x \rangle$, so $t$ induces an inner automorphism on $L$. Solomon derives a contradiction by a comparison of the action of $t$ on $\Lambda$ with the action of $x$ in the corresponding permutation representation of $N_G(L^g)/H^g$. His leverage comes from the fact that $x$ is a *square* in $C_T(t)$ [as (ii) is assumed false].

In the present situation the *generic* case occurs for $n \geqslant 12$. Thus in Propositions 5.76–5.83:

$$\text{We assume } n \geqslant 12. \qquad (5.88)$$

In treating this case, we assume, in addition, that the cases $n = 8, 9, 10, 11$ have already been solved.

Let $D$ be the $A_4$-subgroup of $L$ with support $\{1, 2, 3, 4\}$ on $\Lambda$ and set $I = L\big(C_L(D)\big)$, so that $I \cong A_{n-4}$. Since $C_D \leqslant C_d$ for any $d \in \mathscr{I}(D)$, it follows by $L$-balance and Hypothesis S that $I \leqslant L_{2'}(C_D)$ and its normal closure $J$ in the latter group is semisimple. We fix this notation.

By a careful involution fusion analysis, Solomon first proves the following result:

PROPOSITION 5.76. *We have $I < J$.*

Since $I$ is a component of $C_J(x)$, $I$ is standard in $J\langle x\rangle$ and $\langle x\rangle \in \mathrm{Syl}_2\big(C_{J\langle x\rangle}(x)\big)$. Hence by induction and the previous proposition, he concludes:

PROPOSITION 5.77. *One of the following holds*:

(i) $J \cong A_n$; *or*
(ii) $J \cong A_{n-2} \times A_{n-2}$.

The induction works smoothly if $n \neq 12$ or $14$, but in the latter two cases (whence $n - 4 = 8$ or $10$), the possibilities $J \cong L_4(4)$, $HS$, or $F_5$ must first be excluded to obtain the proposition. This is achieved by considering an element $y$ of order 3 in $D$, setting $I^* = L\big(C_L(y)\big)$ $(\cong A_{n-3})$ and arguing by induction in $\bar{C}_y = C_y/O(C_y)$ that the normal closure $\bar{J}^*$ of $\bar{I}^*$ in $L(\bar{C}_y)$ is isomorphic to $A_{n-1}$ or $A_{n-3} \times A_{n-3}$. But as $I \leqslant I^*$ and $I \leqslant J \leqslant C_d$, it follows that $\bar{J} \leqslant \bar{J}^*$. However, neither $A_{11}$ nor $A_9 \times A_9$ involves $L_4(4)$ or $HS$, while neither $A_{13}$ nor $A_{11} \times A_{11}$ involves $F_5$, so these three exceptional possibilities are indeed excluded.

We first consider the quasisimple case. Using the standard presentation for $\Sigma_{n+2}$, Solomon proves:

PROPOSITION 5.78. *If $J \cong A_n$, then $\langle L, J\rangle \cong A_{n+2}$.*

Indeed, observe first that as $x$ centralizes $I \cong A_{n-2}$, $x$ acts on $J$ like a transposition and consequently $C_J(x) \cong \Sigma_{n-2}$. Thus $D$ centralizes a $\Sigma_{n-2}$-subgroup of $C_x$ and it follows that $C_x/O(C_x)\langle x\rangle \cong \Sigma_n$. Since $C_x$ splits over $\langle x\rangle$, this implies that $C_x$ contains a subgroup $L^* \cong \Sigma_n$ with $L \leqslant L^*$. Moreover, $I^* = C_{L^*}(D) \cong \Sigma_{n-4}$, and the discussion shows that we can choose $I^*$ so that $I^* \leqslant J\langle x\rangle$.

Let $\{t_i,\ 1 \leqslant i \leqslant n-1\}$ be a canonical set of generators of $L^*$ with $t_i$, $4 \leqslant i \leqslant n-1$, a canonical set of generators of $I^*$. Thus

(1)  $t_i$ is a transposition, $1 < i \leqslant n-1$;
(2)  For $i \neq j$, $t_i t_j$ has order 1, 2, or 3 according as $j = i$,
       $j \neq i, i \pm 1$, or $j = i \pm 1$, $1 \leqslant i$, $j \leqslant n-1$; and     (5.89)
(3)  $L^* = \langle t_i \mid 1 \leqslant i \leqslant n-1\rangle$ and $I^* = \langle t_i \mid 4 \leqslant i \leqslant n-1\rangle$.

Note that also $D$ has index 2 in $\langle t_1, t_2, t_3\rangle \cong \Sigma_4$.

Now set $x = t_{n+1}$. Since $x$ centralizes $I^*$, we see that there exists a transposition $t_n$ in $J\langle x \rangle$ such that $t_i$, $4 \leqslant i \leqslant n + 1$, is a canonical set of generators of $J\langle x \rangle$. Thus we similarly have

(1) $t_i$ is a transposition, $4 \leqslant i \leqslant n + 1$;
(2) $t_i t_j$ has order 1, 2, or 3 according as $j = i$, $j \neq i$, $i \pm 1$, or $j = i \pm 1$, $4 \leqslant i, j \leqslant n + 1$; and  (5.90)
(3) $J\langle x \rangle = \langle t_i \mid 4 \leqslant i \leqslant n + 1 \rangle$.

We now set

$$G^* = \langle t_i \mid 1 \leqslant i \leqslant n + 1 \rangle \qquad (5.91)$$

so that

$$\langle L, J \rangle \leqslant \langle L^*, J \rangle \leqslant G^*. \qquad (5.92)$$

To establish the desired conclusion $\langle L, J \rangle \cong A_{n+2}$, it will clearly suffice to show that $G^* \cong \Sigma_{n+2}$; and this will follow provided we show that the generators $t_i$ of $G^*$ satisfy the relations of the canonical presentation of $\Sigma_{n+2}$ ([I: 181, pp. 137–139]). In view of (5.89) and (5.90) and the fact that $x = t_{n+1}$ centralizes $L^*$, (whence $t_i t_{n+1}$ has order 2 for all $i$, $1 \leqslant i \leqslant n - 1$), this reduces to verification of the single relation

$$t_1 t_n \text{ has order 2.} \qquad (5.93)$$

Consider the transposition $t_4$ and set $L_1 = L\big(C_L(t_4)\big) \cong A_{n-2}$, $J_1 = L\big(C_J(t_4)\big) \cong A_{n-4}$ and $I_1 = L\big(C_I(t_4)\big) \cong A_{n-6}$. Then

(1) $I_1 \leqslant L_1 \cap J_1$; and  (5.94)
(2) $x$ does not centralize $J_1$.

Hence by $L$-balance and (5.94), $L_1$ and $J_1$ necessarily lie in the same 2-component $K$ of $C_{t_4}$. Moreover, using the simplicity of $L_1$ and $J_1$ and the involution fusion pattern in $G$, we easily see that $K$ is quasisimple. Now applying induction once again, we conclude that $K \cong A_n$.

Finally $\langle t_1, x \rangle = \langle t_1, t_{n+1} \rangle \leqslant C_{t_4}$, $L_1 \langle t_1 \rangle \cong \Sigma_{n-2}$, $J_1 \langle x \rangle \cong \Sigma_{n-4}$, and $\langle t_1, x \rangle$ centralizes $I_1$. Since $K \cong A_n$, one can now easily argue from these conditions that $t_1$ must centralize $x$, whence $t_1 x = t_1 t_{n+1}$ has order 2, as required.

Now Solomon proves:

PROPOSITION 5.79. *If* $J \cong A_n$, *then* $F^*(G) \cong A_{n+2}$.

Indeed, setting $G^* = \langle L, J \rangle$, then as $G^* \cong A_{n+2}$, it suffices to prove that $G^* \lhd G$. If false, then $M = N_G(G^*) < G$. Note that as $J$ is quasisimple, $O(C_x)$ centralizes $J$ as well as $L$. Hence $O(C_x) \leqslant C_{G^*} \leqslant M$ and as $C_x / O(C_x)\langle x \rangle \cong \Sigma_n$, it follows that $C_x \leqslant M$. Using the previous canonical generators $t_1, t_2, ..., t_{n+1} = x$ of $G^*$, then as $x$ is $G^*$-conjugate to $t_1$, this in turn implies that

$$C_{t_1} \leqslant M. \tag{5.95}$$

Solomon next argues that

$$C_{t_1 t_3} \leqslant M. \tag{5.96}$$

Indeed, $I^* = L\left( C_{G^*}(t_1 t_3) \right) \cong A_{n-2}$, and by Hypothesis $S$ the pump-up $J^*$ of $L(I^* \cap L)$ $(\cong A_{n-4})$ in $L_{2'}(C_{t_1 t_3})$ is semisimple. It follows readily now that the only possibility is $J^* = I^*$. Since $x \in I^* \leqslant M$ and $C_x \leqslant M$, this immediately yields (5.96).

Similarly, using the structure of the centralizers of $t_1$ and $t_1 t_3$, determined above (as well as of their $M$-conjugates), Solomon argues finally that any $G$-conjugate of $t_1$ or $t_1 t_3$ contained in $M$ is necessarily an $M$-conjugate of $t_1$ or $t_1 t_3$, respectively. [For example, if $y = t_1^g \in M$ and $y$ fixes four points in the natural representation of $G^*$, then $y$ centralizes an $M$-conjugate of $t_1 t_3$, which without loss we can assume to be $t_1 t_3$ itself. Hence $y$ normalizes $J^*$ and Solomon argues that $y$ corresponds to a transposition in $J^* \langle y \rangle$, whence $y$ is $M$-conjugate to $t_1$.]

Hence the conditions of Aschbacher's strong embedding criterion [I, 4.31] are satisfied, whence $M$ is strongly embedded in $G$. As a consequence, $M$ possesses only one conjugacy class of involutions, contrary to the fact that $t_1$ and $t_1 t_3$ are clearly not $M$-conjugate.

In the wreathed case, Solomon similarly proves the following result:

PROPOSITION 5.80. *If* $J \cong A_{n-2} \times A_{n-2}$, *then* $\langle L, J \rangle \cong A_n \times A_n$.

Again this is derived by means of the classical presentation of $A_n$ by generators and relations [I, 3.42], the calculations depending in turn on prior information about the structure of the centralizers in $G$ of suitable subgroups of $L$.

Let $d$ be the 3-cycle (123), $d^*$ the 3-cycle $(n-2, n-1, n)$, and set $I = L(C_L(d))$, $I^* = L(C_L(d^*))$, so that $I \cong I^* \cong A_{n-3}$. Solomon argues that $I \leqslant L(C_d)$, $I^* \leqslant L(C_{d^*})$ and if $K, K^*$ denote their respective normal closures in $L(C_d)$, $L(C_{d^*})$, that the following conditions hold:

(1) $K = K_1 \times K_1^x$, where $K_1 \cong A_{n-3}$;

(2) $K^* = K_1^* \times K_1^{*x}$, where $K_1^* \cong A_{n-3}$; and $\qquad$ (5.97)

(3) $K_1 \cap K_1^* \cong A_{n-6}$ and fixes three letters in the natural representations of both $K_1$ and $K_1^*$.

We can therefore identify $K_1, K_1^*$ with the alternating group on the letters $\varLambda_1 = \{4, 5,..., n\}$ and $\varLambda_1^* = \{1, 2, 3,..., n-3\}$, respectively, with $K_1 \cap K_1^*$ moving the letters $\varLambda_1 \cap \varLambda_1^* = \{4, 5,..., n-3\}$ and fixing the letters $\{1, 2, 3\}$ and $\{n-2, n-1, n\}$. Clearly Solomon's aim is to show that $\langle K_1, K_1^* \rangle$ is the alternating group on the set $\{1, 2,..., n\}$.

Setting $t = (123)$ and $t_i = (12)(i, i+1)$, $3 \leqslant i \leqslant n-4$, it follows that

(1) $\{t, t_1, 3 \leqslant i \leqslant n-4\}$ is a canonical set of generators of $K_1$; and

(2) $\{t_4 t_5 = (456), t_4 t_i = (45)(i, i+1), 6 \leqslant i \leqslant n-3\}$ is a canonical set of generators for $K_1 \cap K_1^*$. $\qquad$ (5.98)

Note that $t_4 = (12)(45) \notin K_1^*$. Nevertheless, by considering the alternating subgroup $B$ of $L$ of degree $n-6$, moving the letters $\{4, 5,..., n-3\}$ (and fixing $\{1, 2, 3, n-2, n-1, n\}$) and the pump-up in $C_B$ of $L(C_L(B)) \cong A_6$ (which likewise has the form $A_6 \times A_6$), Solomon is able to show that there exist elements $t_{n-2}$, $t_{n-1}$ in $C_B$ such that $t_4 t_{n-i} \in K_1^*$, $i = 1, 2$, with

$$t_4 t_{n-i} = (45)(n-i, n-i+1). \qquad (5.99)$$

It follows that

(1) $\{t_4 t_5, t_4 t_i, 6 \leqslant i \leqslant n-1\}$ is a canonical set of generators of $K_1^*$; and $\qquad$ (5.100)

(2) $t_{n-i}$ is represented by $(12)(n-i, n-i+1)$, $i = 1, 2$.

This yields the desired conclusion that $\{t, t_i, 3 \leqslant i \leqslant n-1\}$ satisfy the desired canonical defining relations for the alternating group of degree $n$ and hence that

$$L_1 = \langle K_1, K_1^* \rangle \cong A_n. \tag{5.101}$$

It is now an easy matter to see that these results also force

(1) $G^* = \langle K, K^* \rangle = L_1 \times L_1^x \cong A_n \times A_n$; and
(2) $G^* = \langle L, J \rangle$, $\tag{5.102}$

thus establishing the proposition.

Finally Solomon proves:

PROPOSITION 5.81. *If* $J \cong A_{n-2} \times A_{n-2}$, *then* $F^*(G) \cong A_n \times A_n$.

Indeed, as in the quasisimple case, he must derive a contradiction from the assumption that $M = N_G(G^*) < G$. Again this is obtained by appeal to Aschbacher's strong embedding criterion, which requires a prior analysis of the fusion of involutions of G. Solomon, and proves:

PROPOSITION 5.82. *The following conditions hold*:

(i) *M contains a Sylow 2-subgroup S of G*;
(ii) *G contains a normal subgroup* $G_0$ *of index 2 with* $x \notin G_0$; *and*
(iii) *If* $M_0 = M \cap G_0$ *and* $S_0 = S \cap G_0$, *then M controls the* $G_0$-*fusion of involutions of* $S_0$.

As an immediate corollary, one obtains the following result:

LEMMA 5.83. *We have* $M_0/O(M_0) \cong A_n \times A_n$.

Indeed, if not, then $M_0$ and hence $G_0$ has a normal subgroup of index 2 by Proposition 5.82(iii), whence $G_1 = O^2(G) < G_0$. But then $G_1\langle x \rangle < G$. Since $G$ is a minimal counterexample to the desired theorem, it follows that $G^* = F^*(G_1\langle x \rangle) = F^*(G_1) = F^*(G)$, so $M = G$, contrary to assumption.

In particular, it follows that $M_0 = O(M)G^*$, so $S_0 \in \mathrm{Syl}_2(G^*)$ and hence $S_0$ is of type $A_n \times A_n$.

To complete the analysis, Solomon invokes Aschbacher's criterion *twice*. First, if $T_1$ is a root four-subgroup of $L_1$, he uses it to prove

$$C_{G_0}(T_1) \leqslant M_0, \tag{5.103}$$

which immediately yields

> (1) If $t_1 \in T_1^{\#}$, then $C_{G_0}(t_1) \leqslant M_0$; and
> (2) If $t_1^g$ centralizes $t_1$ with $t_1^g \neq t_1$ for $g \in G_0$, then $C_{G_0}(t_1 t_1^g) \leqslant M_0$. $\tag{5.104}$

Note that as $M_0$ controls the fusion of its involutions in $G_0$, $\langle t_1, t_1^g \rangle$ is, in fact, a root four-subgroup of $L_1$ in (2). Moreover, for the same reason, it follows from (5.103) that Aschbacher's criterion is indeed satisfied in $G_0$. Thus $M_0$ is strongly embedded in $G_0$, contrary to the fact that $M_0$ has more than one class of involutions.

There you have the principal moves of Solomon's analysis of the $A_n$ standard component problem for $n \geqslant 12$.

We are therefore left to consider the *exceptional* cases $n \leqslant 12$. Thus for the balance of the section:

$$\text{We assume } n = 8, 9, 10, \text{ or } 11. \tag{5.105}$$

Again by Proposition 5.75, we have $R\langle x \rangle$ and $C_x$ splits over $\langle x \rangle$. In particular, $\overline{C}_x = C_x/O(C_x) \cong A_n \times Z_2$ or $\Sigma_n \times Z_2$.

Solomon first proves the following result:

**PROPOSITION 5.84.** *If $T \in \text{Syl}_2(G)$, then $n = 8$ or $9$ and $F^*(G) \cong A_{n+2}$.*

Indeed, in the contrary case, standard fusion arguments imply that $G$ has a normal subgroup of index 2 not containing $x$, and, moreover, that $T_1 = T \cap L \in \text{Syl}_2(G')$. Since $T_1$ is of type $A_n$, $n \leqslant 11$, $T_1$ has sectional 2-rank $\leqslant 4$; and as $L \leqslant G'$, $F^*(G') = F^*(G)$ is necessarily simple and so its possibilities are determined by the sectional 2-rank $\leqslant 4$ theorem. Since $T \in \text{Syl}_2(G)$ and $L$ is a component of $C_x$, we check that the only solutions are those listed in the proposition.

Next, Solomon proves:

PROPOSITION 5.85. *If* $\bar{C}_x \cong \Sigma_n \times Z_2$, $n = 10$ *or* $11$, *then* $F^*(G) \cong A_{n+2}$ (*with* $n = 10$).

Indeed, let $t$ be an involution of $C_x$ inducing the transposition $(n-1, n)$ on $L$, set $I = L(C_L(t))(\cong A_{n-2})$, and let $J$ be the normal closure of $I$ in $L_t$. Observe that if $n = 10$, then by Hypothesis $S$, $G$ has the $B$-property, so $J$ is semisimple. On the other hand, if $n = 11$, then $I$ contains the four-subgroup $D = \langle (12)(34), (13)(24) \rangle$ and for each $d \in D^\#$, the pump-up of $I_d = L(C_L(d))(\cong A_7)$ in $C_d$ is semisimple by Hypothesis $S$. Since each $L(I_d \cap I) \cong A_5$, this easily forces $J$ to be semisimple in this case as well.

Hence by the minimality of $G$, one of the following holds:

$$
\begin{aligned}
&(1) \ J/Z(J) \cong L_4(4) \text{ or } HS \text{ with } n = 10; \text{ or} \\
&(2) \ J \cong A_{n-2}, A_{n-2} \times A_{n-2}, \text{ or } A_n.
\end{aligned}
\tag{5.106}
$$

In the first two cases, we conclude from (5.84) and the known structure of the centralizers of involutions acting on quasisimple $K$-groups, together with the $B$-property, that $G$ possesses a standard component isomorphic to $L_4(4^m)$, $HS$, $HS/Z_2$, or $F_5$, But then $F^*(G)$ is determined by our assumption (5.84), and the only possibility compatible with the structure of $\bar{C}_x$ is $F^*(G) \cong F_5$.

The case $J \cong A_{n-2}$ easily leads to a fusion contradiction. On the other hand, if $J \cong A_{n-2} \times A_{n-2}$, Solomon argues by induction, using the 3-cycle $(123)$ of $L$ (along the lines of Proposition 5.80) that $G$ possesses a component $K \cong A_n$ such that $m(C_K) \geqslant 2$. If $K$ is standard, $F^*(G)$ is determined from Aschbacher and Seitz's results and the structure of $C_x$ is contradicted, so $K$ is not standard.

Solomon now attempts to pump $K$ up to a standard component $K^*$, which he identifies by the minimality of $G$ and (5.84). The possibilities for $F^*(G)$ are then determined either by (5.84), the previous $n \geqslant 12$ solution, or Aschbacher and Seitz, and again the structure of $C_x$ is contradicted. This procedure works fine if $n = 10$, for then $G$ has the $B$-property. However, if $n = 11$, it may instead lead only to a *nonquasisimple* 2-component $K^*$ with $K^*/O(K^*) \cong A_m$, $m$ odd, $m \geqslant 13$, in which case his earlier theorem (Theorem 4.39) is used to reach the desired contradiction.

Finally if $J \cong A_n$, Solomon pins down the structure of the Sylow 2-subgroup $S$ of $G$ and argues that $G$ has a normal subgroup $G_0$ of index 2 not containing $x$ with $S \cap G_0$ of type $A_{12}$. But then by his prior classification theorem (Theorem 2.72) and the structure of $C_x$, the only possibilities are $F^*(G) \cong A_{12}$ or $A_{13}$, so the proposition holds in this case as well.

In view of the preceding two results, we can assume henceforth:

(1) $T \notin \mathrm{Syl}_2(G)$; and
(2) if $n = 10$ or $11$, then $C_x \cong A_n \times Z_2$.           (5.107)

Now let $z$ be the involution of $Z(T \cap L)$. Under the conditions of (5.107), Solomon next proves:

LEMMA 5.86. *The involutions $x$ and $xz$ are $G$-conjugate.*

This puts him in a position to carry out a 2-local pushing-up analysis. Let $B$ be an $E_8$-subgroup of $T \cap L$ which acts regularly on $\{1, 2, \cdots, 8\}$ and fixes the remaining $n - 8$ letters of $\Lambda$. Set $A = B \times \langle x \rangle (\cong E_{16})$, $N = N_G(A)$, $\overline{N} = N/O(N)$, and $I = N \cap L = N_L(B)$. By the structure of $L$, we have

$$I \cong L_3(2)/E_8.$$           (5.108)

Using Lemma 5.86 and the Finkelstein–Landrock–Solomon Lemma 5.19, Solomon first proves the following result:

PROPOSITION 5.87. (i) $O_2(\overline{N}') \cong Z_4 \times Z_4 \times Z_4$ or $E_{64}$; and
(ii) $\overline{N}$ *is a split extension of* $O_2(\overline{N}')$ *and* $\overline{I} \times \langle \overline{x} \rangle$.

(The possibilities $O_2(\overline{N}')$ of type $L_3(8)$ or $U_3(8)$ are excluded here as such 2-groups do not admit $L_3(2)$ as a group of automorphisms.)

In the $Z_4 \times Z_4 \times Z_4$ case, Solomon argues exactly as in the Griess–Mason–Seitz proof of Proposition 5.21 and concludes with the aid of O'Nan's theorem [I, 2.34, 2.35] and the characterization of $HS$ by its Sylow 2-subgroup (Theorem 2.51):

PROPOSITION 5.88. *If* $O_2(\overline{N}') \cong Z_4 \times Z_4 \times Z_4$, *then we have* $L \cong A_8$ *and* $F^*(G) \cong HS$.

We can therefore assume henceforth that

$$O_2(\overline{N}') \cong E_{64}.$$           (5.109)

Now let $U$ be an $x$-invariant $E_{64}$-subgroup of $N$ mapping on $O_2(\overline{N}')$, so that $N_1 = N_N(U)$ covers $\overline{N}$, and, in particular, contains a Sylow 2-subgroup $V$

of $N$. [Thus $V/U \cong D_8 \times Z_2$ and we see that $U = J(V)$.] By a further involution fusion analysis, Solomon next proves:

LEMMA 5.89. *N does not contain a Sylow 2-subgroup of $G$.*

Hence if we set $M = N_G(U)$ and $\bar{M} = M/O(M)U$, it follows that $\bar{M} > \bar{N}_1 \cong L_3(2) \times Z_2$. But it is also easy to see that $L(\bar{N}_1)$ is standard in $\bar{M}$ with $\langle \bar{x} \rangle = C_{\bar{M}}(L(\bar{N}_1))$. But now, applying the Fritz–Harris–Solomon theorem (Theorem 2.161) and the sectional 2-rank $\leqslant 4$ theorem together with the fact that $\bar{M} \leqslant \text{Aut}(U) \cong GL_6(2)$, he immediately obtains the following result:

PROPOSITION 5.90. *We have $F^*(\bar{M}) \cong SL_3(4)$ or $L_3(2) \times L_3(2)$.*

These two cases are treated separately. Solomon first proves:

PROPOSITION 5.91. *If $F^*(M) \cong SL_3(4)$, then $L \cong A_8$ and we have $F^*(G) = L_4(4)$.*

It is a rather delicate argument. First, it is not difficult to show that $M/O(M)$ splits over $UO(M)/O(M)$ and that a Sylow 2-subgroup $S$ of $M$ is necessarily of type $L_4(4)$. [$L_4(4)$ possesses a maximal parabolic of the form $SL_3(4)/E_{64}$.] Hence if $S \in \text{Syl}_2(G)$, it follows from McBride's theorem (Theorem 5.4) that $F^*(G) \cong L_4(4)$, as asserted, so we can assume that $S \notin \text{Syl}_2(G)$. However, if $\bar{C}_x \cong A_n \times Z_2$, $n = 8$ or 9, Solomon argues to the contrary that $S \in \text{Syl}_2(G)$. Hence $\bar{C}_x$ does not have this form and so by Proposition 5.85,

$$\bar{C}_x \cong \Sigma_n \times Z_2, \ n = 8, 9, \ \text{or} \ A_n \times Z_2, \ n = 10, 11. \qquad (5.110)$$

In these cases, he chooses an appropriate involution $y \in T$, correspondingly a transposition or product of two transpositions, whence $J = L(C_L(t)) \cong A_{n-2}$ or $A_{n-4}$, respectively. Hence, in either case

$$J \cong A_6 \ \text{or} \ A_7. \qquad (5.111)$$

Solomon's aim now is to pin down the pump-up $K$ of $J$ in $L_t$. Since $J$ is a component of $L(C_K(x))$, the complete list of possibilities for $K$ is again given by our earlier results on $L_2(q)$ and $A_7$ components. However, in the present instance he argues that there is only one solution:

$$J \cong A_6, \qquad K \cong PSp_4(4). \tag{5.112}$$

This is achieved by forcing $K$ to contain a 2-local of the form $A_5/E_{64}$. To establish this fact, Solomon is forced to analyze the normalizer in $G$ of a suitable $y$-invariant $E_{2^8}$-subgroup of $M$ [2-groups of type $L_4(4)$ have rank 8]. From the structure of $K$, he also easily concludes that $n = 8$ and that $\langle y \rangle \in \mathrm{Syl}_2(C_K)$.

Since $n = 8$, $G$ has the $B$-property. By the previous paragraph, $K$ is standard in $G$, so by (5.84) and the structure of $C_x$, he concludes that $F^*(G) \cong L_4(4)$ in this case as well.

Thus, to complete the analysis, it remains to eliminate the wreathed possibility for $F^*(M)$.

PROPOSITION 5.92. *We have* $F^*(M) \not\cong L_3(2) \wr Z_2$.

Solomon first argues that $F^*(M)$ acts trivially on $O(M)$, whence $M$ contains an $x$-invariant subgroup

$$J = J_1 \times J_2 \quad \text{with} \quad J_1 \cong J_2 \cong L_3(2)/E_8. \tag{5.113}$$

Clearly $x$ interchanges $J_1$ and $J_2$ under conjugation and $C_J(x) = I$. Also $U_i = U \cap J_i \cong E_8$, $i = 1, 2$.

As usual in the wreathed case, the goal is to produce a subgroup $G^*$ of $G$ of the form

$$G^* = L_1 \times L_2 \quad \text{with} \quad L_1 \cong L_2 \cong A_n. \tag{5.114}$$

In the present situation, this is achieved by arguing first that $\bar{U}_2 \in \mathrm{Syl}_2(C_{C_{\bar{U}_1}}(\bar{U}_2))$, where $C_{\bar{U}_1} = C_{U_1}/U_1$, and then using Harada's self-centralizing $E_8$ theorem (Theorem 2.156) and the sectional 2-rank $\leqslant 4$ theorem to identify the normal closure $L_2$ of $J_2$ in $C_{U_1}$. (Solomon has sufficient fusion information to know that $J_2 < L_2$.) He thus concludes that $L_2 \cong A_m$, $8 \leqslant m \leqslant 11$, $M_{22}$, $M_{23}$, or $Ly$. Similarly he produces $L_1$ in $C_{U_2}$ of the same form and argues that $L_1 L_2 = L_1 \times L_2$ with $x$ interchanging $L_1$ and $L_2$, forcing $L_1 \cong L_2 \cong L$, thus establishing (5.114).

Now he pumps-up $L_1$ and derives a contradiction in the same way as in the $A_{n-2} \times A_{n-2}$ case of Proposition 5.85.

We have finally disposed of alternating components!

## 5.7. The $\mathscr{L}_0(3)$ Case

We must first mention a technical problem connected with the $U_3(3)$ subcase, which, as noted in Section 2.5 Harris has treated only under the assumption that $G$ is simple [60]. Indeed, he shows that $G$ possesses a subgroup $G_0$ of index 2 with Sylow 2-subgroup $S_0$ of order $2^{11}$ or $2^{12}$, containing an abelian normal subgroup $A$ of order $2^8$ isomorphic to $Z_4 \times Z_4 \times Z_4 \times Z_4$ or $Z_8 \times Z_8 \times Z_2 \times Z_2$, and on the way establishes considerable 2-fusion information about $G_0$. Presumably with only modest additional work, Harris could have forced $S_0$ to be of type $G_2(4)$ (in particular, $|S_0| = 2^{12}$ and $A \cong Z_4 \times Z_4 \times Z_4 \times Z_4$), in which case Thomas's theorem (Theorem 5.5) would imply that $F^*(G) \cong G_2(4)$. [Since $U_3(3) \cong G_2(2)'$, $G_2(4)$ was an expected solution.] However, to date this last phase of the analysis has not been carried out.

Even though the $U_3(3)$ problem falls under the domain of the standard component theorem (I), it was placed there because the inductive characterization of the family $\mathscr{L}\mathscr{O}$ requires a solution of the standard form problems for all $L \in \mathscr{L}\mathscr{O}_0(3)$. However, a careful reading of this inductive argument reveals that if one simply *deletes* the group $U_3(3)$ from $\mathscr{L}_0(3)$, one obtains the corresponding characterization of the family $\mathscr{L}\mathscr{O} - \{U_3(3)\}$. In effect, this allows one to *shift* $U_3(3)$ to the standard component theorem (II), in which case a solution with $G$ simple suffices for the noncharacteristic 2 theorem. (It is this observation which explains Harris's termination of his analysis once he had shown that $G$ is not simple.)

However, this is really a minor issue, for Aschbacher has outlined (unpublished) a (short) complete solution of the $U_3(3)$ standard form problem [as well as the $L_3(3)$ and $G_2(3)$ problems] under the assumption that *all proper subgroups of the group $G$ under investigation are $K$-groups*, which, of course, is all that is needed for the classification of the finite simple groups. It is only our insistence on establishing the $B$- and $U$-theorems as independent results that forces us to consider the question at all. In any event, we discuss the $U_3(3)$ case here only under the assumption that $G$ is simple.

We first describe the highlights of Aschbacher's work in the $\mathscr{L}_0(3)$ case. [Although his published results assume a complete solution of the $U_3(3)$ standard form problem, Aschbacher has shown that the argument can be easily modified to avoid the $U_3(3)$ problem completely, thus fitting in with the inductive characterization of the family $\mathscr{L}\mathscr{O} - \{U_3(3)\}$.]

For convenience, we let $\mathscr{F}$ denote the set of quasisimple groups $K$ such that

$$K/O(K) \in \{Psp_4(3), L_4(3), U_4(3), U_4(3)/Z_2, P\Omega_7(3), P\Omega_8^{\pm}(3)\} \qquad (5.115)$$

and we let $\mathscr{F}^*$ denote the set of quasisimple groups $K$ such that

$$K/O(K) \in \mathscr{F} \cup \{L_2(9), L_2(81), L_3(3)\}. \qquad (5.116)$$

In this terminology, Aschbacher's operating assumption (5.14), takes the following form:

> If $J$ is a component of $C_y$ for $y \in \mathscr{I}(G)$ with $J \in \mathscr{F}^*$ and $K$ is a component of the pump-up of $J$ in $C_u$ for some $u \in \mathscr{I}(C_y)$ centralizing $J$, then $K \in \mathscr{F}^*$. (5.117)

Aschbacher's result [5] is as follows:

THEOREM 5.93. *If* $L \in \mathscr{F}$ *and* $G$ *satisfies* (5.117), *then* $L \cong PSp_4(3)$ $\left(\cong U_4(2)\right)$ *and* $F^*(G) \cong L_4(4)$.

In proving the theorem, Aschbacher can assume that $F^*(G)$ is not of Lie type of odd characteristic and hence that $G$ does not possess a classical involution. Under this assumption and using (5.117), he establishes a key general result concerning pump-ups of $SL_2(3)$ components. We state three of its immediate consequences, which incorporate the principal uses of the result in the analysis. We emphasize once again that it is proved by forcing $G$ to otherwise possess a classical involution, in which case $F^*(G)$ is of Lie type of odd characteristic by the classical involution theorem—contradiction.

Aschbacher first fixes a Sylow 2-subgroup $T$ of $C_x$ and a uniquely determined involution $z$ of $Z(T) \cap L$, having the property that $C_L(z)$ contains a subnormal $SL_2(3)$-subgroup $I$ (i.e., a "solvable" component) with $z \in I$ (thus $z$ is a classical involution of $L$).

PROPOSITION 5.94. *Let* $t \in \mathscr{I}(G)$ *and let* $J$ *be a component of* $C_y$ *with* $J \in \mathscr{F} \cup \{L_3(3)\}$. *Then we have*

(i) *If* $y$ *is a classical involution of* $J$ *and* $u$ *is any involution of* $C_y$, *then* $C_{\langle u, y \rangle}$ *has no components in* $\mathscr{F}^*$. (*In particular,* $C_y$ *has no such components*);

(ii) $J$ *is the unique component of* $C_y$ *contained in* $\mathscr{F}^*$; *and*

(iii) *If* $z^g \in T$ *for* $g \in G$, *then either* $L\left(C_L(z^g)\right) \cong 1$ *or* $U_3(3)$.

In his published paper, there is no $U_3(3)$ possibility in (iii). Indeed, if that case occurs, then necessarily $L/O(L) \cong U_4(3)/Z_2$ or $U_4(3)$ (with $z^g$ inducing an outer automorphism on $L$) and Aschbacher derives a contradiction by analyzing the pump-up $J$ of $I = L(C_L(z^g)) \cong U_3(3)$ in $L_{z^g}$. However, as usual, the identification of $J$ assumes a prior solution of the $U_3(3)$ standard form problem *without* the assumption that the corresponding group is simple.

Except for one place in the $U_4(3)/Z_2$ analysis, Aschbacher is able to avoid the $U_3(3)$ problem by reducing the application of (iii) to the case in which either $L(C_L(z^g)) = 1$ or $m(C_{L/Z(L)}(z^g)) \geqslant 3$. This rules out the possibility $I \cong U_3(3)$, for then $m(C_{L/Z(L)}(z^g)) = m(I) = 2$. We defer the remaining application of (iii) to the discussion of the $U_4(3)/Z_2$ standard form problem.

This proposition dominates the analysis. For example, it immediately yields the following result:

PROPOSITION 5.95. *We have* $L/O(L) \ncong P\Omega_8^-(3)$.

Indeed, in this case $C_L(z)$ contains an $L_2(9)$ component, contrary to Proposition 5.94(iii) with $g = 1$.

The basic idea of the proof of Theorem 5.93 is to study the possible conjugacies of both $z$ and $x$ in $T$, forced by the $Z^*$-theorem, which requires, in particular, a detailed knowledge of the centralizers of involutions acting on $L$. The resulting analysis becomes increasingly more difficult as $L$ gets smaller in size. Moreover, Aschbacher treats the individual cases according to a suitable descending ordering of $L$, eliminating each in succession [until the final case $L \cong PSp_4(3)$]. This approach has the following consequence: at each stage, if $y \in \mathscr{I}(G)$ and $C_y$ possesses a component $K$ with

$$K/O(K) \cong L/O(L),$$

the previous case analyses together with the initial hypothesis imply that $K/O(K) \in \mathscr{L}^*(G)$ and that also $K$ is standard in $G$ (whence also $C_K$ has cyclic Sylow 2-subgroups). Similarly, these previous results considerably restrict the pump-ups in $C_t$ of components of $C_L(t)$, $t \in \mathscr{I}(T)$.

The "largest" case is $L \cong P\Omega_8^+(3)$. Here, using Proposition 5.94, Aschbacher first argues that

$$z^g = zx \tag{5.118}$$

for some $g \in G$. Hence $z^g$ centralizes $C_L(z)$, which in the present case contains the central product of four $SL_2(3)$ components $I_i$, $1 \leqslant i \leqslant 4$. Setting $Q_i = O_2(I_i)$, $1 \leqslant i \leqslant 4$, and $Q^* = Q_1 Q_2 Q_3 Q_4$, Aschbacher now derives a contradiction by an argument of Walter. If $\mathcal{D}$ denotes the set of involutions of $Q^*$ that centralize some pair of $Q_i$, he shows from the embedding of $Q^*$ in $L$ and $L^g$ that for $d \in \mathcal{D}$,

$$\begin{array}{l} (1)\ L\big(C_L(d)\big) \cong L\big(C_{L^g}(d)\big) \cong U_4(3)/Z_2 \big(\cong P\Omega_6^-(3)/Z_2\big);\ \text{and} \\ (2)\ L = \big\langle L\big(C_L(d)\big)\ |\ d \in \mathcal{D}\big\rangle\ \text{and}\ L^g = \big\langle L\big(C_{L^g}(d)\big)\ |\ d \in \mathcal{D}\big\rangle. \end{array} \quad (5.119)$$

Furthermore, using Proposition 5.94 and assumption (5.117), it follows that each $L\big(C_L(d)\big)$ has a trivial pump-up in $C_d$, which is, in fact, the unique component of its isomorphic type in $C_d$. A similar statement holds for $C_{L^g}(d)$, and consequently for $d \in \mathcal{D}$, we have

$$L\big(C_L(d)\big) = L\big(C_{L^g}(d)\big), \quad (5.120)$$

whence by (5.119)(2)

$$L = L^g, \quad (5.121)$$

forcing $x^g = zx$ to centralize $L$, which is not the case.

Thus Aschbacher proves:

PROPOSITION 5.96. *We have* $L/O(L) \not\cong P\Omega_8^+(3)$.

Next comes $L/O(L) \cong P\Omega_7(3)$. Here, as in all subsequent cases, Aschbacher focuses on the normalizers $N$ in $G$ of a suitable elementary abelian 2-subgroup $A$ of $T$ with $A$ containing every involution in $C_T(A)$ (and hence, as $x \in A$, in $C_A$ itself). We set $B = A \cap L$, $N_0 = N \cap L$, and $\bar{N} = N/C_A$, and fix this notation.

In the present case, one has

$$\begin{array}{l} (1)\ m(A) = 7;\ \text{and} \\ (2)\ \bar{N}_0 \cong A_7. \end{array} \quad (5.122)$$

Thus $\bar{N}$ is a suitable subgroup of $\text{Aut}(A) \cong GL_7(2) = L_7(2)$, containing $A_7$, and Aschbacher's goal is to show that the $G$-fusion of $z$ and $x$, controlled by $\bar{N}$, is incompatible with the action of $\bar{N}$ on $A$.

From the structure of centralizers of involutions $y$ of $\text{Aut}(P\Omega_7(3))$, $L(C_L(y)) \neq 1$ for all $y \in \mathcal{I}(T)$ that are not $L$-conjugate to $z$. Hence, by Proposition 5.94(iii), Aschbacher knows at the outset:

$$\text{If } z^g \in T \text{ for } g \in G, \text{ then } z^g = z^u \text{ or } (zx)^u \text{ for some } u \in L. \quad (5.123)$$

By a detailed analysis of the fusion of involutions, in conjunction with the structure of the centralizers of involutions acting on $L$, Aschbacher is eventually able to establish the following three results:

(1) $x$ is $N$-conjugate to $xz$;

(2) $B$ is generated by the set of all $G$-conjugates of $z$ contained in $A$ (whence $B \lhd N$); and

(3) For a suitable $b \in B^\#$ with $b$ not $L$-conjugate to $z$, either $C_N(b) \leqslant C_N(x)$ or else $x$ is $N$-conjugate to $xb$ and $C_N(b) \leqslant N_N(\langle x, b \rangle)$. $\quad (5.124)$

Finally, considering the possible orbit lengths of the $\bar{N}$-conjugates of both $x$ and $b$, Aschbacher argues that there is no subgroup of $GL_7(2)$ containing $A_7$ that satisfies the three conditions of (5.124).

Thus he proves:

PROPOSITION 5.97. *We have $L/O(L) \not\cong P\Omega_7(3)$.*

We turn next to the cases $L/O(L) \cong U_4(3)/Z_2$ or $U_4(3)$, which in some sense are the most difficult, due to the complex structure of $\text{Out}(U_4(3)) \cong D_8$. In fact, $U_4(3)$ possesses outer automorphisms of order 2 whose centralizers on $U_4(3)$ have layers isomorphic to $PSp_4(3)$, $U_3(3)$, $A_6$, or 1, respectively. In particular, as $m(PSp_4(3)) = 4$, it follows that

$$m(\text{Aut}(U_4(3))) = 5 > m(U_4(3)) = 4. \quad (5.125)$$

First consider the $L/O(L) \cong U_4(3)/Z_2$ case. Aschbacher's initial fusion analysis shows that for some $g \in G$, $t = x^g$ induces an outer automorphism on $L$ with $L(C_L(t)) \cong PSp_4(3)$ and $C_T(t) \in \text{Syl}_2(C_{C_x}(t))$. It follows that $\langle t, x \rangle$ is contained in an $E_{64}$-subgroup of $T$. It is this subgroup which Aschbacher takes as $A$.

Hence, in this case we have

$$
\begin{array}{ll}
(1) \ m(A) = 6; \text{ and} \\
(2) \ \bar{N}_0 \cong A_6 \text{ or } \Sigma_6.
\end{array}
\qquad (5.126)
$$

Aschbacher now goes on to prove

$$
\bar{N} \cong A_8 \text{ or } \Sigma_8. \qquad (5.127)
$$

[We note that to verify (5.127), Aschbacher must avoid the $U_3(3)$ standard form problem, in the manner described above.]

However, in this case the action of $\bar{N}$ on $A$ *is* compatible with the developed fusion information, so Aschbacher must delve more deeply into the structure of $G$ to reach a contradiction, closely analyzing $C_z$. Indeed, using induction, it follows easily that $O(L) = 1$, which in turn implies that $C_L(z)$ contains a subgroup $I_1 * I_2$ with $z \in I_i \cong SL_2(3)$, $i = 1,2$. Now using the structure of $\bar{N}$, he argues that $C_z$ contains the central product of *four* copies $I_i$ of $SL_2(3)$, $1 \leqslant i \leqslant 4$. Again set $Q_i = O_2(I_i)$ and $Q^* = Q_1 Q_2 Q_3 Q_4$ and put $\tilde{C}_z = C_z / \langle z \rangle$.

Aschbacher's goal is to prove that $Q^*$ is strongly closed in a Sylow 2-subgroup $\tilde{S}$ of $\tilde{C}_z$, In the published paper, he argues with the aid of the $U_3(3)$ standard form problem that $L(C_y) \cong U_4(3)/Z_2$ for *every* involution $y$ of $I_i I_j - \{z\}$, $1 \leqslant i,j \leqslant 4$ (this is automatic for those $y$ which are $G$-conjugate to $x$) and uses this result in the analysis. Hence, to avoid the $U_3(3)$ problem, he must establish strong closure without utilizing this piece of information.

Now invoking Goldschmidt's theorem [I, 4.128] and using the structure of $N_{\tilde{C}_z}(\tilde{Q}^*)$ (which he also determines), it follows that

$$
\begin{array}{ll}
(1) \ \langle \tilde{Q}^{* \tilde{C}_z} \rangle = \tilde{K}_1 \times \tilde{K}_2 \times \tilde{K}_3 \times \tilde{K}_4, \text{ where} \\
(2) \ \tilde{I}_i \leqslant \tilde{K}_i \cong L_2(q_i), \ q_i \equiv 3,5 (\text{mod } 8), \ 1 \leqslant i \leqslant 4.
\end{array}
\qquad (5.128)
$$

[The groups $L_2(q)$, $q \equiv 3,5 (\text{mod } 8)$, are the only quasisimple groups with $Z_2 \times Z_2$ Sylow 2-subgroups]. But then if $K_1$ denotes the preimage of $\tilde{K}_1$ in $C_z$, we conclude that

$$
z \in K_1 \cong SL_2(q_1) \text{ and } K_1 \text{ is subnormal in } C_z. \qquad (5.129)
$$

Hence, $z$ is a classical involution, contrary to assumption.

Thus, Aschbacher proves:

PROPOSITION 5.98. *We have $L/O(L) \not\cong U_4(3)/Z_2$.*

In treating the $L/O(L) \cong U_4(3)$ case, Aschbacher follows, to the extent possible, Finkelstein's analysis of the $M_{23}$ standard component problem [24]. The two problems are closely related since $U_4(3)$ and $M_{23}$ have isomorphic Sylow 2-subgroups. Moreover, Finkelstein obtains his $M_{23}$ solution as a corollary of the following more general result.

THEOREM 5.99. *If $L$ has Sylow 2-subgroups of type $U_4(3)$ and both $Z(L)$ and $N_G(L)/C_L$ have odd order, then $F^*(G)$ is not simple.*

In view of the sectional 2-rank $\leqslant 4$ theorem, the Sylow 2-group assumption on $L$ implies that

$$L/O(L) \cong L_4(q), \ q \equiv 5 (\text{mod } 8), \ U_4(q), \ q \equiv 3 (\text{mod } 8), \ M_{22}, \ M_{23}, \ \text{or} \ Mc,$$
$$(5.130)$$

so Finkelstein's result gives partial information in the $U_4(3)$, $M_{22}$, $M_{23}$, and $Mc$ standard component problems.

In particular, since $\text{Out}(M_{23}) = \text{Out}(Mc) = 1$, it yields as a corollary [under the assumption that $F^*(G)$ is simple] the following theorem:

THEOREM 5.100. *We have $L/O(L) \not\cong M_{23}$ or $Mc$.*

Moreover, we see that it is the "decoration" of $L$ which complicates the general solution in the $U_4(3)$ and $M_{22}$ cases. However, it turns out that large portions of Finekelstein's argument can be adapted to the general cases of these problems, so we describe his proof briefly.

First, $T \cap L$ contains two normal $E_{16}$-subgroups $B_1$ and $B_2$. Setting $X_i = N_L(B_i)/C_L(B_i)$, one can choose the ordering so that we have Table 5.14.

**Table 5.14.** Automizers of $E_{16}$-Subgroups of $L$

| $L$ | $X_1$ | $X_2$ |
|:---:|:---:|:---:|
| $U_4(3)$ | $A_6$ | $A_6$ |
| $M_{22}$ | $A_6$ | $\Sigma_5$ |
| $M_{23}$ | $A_7$ | $N_{A_8}(\langle\langle(123)\rangle\rangle)$ |
| $Mc$ | $A_7$ | $A_7$ |

Taking $A = B_1 \times \langle x \rangle$ [with $N = N_G(A)$] and setting $\tilde{N} = N/O(N)$, Finkelstein first proves

$$O_2(\tilde{N}) = \tilde{U}\langle \tilde{x} \rangle, \text{ where } \tilde{U} \cong E_{2^8} \text{ and } \tilde{U} \triangleleft \tilde{N}. \qquad (5.131)$$

Under Finkelstein's assumptions, we also have $T \leqslant N$. Let $U$ be a $T$-invariant $E_{2^8}$-subgroup of $N$ mapping on $\tilde{U}$ and set $M = N_G(U)$ and $\bar{M} = M/C_U$. Finkelstein argues next that $\bar{K} = L(C_{\bar{M}}(\bar{x})) \cong X_1$, that $\langle \bar{x} \rangle \in \mathrm{Syl}_2(C_{\bar{M}}(\bar{K}))$, and that $\bar{K}$ is not normal in $\bar{M}$. Now applying the Fritz–Harris–Solomon theorem (Theorem 2.161) and the sectional 2-rank $\leqslant 4$ theorem, he concludes that

(1) $\bar{M} \cong X_1 \wr Z_2$; and
(2) $M$ possesses a subgroup of index 2 not containing $x$ with
    Sylow 2-subgroups of type $M_{23} \times M_{23}$.         (5.132)

Using the fact that $T \leqslant M$ and $T \in \mathrm{Syl}_2(C_x)$, it follows now that

$$M \text{ contains a Sylow 2-subgroup of } G. \qquad (5.133)$$

Next, Finkelstein applies Thompson transfer to conclude that

$G$ possesses a normal subgroup $G_0$ of index 2 not containing $x$
with Sylow 2-subgroup $S_0$ of type $M_{23} \times M_{23}$.         (5.134)

Thus $S_0 = S_1 \times S_2$ with $S_1$, $S_2$ of type $M_{23}$ and $S_1$, $S_2$ interchanged by $x$. As his final step, Finkelstein proves

$G$ has product fusion with respect to a suitable choice of $S_1$
and $S_2$.                                          (5.135)

Now Goldschmidt's product fusion theorem [I, 4.148] and the action of $x$ implies that $F^*(G_0) = L_1 \times L_2$ with $L_1$ and $L_2$ interchanged by $x$ and $L = C_{L_1 L_2}(x)$. Thus Finkelstein forces the wreathed possibility as the only solution. In particular, $F^*(G)$ is not simple and Theorem 5.99 is proved.

It would be erroneous to infer from this outline that the complete proof of Theorem 5.99 is easy. On the contrary, it involves very delicate fusion analysis. Moreover, to obtain the desired fusion information, it is necessary for Finkelstein to consider the groups $A_2 = B_2 \times \langle x \rangle$, $N_2 = N_G(A_2)$, and $M_2 = N_G(U_2)$, where again $U_2$ is an $x$-invariant $E_{2^8}$-subgroup of $N_2$ (which he proves exists), and argue that $\overline{M}_2 = M_2/C_{U_2}$ has the analogous structure to $\overline{M}_1$ (namely, $\overline{M}_2 \cong X_2 \int Z_2$).

Now return to the $U_4(3)$ situation (whence $X_1 \cong X_2 \cong A_6$). Aschbacher's analysis splits into two subcases according as there is or is not an involution $t \in T$ inducing an outer automorphism on $L$ with

$$C_{L/O(L)}(t) \cong \Sigma_4 \int Z_2. \tag{5.136}$$

[$\mathrm{Aut}(U_4(3))$ possesses such involutions.]

If no such $t$ exists, Aschbacher follows Finkelstein's proof very closely, arguing again that $M$ satisifes (5.132). [To establish this, he must again avoid the $U_3(3)$ standard form problem, in the way described earlier.] In view of Theorem 5.99, Aschbacher can assume that $T \not\leq M$, otherwise the balance of Finkelstein's argument applies without change.

In this case, Aschbacher proceeds as follows, first arguing that

$$M \text{ does not contain a Sylow 2-subgroup of } G. \tag{5.137}$$

He now espands $T \cap M$ to $T_1 \in \mathrm{Syl}_2(M)$ and produces an involution $y$ of $N_G(T_1) - T_1$ which centralizes $x$ and induces an outer automorphism on $L$ such that

$$I = L(C_L(y)) \cong A_6. \tag{5.138}$$

If $J$ denotes the pump-up of $I$ in $L_y$, Aschbacher's final step is to force

(1) $J/O(J) \cong U_4(3)$; and
(2) $x$ interchanges the two conjugacy classes of $E_{16}$-subgroups of $J$. $\tag{5.139}$

This yields an immediate contradiction, for it follows from the structure of $M$ that $J$ must contain an $x$-invariant $E_{16}$-subgroup.

It thus remains to treat the case in which there is an involution $t \in T$ satisfying (5.136). Then $m(C_L(t)) = 4$ and Aschbacher takes $A$ to contain $\langle x, t \rangle$ with $A \cong E_{64}$. However, this time it is preferable to consider $A_1 = B \langle x \rangle$ (recall $B = A \cap L$) and $N_1 = N_G(A_1)$ rather than $N = N_G(A)$. Following Finkelstein, but with some additional argument, he again produces an $E_{2^8}$-subgroup $U$ of $N_1$ (invariant under $T \cap N_1$) with $O(N_1)U \lhd N_1$ and, as before, investigates $\overline{M} = M/C_U$, where $M = N_G(U)$.

Again $\overline{K} = L\big(C_{\overline{M}}(\bar{x})\big) \cong A_6$. However, this time as $t$ centralizes $[U, x]$ ($\leqslant A_1$), it follows that $C_{\overline{M}}(\overline{K}) < \langle \bar{x} \rangle$. Aschbacher, in fact, argues that

(1) $C_{\overline{M}}(\overline{K}) \cong Z_2 \times Z_2$; and
(2) $\overline{K}$ is a component of $C_{\overline{M}}(\bar{y})$ for every involution $\bar{y}$ of $C_{\overline{M}}(\overline{K})$. \hfill (5.140)

Hence $\overline{K}$ is standard in $\overline{M}$ and so by Foote's maximal $L_2(9)$ component theorem (cf. Lemma 4.58), either $F^*(\overline{M}) \cong A_{10}$ or $\overline{K} \triangleleft \overline{M}$. However, the first possibility is excluded here since $A_{10} \not\leqslant \mathrm{Aut}(U) \cong GL_8(2)$. Thus $\overline{K} \triangleleft \overline{M}$ and it follows at once that

$$M \text{ contains a Sylow 2-subgroup of } G. \tag{5.141}$$

Finally, using the structure of $M$ and of $C_x$, Aschbacher derives an easy fusion contradiction.

Thus he has proved:

PROPOSITION 5.101. *We have $L/O(L) \not\cong U_4(3)$.*

Now consider the case $L/O(L) \cong L_4(3)$. The proof is similar, but considerably easier than the $U_4(3)$ case [this is partly because $\mathrm{Out}\big(L_4(3)\big) \cong Z_2 \times Z_2$ rather than $D_8$, and partly because $L_4(3)$ contains an involution with a non-2-constrained centralizer, while $U_4(3)$ does not]. Again there are two subcases to consider, this time according as there is, or is not an involution $t \in T$ inducing an outer automorphism on $L$ with

$$I = L\big(C_L(t)\big) \cong PSp_4(3). \tag{5.142}$$

[$\mathrm{Aut}\big(L_4(3)\big)$ possesses such involutions.]

For uniformity, put $t = 1$ if no such involution $t$ exists. Aschbacher takes $A$ to contain $\langle t, x \rangle$ with $A \cong E_{32}$ or $E_{64}$, according as $t = 1$ or $t \neq 1$. Then in either case

$$\overline{N}_0 \cong \Sigma_5. \tag{5.143}$$

First suppose $t = 1$. In this case, Aschbacher proves

(1) $B \triangleleft N$; and
(2) $\overline{N} = O_2(\overline{N})\,\overline{N}_0$ with $O_2(\overline{N}) \cong E_{16}$. \hfill (5.144)

Setting $\tilde{N} \cong N/O(N)$, he argues next that $\tilde{N}$ contains a normal subgroup $\tilde{U}$ with

$$\tilde{U} \cong Z_4 \times Z_4 \times Z_4 \times Z_4 \text{ or } E_{2^8}. \tag{5.145}$$

On the other hand, $B$ contains an involution $y$ such that $I_1 = L(C_L(b)) \cong L_2(9)$. If $J_1$ denotes the pump-up of $I_1$ in $L_b$, then with the aid of his basic hypothesis, Aschbacher argues that

$$J_1/O(J_1) \cong L_2(9), \; L_2(81), \; PSp_4(3), \; L_4(3), \text{ or } U_4(3). \tag{5.146}$$

But $\tilde{b} \in \tilde{U}$ and so $C_N(b)$ contains a 2-subgroup which maps on $\tilde{U}$. However, this is easily seen to be incompatible with the 2-component $J_1$ of $C_b$.

Hence $t \neq 1$. Let $J$ be the pump-up of $I$ in $L_t$. In this case, Aschbacher argues that

$$J/O(J) \cong PSp_4(3) \text{ or } L_4(3), \tag{5.147}$$

which implies that

$$\overline{N}_0 = \overline{J \cap N}. \tag{5.148}$$

However, an analysis of fusion shows that (5.148) fails for some choice of $t$.

Thus Aschbacher proves:

PROPOSITION 5.102. *We have* $L/O(L) \not\cong L_4(3)$.

Finally, Aschbacher proves:

PROPOSITION 5.103. *If* $L \cong PSp_4(3)$, *then* $F^*(G) \cong L_4(4)$.

In this case, we have

$$\begin{array}{l} (1) \;\; m(A) = 5; \text{ and} \\ (2) \;\; \overline{N}_0 \cong A_5. \end{array} \tag{5.149}$$

To describe the results of Aschbacher's fusion analysis, let $\Omega$ denote the set of $G$-conjugates of $x$ contained in $A$. Then he proves:

LEMMA 5.104. *The following conditions hold*:

(i) $B \lhd N$;
(ii) $N$ *acts transitively on $\Omega$ (i.e., any G-conjugate of $x$ contained in $A$ is an N-conjugate of $x$)*;
(iii) *Every element of $\Omega$ is contained in the coset $Bx$; and*
(iv) *One of the following holds*:
    (1) $|\Omega| = 6$ *and* $\bar{N} \cong A_6$ *or* $\Sigma_6$; *or*
    (2) $|\Omega| = 16$ *and* $O_2(\bar{N}) \cong E_{16}$.

Aschbacher next eliminates the $|\Omega| = 6$ subcase as in the $U_4(3)/Z_2$ situation: namely, if $I_1$ is an $SL_2(3)$ component of $C_L(z)$, he argues that $O(C_z)I_1$ is subnormal in $C_z$, whence again $z$ is a classical involution—contradiction.

Thus he proves:

LEMMA 5.105. *We have $|\Omega| = 16$ and $O_2(\bar{N}) \cong E_{16}$.*

Since $A_5 \cong L_2(4) = SL_2(4)$, $\bar{N}$ is now beginning to resemble a suitable 2-local subgroup in $L_4(4)\langle\tau\rangle$, where $\tau$ induces a graph automorphism on $L_4(4)$. Indeed, compare equation (5.185) below of Seitz's corresponding analysis in the case $L/O(L) \cong L_m(2^n)$, $m \geqslant 3$. In fact, in the same general way that Seitz deduces (5.186) from (5.185), Aschbacher next proves the following result:

LEMMA 5.106. *If we set $U = [O_2(N), N_0']$, then $U$ is elementary abelian of order $2^8$.*

It is immediate from these conditions that $R = \langle x \rangle$ and $O_2(N) = U\langle x \rangle$.

Now set $M = N_G(U)$ and $\bar{M} = M/C_U$. Then again in the same general way that Seitz deduces (5.188) from (5.186), Aschbacher next proves:

LEMMA 5.107. *Either $F^*(\bar{M}) \cong SL_3(4)$ or $SL_2(4) \times SL_2(4)$.*

Moreover, in either case, he determines the action of $F^*(\bar{M})$ on $U$. [Note that $SL_3(4) \cong P\Omega_4^+(4) = O_4^+(4)$, and in that case $U$ turns out to be the natural $GF(4)$-module for $O_4^+(4)$.]

Now let $S \in \mathrm{Syl}_2(M)$. It is immediate from the structure of $S$ that $U = J(S)$. Hence with the aid of Thompson transfer, we immediately obtain:

LEMMA 5.108. *The following conditions hold*:

(i)  $S \in Syl_2(G)$; *and*
(ii) *G contains a normal subgroup*  $G_0$  *of index* 2 *with*  $x \notin G_0$ .

The next step is to eliminate the wreathed possibility  $F^*(\bar{M}) \cong SL_2(4) \times SL_2(4)$ , which Aschbacher accomplishes by again forcing  $G$  to possess a classical involution. (Note that the expected solution in this case is  $G_0 = F^*(G) \cong PSp_4(3) \times PSp_4(3)$ , whence indeed  $G$  would contain a classical involution.)

To conform to later notation, we set  $P^* = G_0 \cap M$ . Having eliminated the wreathed case, Aschbacher immediately obtains the following result:

LEMMA 5.109. *$P^*$  is a split extension of  $U$  by a subgroup  $J \cong O_4^+(4)$  with  $U$  the natural GF(4)-module for  $J$ .*

The structure of  $P^*$  is indeed that of a maximal parabolic of  $L_4(4) = SL_4(4)$ ; one can identify  $U$  with the set of matrices

$$U = \begin{pmatrix} 1 & 0 & 0 & 0 \\ 0 & 1 & 0 & 0 \\ a & b & 1 & 0 \\ c & d & 0 & 1 \end{pmatrix}, \qquad a, b, c, d \in GF(4). \tag{5.150}$$

In particular, it follows that  $S \cap G_0$  is isomorphic to a Sylow 2-subgroup of  $L_4(4)$ , in which case  $G_0 = F^*(G)$  is determined from McBride's theorem (Theorem 5.6). However, Aschbacher prefers to continue the analysis, reducing instead to Timmeseld's theorem [I, 4.161]. The argument is similar to that used by Miyamoto in the case  $L/O(L) \cong U_4(2^n)$ ,  $n \geqslant 2$  (cf. Propositions 5.122–124 below). We conclude by either method that  $F^*(G) \cong L_4(4)$ , completing the proof of Proposition 5.103.

This completes Aschbacher's contributions to the  $\mathcal{L}_0(3)$  standard form problems.

We next briefly discuss the three remaining cases:  $L/O(L) \cong G_2(3)$ ,  $L_3(3)$ , or  $U_3(3)$ . Again we may assume that  $G$  does not possess a classical involution (although Harris and Yamada do not make this assumption). We begin with Yamada's treatment [139] of the case

$$L/O(L) \cong G_2(3). \tag{5.151}$$

Here one has

$$\begin{aligned}&(1)\ \ m(A) = 4; \text{ and}\\&(2)\ \ \bar{N}_0 \cong L_3(2).\end{aligned} \qquad (5.152)$$

Moreover, $N_0$ does not split over $B$ [as is known, $G_2(3)$ contains a nonsplit extension of $E_8$ by $L_3(2)$].

If $T \in \mathrm{Syl}_2(G)$, it follows easily by Thompson transfer that $G$ contains a normal subgroup $G_0$ of index 2 not containing $x$. But then $B \in \mathrm{Syl}_2\big(C_{G_0}(B)\big)$ and so $G_0$ is determined by Harada's extension (Theorem 2.156) of the sectional 2-rank $\leqslant 4$ theorem. The only possibility turns out to be $F^*(G) \cong G_2(9)$, which is excluded here as then $G$ possesses a classical involution.

Now Yamada's fusion analysis yields that $O_2(\bar{N}) \cong E_8$, and he then obtains the following analogue of Lemmas 5.105 and 5.106: if we set $U = [O_2(N), \bar{N}_0]$, then

$$\begin{aligned}&(1)\ \ U \cong E_{2^6} \text{ or } Z_4 \times Z_4 \times Z_4; \text{ and}\\&(2)\ \ \text{Correspondingly, } F^*(\bar{N}) \cong L_3(2) \times L_3(2) \text{ or } L_3(2).\end{aligned} \qquad (5.153)$$

The first case leads, as expected, to the wreathed solution $F^*(G) = G_2(3) \times G_2(3)$. However, as $G_2(3)$, and hence $G$, contains a classical involution, this case is again excluded.

In the course of deriving Lemma 5.105, Aschbacher easily rules out the homocyclic possibility $U \cong Z_4 \times Z_4 \times Z_4$ [as does Seitz in deriving (5.186)]. However, that argument is not applicable in the present situation and a much more elaborate procedure is required to derive a contraction in the case $U \cong Z_4 \times Z_4 \times Z_4$. Indeed, if $Z$ is a $Z_4$-subgroup of $U$, normal in a Sylow 2-subgroup of $N$, it is necessary to determine the structure of the group $C_Z$.

Setting $\bar{C}_Z = C_Z/O(C_Z)$, $\tilde{C}_Z = \bar{C}_Z/\bar{Z}$, and $\bar{K} = L(\bar{C}_Z)$, Yamada mimics a rather involved general lemma of Gomi in the latter's treatment of the $U_4(2)$ standard component problem [37] and concludes that $\bar{K}$ has the following properties:

(1) $\bar{K}$ is quasisimple with $\bar{Z} = Z(\bar{K})$;
(2) $\tilde{K}$ contains an $E_{16}$-subgroup $\tilde{E}$ with $\tilde{E} \in \mathrm{Syl}_2\big(C_{\bar{K}}(\tilde{E})\big)$ and $N_K(\tilde{E})/C_K(\tilde{E}) \cong A_6$ or $\Sigma_6$; and
(3) There exists an involution $u$ in $N_G(Z) - C_Z$ with $C_{\bar{K}}(u)$ solvable.

$$(5.154)$$

[We note that in establishing (5.154), Yamada at one point produces an involution $y$ acting on $L$ such that $C_L(y)$ contains a component $J \cong U_3(8)$ or $L_3(8)$ with $J$ standard in $L\langle y \rangle$ and $\langle y \rangle \in \mathrm{Syl}_2(C_{L\langle y \rangle}(J))$. Correspondingly he then invokes the Griess–Mason–Seitz Lie rank 1 case or the Seitz linear Lie rank $\geqslant 2$ theorem to reach a contradiction [$G_2(3)$ is not one of the solutions of either theorem].]

By (5.154)(2), the possibilities for $\tilde{K}$ are now determined from Stroth's theorem (Theorem 2.167). But then using (5.154)(1) and (3), we check that there are unique solutions for $\bar{K}$ and $\tilde{K}$: namely,

$$\tilde{K} \cong U_4(3) \quad \text{and} \quad \bar{K} \cong SU_4(3). \tag{5.155}$$

Furthermore, by the structure of $N$, it follows at once that if $z$ is the involution of $Z$, $C_z$ has a component $K$ which maps on $\bar{K}$ [whence $K/O(K) \cong SU_4(3)$] and such that $Z \in \mathrm{Syl}_2(C_K)$. Thus Yamada has reduced the $G_2(3)$ problem to the $SU_4(3)$ standard form problem. However, as $SU_4(3) \notin \mathscr{L}_0(3)$, the latter problem is included in the inductive characterization of $\mathscr{L}\mathscr{O}$ and consequently $F^*(G) \in \mathscr{L}\mathscr{O}$. The only possibility is $F^*(G) \cong G_2(9)$, which does not admit an automorphism of order 2 whose centralizer has an $SU_4(3)$ component.

Thus Yamada proves the following result:

PROPOSITION 5.110. *We have* $L/O(L) \not\cong G_2(3)$.

We turn now to the cases

$$L/O(L) \cong L_3(3) \quad \text{or} \quad U_3(3). \tag{5.156}$$

In these cases we have

$$\begin{array}{ll} (1) \ m(A) = 3 \text{ or } 4; \text{ and} \\ (2) \ \bar{N}_0 \cong \Sigma_3. \end{array} \tag{5.157}$$

The solvability of $\bar{N}_0$ results, unfortunately, in greater indeterminacy in successive steps of the pushing-up process. In fact, the total analysis closely resembles that of the Fritz–Harris–Solomon $L/O(L) \cong L_2(q)$ or $A_7$ theorem (Theorem 2.181), where again (5.157) holds [but with the additional stipulation that $N_L(B)$ has $D_8$ Sylow 2-subgroups]. However, this last condition also holds in the $L_3(3)$ case, so the similarity is even sharper in that case. [Since $\mathrm{Out}(L_3(3)) \cong Z_2$, the analysis, in fact, corresponds to the $C_x / LC_{C_x}(L)$ cyclic subcase of the $L_2(q), A_7$-problem.]

Harris's aim in both problems is to pin down a Sylow 2-subgroup $S$ of $G$. Moreover, in those cases in which $|S \cap O^2(G)| \leqslant 2^{10}$, he again invokes the corresponding general classification theory (Theorem 2.166) to determine $F^*(G)$, although it is clear that with his already developed information on the structure of $S$ and some additional 2-local analysis, he could easily have forced $G$ to possess a classical involution.

The $\leqslant 2^{10}$ and sectional 2-rank $\leqslant 4$ theorems are sufficient for dealing with the $L_3(3)$ case, and thus Harris proves the following result:

PROPOSITION 5.111. *We have* $L/O(L) \not\cong L_3(3)$.

On the other hand, in the $U_3(3)$ case, they enable Harris to establish only the following weaker result.

PROPOSITION 5.112. *If* $L/O(L) \cong U_3(3)$, *then* $G$ *possesses a normal subgroup* $G_0$ *of index* 2 *not containing* $x$. *Moreover, a Sylow* 2*-subgroup* $S_0$ *of* $G_0$ *has the following properties*:

   (i) $|S_0| = 2^{11}$ *or* $2^{12}$; *and*
   (ii) $S_0$ *contains an abelian normal subgroup isomorphic to* $Z_4 \times Z_4 \times Z_4 \times Z_4$ *or* $Z_8 \times Z_8 \times Z_2 \times Z_2$.

The proof involves extremely delicate and lengthy fusion analysis. Moreover, the structure of $S_0$ is much more precisely determined than indicated by the proposition.

We make only one further comment on Harris's $U_3(3)$ analysis: namely, we describe the structure of the Sylow 2-subgroup $T$ of $N_G(L)$ in the case which leads ultimately to the conclusions of Proposition 5.112. Note that $T \cap L$ is wreathed of order 32, which implies that $B \lhd T$ and hence that $T \leqslant N$. By a delicate fusion analysis and using both the $\leqslant 2^{10}$ and sectional 2-rank $\leqslant 4$ theorems, Harris is able to prove:

LEMMA 5.113. *The following conditions hold*:

   (i) $R = \langle x \rangle$;
   (ii) $|T| = 2^7$ (*whence* $T > (T \cap L)\langle x \rangle$);
   (iii) $T$ *splits over* $\langle x \rangle$; *and*
   (iv) *A complement to* $\langle x \rangle$ *in* $T$ *is of type* $M_{12}$.

[Note that $\mathrm{Aut}\big(U_3(3)\big)$ has Sylow 2-subgroups of type $M_{12}$.]

However, it is a long and painful climb from $T$ and $N$ to a 2-local subgroup of $G$ containing $S = S_0\langle x\rangle$ of order $2^{12}$ or $2^{13}$.

Finally, as noted at the beginning of the section, Proposition 5.112 suffices for both the $B$- and $U$-theorems and the noncharacteristic 2 classification theorem, since then $G$ is not simple.

This at last completes our *outline* of the $\mathscr{L}_0(3)$ case!

## 5.8. The Nonlinear Lie Rank 2 Case

As we have noted (Section 5.2.6), the six families of groups that comprise this case have received independent treatment, each involving the same general methods of proof. We shall focus here on the single family

$$L/O(L) \cong U_4(2^n), \qquad n \geqslant 2. \tag{5.158}$$

(We have chosen this family because it is "twisted," and since we shall be considering the "nontwisted" families $L_m(2^n)$, $m \geqslant 3$, in the next section.) However, as Miyamoto, the one who treated this case [54], bypasses the construction of a $(B,N)$-pair subgroup $G^*$ of $G$ in the subcase leading to $F^*(G) \cong L_4(2^{2n})$, we shall add some comments pertaining to this point for the family $L/O(L) \cong {}^3D_4(2^n)$.

We first fix some notation. We set $q = 2^n$, we let $P_1$, $P_2$ be the two maximum parabolic subgroups of $L$ containing $Q$ and set $Q_i = O_2(P_i)$, $i = 1,2$.

The structures of $P_1$ and $P_2$ are given by the following lemma:

LEMMA 5.114. *For a suitable ordering of $P_1$ and $P_2$, we have*

(i) $P_i' = O^{2'}(P_i)$ *(whence $Q_i \leqslant P_i'$) for both $i = 1$ and 2;*
(ii) (1) $Q_1$ *is elementary of order $q^4$;*
   (2) $P_1'/Q_1 \cong L_2(q^2)$;
(iii) (1) $Q_2$ *is special of order $q^5$ with $Z(Q_2) = Z(Q)$ elementary of order $q$;*
   (2) $P_2'/Q_2 \cong L_2(q)$; *and*
(iv) $P_i' = I_i Q_i$, *where $I_i \cap Q_i = 1$, $i = 1,2$ [whence $I_1 \cong L_2(q^2)$ and $I_2 \cong L_2(q)$].*

In particular, (iv) asserts that $P_i'$ *splits* over $Q_i$, $i = 1, 2$. To establish this fact, one uses a general splitting criterion due to Gaschütz [I: 154, Theorem 15.8.6], which asserts that for any prime $p$ a group $X$ splits over an abelian normal $p$-subgroup $Y$, provided a Sylow $p$-subgroup of $X$ splits over $Y$.

Notice also that the structures of $P_1$ and $P_2$ are *asymmetric*; this is typical of the twisted situation.

This time the initial fusion analysis yields:

PROPOSITION 5.115. *The following conditions hold*:

   (i) $T \notin \mathrm{Syl}_2(G)$;
   (ii) $R = \langle x \rangle$;
   (iii) $x$ *is $G$-conjugate to every involution of the coset $Q_1 x$; and*
   (iv) $x$ *is $G$-conjugate to no involution of $Q_1$.*

Miyamoto gives his own proof of the proposition, but it is also a consequence of Seitz's Theorem 5.2.

Now begins the pushing-up process. Set $U = Z(Q)$, $V = UR$, and $N = N_G(V)$. From the $N$-fusion of $x$ determined from Proposition 5.115, one again obtains the split $(B, N)$-pair configuration of the Bender case.

PROPOSITION 5.116. *The following conditions hold*:

   (i) $N$ *induces a doubly transitive permutation group $\overline{N}$ of degree $q$ on the set $\Omega$ of $N$-conjugates of $R$ in $V$;*
   (ii) *The stabilizer $\overline{N}^R$ of $\{R\}$ in $\overline{N}$ contains a cyclic normal subgroup of order $q - 1$; and*
   (iii) $O_2(\overline{N})$ *is elementary of order $q$ and acts regularly on $\Omega$.*

[The exceptional $L_3(2)$ case is easily excluded here.]

This information is now fed into $N_1 = N_G(V_1)$, where $V_1 = Q_1 R$. As usual, let $\Omega_1$ be the set of $N_1$-conjugates of $R$ contained in $V_1$ and let $\overline{N}_1$ be the permutation group induced by $N_1$ on $\Omega_1$, with $\overline{N}_1^R$ the stabilizer of $\{R\}$ in $\overline{N}_1$. Using Proposition 5.116, Miyamoto argues that $\{R\}$ is $N_1$-conjugate to an element of $\Omega_1 - \{R\}$. Since $I_1 \cong SL_2(q^2)$ acts transitively on $U_1^\#$ and $I_1 \cong \overline{I}_1 \leqslant \overline{N}_1^R$, one immediately concludes:

PROPOSITION 5.117. *The following conditions hold*:

   (i) $|\Omega_1| = q^4$;
   (ii) $\overline{N}_1$ *acts doubly transitively on $\Omega_1$; and*
   (iii) $\overline{I}_1$ *is normal in $\overline{N}_1^R$.*

Thus we are again reduced to determining the structure of a doubly transitive permutation group, but now with a "larger" one-point stabilizer than in the Bender situation of Section 3.

In the linear case $L/O(L) \cong L_m(2^n)$, Seitz reaches a more general doubly transitive group problem of the same shape [namely, $|\Omega_1| = 2^{mn}$ and $\bar{N}_1^R$ contains a normal subgroup $\bar{I}_1 \cong SL_m(2^n)$; see Theorem 5.126 below], which specializes to the Miyamoto situation when $m = 2$ and $2^n = q^2$. [Note that $SL_2(2^m) = L_2(2^m)$.] [The additional technical side conditions of Seitz's theorem follow here from the structure $N_G(Q_1) \cap N_G(L)$ and its action on $Q_1$.] Under the corresponding general conditions, Seitz shows that $\bar{N}_1$ necessarily possesses a regular normal subgroup which is elementary abelian of order $2^{mn}$. [That $O_2(\bar{N}_1)$ is elementary follows from the easily verified fact that $\bar{I}_1$ acts transitively on $O_2(\bar{N}_1)^{\#}$. In fact, $Q_1$ and $O_2(\bar{N}_1)$ are isomorphic as $\bar{I}_1$-modules.] Seitz then goes on to analyze $O_2(N_1)$, which is the pre-image of $O_2(\bar{N}_1)$ in the corresponding normalizer $N_1$ and proves that the group $[O_2(N_1), I_1]$ is elementary abelian of order $2^{mn}$.

Seitz's results will be discussed more fully in the next section. Using them, Miyamoto can thus conclude:

PROPOSITION 5.118. *If we set* $U_1 = [O_2(N_1), I_1]$, *then* $U_1$ *is elementary abelian of order* $q^4$.

Now Miyamoto pushes up the two groups $U_1$ and $Q_2$. Set $M_1 = N_G(U_1)$ and $M_2 = N_G(Q_2)$ and put $\bar{M}_1 = M_1/U_1$, $\bar{M}_2 = M_2/Q_2$. Clearly, $I_i \leqslant M_i$, $i = 1,2$.

Using the already developed structural and fusion information, it is not difficult to obtain the following properties of $\bar{I}_i$.

LEMMA 5.119. *For both* $i = 1$ *and* 2, *we have*

  (i) $\bar{I}_i$ *is standard in* $\bar{M}_i$;
  (ii) $\bar{R} \in \mathrm{Syl}_2(C_{\bar{M}_i}(\bar{I}_i))$; *and*
  (iii) $\bar{I}_i$ *is not normal in* $\bar{M}_i$.

Since $\bar{I}_1 \cong L_2(q^2)$ and $\bar{I}_2 \cong L_2(q)$, Miyamoto is thus in a position to apply the Griess–Mason–Seitz results of the Bender case to identify $\bar{J}_i = F^*(\bar{M}_i)$, $i = 1, 2$ {except in the case $q = 4$, $\bar{I}_2 \cong L_2(4)\ (\cong L_2(5))$, where instead he uses the inductive characterization of $\mathcal{LO}$}. We have an extra constraint here which serves to eliminate many of the initial possibilities for $\bar{J}_i$: namely, $\bar{J}_1$ must act nontrivially on $U_1 \cong E_{q^4}$ and $\bar{J}_2$ on $Q_2$ and hence on

$Q_2/\phi(Q_2) \cong E_{q^4}$ (since $I_1$, $I_2$ act nontrivially on $U_1$, $Q_2$, respectively). For example, these conditions exclude $\bar{J}_2 \cong U_3(q)$ since the minimal dimension of a faithful representation of $U_3(q)$ on a vector space over $GF(q)$ is 6.

Miyamoto is also able to show that $\bar{J}_1$ does *not* act irreducibly on $U_1$, which enables him to conclude that the possibilities for $\bar{J}_i$ are as follows:

PROPOSITION 5.120. *The following conditions hold*:

(i) $\bar{J}_1 \cong L_2(q^2) \times L_2(q^2)$; *and*
(ii) $\bar{J}_2 \cong L_2(q^2)$ *or* $L_2(q) \times L_2(q)$.

The two solutions for $\bar{J}_2$ lead, respectively, to the quasisimple and wreathed cases. Note that $\bar{J}_1$ is a direct product in *both* cases.

Let $P_i^*$ be the pre-image of $\bar{J}_i$ in $M_i$, $i = 1,2$. Then it is not difficult to prove:

LEMMA 5.121. *The following conditions hold*:

(i) $P_1^*$ *and* $P_2^*$ *have a common Sylow 2-subgroup*;
(ii) $P_1^* = U_1 J_1$, *where* $U_1 \cap J_1 = 1$; *and*
(iii) $P_2^* = Q_2 J_2$, *where* $Q_2 \cap J_2 = 1$.

Setting $G^* = \langle P_1^*, P_2^* \rangle$, the goal as indicated earlier is to prove that $G^* = F^*(G)$ [$\cong L_4(q^2)$ or $U_4(q) \times U_4(q)$ in the quasisimple and wreathed cases, respectively]. (It will then follow that $P_1^*$ and $P_2^*$ are the two maximal parabolic subgroups of $G^*$ containing a given Sylow 2-subgroup of $G^*$.)

However, as already noted, Miyamoto bypasses this approach in the quasisimple case, moving instead directly to an analysis of fusion. He lets $G_0$ be the normal closure of $L$ in $G$ and chooses a Sylow 2-subgroup $S_0$ of $G_0$ such that $S_0 \cap P_1^* \in \mathrm{Syl}_2(P_1^*)$. (It is easy to see that $P_i^* \leqslant G_0$, $i = 1,2$.) Moreover, he can take $S_0$ so that $S_0 \cap J_1 \in \mathrm{Syl}_2(J_1)$.

By Proposition 5.120, $J_1 = K \times K^*$, where $K \cong K^* \cong SL_2(q)$. Set $B = S_0 \cap K$, so that $B \in \mathrm{Syl}_2(K)$ and finally put

$$A = BC_{U_1}(B). \tag{5.159}$$

This group $A$ is the subgroup which Miyamoto will show satisfies the conditions of Timmesfeld's weakly closed theorem. He carries this out in two steps, first proving the following:

PROPOSITION 5.122. *The following conditions hold*:

(i) $G_0$ *has exactly two conjugacy classes of involutions*;
(ii) *Representatives* $z_1$, $z_2$ *of these classes can be taken in* $U_1$;
(iii) $C_{S_0}(z_i) \in \mathrm{Syl}_2(C_{G_0}(z_i))$, $i = 1,2$; *and*
(iv) $A$ *is elementary abelian and weakly closed in* $S_0$ *with respect to* $G_0$.

He next proves:

PROPOSITION 5.123. *The following conditions hold*:

(i) $C_{G_0}(z_i)$ *is 2-constrained for* $i = 1$ *and* 2; *and*
(ii) $G_0$ *is of characteristic 2 type.*

Note that (ii) follows from (i) and the full balanced, connected theorem [I, 1.40].

As an immediate corollary of these two results, he finally obtains the desired conclusion:

PROPOSITION 5.124. $A$ *is weakly closed in* $C_{G_0}(a)$ *for all* $a \in A^{\#}$.

Thus Timmesfeld's theorem [I, 4.161] is indeed applicable and Miyamoto concludes that $G_0 \cong L_4(q^2)$ [whence also $G^* = F^*(G) = G_0$].

On the other hand, in the wreathed case Miyamoto follows the more standard procedure, first identifying $G^*$ as a $(B, N)$-pair and then applying Seitz's general result, mentioned in Remark 5.1.6, to conclude that $G^* \lhd G$ [whence again $G^* = F^*(G)$]. We shall have more to say concerning this wreathed argument in the next section.

We turn now briefly to the case

$$L/O(L) \cong {}^3D_4(q), \qquad q = 2^n, \tag{5.160}$$

which was treated by Yamada [137]. The pushing-up process involves more steps than in the unitary case. Moreover, as $SL_2(2)$ $(\cong \Sigma_3)$ is solvable, the subcase $n = 1$ is somewhat exceptional and requires additional analysis at certain places in the argument. For simplicity, we assume $n \geqslant 2$. Likewise to preserve the similarity with the unitary case, we again denote the final pushed-up 2-local subgroups of $G$ by $M_1$, $M_2$ and set $U_i = O_2(M_i)$, $\bar{M}_i = M_i/U_i$, and $\bar{J}_i = F^*(\bar{M}_i)$, $i = 1,2$.

Yamada establishes the following analog of the unitary situation.

PROPOSITION 5.125. *The following conditions hold*:

(i) (1) $\bar{J}_1 \cong SL_2(q^6)$ *or* $SL_2(q^3) \times SL_2(q^3)$;
    (2) *Correspondingly* $\bar{J}_2 \cong SL_2(q^2)$ *or* $SL_2(q) \times SL_2(q)$; *and*
(ii) (1) $|U_1| = q^{18}$ *and* $|U_2| = q^{21}$;
    (2) *The general structure of* $U_i$ *is specified as is the action of* $\bar{J}_i$ *on* $U_i$, $i = 1, 2$.

We are interested here in the quasisimple case:

$$\bar{J}_1 \cong SL_2(q^6), \qquad \bar{J}_2 \cong SL_2(q^2). \tag{5.161}$$

Let $P_i^*$ be the pre-image of $\bar{J}_i$ in $M_i$, $i = 1, 2$, and set

$$G^* = \langle P_1^*, P_2^* \rangle. \tag{5.162}$$

The goal is to prove that $G^*$ is a split $(B, N)$-pair, in which case the Fong–Seitz theorem [I, 3.14] will yield the desired identification

$$G^* \cong {}^3D_4(q^2). \tag{5.163}$$

Let $T$ be the common Sylow 2-subgroup of $P_1^*$ and $P_2^*$. Then $N_{P_i^*}(T) = TA_i$, where

$$\begin{gathered} \text{(1) } A_i \text{ is cyclic, } i = 1, 2; \text{ and} \\ \text{(2) } |A_1| = q^6 - 1 \text{ and } |A_2| = q^2 - 1. \end{gathered} \tag{5.164}$$

Moreover, it is not difficult to show that $A_1$ and $A_2$ can be chosen so that

$$\begin{gathered} \text{(1) } A = A_1 A_2 = A_1 \times A_2; \text{ and} \\ \text{(2) } A_j \text{ centralizes } \bar{J}_i \text{ for } J \neq i, \ 1 \leqslant i, j \leqslant 2. \end{gathered} \tag{5.165}$$

The group $B = TA$ is then our candidate for the Borel subgroup of $G^*$.

The key step is to determine the Weyl group $W$ of $G^*$ and its pre-image $N$ in $N_G(A)$. (In the present case, $N$ splits over $A$, so that $W$ can be identified with a subgroup of $N$.) Since $\bar{J}_i$ is itself a split $(B, N)$-pair of rank 1 with respect to its Borel subgroup $\bar{B}_i = \bar{A}_i \bar{T}$, we have

$$\bar{J}_i = \bar{B}_i \cup \bar{B}_i \bar{w}_i \bar{B}_i \tag{5.166}$$

for a suitable involution $\bar{w}_i \in N_{\bar{J}_i}(\bar{A}_i)$, $i = 1, 2$.

From the action of $\bar{J}_i$ on $U_i$, $i = 1,2$, we calculate $C_T(A) = 1$. Since $A_j$ centralizes $\bar{J}_i$ for $i \neq j$, it follows that $N_{G^*}(A)$ contains an involution $w_i$ which maps on $\bar{w}_i$ and centralizes $A_j$ for $i \neq j$, $1 \leqslant i, j \leqslant 2$. Furthermore, as $x$ normalizes both $\bar{J}_1$ and $\bar{J}_2$, we can choose $w_i$ to centralize $x$. We now set

$$N = A \langle w_1, w_2 \rangle, \tag{5.167}$$

so that $A \lhd N$.

Moreover, in view of (5.165), we have $G^* = \langle P_1^*, P_2^* \rangle \leqslant \langle B, N \rangle$. Since clearly $\langle B, N \rangle \leqslant G^*$, this yields

$$G^* = \langle B, N \rangle. \tag{5.168}$$

The next step is to determine the structure of the group $W = N/A$, which is to be the candidate for Weyl group of $G^*$. Since $N/A$ is generated by the two involutions $Aw_1$, $Aw_2$, we know at the outset that $W$ is a dihedral group [I, 2.41]; the task is to determine its order, and this is equivalent to determining the least positive integer $r$ for which $(w_1 w_2)^r \in A$. But $r$ can indeed be computed from the following available data: (a) the fact that $w_1 w_2 \in C_x$; and (b) the known action of $w_i$ on $U_i$, $i = 1,2$. The conclusion is as follows:

(1) $r = 6$ and $(w_1 w_2)^6 = 1$;
(2) $W \cong \langle w_1, w_2 \rangle$ is dihedral of order 12. $\tag{5.169}$

To complete the proof that $G^*$ is a $\langle B, N \rangle$ pair, Yamada must verify the critical double coset rules: namely, for all $w \in W$ and $i = 1,2$,

(1) $BwBw_i B \subseteq BwB \cup Bww_i B$; and
(2) $T \cap T^{(w_1 w_2)^6} = 1$. $\tag{5.170}$

Yamada establishes (5.170) by appealing to a result of Gomi [35, (1L)], who has made the corresponding computations under quite general conditions that guarantee a split $(B, N)$-pair of rank 2. Keeping the same notation $T$, $P_i^*$, $U_i$, and $J_i$, Gomi allows $\bar{J}_i$ to be an *arbitrary* group of Lie rank 1 and characteristic 2 [thus $\bar{J}_i \cong L_2(2^{n_i})$, $Sz(2^{n_i})$, $U_3(2^{n_i})$, or when $n_i$ is odd, $SU_3(2^{n_i})$ $(\cong U_3(2^{n_i})/Z_3)$, $i = 1,2$]. With the corresponding definitions of $B$, $w_1$, $w_2$, and $N$, Gomi shows that $G^* = \langle P_1^*, P_2^* \rangle = \langle B, N \rangle$ is a split $(B, N)$-pair under the following conditions:

(1) $w_1 w_2$ has *even* order $r > 2$;

(2) $T = U_i(T \cap T^{w_3 - i} \cap T^{w_i w_3 - i} \cap \cdots \cap T^{\overbrace{w_3 - i w_i \cdots w_3 - i}^{r-1}})$,
    $i = 1, 2$; and                                                                        (5.171)

(3) $T \cap T^{(w_1 w_2)^r} = 1$.

Since $(5.171)(2)$ is much easier to verify than the full set of inclusions of $(5.171)(1)$, Gomi's result considerably shortens the work required to show that $G^*$ is a $(B, N)$-pair in the Lie rank 2 case. [Of course, as $L_4(2^n)$ has Lie rank 3 (and Weyl group $\Sigma_4$), $(5.171)$ is not applicable in Miyamoto's unitary problem. [This can also be seen from the fact that $\bar{J}_1 \cong L_2(2^{2n}) \times L_2(2^{2n})$ is not quasisimple in his case.] However, it is clear that one could use Miyamoto's 2-local information to prove directly that $G^*$ is a split $(B, N)$-pair in this case as well.

Finally, to prove that $G^* \lhd G$, Yamada shows by transfer that $T \in \mathrm{Syl}_2(O^2(G))$ and then quotes Thomas's ${}^3D_4(2^m)$ Sylow 2-group characterization theorem (Theorem 5.5). However, as we have observed, with some additional work he could undoubtedly have reduced instead to Holt's theorem.

## 5.9. Two General Lie Type Characteristic 2 Theorems of Seitz

In this section, we describe the two general results which underlie Seitz's work on the standard component problem with $L$ of Lie type of characteristic 2.

We begin with his doubly transitive group theorem, whose exact statement is as follows:

THEOREM 5.126. *Let $\Omega$ be a set of cardinality $2^{mn}$, $m \geqslant 2$, $n \geqslant 1$, and let $X$ be a doubly transitive permutation group in which the stabilizer $X_a$ of the point $a \in \Omega$ has the following properties*:

(a) $X_a$ *contains a normal subgroup $Y \cong SL_m(2^n)$*;

(b) $Y C_{X_a}(Y)$ *is isomorphic to a subgroup of $GL_m(2^n)$; and*

(c) *The permutation action of $X_a$ on $\Omega - \{a\}$ is equivalent to the natural action of $X_a$ on the nonzero vectors of an $m$-dimensional vector space over $GF(2^n)$*.

*Then $O_2(X)$ is elementary abelian of order $2^{mn}$ and acts regularly on $\Omega$.*

Observe that these conditions imply that $X_a$ is an extension of $Y C_{X_a}(Y)$

by a group of "field" automorphisms of $Y$ and hence that $X_a$ is isomorphic to a subgroup of the group of "semilinear" transformations of an $m$-dimensional vector space over $GF(2^n)$, so condiction (c) is compatible with (a) and (b).

Seitz invokes a theorem of Ostrom and Wagner on finite affine planes admitting a transitive group of collineations to handle the case $m = 2$ [92]. On the other hand, when $m \geqslant 3$, he identifies $\Omega - \{a\}$ with an $m$-dimensional vector space $V$ over $GF(2^n)$, chooses an appropriate one-dimensional subspace $V_1$ of $V$, and using induction argues in rapid succession that

(1) The stabilizer $X_1 = N_X(V_1)$ of $V_1$ contains a Sylow 2-subgroup of $X$;

(2) $X$ acts doubly transitively on the set $\Delta$ of $X$-conjugates of $X_1$; and

(3) $O_2(X_1) \neq 1$.                                                              (5.172)

But such doubly transitive groups have been completely classified by O'Nan in the case that $X_1$ does not act semiregularly on $\Delta - \{X_1\}$ and by Shult in the contrary case (both results are included in the statement of [I, 3.41]). Given the structure of $X_a$ in the present situation, one easily checks that the only possibility here is that $X$ possesses an elementary abelian normal subgroup acting regularly on $\Omega$.

O'Nan has indicated the following more direct proof of the theorem. By condition (c), the subgroup of $Y$ fixing a second point $b \in \Omega$ fixes precisely $2^n$ points of $\Omega$ elementwise. The collection of all such sets of $2^n$ points as the pair $\{a, b\}$ ranges over $\Omega$ determines a block design $\mathscr{B}$ which is preserved by the action of $X$ on $\Omega$. (By definition, a *block design* on $\Omega$ is a collection of subsets of $\Omega$, called *blocks*, each of the same cardinality, with the property that any two distinct points of $\Omega$ are contained in a unique block.)

O'Nan's key observation is that the action of $X$ on $\mathscr{B}$ *is uniquely determined by* $Y$. Indeed, it follows from (c) that each Sylow 2-subgroup of $Y$ has precisely one orbit of size $2^n$ on $\Omega - \{a\}$; and it is not difficult to see that the elements of $\mathscr{B}$ are exactly the set of all 2-point stabilizers which contain $a$ together with the orbits of size $2^n$ of the Sylow 2-subgroups of $Y$.

On the other hand, if one considers the natural action of $Y \cong SL_m(2^n)$ on an $m$-dimensional vector space $V$ over $GF(2^n)$, then $V$ has a natural associated block design $\mathscr{B}^*$, whose elements are the points of the one-dimensional subspaces of $V$. But then by the uniqueness of $\mathscr{B}$, $\mathscr{B}$ and $\mathscr{B}^*$ must be isomorphic block designs (i.e., there exists a one-to-one map between $\Omega$ and the vectors of $V$ transforming the blocks of $\mathscr{B}$ into those of

$\mathscr{B}^*$). This means that we can identify $X$ with a group of affine transformations of $V$. But by the classical theorem of Hilbert the full group of such affine transformations is the semidirect product of the group of translations of $V$ with the group of linear transformations of $V$. Hence, by the transitive action of $X$ on $V$, $X$ must contain this translation group and the theorem follows.

We turn now to Seitz's proof that $G^*$ is normal in $G$, when $R = \langle x \rangle$, $L$ is of Lie type over $GF(2^n)$ of Lie rank $\geqslant 2$, and $G^*$ is an $x$-invariant subgroup of $G$ such that

(1) $L < G^*$;
(2) Either $G^*$ is quasisimple or $G^*$ is the direct product of two quasisimple groups $L_1$, $L_2$, isomorphic to $L$, and interchanged under conjugation by $x$; and
(3) $N_G(L)$ normalizes $G^*$.                                               (5.173)

For the actual theorem, certain groups of low Lie rank must be excluded when $n = 1$. We shall not attempt to list these here. Moreover, we shall discuss only the general lines of the argument, even though at various points several families require separate treatment, and, in addition, orthogonal groups require a slight variation of the argument throughout [with $P\Omega_8^+(2^n)$ in large measure requiring its own special treatment].

Furthermore, as we have indicated earlier, Seitz does not follow this approach when $L$ is a linear group, and $G^*$ is quasisimple, reducing instead to McBride's and Collins's Sylow 2-group characterization theorems. Likewise in most portions of the argument, he assumes that $L$ has Lie rank $\geqslant 3$. However, it is clear from the nature of the proofs that, again with some small case exceptions, the full argument can be modified to cover these groups as well. Our comments will ignore all these subtleties.

THEOREM 5.127. *Assume that $G$ possesses a subgroup $G^*$ satisfying the conditions of* (5.173). *Moreover, if $G^* = L_1 \times L_2$, assume, in addition, that every 2-local subgroup of $G$ has the B-property. Under these conditions, $G^*$ is normal in $G$.*

Seitz uses the $B$-property to the following extent only:

If $T$ is a nontrivial 2-subgroup of $L_1$ such that
$O(N_G(T))L_2 \lhd N_G(T)$, then $L_2 \cong N_G(T)$.                          (5.174)

[Since $x$ interchanges $L_1$ and $L_2$, (5.174) also holds, of course, with $L_2$, $L_1$ in place of $L_1$, $L_2$, respectively.]

It is very likely that with some additional work the balanced, connected theorem [I, 4.54] can be shown to apply in this situation, thus eliminating the necessity of assuming (5.174). However, as this condition is included in both the assumption of Hypothesis $S$ and the standard component theorem (II), there is no compelling reason to avoid it.

For most of the argument, Seitz must separate the direct product and quasisimple cases. However, the first step applies to both cases. We set $M = N_G(G^*)$ and fix this notation.

PROPOSITION 5.128. *If $x$ leaves $G^{*g}$ invariant for $g \in G$, then $g \in M$. In particular, $M$ contains a Sylow 2-subgroup of $G$.*

In other words, in the permutation representation of $G$ on the set $\Omega$ of conjugates of $G^*$, $x$ fixes a *unique* point: namely, $G^*$.

This conclusion will clearly imply that $x$ interchanges the elements of $\Omega - \{G^*\}$ in pairs, whence $|\Omega|$ is odd. But it is immediate that $|\Omega| = |G : N_G(G^*)| = |G : M|$, so $M$ will thus contain a Sylow 2-subgroup of $G$. Hence the second assertion of the proposition is a direct consequence of the first.

The proof of the proposition depends heavily on the previously determined structures of the centralizers of involutions acting on groups of Lie type of characteristic 2 (Aschbacher and Seitz [I: 21]). Note first, however, that as $C_x \leqslant N_G(L)$, condition (5.173)(3) implies that

$$C_x \leqslant M. \tag{5.175}$$

Suppose now that $x$ leaves $H^* = G^{*g}$ invariant for some $g \in G - M$. By the structure of $C_x$, $L^* = L(C_{H^*}(x)) \leqslant L$. There are then three cases to consider, according as $(a)\, L^* = L$, $(b)\, 1 \neq L^* < L$, and $(c)\, L^* = 1$ [whence $C_{H^*}(x)$ is 2-constrained].

In case (a), for example, observe that $x^g$ acts on $H^*$ and $L(C_{H^*}(x^g)) = L^g \cong L^* = L$. However, Seitz knows then that $x$ and $x^g$ must be $H^*$-conjugate, whence $x^{gh} = x$ for some $h \in H^*$. But then $gh \in C_x \leqslant N_G(G^*)$ and therefore $H^* = H^{*h} = G^{*gh} = G^*$—contradiction.

On the other hand, if, say, $C_{G^*}(x)$ is 2-constrained, Seitz uses the fact that $x \notin C_x^{(\infty)}$ [whence $x \notin C_{H^*}^{(\infty)}$ ] and that a Sylow 2-subgroup of $C_{H^*}(x)$ has order at most that of $C_x$ [and hence of $N_G(L)$] to conclude that one of the following holds:

(1) $L/O(L) \cong PSp_m(2^n)$ and $G*/O(G*) \cong P\Omega^\pm_{m+2}(2^n)$; or

(2) $x$ induces an outer automorphism on $H*$ and either       (5.176)

    (a) $L/O(L) \cong PSp_m(2^n)$ and $G*/O(G*) \cong L_m(2^n)$ or $U_m(2^n)$; or

    (b) $L/O(L) \cong F_4(2^n)$ and $G*/O(G*) \cong E_6(2^n)$ or $^2E_6(2^n)$.

Detailed properties of the centralizers of their involutions are required to eliminate these possibilities.

We now distinguish the quasisimple and direct product cases, discussing the latter possibility first. Thus $G* = L_1 \times L_2$, and we let $S \in \mathrm{Syl}_2(M)$ with $x \in S$ [whence $S \in \mathrm{Syl}_2(G)$] and set $S_i = S \cap C_{L_j}$ for $j \neq i$, $1 \leqslant i, j \leqslant 2$, so that $S_i \in \mathrm{Syl}_2(C_{L_i})$ and $x$ interchanges $S_1$ and $S_2$ under conjugation. Furthermore, $S_i \cap L_i \in \mathrm{Syl}_2(L_i)$, $i = 1,2$ (however, if some 2-element of $S$ induces an outer automorphism on $L_j$, $S_i$ may contain $S_i \cap L_i$ properly). We fix this notation.

PROPOSITION 5.129. *If $N_G(S_1) \leqslant M$, then $G*$ is normal in $G$.*

The proof divides into three parts, the first of which is a variation of Proposition 5.128.

LEMMA 5.130. *If $L_2$ leaves $G*^g$ invariant for $g \in G$, then $g \in M$.*

However, the key to the proposition is the following result.

LEMMA 5.131. *If $T$ is a nontrivial subgroup of $S_1$, then $N_G(T)$ normalizes $L_2$.*

If $T = S_1$, this follows at once from the hypothesis of the proposition. Hence if $T$ is chosen of maximal order violating the conclusion of the lemma, then $T < S_1$ and consequently $T_1 = N_{S_1}(T) > T$ by [I, 1.11]. But now if we set $N = N_G(T)$ and $\bar{N} = N/T$, the maximality of $T$ implies that $\bar{T}_1 \in \mathrm{Syl}_2(C_{\bar{N}}(\bar{L}_2))$. It now follows with the aid of the previous lemma that $\tilde{L}_2$ is standard in $\tilde{N} = \bar{N}/O(\bar{N})$. Hence by Aschbacher and Seitz [in the case $m(\tilde{T}_1) > 1$], or either by induction or previous classification theorems [in the case $m(\tilde{T}_1) = 1$], the structure of $\tilde{L}_2^* = \langle \tilde{L}_2^{\tilde{N}} \rangle$ is determined. If $\tilde{L}_2 = \tilde{L}_2^*$, then $O(N)L_2 \lhd N$ and (5.174) yields that $L_2 \lhd N$, in which case the lemma holds. On the other hand, if $\tilde{L}_2 < \tilde{L}_2^*$, Seitz in essence argues that a Sylow 2-subgroup of $\tilde{L}_2^*$ has a structure incompatible with that of a Sylow 2-subgroup of $G (\leqslant M)$.

As a corollary, we immediately obtain the following result:

LEMMA 5.132. *If* $t \in \mathcal{I}(G)$ *centralizes* $L_2$, *then* $L_2 \lhd C_t$.

Indeed, as $S_1 \in \mathrm{Syl}_2(C_{L_2})$, the result follows from Sylow's theorem and the previous lemma.

We note one other fact.

LEMMA 5.133. *We have* $N_G(L_1) \leqslant M$.

Indeed, $L_2 \leqslant C_{L_1} \leqslant N_G(S_1 \cap L_1)$, whence $L_2 \lhd C_{L_1}$ by the previous lemma, and the result follows immediately.

Note also that as $x$ interchanges $S_1, L_1$ and $S_2, L_2$, Lemmas 5.132 and 5.133 hold for both values of $i = 1, 2$.

Seitz is now in a position to verify the conditions of Aschbacher's 2-component fusion theorem (Theorem 3.39, together with the paragraph following) to complete the proof of Proposition 5.129. [Without the assumption that $F^*(G)$ is quasisimple in the statement of the theorem, one must allow the additional possibility that the given subgroup $K$ is normal in $G$—i.e., $M = G$.]

For example, if $t \in \mathcal{I}(G)$ centralizes $L_i$ and $t^g$ centralizes $L_j$ for $g \in G$, then $L_i \lhd C_t$ and $L_j \lhd C_{t^g}$ by Lemma 5.132. But then also $L_i^g \lhd C_{t^g}$ and it follows from Lemma 5.133 (as $|C_{G^*}|$ is odd) that $L_i^g = L_j$. Hence either $g$ or $gx$ normalizes $L_i$ and we conclude from Lemma 5.133 that $g \in M$, as required.

In view of Proposition 5.129, to establish Theorem 5.127 in the case $G^* = L_1 \times L_2$, it remains for Seitz to prove the following:

PROPOSITION 5.134. *We have* $N_G(S_1) \leqslant M$.

Setting $N = N_G(S_1)$, we have $L_2 \leqslant N$, and the natural approach is to try to show that $O(N)L_2 \lhd N$, whence $L_2 \lhd N$ by condition (c) and the result would follow easily. Setting $\bar{N} = N/S_1$ and $\tilde{N} = \bar{N}/O(\bar{N})$, we are thus required to determine the structure of $\tilde{N}$. Let $\tilde{T} \in N_{\tilde{N}}(\tilde{L}_2)$. Then our conditions imply that $\tilde{T} \in \mathrm{Syl}_2(\tilde{N})$ and that $\tilde{T}$ acts faithfully on $\tilde{L}_2$, so that $\tilde{T} \leqslant \mathrm{Aut}(\tilde{L}_2)$, $\tilde{T} \cap \tilde{L}_2 \in \mathrm{Syl}_2(\tilde{L}_2)$, and $C_{\tilde{N}}(\tilde{L}_2)$ has odd order. However, because of this last condition, one cannot obtain the desired conclusion by an inductive argument, so that, in effect, we are reduced to identifying $\tilde{N}$ solely from the structure of its Sylow 2-subgroup $\tilde{T}$. Given the number of possibilities for $\tilde{L}_2$ and hence of $\tilde{T}$, this would be a very elaborate undertaking.

Seitz finds a very ingenious way of avoiding this problem. Under the assumption that $N \not\leq M$, he first produces an element $g \in C_N(x)$ of *odd* order such that

$$\text{(1) } g \notin M; \text{ and}$$
$$\text{(2) } g \text{ normalizes } S \cap L_2. \tag{5.177}$$

Using (2), the goal of his analysis is now to force $g \in M$, thus contradicting (1). This is achieved in two stages. He first argues that with respect to a suitable root system $\Sigma_i$ of $L_i$, $g$ normalizes each root subgroup of $S \cap L_i$ for both $i = 1$ and 2. (Since $g$ centralizes $x$ and normalizes $S \cap L_2$, it suffices to do this for $i = 2$.)

Seitz now uses this information to construct two $g$-invariant subgroups, which together generate $G^*$ (whence $g$ normalizes $G^*$, giving the desired contradiction).

First (apart from certain low Lie rank cases with $n = 1$), $L_1$ possesses a maximal parabolic subgroup $P_1$ containing $S_1 \cap L_1$ and having a *quasisimple* Levi factor $J_1$. Set $P_2 = P_1^x$, $J_2 = J_1^x$, and put $Q_i = O_2(P_i)$, $i = 1,2$. Then each $Q_i$ is a product of $g$-invariant root groups. Hence, if we set $Q = Q_1 Q_2$ and $H = N_G(Q)$, we have $g \in H$. Furthermore, setting $\bar{H} = H/Q$ and $\bar{J} = C_{\bar{J}_1 \bar{J}_2}(\bar{x})$, it follows that $\bar{J}$ is quasisimple. In addition, it is not difficult to show that $\langle \bar{x} \rangle \in \text{Syl}_2(C_{\bar{H}}(\bar{J}))$, so that $\bar{J}$ is standard in $\bar{H}$. Therefore, either by induction or previously treated standard form problems ($\bar{J}_i$ and hence $\bar{J}$ is of Lie type of characteristic 2), we conclude with the aid of (5.174) that

$$\bar{J}_1 \bar{J}_2 = \langle \bar{J}^{\bar{H}} \rangle = F^*(\bar{H}), \tag{5.178}$$

and consequently

$$g \text{ normalizes } J_1 J_2 Q. \tag{5.179}$$

In general, $L_1$ possesses a second maximal parabolic subgroup $P_1^*$ containing $S_1 \cap L_1$ with quasisimple Levi factor $J_1^*$, and if we set $P_2^* = P_1^{*x}$, $J_2^* = J_1^{*x}$, $Q_i^* = O_2(P_i^*)$, $i = 1,2$, and $Q^* = Q_1^* Q_2^*$, it follows similarly that

$$g \text{ normalizes } J_1^* J_2^* Q^*. \tag{5.180}$$

But it is easily shown that

$$G^* = \langle J_1 J_2 Q, J_1^* J_2^* Q^* \rangle. \tag{5.181}$$

[If $B_1 = N_{L_1}(S_1 \cap L_1)$, then $P_1 = B_1 J_1 Q_1$ with $B_1$ normalizing $J_1 Q_1$, and similarly for $P_1^*$. But $L_1 = \langle P_1, P_1^* \rangle$ by [I, 2.18], so $\langle J_1 Q_1, J_1^* Q_1^* \rangle \lhd L_1$, forcing equality.]

Now (5.179), (5.180), and (5.181) imply that $g$ normalizes $G^*$, as required.

(In order to accommodate cases in which $L_1$ does not possess such a second maximal parabolic subgroup $P_1^*$, Seitz takes as his second subgroup the layer $K_1 K_2$ of the centralizer in $G$ of a suitable $x$-invariant subgroup of $P_1 P_2$ of *odd* order and argues that $K_1 K_2 \leqslant G^*$, that $g$ leaves $\langle J_1 J_2 Q, K_1 K_2 \rangle$ invariant, and that the latter group is $G^*$.)

We now consider the quasisimple case. Again we let $S \in \mathrm{Syl}_2(M)$, so that $S \in \mathrm{Syl}_2(G)$. Since $G^* \lhd M$, $S^* = S \cap G^* \lhd S$ and consequently $S^*$ contains a 2-central involution $z$ of $S$. Moreover, it is easy to see that $z$ can be taken to lie in a root subgroup of $S$.

Here, in succession, are Seitz's principal moves in proving that $G^* \lhd G$.

PROPOSITION 5.135. *One of the following holds*:

(i) *If $z$ is $G$-conjugate to $y \in \mathscr{I}(S)$, then $y \in S^*$ and $z, y$ are $G^*$-conjugate; or*
(ii) *$L/O(L) \cong PSp_{m-2}(2^n)$ or $P\Omega_m^-(2^n)$ with $4 \mid m$ and correspondingly $G^*/O(G^*) \cong P\Omega_m^{\pm}(2^n)$ or $P\Omega_m^+(2^{2n})$.*

The proof depends heavily upon the structural information developed by Aschbacher and Seitz in [I: 21].

The possibilities in (ii) require a modification of the argument which applies in (i). We ignore these exceptional cases.

PROPOSITION 5.136. *One of the following holds*:

(i) *$O_2\big(C_{G^*}(z)\big)$ is normal in $C_z$; or*
(ii) *$G^*/O(G^*) \cong L_m(2^n)$, $U_m(2^n)$, or $PSp_m(2^n)$.*

Set $Q = O_2\big(C_{G^*}(z)\big)$ and let $Q_\alpha$ be the root subgroup of $S$ containing $z$ (whence $Q_\alpha \leqslant Q$). In most cases, $Q_\alpha \lhd C_{G^*}(z)$ and $C_{G^*}(z)$ acts irreducibly on $Q/Q_\alpha$. The proposition is proved by combining this fact with an analysis of the set $\Omega_z$ of fixed points of $z$ on the set $\Omega$ of $G$-conjugates of $G^*$. In particular, Seitz shows that $C_z$ is transitive on $\Omega_z$ and also that $Q_\alpha$ fixes every point of $\Omega_z$.

Finally Seitz proves [again we ignore the exceptional cases of (ii), which require further analysis] the following result:

PROPOSITION 5.137. *We have* $C_z \leqslant M$.

Setting $Q = O_2(C_{G^*}(z))$, it follows from the previous proposition that $\Omega_z = \Omega_Q$, where $\Omega_Q$ denotes the set of fixed points of $Q$ on $\Omega$. Using this fact, Seitz produces $G^*$-conjugates $z_1, z_2$ of $z$ with the following properties:

$$
\begin{aligned}
&(1)\ \langle z_2, z \rangle \cong D_8; \\
&(2)\ \langle z_1 \rangle = Z(\langle z_2, z \rangle); \\
&(3)\ \langle z_2, z \rangle \leqslant O_2(C_{z_1}); \text{ and} \\
&(4)\ G^* = \langle C_{G^*}(z), z_2 \rangle.
\end{aligned}
\qquad (5.182)
$$

Using all his information on the action of various subgroups on $\Omega$, he now argues that $\langle C_{G^*}(z), z_2 \rangle$ stabilizes $\Omega_z$ and hence so does $G^*$. But $z$ acts trivially on $\Omega_z$, while $G^*/O(G^*)$ is simple, whence $G^*$ must act trivially on $\Omega_z$. But his analysis also yields that $\{G^*\}$ is the only element of $\Omega$ fixed by $G^*$, so $\Omega_z = \{G^*\}$. Since $C_z$ leaves $\Omega_z$ invariant, we thus conclude that $C_z$ normalizes $G^*$, whence $C_z \leqslant M$, as asserted.

Propositions 5.135(i) and 5.137 together imply that the conditions of Holt's theorem [I, 4.32] are satisfied when $M < G$. Given the possibilities for $G^*$, we reach a contradiction in each case, so $M = G$ and hence $G^* \lhd G$.

This completes our discussion of Theorem 5.127.

## 5.10. The General Lie Type Characteristic 2 Case

In this section, we describe Seitz's solutions of the standard form problems with $L$ linear of Lie rank $\geqslant 2$ or nonlinear of Lie rank $\geqslant 3$, which as we have already noted in Sections 5.2.7 and 5.2.8, he treats by somewhat different methods.

We first discuss the linear case:

$$
L/O(L) \cong L_m(2^n), \qquad m \geqslant 3 \qquad (5.183)
$$

(with $n \geqslant 2$ if $m = 3$ or 4 and $n \geqslant 1$ otherwise). Again, as we have said, Miyamoto's treatment of the $U_4(2^n)$ case parallels Seitz's analysis, so we shall simply list the structures of the corresponding subgroups at successive stages of the pushing-up process, adding an occasional brief comment.

Again we set $q = 2^n$. As described in Section 5.2.7, Seitz begins with the two maximum parabolic subgroups $P_1, P_2$ of $L$ containing $Q$ which correspond in the natural $2m$-dimensional representation of $SL_m(q)$ on a vector space over $GF(q)$ to the stabilizer of a one-dimensional and $(2m - 1)$-dimensional subspace, respectively. We follow the notation of Section 5.8 as closely as possible. Thus we set $Q_i = O_2(P_i)$, $i = 1,2$.

The analog of Lemma 5.114 is the following result, which holds for *both $i = 1$ and 2:*

$$\begin{aligned} &(1)\ \ Q_i \text{ is elementary abelian of order } q^{m-1}; \text{ and} \\ &(2)\ \ P'_i = Q_i I_i, \text{ where } I_i \cong SL_{m-1}(q) \text{ and } Q_i \cap I_i = 1. \end{aligned} \tag{5.184}$$

Furthermore, the fusion assertions of Propositions 5.115 hold now for *both $i = 1$ and 2.* We shall not repeat their statements.

Set $V_i = Q_i R$ and $N_i = N_G(V_i)$ and let $\Omega_i$ be the set of $N_i$-conjugates of $R$ contained in $V_i$, $i = 1,2$. Again let $\bar{N}_i$ be the permutation group induced by $N_i$ on $\Omega_i$ and let $\bar{N}_i^R$ be the stabilizer of $\{R\}$ in $\bar{N}_i$, $i = 1,2$. Because the initial fusion results hold for both values of $i$, so likewise does the analog of Proposition 5.117:

$$\begin{aligned} &(1)\ \ |\Omega_i| = q^{m-1}; \\ &(2)\ \ \bar{N}_i \text{ acts doubly transitively on } \Omega_i; \text{ and} \\ &(3)\ \ \bar{I}_i \text{ is normal in } \bar{N}_i. \end{aligned} \tag{5.185}$$

Seitz's Theorem 5.126 implies now that $O_2(\bar{N}_i)$ is elementary abelian of order $q^{m-1}$. Setting $U_i = [O_2(N_i), I_i]$, $i = 1,2$, Seitz next proves

$$U_i \text{ is elementary abelian of order } q^{2(m-1)} \tag{5.186}$$

Indeed, as noted in the previous section, $\bar{I}_i$ acts transitively on $O_2(\bar{N}_i)^\#$, which immediately implies that $O_2(N_i)/Q_i$ splits over $V_i/Q_i = Q_i R/Q_i$. Hence

$$O_2(N_i) = U_i R, \text{ where } U_i \text{ is } I_i\text{-invariant.} \tag{5.187}$$

Now consider the map

$$\tau : u_i \rightarrow [u_i, x], \qquad u_i \in U_i.$$

We see at once that $\tau$ induces a homomorphism of $U_i/Q_i = O_2(\bar{N}_i)$ on $Q_i$ that is permutable with the action of $I_i$, so by definition $U_i/Q_i$ and $Q_i$ are isomorphic as $I_i$-modules.

But by the structure of $SL_{m-1}(q)$, $I_i$ contains a cyclic subgroup $A_i$ of order $(q^{m-1} - 1)/(q - 1)$. Since $U_i/Q_i$ and $Q_i$ are isomorphic as $A_i$-modules, one can now prove, by passing to the so-called "associated Lie ring" of $W_i$ that $W_i$ must be abelian (cf. [I, 130, Chapter V, Exercise 19]). Hence by the action of $I_i$ on $W_i$, $W_i$ is necessarily either homocyclic or elementary abelian (of order $q^{2(m-1)}$).

Finally Seitz eliminates the homocyclic possibility by a fusion argument, involving a comparison of the structures of $N_1$ and $N_2$. Thus $U_i$ is elementary abelian and (5.185) holds.

This time one sets $M_1 = N_G(U_1)$ and $M_2 = N_G(U_2)$ and puts $\bar{M}_i = M_i/U_i$, $i = 1,2$. Again it follows that $\bar{I}_i$ is standard in $\bar{M}_i$ with $\bar{R} \in \mathrm{Syl}_2(C_{\bar{M}_i}(\bar{I}_i))$, and $\bar{I}_i$ not normal in $\bar{M}_i$, $i = 1$ and 2, and again the next step is to identify $\bar{J}_i = F^*(\bar{M}_i)$.

If $m = 3$ [whence $I_i \cong L_2(q)$, $q \geqslant 4$ (as $n \geqslant 2$ in this case)], Seitz again uses Griess–Mason–Seitz and the inductive characterization of $\mathcal{LO}$ to make an initial determination of $\bar{J}_i$, while if $m \geqslant 4$, then as $\bar{I}_i \cong SL_{m-1}(q)$ with $m - 1 \geqslant 3$, the desired determination follows by induction, except in the case $m = 5$, $n = 1$, where Solomon's prior $L_4(2)$ ($\cong A_8$) standard form theorem is needed. Again many of the initial possibilities are excluded as $\bar{J}_i$ must act faithfully on $U_i$. [The case $\bar{I}_i \cong L_4(2)$, $\bar{J}_i \cong A_{10}$ requires some additional work since $\Sigma_{10}$ ($\cong \bar{J}_i\bar{R}$) possesses an eight-dimensional $GF(2)$-representation ($U_i \cong E_{2^8}$ in this case), but not one in which an element of order 5 in $\bar{I}_i = C_{\bar{J}_i}(\bar{R})$ acts fixed-point-free on the corresponding module, as is the case here.] The final result for both $i = 1$ and 2 is as follows:

$$\bar{J}_i \cong SL_{m-1}(q^2) \quad \text{or} \quad SL_{m-1}(q) \times SL_{m-1}(q). \tag{5.188}$$

Again the pre-image $P_i^*$ of $\bar{J}_i$ in $M_i$ splits over $U_i$ (whence $P_i^* = U_iJ_i$ with $U_i \cap J_i = 1$ and $J_i \cong \bar{J}_i$), $i = 1, 2$.

Seitz now distinguishes the various cases of (5.188). Suppose first that $\bar{J}_i \cong SL_{m-1}(q^2)$ for some $i$, say $i = 1$. In this case, he first proves

The action of $J_1$ on $U_1$ is that of the standard representation of $SL_{m-1}(q^2)$ on a vector space of dimension $m - 1$ over $GF(q^2)$. (5.189)

In particular, it follows that $P_1^*$ has Sylow 2-subgroups of type $L_m(q^2)$. To be in a position to invoke Theorem 5.127, Seitz would first have to argue that $G^* = \langle P_1^*, P_2^* \rangle$ is a $(B, N)$-pair. However, as we have said several times, he bypasses this approach, opting instead for McBride's and Collins's Sylow 2-subgroup characterization of $L_m(q^2)$ (Theorem 5.4). He thus first argues that $M_1$ contains a Sylow 2-subgroup of $G$ and then by a rather delicate fusion analysis proves

$$G_1 = O^2(G) \text{ has Sylow 2-subgroups of type } L_m(q^2), \qquad (5.190)$$

whence $G_1 = F^*(G) \cong L_m(q^2)$ by their theorems.

In verifying (5.190), it is easy to show by Thompson transfer that $G$ has a normal subgroup $G_0$ of index 2 not containing $x$. The difficulty arises from 2-elements of $M_1$ which induce *graph* automorphisms on $J_1$ (in which case Seitz must argue that $G_0$ itself has a normal subgroup of index 2: namely, $G_1$).

Thus there remains only the wreathed possibility: $\bar{J}_i \cong SL_{m-1}(q) \times SL_{m-1}(q)$ for *both* $i = 1$ and 2. In this case, he does construct an $x$-invariant semisimple subgroup $G^*$ with

$$G^*/O(G^*) = L_1 \times L_2, \text{ where } L_1 = L_2 \cong L_m(q), \text{ and } |Z(G^*)| \text{ is odd.}$$
$$(5.191)$$

The basic idea is as follows. For $i = 1$, 2, write $J_i = K_i \times K_i^*$, where $K_i \cong K_i^* \cong SL_{m-1}(q)$, so that $K_i$ and $K_i^*$ are interchanged by $x$. He first determines the action of $J_i$ on $U_i$, obtaining the following analog of (5.189) for $i = 1, 2,$:

(1) $P_i^* = K_i A_i \times K_i^* A_i^*$, where $U_i = A_i \times A_i^*$ and $A_i \cong A_i^* \cong E_{q^{m-1}}$; and

(2) The action of $K_i, K_i^*$ on $A_i, A_i^*$, respectively, is that     $(5.192)$
of the standard representation of $SL_{m-1}(q)$ on a vector
space of dimension $m - 1$ over $GF(q)$.

From the pushing-up process, Seitz knows that $P_1^*$ and $P_2^*$ contain a common Sylow 2-subgroup. In particular, $A_1^* \leqslant P_2^*$. Clearly $A_1^*$ centralizes $K_1$, and he now argues that $A_1^* \leqslant K_2^* A_2^*$, so that $A_1^*$ also centralizes $K_2$. Thus the group

$$L_1 = \langle K_1, K_2 \rangle \leqslant C_{A_1^*}, \qquad (5.193)$$

and it remains only to determine the structure of $L_1$.

Considering the Weyl group $W$ of $L$ and a representative $w$ of the element of maximal length in a canonical set of generators of $W$, [taken relative to the $(B, N)$-pair representation of $L$ in which the Borel subgroup $B$ is contained in $P_1 \cap P_2$]. Seitz proves that

$$K_2 = K_1^w, \tag{5.194}$$

and then identifies $L_1$ by Curtis's theorem (Theorem 3.57), whence

$$L_1/Z(L_1) \cong L_n(q). \tag{5.195}$$

In addition, $L_1$ is perfect and $Z(L_1)$ is easily seen to have odd order. It follows now that

$$L_2 = L_1^x = \langle K_1^*, K_2^* \rangle \text{ centralizes } L_1, \tag{5.196}$$

and this yields (5.191).

We turn now to the nonlinear Lie rank $\geqslant 3$ case [excluding the cases $PSp_6(2)$, $U_6(2)$, and $P\Omega_8^\pm(2)$], and add some details to the discussion of Section 5.2.8. Again we ignore the slight modifications required to handle the orthogonal case [which begins with an appropriate product of *two* $SL_2(2^n)$ subgroups rather than, as in all other cases, with a single such subgroup]. We again put $q = 2^n$ and let $J$ be the $SL_2(q)$-subgroup of $L$ defined in Section 5.2.8, with $Y$ the cyclic subgroup of $J$ of order $q + 1$ and $X = L(C_L(Y))$.

Our intent is to give some indication of the elaborate case division which Seitz is forced to consider in order to construct the desired $(B, N)$-pair subgroup $G^*$ of $G$. We first list the possibilities for $X$ corresponding to the various families for $L$:

PROPOSITION 5.138. *According as $L/O(L)$ is isomorphic to $PSp_m(q)$, $m$ even, $P\Omega_m^+(q)$, $m$ even, $U_m(q)$, $E_6(q)$, $E_7(q)$, $E_8(q)$, $^2E_6(q)$, or $F_4(q)$, $X/O(X)$ is isomorphic to*

$$PSp_{m-2}(q), \ P\Omega_{m-4}^\pm(q), \ U_{m-2}(q), \ L_6(q), \ P\Omega_{12}^+(q), \ E_7(q), \ U_6(q), \text{ or } PSp_6(q).$$

Again let $X^*$ be the pump-up of $X$ in $C_Y$. As noted in (5.17) and (5.19), $X$ is standard in $C_Y$ and $X$ is not normal in $C_Y$, whence $X^* = F^*(C_Y)$ is determined from the previous proposition, either by induction, the solution of a prior standard form theorem, or the assumed solution of one of the cases within Table 5.6.

The general result is as follows:

PROPOSITION 5.139.   $R = \langle x \rangle$ *and one of the following holds*:

(i) $X^* = X_1 \times X_2$, *where* $X_1 \cong X_2 \cong X$ *and* $x$ *interchanges* $X_1, X_2$ *under conjugation; or*
(ii) $X^*$ *is of Lie type of characteristic* 2 *and* $x$ *induces an outer automorphism on* $X^*$ (*either a field, graph, or graph-field automorphism*).

In fact, there are a total of *thirteen* distinct possibilities for $X^*$ under (ii), *seven* of which arise when $X/O(X)$ is a symplectic group. Here is the list of $X^*/O(X^*)$'s when $X/O(X) \cong PSp_{m-2}(q)$, with corresponding action of $x$:

$$PSp_{m-2}(q^2)(\text{field}),\ L_{m-2}(q)(\text{graph}),\ L_{m-1}(q)(\text{graph}),$$
$$P\Omega_m^+(q)\ (\text{graph}),\ P\Omega_m^-(q)\ (\text{graph–field}),\ U_{m-2}(q)(\text{graph–field}),$$
$$U_{m-1}(q)(\text{graph–field}). \tag{5.197}$$

Note that in all the cases of (5.197), $L/O(L) \cong PSp_m(q)$, except when $m - 2 = 6$, in which case there is the second alternative $L/O(L) \cong F_4(q)$.

The construction of $G^*$ depends on specific properties of the groups $L$, $X$, and $X^*$, forcing the analysis to proceed case-by-case (in some instances several cases can be combined in a single argument; but offsetting this is the fact that certain general arguments break down when the Lie rank is low and these exceptional cases require individual treatment).

The aim, of course, is to identify $G^*$ by means of the appropriate form of the Curtis–Phan conditions (Theorems 3.55, 3.57, or 3.58). The procedure is similar in all cases, but is most easily described when $L$ has an $L_3(q)$- or $U_3(q)$-generating system

$$L = \langle J_1, J_2, ..., J_r \rangle, \tag{5.198}$$

with the $J_i$ satisfying the conditions of Definition 3.56. We ignore the orthogonal case here, so that we can take $J = J_1$. Since we are in the nonlinear case, we have, in fact,

$$L/O(L) \cong U_n(q),\ E_6(q),\ E_7(q),\ \text{or}\ E_8(q). \tag{5.199}$$

Moreover, it is easy to see that

$$X = \langle J_2, J_3, ..., J_r \rangle.$$

Now $X = L(X_{X^*}(x))$, and given the nature of the action of $x$ on $X^*$, one can show, in general, for each $i$, $2 \leqslant i \leqslant r$, that $J_i$ is contained in a subgroup $K_i$ of $X^*$ with

$$K_i \cong SL_2(q^2), \ SL_2(q), \ \text{or} \ SL_2(q) \times SL_2(q) \qquad (5.200)$$

[according as $X^*$ is quasisimple and defined over $GF(q^2)$, quasisimple and defined over $GF(q)$, or the direct product of two components interchanged by $x$]. [In the case that $X^*$ is a unitary group, one must also allow the possibility that $K_i \cong SU_3(q)$.]

Likewise, there are certain additional relationships between the $J_i$'s and $K_i$'s. To describe these, let $S$ be a Sylow 2-subgroup of $L$ such that $S \cap J_i$ is a root subgroup of $S$ for each $i$, $1 \leqslant i \leqslant r$ (such an $S$ exists), set $B = N_L(S)$, and let $N$ be the pre-image of the Weyl group $W$ of $L$, relative to $B$, so that $B \cap N \lhd N$ and $N / B \cap N \cong W$. Then for $2 \leqslant i, j \leqslant r$, we have (in general):

(1) If $y \in N \cap X$ leaves $J_i$ invariant, then $y$
    leaves $K_i$ invariant; and                                      (5.201)
(2) If $[J_i, J_j] = 1$, then $[K_i, K_j] = 1$.

Thus there is a considerable parallel between the structure of $X$ and a portion of $X^*$. The trick is now to extend this connection to one from $L$ to $X^*$. The idea is as follows. In the easiest cases, there exists an element $y \in N$ leaving $J_i$ invariant, $2 \leqslant i \leqslant r - 1$, and transforming $J_r$ into $J_1$. In particular,

$$X^y = \langle J_2, J_3, ..., J_r \rangle^y = \langle J_1, J_2, ..., J_{r-1} \rangle. \qquad (5.202)$$

One now sets

$$G^* = \langle X^*, X^{*y} \rangle \qquad (5.203)$$

and attempts to argue that $G^*$ is a $(B, N)$-pair.

In view of (5.201) and (5.202), one has

$$G^* = \langle K_r^y, K_2, ..., K_r \rangle. \qquad (5.204)$$

Setting $K_1 = K_r^y$, one first determines the structure of the groups $\langle K_i, K_j \rangle$ for all $i, j$ except $i = 1, j = r$, which is obtained directly from the embedding of $X$ and $X^y$ in $X^*$ and $X^{*y}$, respectively. Finally, one must argue that $[K_1, K_r] = 1$ (using the fact that $[J_1, J_r] = 1$). The argument yields ultimately that $G^*$ satisfies the Curtis–Phan conditions for a suitable $(B, N)$-pair, and this enables one to identify $G^*$.

This gives the flavor of the construction; however, to carry it through in all cases involves Seitz in considerable technical subtleties.

Now Seitz is in a position to invoke Theorem 5.127 to prove that $G^* \lhd G$, whence $G^* = F^*(G)$ and so $F^*(G)$ is determined.

## 5.11 The Remaining Cases over $GF(2)$ and $GF(4)$

By now, the reader who has persevered must surely be convinced that, although the process may be long and painful, any given standard form problem can be solved using some combination of the techniques described above. This is fortunate, for no less than 47 individual possibilities for $L/O(L)$ still remain to be considered before the noncharacteristic 2 theorem is completely proved! These include the 6 simple groups over $GF(2)$ of Table 5.6, the 8 nonsimple groups over $GF(2)$ and $GF(4)$ of Table 5.7, the 16 simple and nonsimple sporadic cases to complete the verification of Hypothesis $S$ of Table 5.11 (omitting $Mc$, which has already been covered in Section 5.7), and the remaining 17 simple and nonsimple sporadic cases of Table 5.12 (omitting $M_{23}$, likewise covered in Section 5.7).

For the purposes of an outline, little is to be gained by discussing each of these 47 cases in succession, since the principal methods for solving standard form problems have been fully described above:

(a) Analysis of fusion of involutions.
(b) Pushing-up 2-local subgroups.
(c) In some cases, investigation of centralizers of elements of odd order.
(d) Induction and the solution of previous standard form problems to identify proper sections of $G$.
(e) Thompson transfer when $G$ is not simple.
(f) In most cases, construction of an identifiable subgroup $G^*$ within $G$, by one of the following methods:
   (1) For the alternating groups, either by means of the root four-subgroup graph or by the classical presentation by generators and relations.
   (2) For the groups of Lie type of characteristic 2, as a split $(B, N)$-pair.
   (3) For the sporadic groups, as the centralizer of a 2-central involution of $G$.

(g) In the cases under $(f)(1)$ and $(f)(2)$, either Holt's or Timmesfeld's fusion theorem to prove that $G^* = F^*(G)$.

(h) In the cases not covered by (f)(1), (2), or (3), pinning down a Sylow 2-subgroup of $G_1 = O^2(G)$ and identifying $F^*(G_1) = F^*(G)$ by a prior Sylow 2-group classification theorem.

In addition, we have sketched the solutions in the alternating case, for most families of groups of Lie type of characteristic 2, and for a number of individual groups [including those in $\mathscr{L}_0(3)$ as well as the residual cases of the Aschbacher–Seitz reduction theorem when $m(H) \geqslant 2$].

It would therefore seem to be more profitable to limit the discussion of these remaining 47 cases to a few comments, including, in particular, points that involve some twist or variation not previously encountered. However, in adopting this approach, we emphasize that there is no intent to minimize the importance of any of these theorems, many of which are very difficult and all of whose proofs require a thorough mastery of finite simple group theory.

In this section we comment on the Lie type characteristic 2 cases. First of all, when $L/O(L)$ is simple (Table 5.6), the basic approach is to push-up the parabolic subgroups of $L$. In some cases, control of the process (in particular, determination of the possible involution fusion patterns) is maintained by obtaining auxiliary information about the centralizers of suitable elements of odd order in these parabolics and related odd local subgroups.

We consider only the $PSp_6(2)$ and $^2F_4(2)'$ cases. When $L/O(L) \cong PSp_6(2)$, the table indicates that as in the previous section there are *seven* distinct simple solutions for $F^*(G)$. We shall explain how these arise. First, $L$ contains a parabolic subgroup $P$ with

$$U = O_2(P) \cong E_{64} \quad \text{and} \quad P/U \cong L_3(2). \tag{5.205}$$

Moreover, in the action of $P$ on $U$, the elements of $U^\#$ decompose into four orbits, which we designate as $U_i$, $1 \leqslant i \leqslant 4$.

As usual, set $V = UR$, $N = N_G(V)$, and let $\Omega$ be the set of $G$-conjugates of $R$ which lie in $V$. For a suitable ordering of the $U_i$, Gomi [37] proves the following result:

PROPOSITION 5.140. *We have $R = \langle x \rangle$ and one of the following holds*:

(i) $|\Omega| = 8$ *and either $\Omega = \{x\} \cup U_1$ or $\{x\} \cup U_2$; or*
(ii) $|\Omega| = 64$ *and either $\Omega = Ux$, $\{x\} \cup U_1 x \cup U_4 \cup U_4 x$,*
    $\{x\} \cup U_2 \cup U_4 \cup U_4 x$, *or $\{x\} \cup U_1 x \cup U_2 x \cup U_3 x \cup U_4$.*

Thus there are a total of six possible fusion patterns, which in turn lead to distinct structures for $N$.

Ultimately Gomi proves:

PROPOSITION 5.141. *One of the following holds*:

(i) $|\Omega| = 8$ *and* $F^*(G) \cong L_6(2)$, $U_6(2)$, $P\Omega_8^+(2)$, *or* $P\Omega_8^-(2)$; *or*
(ii) $|\Omega| = 64$ *and* $F^*(G) \cong L_7(2)$, $U_7(2)$, $PSp_6(4)$, *or* $PSp_6(2) \times PSp_6(2)$.

Suppose next that $L/O(L) \cong {}^2F_4(2)'$. Then $O(L) = 1$ and $L$ has index 2 in $\hat{L} = \text{Aut}(L) \cong {}^2F_4(2)$. Thus $\hat{L}$ is a split $(B, N)$-pair of rank 2 and moreover, the maximal 2-local subgroups of $L$ are the intersections with $L$ of the maximal parabolic subgroups of $\hat{L}$.

The primary difficulty in this case arises, as in the earlier Griess–Mason–Sietz Lie rank 1 situation, from the fact that these maximal 2-local subgroups do not possess a very rich structure (indeed, $L$ is a *thin* group). More precisely, if $P_1, P_2$ denote the two maximal subgroups of $L$ containing $Q$, then for a suitable ordering, one has

$$P_1/O_2(P_1) \cong \Sigma_3 \text{ and } P_2/O_2(P_2) \text{ is a Frobenius group}$$
$$\text{of order 20.} \tag{5.206}$$

In particular, $|P_1 : Q| = 3$ and $|P_2 : Q| = 5$, so that the structure of $P_1$ and $P_2$ are completely dominated by $Q$.

Furthermore, if $z$ is the 2-central involution of $Q$, we also have

$$
\begin{aligned}
&(1) \ P_2 = C_L(z); \\
&(2) \ U = [O_2(P_2), O_2(P_2)] \cong E_{32}; \text{ and} \\
&(3) \ \langle z \rangle = [U, O_2(P_2)].
\end{aligned}
\tag{5.207}
$$

In particular, $O_2(P_2)$ has class 3.

The goal, of course, is to force $F^*(G) \cong F_4(2)$. However, Aschbacher [6] settles for less; namely, he pins down the structure of a Sylow 2-subgroup of $G$ sufficiently to show that $G$ possesses a normal subgroup $G_0$ of index 2, so that $G$ is not simple. As in Harris's $U_3(3)$ problem, he could undoubtedly have continued the analysis until he either forced $F^*(G_0) \cong F_4(2)$ or reached the wreathed solution $F^*(G_0) \cong {}^2F_4(2)' \times {}^2F_4(2)'$.

Note that a Sylow 2-subgroup of $F_4(2)$ has order $2^{24}$, while that of $L$ and $\text{Aut}(L)$ have orders $2^{11}$ and $2^{12}$, respectively, so that it is again a long climb to reach a Sylow 2-subgroup of $G$.

Aschbacher's first step is to determine the $G$-fusion of $x$ in $T$ (note that $L$ has precisely two conjugacy classes of involutions). He proves the following:

PROPOSITION 5.142. *The involution* $t \in T$ *is* $G$-*conjugate to* $x$ *if and only if* $t = yx$ *for some involution* $y \in T$.

To establish this result, he requires the following 5-local information. Let $D \in \mathrm{Syl}_5(P_2)$, let $E$ be an $x$-invariant Sylow 5-subgroup of $O(C_D)$ and set $E_1 = \Omega_1(E)$, $N_1 = N_G(E)$, and $\bar{N}_1 = N_1/C_{E_1}$.

PROPOSITION 5.143. *The following conditions hold*:

(i) $[O(C_D), z] \cong Z_5 \times Z_5$ *and is contained in* $E_1$; *and*
(ii) *If* $K$ *denotes the normal closure of* $x$ *in* $N_1$, *then* $\bar{K} \cong U_3(3) \times Z_2$.

With this fusion information available, Aschbacher is ready to begin the pushing-up process, taking $V = UR$ and $N = N_G(V)(U = [O_2(P_2), O_2(P_2)])$. By a long argument, he eventually proves:

PROPOSITION 5.144. *The following conditions hold*:

(i) $N \leqslant C_z$; *and*
(ii) $N$ *contains a* $D\langle x \rangle$-*invariant elementary* 2-*subgroup* $U_1 \cong E_{2^{10}}$ *with* $U \leqslant U_1$.

Finally, he analyzes the 2-local subgroup $M = N_G(U_1)$, arguing that $M$ contains a Sylow 2-subgroup of $G$ and determining sufficient fusion information to prove the existence of the desired normal subgroup $G_0$ of index 2 in $G$.

We turn now to the nonsimple cases, which we divide into four subcases.

PROPOSITION 5.145. *If* $L/O(L) \cong L_3(4)/Z_4$, *or* $PSp_6(2)/Z_2$, *then correspondingly* $F^*(G) \cong ON$ *or* $.3$.

In the first case, $T$ contains a normal $Z_4 \times Z_4 \times Z_4$ subgroup $U$ [with $x \in O_2(Z(L))) \leqslant U$]. Setting $N = N_G(U)$ and $\bar{N} = N/O(N)$, Nah [88] argues, on the basis of her initial fusion analysis, that $\bar{N}/\bar{U} \cong L_3(2)$ and then quotes O'Nan's theorem [I, 2.34, 2.35] to conclude that $F^*(G) \cong ON$.

In the second case, one argues immediately that $R = \langle x \rangle$. Since

$\text{Out}(PSp_6(2)) = 1$, it follows that $C_x = L \times O(C_x)$. If $O(C_x) = 1$, then $C_x = L$ is isomorphic to the centralizer of a 2-central involution of .3 and Fendel's theorem [I, 3.48] implies that $F^*(G) \cong .3$. On the other hand, if $O(C_x) \neq 1$, and we set $M \times N_G(O(C_x))$, it is easily shown first that $M/O(M) \cong .3$ and then that $M$ is strongly embedded in $G$—contradiction. However, Seitz, who treated this case [107], prefers to avoid this last argument by observing that $T \in \text{Syl}_2(G)$ [inasmuch as $\langle x \rangle = Z(T)$] and then quoting Solomon's theorem (Theorem 2.72).

PROPOSITION 5.146. *If* $L/O(L) \cong U_6(2)/Z_2$, *then* $F^*(G) \cong M(22)$.

In this case, Davis and Solomon [18] assume that $G$ is a minimal counterexample, so that by Lemma 5.10 $G$ has no normal subgroups of index 2. Their goal is to prove that $C_x = L$, for then $C_x$ will be isomorphic to the exact centralizer of an involution of $M(22)$, in which case $G \cong M(22)$ by [I, 3.50] (cf., Hunt [66]).

As usual, it is easily shown that $R = \langle x \rangle$ and that $O(C_x) = 1$, so if the desired assertion is false, then $C_x/\langle x \rangle \cong \text{Aut}(U_6(2))$, which contains $U_6(2)$ as a subgroup of index 2. By the structure of $\text{Aut}(U_6(2))$, $L$ contains an $E_{2^{10}}$-subgroup $U$ (with $x \in U$) and if we set $N = N_G(U)$, $\bar{N} = N/C_U = N/U$, and $N_0 = C_N(x)$, then

$$\bar{N}_0 \cong L_3(4)\langle t \rangle, \text{ where } \bar{t} \text{ is an involution inducing}$$
a field automorphism of period 2. (5.208)

On the basis of their fusion analysis, Davis and Solomon argue that the set $\Omega$ of $\bar{N}$-conjugates of $x$ has cardinality 22. Viewing $V/\langle x \rangle$ as a projective plane $\pi$ over $GF(4)$, there is associated a natural block design $\mathcal{B}$ on $\Omega - \{x\}$, preserved by $\bar{N}_0$, whose blocks correspond to the points of a line of $\pi$. Thus $\mathcal{B}$ is a Steiner triple system $S(2, 5, 21)$ (see [I, p. 80, 81]). Moreover, it follows that one can define a Steiner triple system $S(3, 6, 22)$ on $\Omega$ which is preserved by $\bar{N}$ and has $\bar{N}_0$ as one-point stabilizer. However, it is well known that there exists a unique such triple system (cf. [77]). We therefore conclude from [I, 2.27] that

$$\bar{N} \cong \text{Aut}(M_{22}). \quad (5.209)$$

In particular, $\bar{N}$ and hence $N$ has a normal subgroup of index 2.

Finally Davis and Solomon argue that $U$ is weakly closed (with respect to $G$) in a Sylow 2-subgroup $S$ of $N$ [whence $S \in \text{Syl}_2(G)$]. A transfer theorem of Yoshida [143] then yields that $G$ has a normal subgroup of index 2, which is not the case.

Finally a word about Hunt's characterization of $M(22)$. Following the usual centralizer of involution approach, he first argues that $x$ is not 2-central, that a Sylow 2-subgroup of $G$ has order $2^{17}$, that $G$ has precisely three conjugacy classes of involutions, and if $N = N_G(U)$ and $\bar{N} = N/U$ are defined as above, then

$$\bar{N} \cong M_{22}. \tag{5.210}$$

Furthermore, he determines the order of the centralizer of every involution of $G$. However, we note that he does not explicitly determine the structure of the centralizer of a 2-central involution.

At this point he is in a position to pin down the order of $G$, using the general form of the Thompson order formula, stated in [I, 2.43] for groups with exactly two classes of involutions.

LEMMA 5.147. *If $X$ is a group with $m \geqslant 2$ conjugacy classes of involutions, represented by $x_i$, $1 \leqslant i \leqslant m$, and if $a(x_i)$ denotes the number of ordered pairs of involutions $(u, v)$ of $X$ such that $u \sim x_1$ and $v \sim x_2$ in $X$ and $x_i = (uv)^j$ for some integer $j$, then we have*

$$|X| = |C_X(x_2)| a(x_1) + |C_X(x_1)| a(x_2)$$

$$+ \sum_{i=3}^{m} \frac{|C_X(x_1)| |C_X(x_2)|}{C_X(x_i)} a(x_i).$$

Let $x$, $z$, and $t$ be representatives of the three conjugacy classes of involutions of $G$ with $z$ 2-central. Hunt's analysis pins down $a(x)$ and $a(t)$ rather quickly $[a(x) = 513 \cdot 693$ and $a(t) = 6]$, but leads to *five* initial possibilities for $a(z)$ [the correct value being $a(z) = 2304$]. This in turn yields five possible orders for $G$. One of these is easily eliminated by Sylow's theorem (for the prime 1987). However, the remaining three incorrect solutions require character-theoretic arguments to reach contradictions. Thus, ultimately Hunt proves

$$|G| = |M(22)|. \tag{5.211}$$

Moreover, the same character analysis shows that in the action of $G$ by conjugation on the elements of the conjugacy class $D$ containing $x$, $C_x$ has precisely three orbits. But if $d \in D$, then for any $y \in C_x$, $(xd)^y = xd^y$, so $xd$ and $xd^y$ have the same order. Taking $d$ in each of the $C_x$-orbits of $D$, it follows that there are at most three distinct possibilities for $|xd|$. If $d = x$, then $|xd| = 1$, while if $d \in C_x - \langle x \rangle$, then $|xd| = 2$. Furthermore, in the remaining case, it can be shown that $x$ and $d$ can be taken to lie in a $G$-conjugate of $C_x (\cong U_6/Z_2)$ with $|xd| = 3$. Thus $G$ is, in fact, a 3-transposition group and we conclude from Fischer's classification theorem [I, 2.58] that $G \cong M(22)$.

Finally, G. Stroth [117] proves the following result:

PROPOSITION 5.148. *If* $L/O(L) \cong {}^2E_6(2)/Z_2$, *then* $F^*(G) \cong F_2$.

Stroth's analysis is similar to Hunt's characterization of $M(22)$, but technically more difficult. His aim is to reach an analogous characterization of $F_2$ as a $\{3, 4\}$-transposition group, due to Fischer [30].

THEOREM 5.149. *Assume $G$ satisfies the following conditions*:

(a) $C_x = L\langle t \rangle$, *where $t$ is an involution inducing a field automorphism on* $L/\langle x \rangle \cong {}^2E_6(2)$;
(b) $|G| = |F_2| = 2^{41} \cdot 3^{13} \cdot 5^6 \cdot 7^2 \cdot 11 \cdot 17 \cdot 19 \cdot 23 \cdot 31 \cdot 47$;
(c) *The conjugacy class $D$ containing $x$ is a class of $\{3, 4\}$-transpositions; and*
(d) $C_x$ *has precisely 5 orbits in its action on the set $D$ under conjugation.*

*Under these conditions,* $G \cong F_2$.

Again it is easy to prove that $R = \{x\}$ and that $O(C_x) = 1$. Moreover, there exists a parabolic subgroup of $L/\langle x \rangle$ whose pre-image $P$ in $L$ has the following structure:

$$\begin{aligned} &\text{(1) } O_2(P) = E \times \langle x \rangle, \text{ where } E \text{ is extraspecial of} \\ &\qquad \text{order } 2^{21}; \text{ and} \\ &\text{(2) } P/O_2(P) \cong U_6(2). \end{aligned} \qquad (5.212)$$

Let $z$ be the involution $Z(E)$. It is immediate that $z$ is 2-central. Stroth argues that $x$ is not 2-central and that $O_2(P)$ is contained in an extraspecial group $U$ of $G$ of order $2^{23}$. Setting $N = N_G(U)$ and $\bar{N} = N/U$, he then goes on to show that $\bar{N}$ is simple and if $\bar{y}$ is a 2-central involution of $\bar{N}$, then $\bar{C} = C_{\bar{N}}(\bar{y})$ is 2-constrained with the following properties:

$$
\begin{align}
&\text{(1) } O_2(\bar{C}) \text{ is extraspecial of order } 2^9; \text{ and} \\
&\text{(2) } \bar{C}/O_2(\bar{C}) \cong PSp_6(2).
\end{align}
\tag{5.213}
$$

Hence by F. Smith's theorem [I, 3.45], he concludes that

$$
\bar{N} \cong .2.
\tag{5.214}
$$

Next, using Goldschmidt's results on strongly closed abelian 2-groups [I, 4.128], Stroth argues that $U \lhd C_z$, whence

$$
N = C_z.
\tag{5.215}
$$

In particular, it follows that a Sylow 2-subgroup of $G$ has order $2^{41}$.

Stroth also proves that $G$ has exactly *four* conjugacy classes of involutions and determines the orders of their centralizers, so that as with Hunt, he is now in a position to apply the Thompson order formula to evaluate $|G|$. This time the initial analysis yields six possibilities. However, in contrast to Hunt, Stroth eliminates the five incorrect solutions solely by a combination of further *odd* local analysis and Sylow's theorem. Thus, ultimately he proves

$$
|G| = |F_2|.
\tag{5.216}
$$

Moreover, this analysis provides him with considerable information about the conjugacy class $D$ containing $x$ and the action of $C_x$ on $D$:

$$
\begin{align}
&\text{(1) Three } C_x\text{-orbits of } D \text{ are contained} \\
&\quad \text{in } C_x; \text{ and} \\
&\text{(2) } C_z \text{ contains an element } d \in D \text{ such that} \\
&\quad (xd)^2 = z.
\end{align}
\tag{5.217}
$$

[Note that if $d$ is in one of the orbits of (1), then $|xd| \leqslant 2$, while if $d$ is as in (2), then $|xd| = 4$.]

Stroth now argues that $x$ inverts an element $y$ of order 3 such that $C_y$ has a normal subgroup of index 2 isomorphic to $M(22)/Z_3$. In particular, he uses Fendel's theorem [I, 3.48] to identify $C_y/\langle y \rangle$.

This in turn implies

> There exists an element $d \in D$ with $d \in C_y\langle x \rangle$
> and $xd = y$.                                                            (5.218)

(whence $|xd| = 3$).

In particular, it follows that $C_x$ possesses at least 5 orbits in its action on $D$. On the other hand, counting involutions, he argues that the reverse inequality also holds, thus proving that $C_x$ has precisely 5 orbits on $D$. But now (5.217) and (5.218) yield

> If $d \in D$, then $|xd| = 1, 2, 3,$ or 4.                                (5.219)

Thus $G$ is indeed a $\{3, 4\}$-transposition group and Theorem 5.149 gives the desired conclusion $G \cong F_2$.

## 5.12. The Sporadic Cases

In this final section we give a similarly abbreviated outline of the cases in which $L/O(L)$ is a sporadic group (excluding $J_1$, $Mc$, and $M_{23}$, which have been previously treated).

We prefer to first discuss those cases which possess simple solutions for $F^*(G)$ (with $R$ cyclic): namely,

$$L/O(L) \cong M_{11}, M_{12}, J_2, HS/Z_2, M(22)/Z_2, M(23), \text{ or } F_2/Z_2; \qquad (5.220)$$

and correspondingly

$$F^*(G) \cong Mc, .3, Suz, F_5, M(23) \text{ or } M(24)', M(24)', \text{ or } F_1. \qquad (5.221)$$

However, in carrying out the analysis, there are instances in which one considers an involution $y$ of $L$ (with $y \neq x$), a component $I$ of $L(C_L(y))$, and proceeds to pump-up $I$ in the usual fashion until one reaches a second standard component $K$ in $G$. In this process, it may well happen that $K$ is a sporadic group not listed in (5.220) (nor $J_1$, $Mc$, or $M_{23}$)—for example, if $L \cong M_{12}$, then $I \cong A_5$, which can pump-up to $J_2$. Hence, as $Suz$ is a pump-up of $J_2$, a possibility for $K$ in this case is $K \cong Suz$.

For expository reasons, we shall therefore discuss the cases of (5.220) under the following assumption:

> If $F^*(G)$ is simple with $K$ standard in $G$
> and $K/Z(K)$ is sporadic, then either $K/O(K) \cong J_1$                    (5.222)
> or is isomorphic to one of the groups of (5.220).

We make one further simplifying assumption. If $O(C_x) \neq 1$ [whence $L/O(L)$ is one of the groups of the standard component theorem (I)], we let $G$ be a minimal counterexample to the proposed theorem and consider $M = N_G(O(C_x))$, $\overline{M} = M/O(M)$. On the basis of the initial fusion and 2-local analysis, one can show that $\overline{L} < \overline{M}$, so that $\overline{M}$ is determined by the minimality of $G$. [If we are in a case in which there is no solution for $F^*(G)$, then necessarily $F^*(\overline{M})$ is not quasisimple and we have the wreathed solution for $\overline{M}$.] If $F^*(\overline{M})$ is quasisimple, one argues as in previous examples that $M$ is strongly embedded—contradiction. In the contrary case, one pumps up a component of $M$ until one reaches a second standard component $K$, which will necessarily have greater order than $L$, so that if one is proceeding in descending order of $L$, the resulting standard form problem will already have been treated.

Thus we assume throughout this section:

$$O(C_x) = 1. \tag{5.223}$$

We begin with the case $L \cong J_2$, treated by Finkelstein [25]. However, as $J_2$ and $J_3$ have isomorphic Sylow 2-subgroups and, in fact, isomorphic centralizers of 2-central involutions ($\cong A_5/Q_8 * D_8$), it was natural for Finkelstein to handle the two problems together.

PROPOSITION 5.150. *If $L \cong J_2$ or $J_3$, then $F^*(G) \cong Suz$ (and $L \cong J_2$).*

Finkelstein easily argues that $R = \langle x \rangle$. Since $\mathrm{Out}(J_2) \cong \mathrm{Out}(J_3) \cong Z_2$, there are then two cases to consider according as $C_x/\langle x \rangle \cong L$ or $\mathrm{Aut}(L)$.

Suppose first that $C_x = L \times \langle x \rangle$. If $U$ denotes a $Q_8 * D_8$ subgroup of $L$ and $U_1$ a Sylow 2-subgroup of $UC_U$, Finkelstein first proves that

$$U_1 = U * U_0, \text{ where } U_0 \cong D_8. \tag{5.224}$$

Setting $N_1 = N_G(U_1)$, he next argues that

(1) $N_1/U_1 \cong A_5/E_{16}$; and
(2) $N_1$ contains a normal subgroup $W = W_1 \times W_2$,                (5.225)
where $W_1 \cong W_2 \cong Q_8 * D_8$.

Finally, setting $M = N_G(W)$, he proves

(1) $M/W \cong A_5 \int Z_2$; and
(2) $M$ contains a Sylow 2-subgroup of $G$.                                    (5.226)

The proofs of (5.224), (5.225), and (5.226) involve a delicate analysis of the fusion of involutions of $G$. Control of this fusion depends on a prior determination of the centralizer in $G$ of $D \in \mathrm{Syl}_3(C_L(Z(U)))$:

$$C_D/O(C_D) \cong A_6 \int Z_2.$$                                             (5.227)

[Note that by properties of $J_2$ and $J_3$, $C_L(D) \cong Z_3 \times A_6$.]
In particular, this fusion analysis yields:

(1) $G$ contains a normal subgroup $G_0$ with $x \notin G_0$; and
(2) $M \cap G_0 = M_1 \times M_2$, where $M_1 \cong M_2 \cong A_5/Q_8 * D_8$   (5.228)
and $M_1, M_2$ are interchanged by $x$.

Let $S_0 = S_1 \times S_2 \in \mathrm{Syl}_2(G_0)$ with $S_i \in \mathrm{Syl}_2(M_i)$, $i = 1, 2$. The final step is to show that $S_0$ has product fusion with respect to the given decomposition, in which case Goldschmidt's theorem [I, 4.148] will imply that $O^{2'}(G_0) = L_1 \times L_2$, where $L_1$ and $L_2$ are interchanged by $x$. But then $L_1 \cong L_2 \cong L$, and we have the wreathed solution. Again the argument involves delicate fusion analysis. Furthermore, because $J_2$ and $J_3$ have different involution fusion patterns, Finkelstein is forced to treat the two possibilities for $L$ separately.

Thus it remains to treat the case $C_x/\langle x \rangle \cong \mathrm{Aut}(L)$. It is not difficult to show that $C_x$ splits over $\langle x \rangle$, so that

$$C_x \cong \mathrm{Aut}(L) \times Z_2.$$                                      (5.229)

With $U, U_1, U_0, N_1, M$, and $D$ as above, this time Finkelstein's analysis yields

(1) $U_0$ is dihedral of order 8 or 16; and
(2) correspondingly, $L(C_D/O(C_D)) \cong A_6 \times A_6$        (5.230)
    or $U_4(3)$.

If $|U_0| = 8$, he closely follows the argument of the $C_x = L \times \langle x \rangle$ case, first producing a normal subgroup $W$ of $N_G(U_1)$ of the same shape as before and then analyzing $M = N_G(W)$. From the resulting structure of $M$, he concludes that $G$ possesses a normal subgroup $G_0$ of index 2 with $x \notin G_0$ and

$$C_{G_0}(x) = L \times \langle x \rangle. \tag{5.231}$$

Hence by the results of the previous case, $F^*(G) = F^*(G_0) = L_1 \times L_2$, where $L_1 \cong L_2 \cong L$, and again we have the wreathed solution.

On the other hand, when $|U_0| \cong 16$, Finkelstein argues first that

$$N_1/U_1 \cong \Sigma_6. \tag{5.232}$$

Now considering an appropriate subgroup $W_1$ of index 2 in $U_1$ with $U \leqslant W_1 \cong Q_8 * D_8 * D_8 (\cong Q_8 * Q_8 * Q_8)$ and setting $M_1 = N_G(W_1)$, he proves next that

(1) $M_1/W_1 \cong O_6^-(2)$; and
(2) $M_1$ contains a Sylow 2-subgroup of $G$.        (5.233)

This in turn yields

(1) $G$ contains a subgroup $G_0$ of index 2
    with $x \notin G_0$; and        (5.234)
(2) $M_0 = M \cap G_0 \cong P\Omega_6^-(2)$

If $z$ denotes the 2-central involution of $M_0$, Finkelstein's final step is to prove

$$W_1 \text{ is normal in } C_{G_0}(z). \tag{5.235}$$

Again all these results involve elaborate involution fusion analyses. In particular, $L \cong J_2$ is forced along the way. Moreover, the proof of (5.235) utilizes, as usual, Goldschmidt's strongly closed Abelian theorem [I, 4.128] to prove that $W_1/\langle z \rangle \lhd C_{G_0}(z)/\langle z \rangle$.

By (5.235), $M_0 = C_{G_0}(z)$ and so by (5.234),

$$C_{G_0}(z) \cong P\Omega_6^-(2)/Q_8 * Q_8 * Q_8,$$

the precise centralizer of a 2-central involution of $Suz$. Now the Patterson–Wong theorem [95] yields that $G_0 \cong Suz$ and Proposition 5.150 is proved.

PROPOSITION 5.151.  *If  $L \cong M_{11}$  or  $M_{12}$,  then  correspondingly $F^*(G) \cong Mc$ or .3.*

In the first case, one argues immediately that $R = \langle x \rangle$, and as $\mathrm{Aut}(M_{11}) = M_{11}$, it follows that $C_x$ contains an $E_8$ subgroup $V$ with $x \in V$ such that $V \in \mathrm{Syl}_2(C_v)$. Hence by Harada's theorem (Theorem 2.156), $G$ has sectional 2-rank $\leqslant 4$. Checking the simple groups of sectional 2-rank $\leqslant 4$, admitting an automorphism of period 2 whose centralizer has an $M_{11}$ component, we conclude at once that $F^*(G) \cong Mc$.

In the $M_{12}$ case, the goal is to reduce to the single possibility $C_x \cong M_{12} \times Z_2$ in which case Yoshida's theorem [142] will yield that $F^*(G) \cong .3$.

The reduction was carried out by Finkelstein and Solomon [29]. The usual fusion analysis forces $R = \langle x \rangle$, so that they can therefore assume that $C_x/\langle x \rangle \cong \mathrm{Aut}(M_{12})$ (otherwise $C_x \cong M_{12} \times Z_2$, as required). Likewise the fusion analysis implies that $C_x$ splits over $\langle x \rangle$, whence, in particular, $x$ is not a square in $C_x$.

They now consider a non-2-central involution $y$ of $L$ [whence $I = L(C_L(y)) \cong A_5$] and the pump-up $J$ of $I$ in $L_y$. If $J_1$ denotes a component of $J$, they then pump-up $J_1$ until they reach a second standard component $K$ of $G$. Using the known structure of the centralizers of involutions on quasisimple $K$-groups, one checks that $K/Z(K)$ is either an alternating group of odd degree, a group of Lie type of characteristic 5, $L_2(4^m)$, $U_3(4^m)$, $L_3(4^m)$, $M_{12}$, $J_1$, $J_2$, $Ly$, $ON$, $He$, $Suz$, or .3.

However, apart from the case in which $K/O(K) \cong M_{12}$, each of the resulting standard form problems either has been previously solved (with solutions incompatible with the structure of $C_x$) or else contradicts (5.222).

Thus we have

$$K \cong M_{12}. \qquad (5.236)$$

Furthermore, by Aschbacher and Seitz's reduction theorem, it also follows that $m(C_K) = 1$. This latter condition implies in turn that $J$ is a single component, so that $J = J_1$.

We see then that there are precisely two possibilities:

$$(1) \ J = K; \text{ or} \\ (2) \ J = I \text{ and } L = K. \qquad (5.237)$$

In the first case, $C_y$ has the same general structure as $C_x$. In particular, $\langle y \rangle \in \text{Syl}_2(C_J)$ and $y$ is not a square in $C_y$ [this is obvious if $C_y/O(C_y) \cong M_{12} \times Z_2$]. However, as $C_x \cong \text{Aut}(M_{12}) \times Z_2$, it is immediate that $y$ is a square in $C_x$ and hence in $C_y$—contradiction.

In the second case, for any involution $a \in I$, if we set $A = \langle a, y \rangle$, it follows from the structure of $M_{12}$ that $L = \langle L(C_L(a')) | a' \in A^{\#} \rangle$. Using this fact and an analysis of $A_5$ components of $C_{a'}$ for $a' \in A^{\#}$, Finkelstein and Solomon argue that for any such four-subgroup $A$,

$$N_G(A) \leqslant N_G(L). \qquad (5.238)$$

With this information they quickly derive a fusion contradiction. Hence $C_x = L \times \langle x \rangle$, as desired.

We briefly describe Yoshida's analysis of this case. First, with $y$, $I$, $J$, and $K$, as above, the same argument again forces $J = K \cong M_{12}$. With this information, if $U$ denotes an $E_8$-subgroup of $L$, $V = U \times \langle x \rangle$ and $N = N_G(U)$, then by considering $N \cap L$ and $N \cap J$, it is not difficult to show that

$$N \text{ is the semidirect product of a normal} \\ \text{subgroup } W \cong D_8 * D_8 * D_8 \text{ by } \Sigma_4. \qquad (5.239)$$

Setting $M = N_G(W)$, Yoshida argues next that

$$(1) \ M/W \cong L_3(2); \text{ and} \\ (2) \ M \text{ contains a Sylow 2-subgroup of } G. \qquad (5.240)$$

Finally if $z$ is the central involution of $W$, he analyzes $\tilde{C}_z = C_z/\langle z \rangle$, arguing first that $\tilde{C}_z$ is simple. But by the structure of $M$, $\tilde{C}_z$ has Sylow 2-subgroups of type $A_{12}$. Since $\tilde{W} \cong E_{64}$ with $N_{\tilde{C}_z}(\tilde{W})/\tilde{W} \cong L_3(2)$, Solomon's theorem (Theorem 2.72) yields that $\tilde{C}_z \cong PSp_6(2)$. Thus

$$C_z \cong PSp_6(2)/Z_2, \tag{5.241}$$

and now Yoshida concludes from Fendel's theorem [I, 3.48] that $G \cong .3$.

PROPOSITION 5.152. If $L \cong HS/Z_2$, then $F^*(G) \cong F_5$.

As usual, the goal is to force $R = \langle x \rangle$, in which case Harada's characterization of $F_5$ [I: 163] will yield the desired conclusion. However, assuming false, one cannot derive an immediate contradiction from Finkelstein's result (Proposition 5.3) since the centralizer of a 2-central involution of $HS$ does contain a cyclic normal subgroup of order 4. On the contrary, in the present case a modification of Finkelstein's argument yields the second possibility

$$R \cong Z_4. \tag{5.242}$$

To rule it out, Solomon [113] considers a subgroup $U$ of $L$ with $U \cong Z_4 \times Z_4 \times Z_4 \times Z_2$, sets $V = UR = U * R \cong Z_4 \times Z_4 \times Z_4 \times Z_4$ and analyzes the structure of $N = N_G(V)$. [Setting $\bar{N} = N/V$, and $N_0 = C_N(x)$, we know from the structure of $L$ that $\bar{N}_0 \cong L_3(2)$.] Solomon proves that $N$ contains a Sylow 2-subgroup $S$ of $G$ and determines a good deal about the structure of $S$ and the $G$-fusion of its elements. Ultimately he argues by transfer that

$$\begin{array}{l} G \text{ contains a normal subgroup } G_0 \text{ of index 2} \\ \text{with } R \nleq G_0. \end{array} \tag{5.243}$$

Hence $\langle x \rangle \in \mathrm{Syl}_2(C_{G_0}(L))$ and so Harada's theorem applies in $G_0$ to yield $G_0 = F^*(G) \cong F_5$.

Harada's analysis of the $R = \langle x \rangle$ case is an elaborate application of the centralizer of involution method, following the same general pattern as in all standard form problems. Thus starting from the given centralizer, he determines the fusion of involutions in $G$ and a considerable amount of the local structure. In particular, $x$ is not 2-central, $G$ has exactly two conjugacy classes of involutions, and if $z$ is a 2-central involution of $G$,

$$C_z \cong (A_5 \int Z_2)/Q_8 * Q_8 * Q_8 * Q_8, \tag{5.244}$$

this last result, as usual requiring Goldschmidt's strongly closed abelian theorem.

Furthermore, his analysis yields the existence of a *nonlocal* subgroup $K$ of $G$ with

$$K \cong A_{12}, \tag{5.245}$$

which he identifies from the standard presentation of $A_{12}$ by generators and relations. With the help of the subgroup $K$, he next determines the orders of the Sylow $p$-subgroups of $G$ for $p = 3, 5, 7, 11$. Now he has sufficient information to use the Thompson order formula to prove

$$|G| = 2^{14} \cdot 3^6 \cdot 5^6 \cdot 7 \cdot 11 \cdot 19. \tag{5.246}$$

Harada then goes on to determine representatives of all the conjugacy classes (54) of elements of $G$ and that

The minimal degree of a faithful irreducible
representation of $G$ is 133.
$$\tag{5.247}$$

[Together with Conway, Curtis, Norton, and P. Smith, Harada has determined the full character table of $G$. The analysis involves some computer calculations.]

Existence of a group of "type" $F_5$ can now be obtained as a subgroup of $F_1$ (as a direct factor in the centralizer of a suitable element of $F_1$ of order 5; see [I, p. 93]). However, Norton [89] constructed $F_5$ prior to Griess's proof of the existence of $F_1$ as a complex linear group of degree 133, by the same general method that Conway and Wales [I: 71] used for their construction of $Ru$—namely, working with a complex vector space $V$ of dimension 133, Norton determines a suitable set of vectors in $V$, which with the aid of computer calculations, is shown to be left invariant by the given group $G$ of type $F_5$.

Likewise, using the existence of an $A_{12}$ subgroup $K$ in a group of type $F_5$, and working with the permutation representation of $G$ on the cosets of $K$, Norton proves the uniqueness of such a group $G$.

PROPOSITION 5.153. *If* $L \cong M(22)/Z_2$ *or* $M(23)$, *then* $F^*(G) \cong M(23)$ *or* $M(24)'$.

[As noted in Table 5.12, $M(24)'$ is a solution in both cases.]

The arguments resemble those of the $U_6(2)/Z_2$ case (Proposition 5.146). In both cases, one easily forces $R = \langle x \rangle$. Suppose first that $L \cong M(22)/Z_2$. Hunt [67] treated the case $C_x/\langle x \rangle \cong M(22)$ and the combined work of Davis and Solomon [18] and Parrott [93] covers the case $C_x/\langle x \rangle \cong \text{Aut}(M(22))$.

This time there is $U \leqslant L$ with $U \cong E_{2^{11}}$ and if $N$, $\bar{N}$, and $N_0$ have the usual meanings, then according to the two possibilities for $C_x/\langle x \rangle$, one has

$$\bar{N}_0 \cong M_{22} \text{ or } \text{Aut}(M_{22}). \tag{5.248}$$

Correspondingly the fusion analysis yields

$$\bar{N} \cong M_{23} \text{ or } M_{24}. \tag{5.249}$$

The first possibility is obtained as in Proposition 5.146, by forcing $\bar{N}$ to act as an automorphism group of the Steiner triple system $S(4, 7, 23)$, the unique transitive extension of the Steiner system $S(3, 6, 22)$ of $\bar{N}_0$. In the second case, assuming $G$ is a minimal counterexample (whence $G$ has no normal subgroups of index 2), Davis and Solomon force $|\bar{N}| \cong |M_{24}|$ and then quote Stanton's theorem [I, 3.44] to conclude that $\bar{N} \cong M_{24}$.

If $\bar{N} \cong M_{23}$, Hunt proceeds as in the earlier $M(22)$ case, arguing that $G$ has exactly three classes of involutions, then computing from the Thompson order formula that $|G| = |M(23)|$, and finally arguing that $C_x$ has exactly three orbits on the conjugacy class $D$ containing $x$. Thus, again $G$ is a 3-transposition group and Fischer's theorem yields the desired conclusion $G \cong M(23)$.

On the other hand, if $\bar{N} \cong M_{24}$, Davis and Solomon show that $N$ contains a subgroup $W \cong (Q_8)^6$ such that $N_N(W)/W$ possesses a normal subgroup of index 2 isomorphic to $A_6/Z_3$. Set $M = N_G(W)$. Since $W/Z(W)$ can be identified with a 12-dimensional orthogonal space of "+ type", $M/W$ is isomorphic to a subgroup of $O_{12}^+(2)$. With this information, Davis and Solomon prove by further local analysis that

$$\begin{array}{ll} (1) & O^2(M)/W \cong U_4(3)/Z_2; \text{ and} \\ (2) & M/O^2(M) \cong Z_2 \text{ or } Z_2 \times Z_2. \end{array} \tag{5.250}$$

Let $z$ be the involution of $Z(W)$. The next step in the argument is to prove that

$$C_z = M. \tag{5.251}$$

To simplify their analysis, Davis and Solomon verify (5.251) only under the assumption that $C_z$ is a *K-group*. However, it is clear that with some additional effort, they could have avoided this assumption by appealing instead, as usual, to Goldschmidt's strongly closed abelian 2-group theorem.

Thus the structure of $C_z$ is given by (5.250), which are precisely the conditions under which Parrott [93] has characterized the group $M(24)'$. Indeed, if $M/O^2(M) \cong Z_2 \times Z_2$, he uses Thompson transfer to show that $G$ has a normal subgroup of index 2, contrary to the minimality of $G$. Hence $M/O^2(M) \cong Z_2$. In this case he argues that $G$ has exactly two conjugacy classes of involutions (represented by $x$ and $z$) and once again uses the Thompson order formula to show that

$$|G| = |M(24)|'. \tag{5.252}$$

Since the class of 3-transpositions of $M(24)$ induce *outer* automorphisms on $M(24)'$, this class is not "visible" in the group $G$. Hence Parrott must use a more roundabout procedure to identify $G$, involving a proof of the uniqueness of a certain graph associated with $G$. The first step is to prove that

$$G \text{ possesses a subgroup } X \cong M(23). \tag{5.253}$$

The point is that $N$ contains a subgroup $N_1$ with $U \leqslant N$ and $N_1/U \cong M_{23}$, and it is possible to prove the existence of a 3-transposition group $X$ in $G$ with $N_1 < X$ and then identify $X$ from Fischer's theorem as $M(23)$.

Parrott next proves

(1) $G$ is a rank 3 transitive permutation group on the right cosets of $X$ with subdegrees 31,671 and 275,264; and
(2) If $X_1$ and $X_2$ denote one-point stabilizers of $X$ on the corresponding orbits, then we have $X_1 \cong M(22)/Z_2$ and $X_2' \cong P\Omega_8^+(3)$ with $X/X' \cong \Sigma_3$. $\qquad$ (5.254)

Now consider the graph $\Lambda(G)$ whose vertex set consists of the right cosets of $X$ with two distinct vertices connected by an edge if and only if they intersect in a subgroup isomorphic to $X_1$ ($\cong M(22)/Z_2$). Parrott proves

$$\Lambda(G) \text{ is uniquely determined up to isomorphism.} \tag{5.255}$$

More specifically, he proves that $\Lambda(G)$ is equivalent to a certain abstractly defined "extension" $\mathscr{D}*$ of the commuting graph $\mathscr{D}$ associated with the class $D$ of 3-transpositions of $X$. From its construction, one sees that the group $X$ acts on $\mathscr{D}*$, so that $X \leqslant \text{Aut}(\mathscr{D}*)$. In [I: 99, 100], Fischer proves that $\mathscr{D}*$ has a *transitive* automorphism group and that

$$\text{Aut}(\mathscr{D}*) \cong M(24). \tag{5.256}$$

Since $\mathscr{D}*$ is isomorphic to $\Lambda(G)$ and $G \leqslant \text{Aut}\big(\Lambda(G)\big)$, $G$ is thus isomorphic to a subgroup of $M(24)$. But it is also easy to see that $G$ is simple, so that $G \leqslant M(24)'$, whence $G \cong M(24)'$ by (5.252).

Finally consider the case $L \cong M(23)$, which Solomon studies in [114]. Then $L$ contains an $E_{2^{11}}$-subgroup $U$ and we set $V = U\langle x \rangle$, $N = N_G(V)$, $\bar{N} = N/V$, and $N_0 = C_N(x)$. By the structure of $M(23)$,

$$\bar{N}_0 \cong M_{23}, \tag{5.257}$$

and Solomon argues by the usual fusion analysis that again

$$\bar{N} \cong M_{24}. \tag{5.258}$$

Thus the situation is very similar to that of the preceding case [(5.248) and (5.249)], except that now $|O_2(N)| = 2^{12}$, whereas there $|O_2(N)| = 2^{11}$. In the present situation, it is not difficult to show from the action of $\bar{N}$ on $V$ that $N$ possesses a normal subgroup of index 2 not containing $x$ and that $V$ is weakly closed with respect to $G$ in a Sylow 2-subgroup $S$ of $N$ [whence $S \in \text{Syl}_2(G)$]. Since $V$ is abelian, Yoshida's transfer theorem [136] again yields that $G$ has a normal subgroup $G_0$ of index 2 not containing $x$. In particular, $G$ is not simple.

Solomon stops at this point [which suffices for the standard component theorem (II)]. However, as the structure of $G_0 \cap N$ is isomorphic to that of the subgroup $N$ in (5.249), he could clearly have continued as before considering $W \cong (Q_8)^6$ with $W \leqslant G_0 \cap N$ and determining the structure of $C = C_{G_0}\big(Z(W)\big)$ by the same general procedure $[O^2(C)/W \cong U_4(3)/Z_3$ and $C/O^2(C) \cong Z_2]$, in which case Parrott's theorem would again yield $F^*(G) \cong M(24)'$.

Finally we have the following result:

PROPOSITION 5.154. *If* $L \cong F_2/Z_2$, *then* $G \cong F_1$.

The aim of Davis and Solomon's analysis [18] is to pin down the structure of the centralizer of a 2-central involution $z$ of $G$:

$$C_z \cong .1/(Q_8)^{12}, \qquad (5.259)$$

in which case Griess's and Norton's theorems [I, 2.89] will yield the desired conclusion. However, in their paper they again determine $C_z$ only under the assumption that $C_z$ is a $K$-group, which could undoubtedly have been avoided by once again using Goldschmidt's strongly closed abelian 2-group theorem.

By the structure of $F_2$, $L$ possesses a 2-constrained maximal 2-local subgroup $P$ containing $Q = T \cap L$ with

$$P \cong .2/[(D_8)^{11} \times Z_2]. \qquad (5.260)$$

Davis and Solomon argue first that

$$W = O_2(N_G(O_2(P))) \cong (D_8)^{12} \cong (Q_8)^{12}. \qquad (5.261)$$

Because of their $K$-group assumption, they can now move directly to an analysis of $C_z$, where $\langle z \rangle = Z(W)$. However, to avoid it, one would first have to determine the structure of $M = N_G(W)$: namely,

$$M/W \cong .1. \qquad (5.262)$$

Proving that $W = O_2(M)$ is not difficult, following once one knows the structure of $C_D$, where $D$ is a subgroup of $P$ of order 3 such that $C_P(D)/O_2(C_P(D)) \cong Z_3 \times \mathrm{Aut}(U_4(2))$ and $C_L(D)/D$ contains a normal subgroup of index 2 isomorphic to $M(22)/Z_2$. ($F_2/Z_2$ possesses such a subgroup $D$.) Using the structure of $C_L(D)$, Davis and Solomon argue that $C_D$ contains a standard component $K \cong M(22)/Z_2$ with $N_{C_D}(K)/C_{C_D}(K)) \cong \mathrm{Aut}(M(22))$, and $K$ not normal in $C_D$. It follows therefore from the previous proposition that

$$C_D/D \cong M(24)'. \qquad (5.263)$$

Thus $C_D$ and hence $W$ are determined.

On the other hand, to prove that $\bar{M} = M/W \cong .1$, one would presumably begin with a 2-central involution $\bar{t}$ of $\bar{P}$ ($\cong .2$) and argue that $C_{\bar{M}}(\bar{t}) \cong P\Omega_8^+(2)/(Q_8)^4$, in which case Patterson's theorem [I.3.48] would yield the desired conclusion.

Finally Goldschmidt's theorem would enable one to conclude 'that $W \lhd C_z$, whence $M = C_z$, thus verifying (5.259).

We have now covered all sporadic cases in which there exists a solution with $F^*(G)$ simple (plus the $M_{23}$, $J_1$, and $Mc$ cases). We let $\mathscr{S}_0$ denote the set of remaining possibilities for $L$. We divide $\mathscr{S}_0$ into three subsets according as

A. $L$ is simple of component type;
B. $L$ is simple of characteristic 2 type;
C. $L$ is nonsimple.

The possibilities for $L \in \mathscr{S}_0$ under (A) are covered by the following result.

PROPOSITION 5.155. *We have* $L \ntrianglelefteq Ly$, *HS*, *He*, *Suz*, *ON*, *.3*, *Ru*, $M(22)$, $M(24)'$, $F_2$, *or* $F_1$.

Most of these cases were treated by Solomon [113, 114 and I: 264] by an essentially uniform method. The general line of argument has already been described in Proposition 5.151 in discussing the $M_{12}$ case. (Moreover, he also covers the $Mc$ case in [I: 264], which as we have seen, was independently treated by Finkelstein. Also the $He$ case was done jointly with Griess [48].)

Let $y$ be an involution of $L$ such that $I = L\big(C_L(y)\big) \neq 1$ and let $J$ be the pump-up of $I$ in $L_y$. As in the $M_{12}$ case, one pumps-up a component $J_1$ of $J$ until one reaches a second standard component $K$ of $G$. In view of all our previous solutions of standard form problems together with the minimality of $G$, it follows that each successive component in the pumping-up process is a $K$-group. In particular, $K$ is a $K$-group. But then by Aschbacher and Seitz, we again have $m(C_K) = 1$. Likewise if $K/O(K) \notin \mathscr{S}_0$, then $F^*(G)$ is determined by our previous results. However, in none of these cases does $G$ contain an involution $x$ whose centralizer has a component isomorphic to $L$. Hence $K/O(K) \in \mathscr{S}_0$. Furthermore, examining the structure of $I$ in each case, we check that these conditions again force $J_1 = J$ and one of the following:

$$(1) \ J = K \text{ with } K/O(K) \cong L; \text{ or}$$
$$(2) \ J = I \text{ and } L = K. \tag{5.264}$$

Note that if $y \in I$ (that is, $I$ is nonsimple), the first possibility is excluded. This occurs for $L \cong Ly$, $He$, $ON$, $.3$, $M(22)$, $M(24)'$, $F_5$, $F_2$, or $F_1$ (with correspondingly $I \cong \hat{A}_{11}$, $L_3(4)/Z_2 \times Z_2$, $L_3(4)/Z_4$, $U_6(2)/Z_2$, $M(22)/Z_2$, $HS/Z_2$, $^2E_6(2)/Z_2$, or $F_2/Z_2$).

In the first case, taking $U$ to be an appropriate elementary abelian 2-subgroup of $L$ of maximal rank containing $y$ and setting $V = U \times \langle x \rangle$, one can determine the precise embedding of $V \cap K$ in $K$. But then setting $N = N_G(V)$ and $\bar{N} = N/V$, one knows the structure of both $N \cap L$ and $N \cap K$, which gives strong information on the structure of $\bar{N}$ and of the consequent fusion of $x$ in $V$, which can be shown to lead to a contradiction. For example, in most cases it follows that $x \in C_N(x)'$. However, by the structure of $\bar{N}$, $R = \langle x \rangle$ is forced and, as usual, a fusion argument then yields that $C_x$ splits over $\langle x \rangle$, whence $x \notin (C_x)'$—contradiction. (This is similar to the contradiction derived in the previous $M_{12}$ case.)

On the other hand, in the second case the argument is again similar to that of the $M_{12}$ case. Indeed, one argues that $N_G(L)$ controls a considerable amount of the $G$-fusion of involutions of $T$, which is then shown to be incompatible with previously derived fusion information about the $G$-conjugates of $x$ contained in $T$, forced by the $Z^*$-theorem.

The next result covers the possibilities for $L$ under (B).

PROPOSITION 5.156. *We have $L \not\cong M_{22}$, $M_{24}$, $.2$, $J_4$, or $F_3$.*

In all cases, the analysis involves pushing-up 2-locals and either forcing the wreathed solution or a contradiction.

Note first note that $M_{22}$ and $M_{23}$ have isomorphic Sylow 2-subgroups (as $M_{22} \leqslant M_{23}$ with index 23). Hence Finkelstein's analysis, which we have described in Section 5.7, covers the case that $C_x/R \cong M_{22}$, and so when $L \cong M_{22}$, it suffices for him to consider the subcase $C_x/R \cong \mathrm{Aut}(M_{22})$ [26]. The proof follows the $M_{23}$ analysis very closely. He again produces an $E_{2_8}$-subgroup $W$ of $G$ and determines the possibilities for the structure of $M = N_G(W)$ and $\bar{M} = M/W$. In the present situation, there turn out to be two possibilities:

$$L(\bar{M}) \cong A_6 \times A_6 \text{ or } PSp_4(4). \tag{5.265}$$

In the first case, Goldschmidt's general strong closure results force, as before, $F^*(G) \cong M_{22} \times M_{22}$, while in the second case a very delicate comparison of two further 2-local subgroups, one pushed up from $M$, the other from $C_x$, is needed to reach a contradiction.

Similarly when $L \cong M_{24}$, Egawa [21] immediately forces $R = \langle x \rangle$ and then analyzes $N = N_G(V)$, where $V = U \times \langle x \rangle$ with $E_{64} \cong U \leqslant L$ and $N_L(U)/U \cong \Sigma_6/Z_3$ ($L$ contains such a subgroup $U$), arguing that $N$ possesses a normal subgroup $W \cong E_{2^{10}}$. Setting $M = N_G(W)$ and $\bar{M} = M/W$, he now forces the unique possibility

$$L(\bar{M}) \cong (A_6/Z_3) \times (A_6/Z_3), \tag{5.266}$$

which leads, as usual, to the wreathed solution.

The remaining three cases, treated, respectively, by Manfredelli [79], Finkelstein [28], and Syskin [120], are entirely similar. With $U$, $V$, and $N$, as above, then according as $L \cong .2$, $J_4$, or $F_3$, they correspondingly take

$$\begin{array}{l} (1) \ \ U \cong E_{2^{10}}, E_{2^{11}}, \text{ or } E_{2^{10}}; \text{ and} \\ (2) \ \ N_L(U)/U \cong \mathrm{Aut}(M_{22}), M_{24}, \text{ or } L_5(2), \end{array} \tag{5.267}$$

arguing first that $N$ contains an elementary normal subgroup $W$ of order $|U|^2$ and then analyzing the structure of $M = N_G(W)$, first determining the possibilities for $L(M/W)$.

Finally we discuss the sporadic cases under (C).

PROPOSITION 5.157. *We have* $L \ncong M_{12}/Z_2$, $M_{22}/Z_2$, $M_{22}/Z_4$, $J_2/Z_2$, $Suz/Z_2$, $Ru/Z_2$, *or* $.0 \ (\cong .1/Z_2)$.

The pattern of argument is essentially the same in all cases, the goal being to derive a contradiction to the fusion of $x$ in $C_x$. Furthermore, the overall arguments are considerably shorter than in previous cases because large portions depend only on properties of $L$.

As usual, one argues first that $R = \langle x \rangle$, except in the case of $M_{22}/Z_4$, where instead Griess [46] proves that $R = Z(L) \cong Z_4$. Since $|\mathrm{Out}(L)| \leqslant 2$ by [I, 4.239], it follows that $|C_x/L| \leqslant 2$. Now the $Z^*$-theorem implies that $t = x^g \in T - \langle x \rangle$ for some $g \in G$. There are then two cases to consider according as $t \in L$ or $t \notin L$.

The first case is eliminated either directly from the structure of a Sylow 2-subgroup of $L$ or combined with a short fusion argument. For example, in the case of $Ru/Z_2$ [whence $C_x = L$ as $\mathrm{Out}(Ru) = 1$], every involution of $L - \langle x \rangle$ is a square, so $t$ and hence $x$ is a square. However, $x$ is not a square in $L$—contradiction.

On the other hand, in the $.0$ case (where again $C_x = L$), Finkelstein

[27] rules out all but a single possible conjugacy class of $L$ for $t$ by the same type of argument. Moreover, in that case $\langle t, x \rangle \leqslant U \cong E_{2^{12}}$ with $N_L(U)/U \cong M_{24}$, and he argues that $g$ can be chosen to normalize $U$. But now from the action of $N = N_G(U)$ on $U^{\#}$, it follows that $N$ has an orbit of length $2{,}576 = 3 \cdot 859$, whence $\bar{N} = N/U$ has order divisible by 859. However, this is impossible as $N \leqslant GL_{12}(2)$, whose order is not divisible by the prime 859.

Similarly in the $M_{12}/Z_2$ and $J_2/Z_2$ cases, Finkelstein and Solomon [29] show that there is only one possibility for $t$: namely, $t \in Z(T)$ [whence $\langle t, x \rangle = Z(T) \cong Z_2 \times Z_2$]. An elementary argument of Burnside implies now that $g$ can be taken to lie in $N_G(T)$. However, Finkelstein and Solomon show in either case that $\mathrm{Aut}(T)$ acts trivially on $Z(T)$. [Note that $\mathrm{Out}(M_{12}) \cong \mathrm{Out}(J_2) \cong Z_2$, so that they must establish this fact both when $T \in \mathrm{Syl}_2(L)$ and when $|T : T \cap L| = 2$.]

Once the subcase $t \in L$ is eliminated, it completes the proof in those cases in which $\mathrm{Out}(L) = 1$ (namely, $L \cong Ru/Z_2$ or $.0$) and also when $L \cong M_{22}/Z_4$ (one shows here that $g$ can be chosen so that $R^g \leqslant C_x$, whence $t$ is a square in $C_x$, forcing $t \in L$). We note also that the subcase $C_x = L \cong Suz/Z_2$ was treated by Wright [133].

Suppose then that $t \notin L$. In the $M_{12}/Z_2$ and $J_2/Z_2$ cases, for example, Finkelstein and Solomon argue first that $g$ can be taken to be an element of order 3 in $N_G(\langle x, t \rangle)$. However, $I = L(C_{\langle x, t \rangle}) = L(C_L(t)) \cong A_5$ or $L_3(2)$, respectively, and we see that $g$ can be chosen so that, in addition, $g$ centralizes $I$. But if $z$ is a 2-central involution of $C_x$ contained in $I$, then by the structure of $C_x$, $t \sim tz$ in $C_x$. On the other hand, $tz = x^g z = (xz)^g$ and so $x$ is $G$-conjugate to $xz \in L - \langle x \rangle$, and we are reduced to the previous case.

Similarly in the case $L \cong Suz/Z_2$, Finkelstein and Solomon [29] argue that $g$ can be chosen to normalize $I = L(C_L(t))$. On the other hand, $C_x - L$ contains only one conjugacy class $D$ of involutions, so $t \in D$. However, $Suz/Z_2$ possesses *two* nonisomorphic extensions by a group of order 2 and so there are two possibilities for the images in $\mathrm{Aut}(L)$ of elements of $D$. According to the two possibilities for $C_x$, one computes that

$$I \cong M_{12}/Z_2 \text{ or } J_2/Z_2 \text{ with } x \in I. \tag{5.268}$$

But then as $g$ normalizes $I$, it follows that $x^g = x$, contrary to $x^g = t \neq x$.

At long last we have completed our discussion of the standard component theorems (I) and (II) and with them the $B$- and $U$-theorems as well as the noncharacteristic 2 theorem.

Despite the sheer number and enormous technical complexity of these standard form problems, I have always felt that taken together their solutions represent one of the most beautiful chapters of finite simple group theory. I pictured each centralizer of an involution as a bud, blooming gradually as the local analysis unfolds, and finally revealing the full flower of the internal structure of $G$.

# Conclusion

We have now obtained our objective, outlining in detail a proof of the noncharacteristic 2 theorem:

> *A minimal counterexample to the classification of finite simple groups is a group of characteristic 2 type in which all proper subgroups are K-groups.*

Along the way we have established a number of additional properties of such a minimal counterexample $G$: it is connected, and has Sylow 2-subgroups of sectional rank $\geqslant 5$, class $\geqslant 3$, and containing no nontrivial strongly closed abelian subgroups, etc. However, the principal conclusion is that $G$ is of characteristic 2 type.

In the sequel to this volume we shall present a similar outline of the classification of simple groups of characteristic 2 type in which all proper subgroups are $K$-groups.

# Bibliography

This bibliography is limited to those articles and books referred to in the text, but not included in the bibliography of [I].

1. Alward, L. Standard subgroups of type $O^-(8, 2)$, Ph.D. thesis, University of Oregon, 1979.
2. Aschbacher, M. 2-components in finite groups, *Commun. Algebra* **3** (1975), 901–911.
3. Aschbacher, M. A homomorphism theorem for finite graphs, *Proc. A.M.S.* **54** (1976), 468–470.
4. Aschbacher, M. Standard components of alternating type centralized by a 4-group (preprint).
5. Aschbacher, M. A characterization of some finite groups of characteristic 3, *J. Algebra* **76** (1982), 400–441.
6. Aschbacher, M. The Tits group as a standard component, *Math. Zeit.* **181** (1982), 229–252.
7. Aschbacher, M., and Seitz, G. On groups with a standard component of known type II, *Osaka J. Math* **18** (1981), 703–723.
8. Beisiegel, B. Uber einfache endliche Gruppen mit Sylow 2-Gruppen der Ordnung hochsten $2^{10}$, *Commun. Algebra* **5** (1977), 113–170.
9. Bender, H. Finite groups with dihedral Sylow 2-subgroups, *J. Algebra* **70** (1981), 216–228.
10. Bender, H., and Glauberman, G. Characters of finite groups with dihedral Sylow 2-subgroups, *J. Algebra* **76** (1981), 200–215.
11. Brauer, R., and Wong, W. Some properties of finite groups with wreathed Sylow 2-subgroups, *J. Algebra* **19** (1971), 263–273.
12. Buekenhout, F., and Shult, E. On the foundations of polar geometry, *Geom. Dedicata* **3** (1974), 155–170.
13. Collins, M. The characterization of the Suzuki groups by their Sylow 2-subgroups, *Math. Z.* **123** (1971), 32–48.
14. Collins, M. The characterization of the unitary groups $U_3(2^n)$ by their Sylow 2-subgroups, *Bull. London Math. Soc.* **4** (1972), 49–53.
15. Collins, M. The characterization of finite groups whose Sylow 2-subgroups are of type $L_3(q)$, $q$ even, *J. Algebra* **25** (1973), 490–512.
16. Collins, M., and Solomon, R. The identification of finite groups of $PSL(5, q)$-type and $PSU(5, q)$-type.

17. Dade, E. Lifting group characters, *Ann. Math.* **79** (1964), 590–596.

18. Davis, S., and Solomon, R. Some sporadic characterizatior.s, *Commun. Algebra* **9** (1981), 1725–1742.

19. Dempwolff, U. A characterization of the Rudvalis simple group or order $2^{14} 3^3 5^3 7$ 13 29 by the centralizers of noncentral involutions, *J. Algebra* **32** (1974), 53–88.

20. Dempwolff, U., and Wong, S. K. On finite groups whose centralizer of an involution has normal extra special and Abelian subgroups I, II, *J. Algebra* **45** (1977), 247–253; 52 (1978), 210–217.

21. Egawa, Y., and Yoshida, T. Standard components of type $M_{24}$ and $\Omega^+(8, 2)$, Ph. D. Thesis, Ohio State University, 1980.

22. Egawa, Y. Standard subgroups of type $\Omega_8^+(2)$, *Hokaido Math. J.* **14** (1982), 279–285.

23. Finkelstein, L. Finite groups with a standard component of type Janko–Ree, *J. Algebra* **36** (1975), 416–426.

24. Finkelstein, L. Finite groups with a standard component isomorphic to $M_{23}$, *J. Algebra* **40** (1976), 541–555.

25. Finkelstein, L. Finite groups with a standard component of type *HJ* or *HMJ*, *J. Algebra* **43** (1976), 61–114.

26. Finkelstein, L. Finite groups with a standard component isomorphic to $M_{22}$, *J. Algebra* **44** (1977), 558–572.

27. Finkelstein, L. Finite groups with a standard component whose centralizer has cyclic Sylow 2-subgroups, *Proc. Amer. Math. Soc.* **62** (1977), 237–241.

28. Finkelstein, L. Finite groups with a standard component of type $J_4$, *Pacific J. Math.* **71** (1977), 237–241.

29. Finkelstein, L., and Solomon, R. Standard components of type $M_{12}$ and .3, *Osaka J. Math.* **16** (1979), 759–774.

30. Fischer, F. A characterization of the $\{3, 4\}$-transposition group $F_2$ (unpublished).

31. Foote, R. Finite groups with 2-components of 2-rank 1, I, II, *J. Algebra* **41** (1976), 16–46, 47–57.

32. Foote, R. Finite groups with Sylow 2-subgroups of type $L_6(q)$, $q \equiv 3 \pmod 4$, *J. Algebra* **68** (1981), 378–389.

33. Fritz, F. On centralizers of involutions with components of 2-rank two, I, II, *J. Algebra* **47** (1977), 323–399.

34. Gilman, R., and Solomon, R. Finite groups with small unbalancing triples, *Pacific J. Math.* **83** (1979), 55–107.

35. Gomi, K. Finite groups with a standard subgroup isomorphic to $Sp(4, 2^n)$, *Jpn. J. Math.* **4** (1978), 1–76.

36. Gomi, K. Finite groups with a standard subgroup isomorphic to $PSU(4, 2)$, *Pacific J. Math.* **79** (1978), 399–462.

37. Gomi, K. Standard subgroups of type $Sp_6(2)$, I, II, *J. Fac. Sci. Univ. Tokyo* **27** (1980), 87–107; 109–156.

38. Gomi, K., Finite groups with a standard component isomorphic to $PSp_4(3)$ (to appear).

39. Gorenstein, D. An outline of the classification of finite simple groups, *Proc. Symposia Pure Math. 37*, Santa Cruz, California, Amer. Math. Soc. (1979), 1–28.

40. Gorenstein, D., and Harada, K. A characterization of Janko's two new simple groups, *J. Fac. Sci. Univ. Tokyo* **16** (1970), 331–406.

41. Gorenstein, D., and Harada, K. On finite groups with Sylow 2-subgroups of type $\hat{A}_n$, $n = 8, 9, 10, 11$, *J. Algebra* **19** (1971), 185–227.

42. Gorenstein, D., and Harada, K. Finite groups with Sylow 2-subgroups of type $PSp(4, q)$, $q$ odd, *J. Fac. Sci. Univ. Tokyo* **20** (1973), 341–372.

43. Gorenstein, D., and Lyons, R. Nonsolvable finite groups with solvable 2-local subgroups, *J. Algebra* **38** (1976), 453–522.

44. Gorenstein, D., and Lyons, R. Signalizer functors, proper 2-generated cores, and nonconnected finite groups, *J. Algebra* **75** (1982), 10–22.

45. Griess, R. A characterization of $U_3(2^n)$ by its Sylow 2-subgroups, *Trans. Amer. Math. Soc.* **175** (1973), 181–186.

46. Griess, R. Finite groups with standard component a 4-fold cover of $M_{22}$ (unpublished).

47. Griess, R., Mason, D., and Seitz, G. Bender groups as standard components, *Trans. Amer. Math. Soc.* **235** (1978), 179–211.

48. Griess, R., and Solomon, R. Finite groups with unbalancing 2-components of $L_3(4)$, *He* type, *J. Algebra* **60** (1979), 96–125.

49. Hall, J. Fusion and dihedral Sylow 2-subgroups, *J. Algebra* **40** (1976), 203–228.

50. Hall, M. Computers in algebra and number theory (*Proc. Symp. Appl. Math.*) New York (1970), 109–134.

51. Harada, K. Finite simple groups whose Sylow 2-subgroups are of order $2^7$, *J. Algebra* **14** (1970), 386–404.

52. Harada, K. On some 2-groups of normal 2-rank 2, *J. Algebra* **20** (1972), 90–93.

53. Harada, K. Finite groups with standard component of type $M_{22}$ (preprint).

54. Harada, K. Groups with nonconnected Sylow 2-subgroups revisited, *J. Algebra* **70** (1981), 339–349.

55. Harris, M. A characterization of odd order extensions of the finite simple groups $PSp(4, q)$, $G_2(q)$, $D_4^2(q)$, *Nagoya Math. J.* **45** (1972), 79–96.

56. Harris, M. A characterization of odd order extensions of the finite projective symplectic groups $PSp(4, q)$, *Trans. Amer. Math. Soc.* **163** (1972), 311–327.

57. Harris, M. Finite groups having an involution centralizer with a 2-component of dihedral type II, *Ill. J. Math.* **21** (1977), 621–647.

58. Harris, M. Finite groups having an involution centralizer with a 2-component of type $PSL(3, 3)$, *Pacific J. Math.* **87** (1980), 69–74.

59. Harris, M. $PSL(2, q)$ type 2-components and the unbalanced group conjecture, *J. Algebra* **68** (1981), 190–235.

60. Harris, M. Finite groups having an involution centralizer with a $PSU(3, 3)$ component, *J. Algebra* **72** (1981), 426–455.

61. Harris, M. Finite groups containing an intrinsic 2-component of Chevalley type over a field of odd order, *Trans. Am. Math. Soc.* **272** (1982), 319–331.

62. Harris, M., and Solomon, R. Finite groups having an involution centralizer with a 2-component of dihedral type I, *Ill. J. Math.* **21** (1977), 575–620.

63. Held, D. A characterization of the alternating groups of degree eight and nine, *J. Algebra* **7** (1967), 218–237.

64. Held, D. Eine Kennzeichnung der Mathieu-Gruppe $M_{22}$ und der alternierenden Gruppe $A_{10}$, *J. Algebra* **8** (1968), 436–449.

65. Held, D. A characterization of some multiply transitive permutation groups, I, II, *J. Math.* **13** (1969), 224–240; II, *Arch. Math.* (*Basel*) **19** (1968), 373–382.

66. Hunt, D. A characterization of the finite simple group $M(22)$, *J. Algebra* **21** (1972), 103–112.

67. Hunt, D. A characterization of the finite simple group $M(23)$, *J. Algebra* **26** (1973), 431–439.

68. Ito, N. On a class of doubly, but not triply transitive groups, *Arch. Math.* **18** (1967), 564–570.

69. Janko, Z. Nonsolvable finite groups all of whose 2-local subgroups are solvable I, *J. Algebra* **21** (1972), 458–517.

70. Janko, Z. A characterization of the simple group $G_2(3)$, *J. Algebra* **12** (1969), 360–371.

71. Janko, Z., and Thompson, J. On finite simple groups whose Sylow 2-subgroups have no normal elementary subgroups of order 8, *Math. Z.* **113** (1970), 385–397.

72. Janko, Z., and Wong, S. A characterization of the McLaughlin simple group.

73. Kondo, T. On the alternating groups I, *J. Fac. Sci., Univ. Tokyo* **15** (1968), 87–97; II, *J. Math. Soc. Jpn.*, **21** (1969), 116–139.

74. Kondo, T. A characterization of the alternating group of degree eleven, *Ill. J. Math.* **13** (1969), 528–541.

75. Landrock, P., and Solomon, R. A characterization of the Sylow 2-subgroups of $PSU(3, 2^n)$ and $PSL(3, 2^n)$, Aarhus Universitet Preprint series 13 1974/75.

76. Lindsey, J. Finite linear groups of degree six, *Can. J. Math.* **23** (1971), 771–790.

77. Luneburg, H. *Transitive Erweiterungen endlicher Permutations-gruppen*, Springer-Verlag Lecture Notes #84, Berlin (1969).

78. Lyons, R, A characterization of $PSU(3, 4)$, *Trans. Amer. Math. Soc.* **164** (1972), 371–387.

79. Manferdelli, J. Standard components of type .2, Ph.D. thesis, University of California, Berkeley, 1979.

80. Mason, D. Finite simple groups with Sylow 2-subgroup dihedral wreath $Z_2$, *J. Algebra* **26** (1973), 10–68.

81. Mason, D. Finite simple groups with Sylow 2-subgroups of type $PSL(4, q)$, $q$ odd, *J. Algebra* **26** (1973), 75–97.

82. Mason, D. Finite simple groups with Sylow 2-subgroups of type $PSL(5, q)$, $q$ odd, *Math. Proc. Camb. Phil. Soc.* **79** (1976), 251–269.

83. Mason, G. Quasi-thin finite groups (to appear).

84. Miyamoto, I. Finite groups with a standard subgroup isomorphic to $U_4(2^n)$. *Jpn. J. Math.* **5** (1979), 209–244.

85. Miyamoto, I. Finite groups with a standard subgroup of type $U_5(2^n)$, $n > 1$, *J. Algebra* **64** (1980).

86. Miyamoto, I. Standard subgroups isomorphic to ${}^2F_4(2^n)$, *J. Algebra* **77** (1982), 261–273.

87. McBride, P. A classification of groups of type $A_n(q)$, for $n \geqslant 3$ and $q = 2^k \geqslant 4$, *J. Algebra* **46** (1977), 220–267.

88. Nah, C. K. Über endlichen einfach Gruppen die eine standard Untergruppe $A$ besitzen derart, dass $A/Z(A)$ zu $L_3(4)$ isomorphist. Ph.D. dissertation, Johannes Gutenberg Universität, Mainz (1975).

89. Norton, S. Existence and uniqueness of Harada's group $F_5$ (unpublished).

90. Olsson, J. Odd-order extensions of some orthogonal groups, *J. Algebra* **28** (1974), 573–596.

91. O'Nan, M. Finite simple groups of 2-rank 3 with all 2-local subgroups 2-constrained, *Ill. J. Math.* **20** (1976), 155–170.

92. Ostrom, J., and Wagner, A. On projective and affine planes with transitive collineation groups, *Math. Z.* **71** (1959), 186–199.

93. Parrott, D. Characterizations of the Fischer groups II, III, *Trans. Amer. Math. Soc.*

94. Parrott, D. A characterization of the Rudvalis simple group, *Proc. London Math. Soc.* **32** (1976), 25–51.

95. Patterson, N., and Wong, S. K. A characterization of the Suzuki sporadic simple group of order 448, 345, 497, 600. *J. Algebra* **39** (1976), 277–282.

96. Phan, K. A characterization of the four-dimensional unimodular group, *J. Algebra* **15** (1970), 252–279.

97. Phan, K. On groups generated by $SU(3, q)$'s, I, II, *J. Austral. Math. Soc.* **23** (1977), 66–76; 129–146.

98. Phan, K. A characterization of the unitary groups $PSU(4, q^2)$, $q$ odd, *J. Algebra* **17** (1971), 132–148.

99. Phan, K. A characterization of the finite groups $PSL(n, q)$, *Math. A.* **124** (1972), 169–185.

100. Phan, K. A characterization of the finite simple groups $PSU(n, q)$, *J. Algebra* **37** (1975), 313–339.

101. Powell, M., and Thwaites, G. On the nonexistence of certain types of subgroups in simple groups, *Quart. J. Math. Oxford Series* (2).

102. Reifart, A. A characterization of the sporadic simple group of Suzuki, *J. Algebra* **33** (1975), 288–305.

103. Schoenwalder, U. Finite groups with a Sylow 2-subgroup of type $M_{24}$, I, II, *J. Algebra* **28** (1974), 20–56.

104. Schur, I. Über eine Klasse von endlichen Gruppen die linearer Substitutionen, *S.-B. Preuss. Akad. Wiss. Berlin*, (1905), 77–91.

105. Seitz, G. Standard subgroups of type $L_n(2^a)$, *J. Algebra* **48** (1977), 417–438.

106. Seitz, G. Chevalley groups as standard subgroups, *Ill. J. Math.*, **23** (1979), I, pp. 36–57; II, pp. 516–553; III, pp. 554–578.

107. Seitz, G. Some standard subgroups, *J. Algebra* **70** (1981), 299–302.

108. Sibley, D. Coherence in finite groups containing a Frobenius section, *Ill. J. Math.* **20** (1976), 434–442.

109. Smith, F. Finite groups all of whose 2-local subgroups are solvable, *J. Algebra* **34** (1975), 481–520.

110. Smith, F., On finite groups with large extra special subgroups, *J. Algebra* **44** (1977), 477–487.

111. Solomon, R., Finite groups with Sylow 2-subgroups of type $A_{12}$, Ph.D. thesis, Yale, 1971.

112. Solomon, R. Finite groups with Sylow 2-subgroups of type $\Omega(7, q)$, $q \equiv 3(\mathrm{mod}\ 8)$, *J. Algebra* **28** (1974), 20–56.

113. Solomon, R. Standard components of alternating type I, II. *J. Algebra* **41** (1976), 496–514; **47**(1977), 162–179.

114. Solomon, R. Some sporadic components of sporadic type, *J. Algebra* **53** (1978), 93–124.

115. Stingl, V. Endliche, einfache Component-Type–Gruppen, deren Ordnung nicht durch $2^{11}$ geteilt wird, Dissertation, Mainz, (1976).

116. Stroth, G. Über Gruppen mit 2-Sylow-Durchschnitten vom Rang $\leqslant 3$, I, II, *J. Algebra* **43** (1976), 457–505.

117. Stroth, G. On standard subgroups of type $^2E_6(2)$, *Proc. Am. Math. Soc.* **81** (1981), 365–368.

118. Suzuki, H. Standard component of type $PSL(4, 3)$, Ph.D. thesis, Ohio State University (1980).

119. Suzuki, M. On finite groups containing an element of order four which commutes only with its powers, *Ill. J. Math.* **3** (1959), 255–271.

120. Syskin, S. Standard components of type $F_3$ (to appear).

121. Thomas, G. A characterization of the groups $G_2(2^n)$, *J. Algebra* **13** (1969), 1043–1091.

122. Thomas, G. A characterization of the Steinberg groups $D_4^2(q^3)$, $q = 2^n$, *J. Algebra* **14** (1970), 373–385.

123. Thompson, J. Two results about finite groups, *Proc. Int. Congress of Mathematicians*, Moscow (1962), 296–299.

124. Wales, D. Finite linear groups of degree 7, I, *Can. J. Math.* **21** (1969), 1042–1056; II, *Pac. J. Math.* **34** (1970), 207–235.

125. Wales, D. Uniqueness of the graph of a rank three group, *Pac. J. Math.* **30** (1969), 271–276.

126. Walter, J. Centralizers of involutions in finite groups and the classification problem, *Proc. Gainesville Conference*, North Holland, Amsterdam (1973), 147–155.

127. Walter, J. A characterization of Chevalley groups I, *Proceedings of the International Symposium on Theory of Finite Groups*, Sapporo, Japan, Japan Society of the Promotion of Science 1974, 117–141.

128. Walter, J. Characterizations of Chevalley groups, *Proc. Taniguichi Inter. Symp. Div. Math.* Tokyo 1976, 117–141.

129. Walter, J. The *B*-conjecture; 2-components in finite simple groups (to appear).

130. Wong, W. A characterization of the alternating group of degree eight, *Proc. London Math. Soc.* **13** (1963), 359–383.

131. Wong, W. A characterization of the simple groups $PSp(4, q)$, $q$ cdd, *Trans. Amer. Math. Soc.* **139** (1969), 1–35.

132. Wong, W. Generators and relations for classical groups, *J. Algebra* **32** (1974), 529–553.

133. Wright, D. The non-existence of a certain type of finite simple group, *J. Algebra* **29** (1974), 417–420.

134. Yamada, H. Finite groups with a standard subgroup isomorphic to $U_5(2)$, *J. Algebra* **58** (1979), 527–562.

135. Yamada, H. Standard subgroups isomorphic to $PSU(6, 2)$ and $SU(6, 2)$, *J. Algebra* **61** (1979), 82–111.

136. H. Finite groups with a standard subgroup isomorphic to $G_2(2^n)$, *J. Fac. Sci. Univ. Tokyo* **26** (1979), 1–52.

137. Yamada, H. Finite groups with a standard subgroup isomorphic to $^3D_4(2^n)$, *J. Fac. Sci. Univ. Tokyo* **26** (1979), 255–278.

138. Yamada, H. A remark on the standard form problem $^2F_4(2^{2n+1})$, $n > 1$, *J. Algebra* (to appear).

139. Yamada, H. Standard subgroups of type $G_2(3)$, *Tokyo J. Math.* **5** (1982), 49–84.

140. Yamaki, H. A characterization of the alternating groups of degrees 12, 13, 14, 15, *J. Math. Soc. Jpn.* **20** (1968), 673–695.

141. Yamaki, H. A characterization of the simple group $S_p(6, 2)$, *J. Math. Soc. Jpn* **21** (1969), 334–356.

142. Yoshida, T. A characterization of Conway's group $C_3$, *Hokkaido Math. J.* **3** (1974), 232–242.

143. Yoshida, T. Character-theoretic transfer, *J. Algebra* **52** (1978), 1–38.

144. Yoshida, T. Transfer theorems, *Proc. Symposia Pure Math.* **37**, Santa Cruz, California, Amer. Math. Soc. (1979), 213–216.

# Index